Satellites are increasingly used in telecommunications, sc
veillance, and meteorology. These satellites rely heavily o
complex onboard control systems.

The aim of this book is to explain the basic theory of spacecraft dynamics and control and the practical aspects of controlling a satellite. The emphasis throughout is on analyzing and solving real-world engineering problems. For example, the author discusses orbital and rotational dynamics of spacecraft under a variety of environmental conditions, along with the realistic constraints imposed by available hardware. Among the topics covered are orbital dynamics, attitude dynamics, gravity gradient stabilization, single- and dual-spin stabilization, attitude maneuvers, attitude stabilization, and structural dynamics and liquid sloshing.

Spacecraft Dynamics and Control reflects Dr. Sidi's experience as a university instructor and as an engineer working on spacecraft control systems. This book will be useful as a reference for engineers and as a text for students.

CAMBRIDGE AEROSPACE SERIES 7

General editors
MICHAEL J. RYCROFT & ROBERT F. STENGEL

Spacecraft Dynamics and Control

Cambridge Aerospace Series

1. J. M. Rolfe and K. J. Staples (eds.): *Flight Simulation*
2. P. Berlin: *The Geostationary Applications Satellite*
3. M. J. T. Smith: *Aircraft Noise*
4. N. X. Vinh: *Flight Mechanics of High-Performance Aircraft*
5. W. A. Mair and D. L. Birdsall: *Aircraft Dynamics*
6. M. J. Abzug and E. E. Larrabee: *Airplane Stability and Control*
7. M. J. Sidi: *Spacecraft Dynamics and Control*

Spacecraft Dynamics and Control

A Practical Engineering Approach

MARCEL J. SIDI
Israel Aircraft Industries Ltd.
and
Tel Aviv University

PUBLISHED BY THE PRESS SYNDICATE OF THE UNIVERSITY OF CAMBRIDGE
The Pitt Building, Trumpington Street, Cambridge, United Kingdom

CAMBRIDGE UNIVERSITY PRESS
The Edinburgh Building, Cambridge CB2 2RU, UK
40 West 20th Street, New York, NY 10011-4211, USA
10 Stamford Road, Oakleigh, VIC 3166, Australia
Ruiz de Alarcón 13, 28014 Madrid, Spain
Dock House, The Waterfront, Cape Town 8001, South Africa

http://www.cambridge.org

© Cambridge University Press 1997

This book is in copyright. Subject to statutory exception and
to the provisions of relevant collective licensing agreements,
no reproduction of any part may take place without
the written permission of Cambridge University Press.

First published 1997
First paperback edition 2000
Reprinted 2001

Printed in the United States of America

A catalog record for this book is available from the British Library

Library of Congress Cataloging in Publication Data is available

ISBN 0 521 55072 6 hardback
ISBN 0 521 78780 7 paperback

To the memory of my parents, Jacob and Sophie,
who dedicated their lives to my education

and

to my wife Raya and children Gil, Talia, Michal and Alon,
who were very patient with me during the preparation of this book

Contents

Preface	*page* xv
Acknowledgments	xvii

Chapter 1 Introduction — 1
- 1.1 Overview — 1
- 1.2 Illustrative Example — 1
  - 1.2.1 Attitude and Orbit Control System Hardware — 2
  - 1.2.2 Mission Sequence — 2
- 1.3 Outline of the Book — 5
- 1.4 Notation and Abbreviations — 7
- *References* — 7

Chapter 2 Orbit Dynamics — 8
- 2.1 Basic Physical Principles — 8
  - 2.1.1 The Laws of Kepler and Newton — 8
  - 2.1.2 Work and Energy — 9
- 2.2 The Two-Body Problem — 10
- 2.3 Moment of Momentum — 11
- 2.4 Equation of Motion of a Particle in a Central Force Field — 12
  - 2.4.1 General Equation of Motion of a Body in Keplerian Orbit — 12
  - 2.4.2 Analysis of Keplerian Orbits — 15
- 2.5 Time and Keplerian Orbits — 18
  - 2.5.1 True and Eccentric Anomalies — 18
  - 2.5.2 Kepler's Second Law (Law of Areas) and Third Law — 19
  - 2.5.3 Kepler's Time Equation — 20
- 2.6 Keplerian Orbits in Space — 22
  - 2.6.1 Definition of Parameters — 22
  - 2.6.2 Transformation between Cartesian Coordinate Systems — 24
  - 2.6.3 Transformation from $\alpha = [a\ e\ i\ \Omega\ \omega\ M]^T$ to $[\mathbf{v}, \mathbf{r}]$ — 26
  - 2.6.4 Transformation from $[\mathbf{v}, \mathbf{r}]$ to $\alpha = [a\ e\ i\ \Omega\ \omega\ M]^T$ — 27
- 2.7 Perturbed Orbits: Non-Keplerian Orbits — 28
  - 2.7.1 Introduction — 28
  - 2.7.2 The Perturbed Equation of Motion — 29
  - 2.7.3 The Gauss Planetary Equations — 30
  - 2.7.4 Lagrange's Planetary Equations — 33
- 2.8 Perturbing Forces and Their Influence on the Orbit — 33
  - 2.8.1 Definition of Basic Perturbing Forces — 33
  - 2.8.2 The Nonhomogeneity and Oblateness of the Earth — 34

	2.8.3	A Third-Body Perturbing Force	39
	2.8.4	Solar Radiation and Solar Wind	41
2.9	Perturbed Geostationary Orbits		42
	2.9.1	Redefinition of the Orbit Parameters	42
	2.9.2	Introduction to Evolution of the Inclination Vector	43
	2.9.3	Analytical Computation of Evolution of the Inclination Vector	45
	2.9.4	Evolution of the Eccentricity Vector	50
	2.9.5	Longitudinal Acceleration Due to Oblateness of the Earth	56
2.10	Euler–Hill Equations		57
	2.10.1	Introduction	57
	2.10.2	Derivation	58
2.11	Summary		62
	References		62

Chapter 3 Orbital Maneuvers — 64

3.1	Introduction		64
3.2	Single-Impulse Orbit Adjustment		65
	3.2.1	Changing the Altitude of Perigee or Apogee	65
	3.2.2	Changing the Semimajor Axis a_1 and Eccentricity e_1 to a_2 and e_2	65
	3.2.3	Changing the Argument of Perigee	68
	3.2.4	Restrictions on Orbit Changes with a Single Impulsive $\Delta \mathbf{V}$	69
3.3	Multiple-Impulse Orbit Adjustment		70
	3.3.1	Hohmann Transfers	70
	3.3.2	Transfer between Two Coplanar and Coaxial Elliptic Orbits	71
	3.3.3	Maintaining the Altitude of Low-Orbit Satellites	72
3.4	Geostationary Orbits		73
	3.4.1	Introduction	73
	3.4.2	GTO-to-GEO Transfers	73
	3.4.3	Attitude Errors During GTO-to-GEO Transfers	76
	3.4.4	Station Keeping of Geostationary Satellites	78
3.5	Geostationary Orbit Corrections		80
	3.5.1	North–South (Inclination) Station Keeping	81
	3.5.2	Eccentricity Corrections	84
	3.5.3	Fuel Budget for Geostationary Satellites	84
3.6	Summary		86
	References		86

Chapter 4 Attitude Dynamics and Kinematics — 88

4.1	Introduction		88
4.2	Angular Momentum and the Inertia Matrix		88
4.3	Rotational Kinetic Energy of a Rigid Body		90
4.4	Moment-of-Inertia Matrix in Selected Axis Frames		90
	4.4.1	Moment of Inertia about a Selected Axis in the Body Frame	90

	4.4.2	Principal Axes of Inertia	91
	4.4.3	Ellipsoid of Inertia and the Rotational State of a Rotating Body	93
4.5	Euler's Moment Equations		95
	4.5.1	Solution of the Homogeneous Equation	95
	4.5.2	Stability of Rotation for Asymmetric Bodies about Principal Axes	96
	4.5.3	Solution of the Homogeneous Equation for Unequal Moments of Inertia	97
4.6	Characteristics of Rotational Motion of a Spinning Body		98
	4.6.1	Nutation of a Spinning Body	98
	4.6.2	Nutational Destabilization Caused by Energy Dissipation	99
4.7	Attitude Kinematics Equations of Motion for a Nonspinning Spacecraft		100
	4.7.1	Introduction	100
	4.7.2	Basic Coordinate Systems	101
	4.7.3	Angular Velocity Vector of a Rotating Frame	102
	4.7.4	Time Derivation of the Direction Cosine Matrix	104
	4.7.5	Time Derivation of the Quaternion Vector	104
	4.7.6	Derivation of the Velocity Vector ω_{RI}	105
4.8	Attitude Dynamic Equations of Motion for a Nonspinning Satellite		107
	4.8.1	Introduction	107
	4.8.2	Equations of Motion for Spacecraft Attitude	107
	4.8.3	Linearized Attitude Dynamic Equations of Motion	108
4.9	Summary		111
	References		111

Chapter 5 Gravity Gradient Stabilization — 112

5.1	Introduction	112
5.2	The Basic Attitude Control Equation	113
5.3	Gravity Gradient Attitude Control	114
	5.3.1 Purely Passive Control	114
	5.3.2 Time-Domain Behavior of a Purely Passive GG-Stabilized Satellite	117
	5.3.3 Gravity Gradient Stabilization with Passive Damping	122
	5.3.4 Gravity Gradient Stabilization with Active Damping	126
	5.3.5 GG-Stabilized Satellite with Three-Axis Magnetic Active Damping	129
5.4	Summary	129
	References	130

Chapter 6 Single- and Dual-Spin Stabilization — 132

6.1	Introduction	132
6.2	Attitude Spin Stabilization during the ΔV Stage	132
6.3	Active Nutation Control	135
6.4	Estimation of Fuel Consumed during Active Nutation Control	137

6.5	Despinning and Denutation of a Satellite	139
	6.5.1 Despinning	140
	6.5.2 Denutation	141
6.6	Single-Spin Stabilization	144
	6.6.1 Passive Wheel Nutation Damping	144
	6.6.2 Active Wheel Nutation Damping	146
6.7	Dual-Spin Stabilization	148
	6.7.1 Passive Damping of a Dual-Spin–Stabilized Satellite	148
	6.7.2 Momentum Bias Stabilization	150
6.8	Summary	151
	References	151

Chapter 7 Attitude Maneuvers in Space — 152

7.1	Introduction	152
7.2	Equations for Basic Control Laws	152
	7.2.1 Control Command Law Using Euler Angle Errors	152
	7.2.2 Control Command Law Using the Direction Cosine Error Matrix	153
	7.2.3 Control Command Law about the Euler Axis of Rotation	155
	7.2.4 Control Command Law Using the Quaternion Error Vector	156
	7.2.5 Control Laws Compared	156
	7.2.6 Body-Rate Estimation without Rate Sensors	158
7.3	Control with Momentum Exchange Devices	160
	7.3.1 Model of the Momentum Exchange Device	161
	7.3.2 Basic Control Loop for Linear Attitude Maneuvers	164
	7.3.3 Momentum Accumulation and Its Dumping	165
	7.3.4 A Complete Reaction Wheel-Based ACS	167
	7.3.5 Momentum Management and Minimization of the $\|\mathbf{h}_w\|$ Norm	169
	7.3.6 Effect of Noise and Disturbances on ACS Accuracy	172
7.4	Magnetic Attitude Control	185
	7.4.1 Basic Magnetic Torque Control Equation	185
	7.4.2 Special Features of Magnetic Attitude Control	186
	7.4.3 Implementation of Magnetic Attitude Control	188
7.5	Magnetic Unloading of Momentum Exchange Devices	189
	7.5.1 Introduction	189
	7.5.2 Magnetic Unloading of the Wheels	190
	7.5.3 Determination of the Unloading Control Gain k	192
7.6	Time-Optimal Attitude Control	195
	7.6.1 Introduction	195
	7.6.2 Control about a Single Axis	197
	7.6.3 Control with Uncertainties	201
	7.6.4 Elimination of Chatter and of Time-Delay Effects	201
7.7	Technical Features of the Reaction Wheel	206
7.8	Summary	208
	References	208

Chapter 8 Momentum-Biased Attitude Stabilization — 210

- 8.1 Introduction — 210
- 8.2 Stabilization without Active Control — 210
- 8.3 Stabilization with Active Control — 214
  - 8.3.1 Active Control Using Yaw Measurements — 215
  - 8.3.2 Active Control without Yaw Measurements — 217
- 8.4 Roll-Yaw Attitude Control with Magnetic Torques — 222
- 8.5 Active Nutation Damping via Products of Inertia — 225
- 8.6 Roll-Yaw Attitude Control with Solar Torques — 229
  - 8.6.1 Dynamic Equations for Solar Panels and Flaps — 230
  - 8.6.2 Mechanization of the Control Algorithm — 233
- 8.7 Roll-Yaw Attitude Control with Two Momentum Wheels — 237
  - 8.7.1 Introduction — 237
  - 8.7.2 Adapting the Equation of Rotational Motion — 238
  - 8.7.3 Designing the Control Networks $G_Y(s)$ and $G_Z(s)$ — 240
  - 8.7.4 Momentum Dumping of the MW with Reaction Thrust Pulses — 241
- 8.8 Reaction Thruster Attitude Control — 242
  - 8.8.1 Introduction — 242
  - 8.8.2 Control of ϕ (Roll) and ψ (Yaw) — 244
  - 8.8.3 Immunity to Sensor Noise — 246
  - 8.8.4 Determining the Necessary Momentum Bias h_w — 247
  - 8.8.5 Active Nutation Damping via Products of Inertia — 248
  - 8.8.6 Wheel Momentum Dumping and the Complete Attitude Controller — 250
  - 8.8.7 Active Nutation Damping without Products of Inertia — 251
- 8.9 Summary — 256
- *References* — 257

Chapter 9 Reaction Thruster Attitude Control — 260

- 9.1 Introduction — 260
- 9.2 Set-Up of Reaction Thruster Control — 260
  - 9.2.1 Calculating the Torque Components of a Single Thruster — 261
  - 9.2.2 Transforming Torque Commands into Thruster Activation Time — 263
- 9.3 Reaction Torques and Attitude Control Loops — 265
  - 9.3.1 Introduction — 265
  - 9.3.2 Control Systems Based on PWPF Modulators — 266
  - 9.3.3 Control Loop Incorporating a PWPF Modulator — 270
- 9.4 Reaction Attitude Control via Pulse Width Modulation — 273
  - 9.4.1 Introduction — 273
  - 9.4.2 Feedback Control Loop of a Pulsed Reaction System — 273
- 9.5 Reaction Control System Using Only Four Thrusters — 287
- 9.6 Reaction Control and Structural Dynamics — 289
- 9.7 Summary — 289
- *References* — 289

Chapter 10 Structural Dynamics and Liquid Sloshing — 291

- 10.1 Introduction — 291
- 10.2 Modeling Solar Panels — 291
  - 10.2.1 Classification of Techniques — 291
  - 10.2.2 The Lagrange Equations and One-Mass Modeling — 292
  - 10.2.3 The Mass–Spring Concept and Multi-Mass Modeling — 296
- 10.3 Eigenvalues and Eigenvectors — 299
- 10.4 Modeling of Liquid Slosh — 301
  - 10.4.1 Introduction — 301
  - 10.4.2 Basic Assumptions — 301
  - 10.4.3 One-Vibrating Mass Model — 302
  - 10.4.4 Multi-Mass Model — 308
- 10.5 Generalized Modeling of Structural and Sloshing Dynamics — 309
  - 10.5.1 A System of Solar Panels — 309
  - 10.5.2 A System of Fuel Tanks — 310
  - 10.5.3 Coupling Coefficients and Matrices — 310
  - 10.5.4 Complete Dynamical Modeling of Spacecraft — 311
  - 10.5.5 Linearized Equations of Motion — 312
- 10.6 Constraints on the Open-Loop Gain — 313
  - 10.6.1 Introduction — 313
  - 10.6.2 Limitations on the Crossover Frequency — 313
- 10.7 Summary — 316
- *References* — 316

Appendix A Attitude Transformations in Space — 318

- A.1 Introduction — 318
- A.2 Direction Cosine Matrix — 318
  - A.2.1 Definitions — 318
  - A.2.2 Basic Properties — 319
- A.3 Euler Angle Rotation — 320
- A.4 The Quaternion Method — 322
  - A.4.1 Definition of Parameters — 322
  - A.4.2 Euler's Theorem of Rotation and the Direction Cosine Matrix — 323
  - A.4.3 Quaternions and the Direction Cosine Matrix — 324
  - A.4.4 Attitude Transformation in Terms of Quaternions — 325
- A.5 Summary — 326
- *References* — 326

Appendix B Attitude Determination Hardware — 328

- B.1 Introduction — 328
- B.2 Infrared Earth Sensors — 329
  - B.2.1 Spectral Distribution and Oblateness of the Earth — 329
  - B.2.2 Horizon-Crossing Sensors — 330
  - B.2.3 IRHCES Specifications — 339
  - B.2.4 Static Sensors — 343

B.3	Sun Sensors		345
	B.3.1	Introduction	345
	B.3.2	Analog Sensors	345
	B.3.3	Digital Sensors	351
B.4	Star Sensors		353
	B.4.1	Introduction	353
	B.4.2	Physical Characteristics of Stars	357
	B.4.3	Tracking Principles	366
B.5	Rate and Rate Integrating Sensors		373
	B.5.1	Introduction	373
	B.5.2	Rate-Sensor Characteristics	375
	References		376

Appendix C   Orbit and Attitude Control Hardware 379

C.1	Introduction		379
C.2	Propulsion Systems		379
	C.2.1	Cold Gas Propulsion	381
	C.2.2	Chemical Propulsion – Solid	381
	C.2.3	Chemical Propulsion – Liquid	382
	C.2.4	Electrical Propulsion	385
	C.2.5	Thrusters	387
C.3	Solar Pressure Torques		388
	C.3.1	Introduction	388
	C.3.2	Description	388
	C.3.3	Maximization	392
C.4	Momentum Exchange Devices		393
	C.4.1	Introduction	393
	C.4.2	Simplified Model of a RW Assembly	393
	C.4.3	Electronics	396
	C.4.4	Specifications	396
C.5	Magnetic Torqrods		397
	C.5.1	Introduction	397
	C.5.2	Performance Curve	398
	C.5.3	Specifications	401
	References		401

Index   403

Preface

The goal of this book is to provide the reader with the basic engineering notions of controlling a satellite. In the author's experience, one of the most important facts to be taught from the beginning is practical engineering reality. Theoretical, "nice" control solutions are seriously hampered when practical problems (e.g., sensor noise amplification, unexpected time delays, control saturation effects, structural modes, etc.) emerge at a later stage of the design process. The control algorithms must then be redesigned, with the inevitable loss of time and delay of the entire program. Early anticipation of these effects shortens the design process considerably. Hence it is of utmost importance to analyze different concepts for engineering solutions of spacecraft control tasks in the preliminary design stages, so that the correct one will be selected at the outset. This is why several approaches may be suggested for a given control task.

Part of the material in this textbook has been used as background for a single-semester course on "Spacecraft Dynamics and Control" – offered since 1986 at the Tel Aviv University and also more recently at the Israel Institute of Technology, the Technion, Haifa. All the material in this book is appropriate for a course of up to two semesters in length. The book is intended for introductory graduate-level or advanced undergraduate courses, and also for the practicing engineer. A prerequisite is a first course in automatic control, continuous and sampled, and a first course in mechanics. This, in turn, assumes knowledge of linear algebra, linear systems, Laplace transforms, and dynamics.

A sequential reading of the book is advised, although the chapters are for the most part self-contained. A preliminary overview is recommended in order to acquire a feeling for the book's contents; this will help enormously in the second, and deeper, reading.

Modern spacecraft control concepts are based on a vast choice of physical phenomena: single- and dual-spin stabilization; gravity gradient attitude control; three-axis stabilization; momentum-bias stabilization; and solar, magnetic, or reaction torque stabilization. It is important to master the essential qualities of each before choosing one as an engineering solution. Therefore, the various concepts are treated, analyzed, and compared in sufficient depth to enable the reader to make the correct choices.

Appendix B and Appendix C detail the space onboard hardware that is essential to any practical engineering solution. Technical specifications of various control items are listed for easy reference.

Acknowledgments

I wish to express my sincere gratitude to the System & Space Technology – MBT, a subsidiary of Israel Aircraft Industries Ltd., on whose premises this textbook was partially prepared. I give special thanks to its manager, Dr. M. Bar-Lev, who encouraged my writing.

Part of the material included here is the result of mutual efforts of the Control and Simulation Department engineers and scientists to develop, design, evaluate, and build the attitude and orbit control systems of the *Offeq* series of low-orbit satellites and of the Israeli geostationary communication satellite *Amos 1*.

It took me more than ten years of effort to study and master, at least partially, the nascent field of space technology. In this context I would like to thank my colleagues, especially P. Rosenbaum, A. Albersberg, E. Zemer, D. Verbin, R. Azor, A. Ben-Zvi, Y. Efrati, Y. Komen, Y. Yaniv, F. Dellus, and others with whom I had long and fruitful discussions and who carefully read parts of the manuscript.

I also wish to express my gratitude to Professor S. Merhav, former head of the aeronautical and astronautical engineering department at the Israel Institute of Technology, and Professor R. Brodsky, former head of the aeronautical engineering department at Iowa State University, for reading the entire manuscript and for their constructive remarks.

During my own education in the field of space dynamics and control, I took advantage of many works written by such excellent scientists as Agrawal, Alby, Balmino, Battin, Bernard, Bittner, Borderies, Bryson, Campan, Deutsch, Donat, Duret, Escobal, Foliard, Frouard, Gantous, Kaplan, Legendre, Pocha, Pritchard, Robert, Sciulli, Soop, Thomson, Wertz, and others. To these scientists I owe my deep gratitude.

Last but not least, I am deeply indebted to Mrs. Florence Padgett, physical sciences editor for Cambridge University Press, who has helped significantly in improving the style and overall presentation of the book.

CHAPTER 1

Introduction

1.1 Overview

Space technology is relatively young compared to other modern technologies, such as aircraft technology. However, in only forty years this novel domain has achieved a tremendous level of complexity and sophistication. The reason for this is simply explained: most satellites, once in space, must rely heavily on the quality of their onboard instrumentation and on the design ingenuity of the scientists and engineers who produced them. Recent achievements of repairing satellites while in orbit testify to the complexity involved in space technology. The desire of humans to conquer space within the solar system will surely encourage new technological achievements that are not yet imagined.

The technical fields in which satellites are used are numerous – telecommunications, scientific research, meteorology, and others. According to the specific task for which they are designed, satellites are very different from one another. They may be in orbits as low as 200 km or as high as 40,000 km above the earth; other spacecraft leave the earth toward planets in the solar system. Satellites may be very heavy: an inhabited space station, for example, could weigh several tons or more. There also exist very light satellites, weighing 20 kg or less. Small satellites may be relatively cheap, of the order of a million dollars apiece. Despite their differences, satellites possess fundamental features that are common to all. The physical laws that govern their motion in space and their dynamics are the same for all spacecraft. Hence, the fundamental technologies that evolved from these laws are common to all.

A satellite's life begins with the specific booster transferring it to some initial orbit, called a *transfer orbit,* in which the satellite is already circling the earth. For a satellite that will stay near earth, the next stage will be to "ameliorate" the orbit; this means that the satellite must be maneuvered to reach the precise orbit for which the satellite was designed to fulfill its mission. Next, the satellite's software must check for the proper functioning of its instrumentation and its performance in space, as well as calibrate some of the instruments before they can be used to control the satellite. The final stage is the one for which the satellite was designed and manufactured.

These stages will be discussed in the next section. Understanding the meaning of each stage will help one to understand the infrastructure of the control system of any satellite. Throughout the text, the terms "satellite" and "spacecraft" (s/c for short) will be used interchangeably. The terms "geosynchronous" and "geostationary" will be used interchangeably to describe the orbit of a satellite whose period can be made exactly equal to the time it takes the earth to rotate once about its axis.

1.2 Illustrative Example

In this section, a geosynchronous communications satellite will be described in its different life stages. The U.S. *Intelsat V* and the European *DFS Kupernikus*

(Bittner et al. 1987) are good examples of a common, medium-sized satellite. Satellites of this type consist of the following main structural parts.

(1) A central body consisting of a cubelike structure with dimensions of about 1.5×2 m.
(2) Solar arrays extended in the N–S direction (Y_B axis), with panel dimensions of about 1.5×7 m.
(3) An antenna tower directed toward the earth (Z_B direction) carrying different communication payloads such as global and beacon horns, feed systems for communication, hemi/zone and spot reflectors, TM/TC (telemetry/telecommand) antenna, and others.
(4) Controllers (such as reaction thrusters) and attitude sensors (such as sun sensors) located over the central body and the solar panels.

1.2.1 *Attitude and Orbit Control System Hardware*

It is important to list the typical attitude and orbit control system (AOCS) hardware of a geostationary satellite in order to understand and perceive from the beginning the complexity of the problems encountered. This hardware may include:

(1) a reaction bipropellant thrust system, consisting of one 420-N thruster used for orbit transfer and two independent (one redundant) low-thrust systems consisting of eight 10-N thrusters each;
(2) two momentum wheels (one redundant) of 35 N-m-sec each;
(3) two infrared horizon sensors (one operating and one redundant);
(4) four fine sun sensors (two redundant);
(5) twelve coarse sun sensors for safety reasons (six redundant);
(6) two three-axis coarse rate gyros; and
(7) two three-axis integrating gyros.

An illustration of the partial control hardware of a typical geostationary communications satellite is given in Figure 1.1. Much of the control hardware is redundant in order to guarantee a reliable control system despite potential hardware failures.

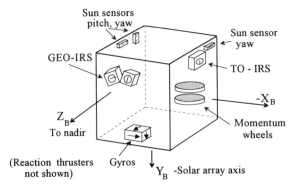

Figure 1.1 Principal arrangement of AOCS equipment; adapted from Bittner et al. (1987) by permission of IFAC.

1.2 / Illustrative Example

Table 1.1 *Typical sequence stages up to the normal mode acquisition stage*
(adapted from Bittner et al. 1987 by permission of IFAC)

Event No.	Stage	Time	Event
1	Launch Stage	T_0	ARIANE lift-off, ignition of first stage
2		$T_0 + 1009$ s	Reorientation for payload separation
3		$T_{SEP} = T_0 + 1122$ s	S/C separation; start of S/C sequence
4	GTO Stage	$T_{SEP} + 9$ min	Start of AOCS Sequences
5	Preparation to ABM Stage (Apogee Boost Motor)	$T_{SEP} + 10$ m $+ 20$ s	Automatic beginning of Sun-acquisition,
6		$T_{SEP} + 11$m to $T_{SEP} + 34$ min	Sun pointing, x-axis pointing to the Sun, roll rate=$0.5°$/s
7		$T_{SEP} + 1$h$+5$m	Solar panels deployment
8		$T_1 = T_{SEP} + 37$h$+11$m$+34$s	Apogee No.4 passage
9		$T_1 - 250$m	Start gyro calibration
10		$T_1 - 160$m	Gyrocalibration finished
11		$T_1 - 160$m	Start GTO Earth acquisition
12		$T_1 - 60$m	SS-bias functions loaded from ground
13	1st ABM	$T_1 - 24$m	Apogee Boost Motor (ABM) ignition
14		$T_1 + 25$m	Apogee boost termination
15		$T_1 + 28$m	Ground comand of Sun acquisition mode
16 -30	2nd and 3rd ABM	$T_{SEP} + 129$h$+48$m	Repetition of GTO Earth capture for second and fourth apogee boost maneuver until S/C in quasi-synchronous orbit.
31	Preparation to Normal Mode (Mission Stage)	$T_2 = T_{SEP} + 129$h$+48$m$+ 18$h	Start of Geosynchronous Orbit Earth acquisition
32		T_2	GEO Earth Acquisition command, finally Earth pointing takes place; y-axis perpendicular to orbital plane.
33		$T_2 + 30$m	Command for wheel run-up
34		$T_2 + 45$m	Wheel at nominal speed.
35		$T_2 + 50$m	Station keeping to reduce initial errors of acquisition loops
36		$T_2 + 55$m	Transition to normal mode as soon as angular and rate values within prescribed limits.

1.2.2 Mission Sequence

The mission events – from launch to in-orbit operation – may be summarized as follows. First is the launch into a geosynchronous transfer orbit (GTO), with perigee and apogee (low and high altitude) of 200 km and 35,786 km, respectively. This is followed by the transfer from GTO to geostationary orbit (GEO), where perigee and apogee both are 35,786 km and the orbit inclination and eccentricity are close to null. Next is the preparation and calibration of the AOCS before the useful GEO mission can start, followed by the actual GEO mission stage. Table 1.1 contains an outline of the typical major events and times related to the pre-mission stages. The significance of each event will become clear in the chapters to follow.

Figure 1.2 illustrates some of the principal stages in the geostationary transfer orbit. After separation from the launcher, the satellite is commanded into a sun acquisition mode with the $-\mathbf{X}_B$ axis pointing toward the sun. After completion of

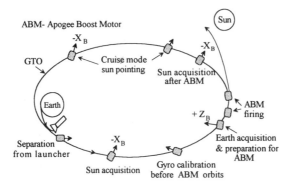

Figure 1.2 Sequence for injecting a satellite into the geostationary orbit.

this stage, the solar panels are partially or fully deployed. If fully deployed, they can be rotated about their axis of rotation toward the sun in order to maximize power absorption.

The satellite stays in this cruise mode until the first apogee boost motor (ABM) orbit is approached. In the first and the subsequent ABM orbits, several hours before the ABM firing at the apogee, the gyros' calibration maneuvers are initiated. Less than an hour before any ABM firing, earth acquisition is initiated with the $+\mathbf{Z}_B$ axis now pointed toward the earth, followed by preparation for the ABM firing stages. After ABM firings ranging from several to more than 30 minutes, the satellite is commanded to GTO cruise, sun-pointing. After the last ABM firing, the satellite is prepared for GEO operation. Some of the first maneuvers in GEO are shown in Figure 1.3. See also Bittner et al. (1989).

In the first GEO, earth acquisition is performed, meaning that the $+\mathbf{Z}_B$ axis of the satellite is directed toward the earth center of mass, thus allowing the normal GEO cruise. The momentum wheel is spun to its nominal angular velocity to provide momentum bias attitude control. In this stage, the satellite is brought to its nominal geographical longitude. The orbit is then corrected for any remaining inaccuracies in inclination and eccentricity (to be explained in Chapters 2 and 3).

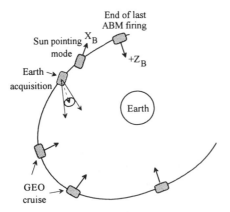

Figure 1.3 Principal stages in the first geostationary orbit (GEO).

When in the mission orbit, the following tasks are fulfilled in *normal mode:*

(1) pitch control by momentum wheel in torque mode control;
(2) roll/yaw control by the WHECON principle (to be explained in Chapter 8) – roll control by horizon sensor and yaw control by momentum bias (see Dougherty, Scott, and Rodden 1968); and
(3) momentum management of the wheel, which keeps the momentum of the wheel inside permitted bounds.

In addition, "station keeping" maintains the s/c within prescribed limits (of the order of $\pm 0.05°$) about the nominal longitude station position, and also within the same permitted deviation in the inclination of the mission orbit. Station keeping involves both north–south and east–west correction maneuvers.

1.3 Outline of the Book

The chapters of this book have been arranged to give the reader an integrated view of the subject of attitude and orbit control. Chapters 2 and 3 deal with the satellite *orbit* dynamics and control. The remaining chapters treat the *attitude* dynamics and control of satellites.

Chapter 2 develops the classical equations of motion of ideal Keplerian orbits. It then presents Gauss's and Lagrange's planetary equations, with which the perturbed orbit motion of a satellite can be analyzed. Chapter 3 covers basic orbital control concepts including control and station keeping of geostationary satellites.

Chapter 4 is devoted to the basic equations of rotational motion about some axis through its center of mass. The usual notions of angular momentum and rotational kinetic energy are introduced, defining the rotational state of a body. Next, Euler's moment equations are stated as a preliminary to analyzing the angular stability of a rotating body with or without the existence of internal energy dissipation. This chapter also develops the linearized angular equations of motion of a nonspinning spacecraft, which are necessary when designing a feedback attitude control system.

Chapter 5 deals with gravity gradient stabilization of a s/c. Gravity gradient control is a passive means of stabilizing the attitude of the satellite. In principle, gravity gradient attitude control is undamped. This chapter analyzes passive and active damping, and emphasizes the inaccuracies in attitude stabilization that arise in response to environmental conditions.

Chapter 6 deals with single- and dual-spin stabilization. The single-spin stabilization mode is frequently used to keep the direction of the thrust vector constant in space during the orbit change process. This chapter discusses the minimum spin rate needed to keep the thrust direction within permitted bounds, despite parasitic disturbing torques acting on the s/c. Also analyzed are active nutation control and despinning of the satellite at the end of the orbit change process together with the denutation stage. The mass of fuel consumed is evaluated analytically for both active nutation control and despin–denutation control. A design example is included.

The single-spin property is also used in the context of attitude stabilizing the spin axis of the s/c perpendicular to the orbit plane, thus allowing an attached communications payload to scan the earth continuously and so provide the communications link. Due to parasitic disturbing torques acting on the s/c, nutational motion is excited and must be constantly damped. Depending on the specific moments of inertia of the s/c, either passive or active damping is added to the attitude control of the satellite; various damping schemes are analyzed.

Dual-spin stabilization was developed in order to increase the efficiency of communication of spinning satellites by enabling the communications antenna to be continuously directed toward the earth. In this control concept, passive nutation damping can be achieved by energy dissipation. The stabilizing conditions are explicitly stated.

Chapter 7 is concerned with attitude stabilization and maneuvering of spacecraft stabilized in three axes. In these satellites, there is no constant angular momentum added to the s/c to keep the direction of one of its axes stabilized in space, so attitude control is achieved by simultaneously controlling the three body axes. For small attitude-angle maneuvering, the common Euler angles are a clear way to express the attitude of the satellite with reference to some defined frame in space. However, for larger attitude changes, the attitude kinematics are expressed much more effectively with the direction cosine matrix and the quaternion vector. The chapter begins with a thorough discussion of control laws for attitude control.

Momentum exchange devices are used to provide the control torques for accurate attitude control. These devices, called *reaction* and *momentum wheels,* are introduced and modeled for use in this and subsequent chapters. If an external inertial disturbance acts on an attitude-controlled satellite, then excess angular momentum is accumulated in the controlling wheels. A control scheme using *magnetic torqrods* to dump this momentum from the wheels is analyzed and simulated.

Attitude sensors and controllers are inherently noisy. When designing a control loop, these noises must be taken into consideration using statistical linear control theory. Tradeoffs in the design process due to such noises are stated. In order to augment the reliability and the control capability of a complete three-axis ACS, more than three reaction wheels are sometimes used. Chapter 7 analyzes optimal distribution of the computed control torques among the different wheels. Time-optimal attitude maneuvers about a single body axis are also analyzed. The last section of Chapter 7 deals with specifying technical characteristics of the reaction wheel based on mission requirements of the attitude control system (ACS).

Chapter 8 is concerned with momentum-biased satellites. A momentum wheel added to the satellite provides inertial stabilization to the three-axis stabilized s/c about one of its axes. The inertial stabilizing torque is achieved with the momentum produced in the momentum wheel. Unfortunately, environmental disturbances tend to destabilize the s/c by increasing the nutational motion, which thus must be actively controlled. There are three essential schemes for controlling nutational motion: magnetic damping, reaction propulsion damping, and – for high-altitude–orbit satellites such as geostationary satellites – solar torque control. These schemes are analyzed and compared.

Chapter 9 reviews the use of propulsion reaction hardware for attitude control. Only reaction thrusters can provide the high torques necessary in different attitude control tasks during orbit changes. The attitude stabilization scheme using reaction thrusters is stated and analyzed. Attitude maneuvering can likewise use reaction thrust torques. The achievable accuracies depend largely on the minimal impulse bit that a thruster can deliver. Also, since the torques delivered are with constant amplitude, the reaction pulses must be width- or frequency-modulated. Both modulation schemes are analyzed, and design examples are given.

Chapter 10 introduces the dynamics of structural modes and fuel sloshing dynamics. The chapter provides simplified analyses of solar panels and fuel sloshing, as well as rules-of-thumb for obtaining the simplified models so necessary in the initial design stage of an ACS. Also, given these initial models, the reader is shown how to approximate the maximum obtainable bandwidths of the system.

Appendix A is a short introduction to attitude transformations in space. It deals with Euler angle transformations, the direction cosine matrix, the quaternion vector, the relations among them, and attitude kinematics in general. Appendix B is a concise introduction to attitude measurement hardware. It is of the utmost importance to have a clear knowledge of such sensor characteristics, as their noise behavior influences achievable accuracies. The hardware treated includes horizon sensors (static or scanning), analog and digital sun sensors, star sensors, and angular rate sensors; characteristics data sheets are shown for various existing products. Appendix C describes a variety of control hardware, such as propulsion systems, magnetic torqrods, reaction wheels, and solar panels and flaps for achieving solar control torques.

1.4 Notation and Abbreviations

Vectors will be expressed by bold letters: **V**, **γ**. Matrices will be denoted by square brackets, and the name of the matrix inside the brackets in capital bold letters: [**A**]. Scalar variables are expressed using italicized letters: V, γ. The scalar "dot" product of two vectors will be expressed by a solid dot: **a**·**b**. The vector "cross" product will be denoted by a boldface cross: **a**×**b**. Multiple products will likewise be denoted by solid dots and boldface crosses: **a**·(**b**×**c**); **a**×(**b**×**c**). The MKS system of units is used throughout the book.

The following abbreviations will be used: ACS ≡ attitude control system; AOCS ≡ attitude and orbit control system; cm ≡ center of mass; ES ≡ earth sensor; LP ≡ low pass; MW ≡ momentum wheel; RW ≡ reaction wheel; s/c ≡ spacecraft; ss ≡ steady state.

References

Bittner, H., Fisher, H., Miltenberger, K., Roche, Z. C., Scheit, A., Surauer, M., and Vieler, H. (1987), "The Attitude and Orbit Control Subsystem of the DFS Kopernikus," Automatic Control World Congress, IFAC (27–31 July, Munich). Oxford: Pergamon.

Bittner, H., Fisher, H., Froeliger, J., Miltenberger, K., Popp, H., Porte, F., and Surauer, M. (1989), "The Attitude and Orbit Control Subsystem of the EUTELSAT II Spacecraft," 11th IFAC Symposium on Automatic Control in Space (17–21 July, Tsukuba, Japan).

Dougherty, H., Scott, E., and Rodden, J. (1968), "Analysis and Design of WHECON – An Attitude Control Concept," Paper no. 68-461, AIAA 2nd Communications Satellite System Conference (8–10 April, San Francisco). New York: AIAA.

CHAPTER 2

Orbit Dynamics

2.1 Basic Physical Principles

Orbital mechanics, as applied to artificial spacecraft, is based on celestial mechanics. In studying the motion of satellites, quite elementary principles are necessary. In fact, Kepler provided three basic empirical laws that describe motion in unperturbed planetary orbits. Newton formulated the more general physical laws governing the motion of a planet, laws that were consistent with Kepler's observations.

In this chapter, the dynamical equations of motion for ideal, unperturbed Keplerian orbits – and subsequently for realistic, perturbed orbits – will be analyzed. Kepler's laws of motion describe ideal orbits that do not exist in nature. Perturbing forces and physical anomalies cause spacecraft orbits to have strange properties; in most cases these cause difficulties for the space control engineer, but in other cases these properties may be of enormous help.

Keplerian orbits are treated in Sections 2.1–2.6. For further reading on this subject, Kaplan (1976) or Thomson (1986) may be consulted. Perturbed non-Keplerian orbits are treated in Sections 2.7–2.9 (see also Deutsch 1963, Alby 1983, and Battin 1990).

2.1.1 *The Laws of Kepler and Newton*

Kepler provided three empirical laws for planetary motion, based on Brahe's planetary observations. First, the orbit of each planet is an ellipse with the sun located at one focus. Second, the radius vector drawn from the sun to any planet sweeps out equal areas in equal time intervals (the law of areas). Third, planetary periods of revolution are proportional to the [mean distance to sun]$^{3/2}$.

Newton provided three laws of mechanics and one for gravitational attraction. Most analysis of celestial and spacecraft orbit dynamics is based on Newton's laws, formulated as follows.

(1) Every particle remains in a state of rest, or of uniform motion in a straight line with constant velocity, unless acted upon by an external force.
(2) The rate of change of linear momentum of a body is equal to the force **F** applied on the body, where **p** = m**v** is the linear momentum and

$$\mathbf{F} = \frac{d\mathbf{p}}{dt} = \frac{d(m\mathbf{v})}{dt}. \qquad (2.1.1)$$

In this equation, m is the mass of the body and **v** is the velocity vector. For a constant mass, this law takes the simplified form

$$\mathbf{F} = m\mathbf{a}, \qquad (2.1.2)$$

where **a** = $d\mathbf{v}/dt$ is the familiar linear acceleration.

2.1 / Basic Physical Principles

(3) For any force $\mathbf{F}_{12}$ exerted by particle 1 on a particle 2, there must likewise exist a force $\mathbf{F}_{21}$ exerted by particle 2 on particle 1, equal in magnitude and opposite in direction:

$$\mathbf{F}_{12} = -\mathbf{F}_{21}. \tag{2.1.3}$$

(4) Any two particles attract each other with a force given by the expression

$$\mathbf{F} = \frac{Gm_1 m_2 \mathbf{r}}{r^3}, \tag{2.1.4}$$

where $\mathbf{r}$ is a vector of magnitude r along the line connecting the two particles with masses m_1 and m_2, and $G = 6.669 \times 10^{-11}$ m^3/kg-s^2 is the universal constant of gravitation. This is the famous *inverse square law of force;* the magnitude of the force is $F = Gm_1 m_2/r^2$.

2.1.2 Work and Energy

If a force $\mathbf{F}$ acting on a body causes its displacement by a distance $d\mathbf{r}$, then the incremental work done by the force on the body is defined as

$$dW = \mathbf{F} \cdot d\mathbf{r}, \tag{2.1.5}$$

where $\mathbf{F} \cdot d\mathbf{r}$ is a scalar "dot" product. This illustrates that only the component of $\mathbf{F}$ in the direction of $d\mathbf{r}$ is effective in doing the work. The total work done by the force on the body is equal to the line integral

$$W_{12} = \int_c \mathbf{F} \cdot d\mathbf{r} = \int_{\mathbf{r}_1}^{\mathbf{r}_2} \mathbf{F} \cdot d\mathbf{r} \tag{2.1.6}$$

(see Figure 2.1.1).

The work done on a body changes its kinetic and potential energies. With respect to kinetic energy, the total work done on a body by moving it along the line c from P_1 to P_2 in Figure 2.1.1 is given by

$$W_{12} = \int_c \mathbf{F} \cdot d\mathbf{r} = \int_c m \frac{d\mathbf{v}}{dt} \cdot d\mathbf{r} = \int_{\mathbf{r}_1}^{\mathbf{r}_2} m \, d\mathbf{v} \cdot \mathbf{v} = \int_{\mathbf{r}_1}^{\mathbf{r}_2} \frac{m}{2} d(\mathbf{v}^2)$$

$$= \frac{m}{2}(\mathbf{v}_2^2 - \mathbf{v}_1^2) = T_2 - T_1, \tag{2.1.7}$$

which is the difference in kinetic energies at $\mathbf{r}_2$ and at $\mathbf{r}_1$; $T = (mv^2)/2$, and

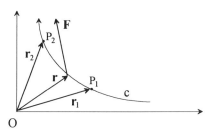

Figure 2.1.1 Line integral of force and work.

$$dW = dT. \tag{2.1.8}$$

With respect to potential energy, in conservative force fields there exists a scalar function U such that $\mathbf{F} = -\operatorname{grad} U(\mathbf{r})$. In such fields, if the work is done from P_1 to P_2 then

$$W_{12} = \int_{\mathbf{r}_1}^{\mathbf{r}_2} \mathbf{F} \cdot d\mathbf{r} = \int_{\mathbf{r}_1}^{\mathbf{r}_0} \mathbf{F} \cdot d\mathbf{r} + \int_{\mathbf{r}_0}^{\mathbf{r}_2} \mathbf{F} \cdot d\mathbf{r}$$

$$= \int_{\mathbf{r}_1}^{\mathbf{r}_0} \mathbf{F} \cdot d\mathbf{r} - \int_{\mathbf{r}_2}^{\mathbf{r}_0} \mathbf{F} \cdot d\mathbf{r} = U(\mathbf{r}_1) - U(\mathbf{r}_2), \tag{2.1.9}$$

where the scalar $U(\mathbf{r})$ is defined as the *potential energy* at $\mathbf{r}$. Hence

$$dW = -dU. \tag{2.1.10}$$

As is well known, the work done in a conservative force field is independent of the path taken by the force, and is a function only of the final position.

From Eq. 2.1.8 and Eq. 2.1.10 follows the law of conservation of energy:

$$dT + dU = 0 \quad \text{and} \quad T + U = \text{const} = E; \tag{2.1.11}$$

E is called the *total energy*. For a conservative force field, the total energy is constant. This is the principle of *conservation of energy*.

2.2 The Two-Body Problem

The two-body problem is an idealized situation in which only two bodies exist that are in relative motion in a force field described by the inverse square law (Eq. 2.1.4). In order to obtain simple analytical results for the motion of celestial bodies or spacecraft, it is assumed that additional bodies are situated far enough from the two-body system, thus no appreciable force is exerted on them from a third body.

In Figure 2.2.1, m_2 exerts an attraction force $\mathbf{F}_1 = m_1 \ddot{\mathbf{r}}_1$ on m_1, and m_1 exerts a force $\mathbf{F}_2 = m_2 \ddot{\mathbf{r}}_2$ on m_2:

$$\mathbf{F}_1 = m_1 \ddot{\mathbf{r}}_1 = G m_1 m_2 \frac{\mathbf{r}_2 - \mathbf{r}_1}{|\mathbf{r}_2 - \mathbf{r}_1|^3}; \tag{2.2.1}$$

$$\mathbf{F}_2 = m_2 \ddot{\mathbf{r}}_2 = G m_1 m_2 \frac{\mathbf{r}_1 - \mathbf{r}_2}{|\mathbf{r}_1 - \mathbf{r}_2|^3} = -\mathbf{F}_1. \tag{2.2.2}$$

From Eq. 2.2.1 and Eq. 2.2.2 we find that

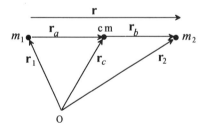

Figure 2.2.1 Displacement vectors in a two-body system.

2.3 / Moment of Momentum

$$\ddot{\mathbf{r}}_2 - \ddot{\mathbf{r}}_1 = -G(m_1 + m_2)\frac{\mathbf{r}_2 - \mathbf{r}_1}{r^3}$$

and, since $\mathbf{r} = \mathbf{r}_2 - \mathbf{r}_1$,

$$\ddot{\mathbf{r}} + G(m_1 + m_2)\frac{\mathbf{r}}{r^3} = 0. \tag{2.2.3}$$

Equation 2.2.3 is the basic equation of motion for the two-body problem. Some properties of the two-body system will be discussed next.

The center of mass (cm) of the two-body system can be found from the equation $\sum m_j \mathbf{r}_j = 0$; it follows that $\mathbf{r}_a m_1 - \mathbf{r}_b m_2 = 0$. In Figure 2.2.1, $\mathbf{r}_c$ is the radius vector from the origin of the reference frame to the cm of the two-body system, and $\mathbf{r}_a$ and $\mathbf{r}_b$ are (respectively) the distances of m_1 and m_2 from the cm. We observe that $\mathbf{r}_a = \mathbf{r}_c - \mathbf{r}_1$ and $\mathbf{r}_b = \mathbf{r}_2 - \mathbf{r}_c$, or (equivalently) that $m_1(\mathbf{r}_c - \mathbf{r}_1) - m_2(\mathbf{r}_2 - \mathbf{r}_c) = 0$. Hence

$$\mathbf{r}_c(m_1 + m_2) = m_1 \mathbf{r}_1 + m_2 \mathbf{r}_2. \tag{2.2.4}$$

After two differentiations of Eq. 2.2.4 with time, and taking into consideration Eq. 2.2.1 and Eq. 2.2.2, we find that

$$\ddot{\mathbf{r}}_c = 0; \quad \dot{\mathbf{r}}_c = \text{const.} \tag{2.2.5}$$

The last equation means that although the cm is not accelerated, the system can be in rectilinear motion with constant velocity.

Using once more the definition of the center of mass of the two-body system, we find that $\mathbf{r}_b = \mathbf{r}_a(m_1/m_2)$ and $\mathbf{r} = \mathbf{r}_a(1 + m_1/m_2)$. After differentiation, we obtain

$$\ddot{\mathbf{r}}_a = \ddot{\mathbf{r}}\frac{m_2}{m_1 + m_2}; \quad \ddot{\mathbf{r}}_b = \ddot{\mathbf{r}}\frac{m_1}{m_1 + m_2}. \tag{2.2.6}$$

If $m_1 \gg m_2$, then $\ddot{\mathbf{r}}_a = \ddot{\mathbf{r}}(m_2/m_1) \to 0$ and $\ddot{\mathbf{r}}_b = \ddot{\mathbf{r}}$. The self-evident conclusion is that the much smaller body m_2 has no influence on the motion of the much larger body m_1, which can be seen as an inertial body as far as the small body is concerned.

2.3 Moment of Momentum

In Figure 2.3.1, $\mathbf{r}$ is the position vector of a particle m that moves in a force field $\mathbf{F}$. The moment of the force $\mathbf{F}$ about the origin O is

$$\mathbf{M} = \mathbf{r} \times \mathbf{F}. \tag{2.3.1}$$

The *moment of momentum* (also called *angular momentum*) about O is defined as

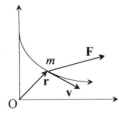

Figure 2.3.1 The moment produced by a force.

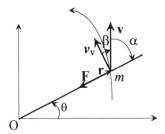

Figure 2.3.2 Radial and transverse components of a body velocity.

$$\mathbf{h} = m(\mathbf{r} \times \mathbf{v}) = \mathbf{r} \times (m\mathbf{v}) = \mathbf{r} \times \mathbf{p}, \tag{2.3.2}$$

where $m\mathbf{v} = \mathbf{p}$ is the linear momentum of the particle. Next, we have

$$\frac{d\mathbf{h}}{dt} = \frac{d}{dt}(\mathbf{r} \times m\mathbf{v}) = \dot{\mathbf{r}} \times (m\mathbf{v}) + \mathbf{r} \times \frac{d}{dt}(m\mathbf{v})$$

$$= \mathbf{v} \times (m\mathbf{v}) + \mathbf{r} \times \frac{d}{dt}(m\mathbf{v}) = 0 + \mathbf{r} \times \mathbf{F} = \mathbf{M}. \tag{2.3.3}$$

The last equation states the very important fact that the moment acting on a particle equals the time rate of change of its angular momentum. This statement is also true if the mass m is variable or if the force is nonconservative.

If the motion of a body takes place in a force field characterized by the inverse square law, then the moment of momentum of the body remains constant. To show this, consider Figure 2.3.2. The central force is located at the origin O. The central force $\mathbf{F}$ acts along the radius vector $\mathbf{r}$ from O to the body with mass m. Since $\mathbf{r}$ is collinear with $\mathbf{F}$, $\mathbf{r} \times \mathbf{F} = \mathbf{M} = 0$. Using Eq. 2.3.3, for a mass of unity we find that $d\mathbf{h}/dt = 0$ and $\mathbf{h} = \mathbf{r} \times \mathbf{v} = \text{const}$. The vector $\mathbf{h}$ is called the *specific* angular momentum; it is perpendicular to both $\mathbf{r}$ and $\mathbf{v}$, and is also constant in space. This means that the motion of the particle takes place in a plane.

In Figure 2.3.2 we define α as the direction of the vector $\mathbf{v}$ relative to $\mathbf{r}$, and β as the direction of $\mathbf{v}$ relative to the local horizontal. Since $\mathbf{h} = \mathbf{r} \times \mathbf{v}$ is a vector cross product, $h = rv \sin(\alpha) = rv \cos(\beta)$. However, since $v_v = v \cos(\beta)$ is the component of $\mathbf{v}$ perpendicular to $\mathbf{r}$, the absolute value of the angular momentum is

$$h = rv \cos(\beta) = rv_v = r\left(r\frac{d\theta}{dt}\right) = r^2 \frac{d\theta}{dt}. \tag{2.3.4}$$

2.4 Equation of Motion of a Particle in a Central Force Field

2.4.1 *General Equation of Motion of a Body in Keplerian Orbit*

Because the motion takes place in a plane, it is easier to solve the equation of motion in polar form. In Figure 2.4.1, $\mathbf{i}$ and $\mathbf{j}$ are unit vectors in the directions (respectively) of $\mathbf{r}$ and of v_v, the component of the velocity vector perpendicular to the radius vector $\mathbf{r}$. Since $\mathbf{r} = \mathbf{i}r$, we find that

2.4 / Equation of Motion of a Particle in a Force Field

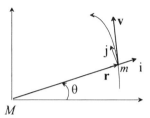

Figure 2.4.1 Radial and transverse components of motion in a plane.

$$\frac{d\mathbf{i}}{dt} = \frac{d\mathbf{i}}{d\theta}\frac{d\theta}{dt} = \mathbf{j}\dot{\theta}, \qquad \frac{d\mathbf{j}}{dt} = \frac{d\mathbf{j}}{d\theta}\frac{d\theta}{dt} = -\mathbf{i}\frac{d\theta}{dt},$$

$$\frac{d\mathbf{r}}{dt} = \frac{d\mathbf{i}}{dt}r + \mathbf{i}\frac{dr}{dt} = r\frac{d\theta}{dt}\mathbf{j} + \mathbf{i}\frac{dr}{dt}, \quad \text{and}$$

$$\frac{d^2\mathbf{r}}{dt^2} = \frac{dr}{dt}\left(\mathbf{j}\frac{d\theta}{dt}\right) + r\left\{-\mathbf{i}\left(\frac{d\theta}{dt}\right)^2 + \mathbf{j}\frac{d^2\theta}{dt^2}\right\} + \frac{d^2r}{dt^2}\mathbf{i} + \frac{dr}{dt}\frac{d\theta}{dt}\mathbf{j}$$

$$= \mathbf{i}\left\{\frac{d^2r}{dt^2} - r\left[\frac{d\theta}{dt}\right]^2\right\} + \mathbf{j}\left\{2\frac{dr}{dt}\frac{d\theta}{dt} + r\frac{d^2\theta}{dt^2}\right\} = -\frac{\mu}{r^2}\mathbf{i} = F(\mathbf{r}) \qquad (2.4.1)$$

with $\mu = GM$. It follows that

$$2\frac{dr}{dt}\frac{d\theta}{dt} + r\frac{d^2\theta}{dt^2} = \frac{1}{r}\frac{d}{dt}\left(r^2\frac{d\theta}{dt}\right) = 0$$

and hence

$$h = r^2\frac{d\theta}{dt} = \text{const} \qquad (2.4.2)$$

(see also Eq. 2.3.4). It also follows that

$$\frac{d^2r}{dt^2} - r\left(\frac{d\theta}{dt}\right)^2 = -\frac{\mu}{r^2}. \qquad (2.4.3)$$

Equation 2.4.3 is nonlinear and cannot be solved directly, but substitution of the variable r by $1/u$ allows an analytical closed-form solution. If $r = 1/u$ then

$$\frac{dr}{dt} = -\frac{1}{u^2}\frac{du}{dt} = -\frac{1}{u^2}\frac{du}{d\theta}\frac{d\theta}{dt}. \qquad (2.4.4)$$

From Eq. 2.4.2, $h = r^2(d\theta/dt)$, hence also $d\theta/dt = hu^2$. It follows that

$$\frac{dr}{dt} = -\frac{1}{u^2}\frac{d\theta}{dt}\frac{du}{d\theta} = -h\frac{du}{d\theta}; \qquad (2.4.5)$$

$$\frac{d^2r}{dt^2} = -h\frac{d}{dt}\frac{du}{d\theta} = -h\frac{d}{d\theta}\frac{du}{d\theta}\frac{d\theta}{dt} = -h\frac{d^2u}{d\theta^2}\frac{d\theta}{dt} = -h^2u^2\frac{d^2u}{d\theta^2}. \qquad (2.4.6)$$

Use Eq. 2.4.6 in Eq. 2.4.3 to obtain

$$-h^2u^2\frac{d^2u}{d\theta^2} - \frac{1}{u}h^2u^4 = -\mu u^2 \quad \text{or}$$

$$\frac{d^2u}{d\theta^2} + u = \frac{\mu}{h^2}, \tag{2.4.7}$$

which is a second-order linear equation for u whose solution is harmonic in θ:

$$u = \frac{\mu}{h^2} + c\cos(\theta - \theta_0). \tag{2.4.8}$$

If $\theta = \theta_0$, then $u = u_{max}$ and $r = r_{min} = 1/u_{max}$.

To find the integration constant c, we use the energy equation (Eq. 2.1.11) for a unit mass $m = 1$. In this case $E = v^2/2 - \mu/r$. This relation for E is called the *total energy per unit mass*. The terms $v^2/2$ and μ/r are identified (respectively) as the kinetic and potential energy of the unit mass. Observing Figure 2.4.1, we can write

$$v^2 = \left(\frac{dr}{dt}\right)^2 + \left(r\frac{d\theta}{dt}\right)^2 = h^2\left(\frac{du}{d\theta}\right)^2 + \left(\frac{hu^2}{u}\right)^2 = h^2\left[\left(\frac{du}{d\theta}\right)^2 + u^2\right]. \tag{2.4.9}$$

Taking a derivative of Eq. 2.4.8 yields $du/d\theta = -c\sin(\theta - \theta_0)$, so that together with Eq. 2.4.9 we have

$$v^2 = \left[c^2 + \frac{2\mu}{h^2}c\cos(\theta - \theta_0) + \left(\frac{\mu}{h^2}\right)^2\right]h^2,$$

and E becomes:

$$E = \frac{h^2}{2}c^2 - \frac{1}{2}\frac{\mu^2}{h^2}. \tag{2.4.10}$$

From the last equation, it follows that

$$c = \frac{\mu}{h^2}\sqrt{1 + 2E\frac{h^2}{\mu^2}}. \tag{2.4.11}$$

If we define the *eccentricity* as $e = \sqrt{1 + 2E(h^2/\mu^2)}$ then

$$E = (e^2 - 1)\frac{\mu^2}{2h^2}, \tag{2.4.12}$$

which is an important relationship between the eccentricity and the total energy of a Keplerian orbit.

Substitution of u for $1/r$ leads to the final equation for Keplerian orbits:

$$r = \frac{h^2/\mu}{1 + e\cos(\theta - \theta_0)} = \frac{p}{1 + e\cos(\theta - \theta_0)}; \tag{2.4.13}$$

here $p = h^2/\mu$ is a geometrical constant of the orbit called the *semi-latus rectum* or the *parameter*. Equation 2.4.13 is the equation of a conical section. This is the general orbit equation from which different kinds of orbits evolve – namely, circular, elliptic, parabolic, and hyperbolic. Motion under a central force results in orbits that are one of these conical sections. Such Keplerian orbits will be analyzed next.

2.4.2 Analysis of Keplerian Orbits

Circular Orbits

For circular orbits, the eccentricity is null ($e = 0$) and r, the magnitude of the radius vector of the orbit from the focus, is constant: $r = p = h^2/\mu = [rv\cos(\beta)]^2/\mu$. But in a circular orbit $\beta = 0$ (the velocity of the body is perpendicular to the radius vector **r**), so it follows that

$$v^2 = \mu/r \tag{2.4.14}$$

and the velocity is also constant. The energy is then $E = -\mu^2/2h^2 < 0$.

Elliptic Orbits

For elliptic orbits $0 < e < 1$, and the energy is $E = (e^2-1)\mu^2/2h^2 < 0$. The point on the ellipse at $\theta = 0°$ (point A in Figure 2.4.2) is called the *periapsis*, and the radius vector from the prime focus F of the ellipse to the periapsis is the minimum radius vector from the focus to any other point on the ellipse. Its value is found from Eq. 2.4.13 to be

$$r_p = p/(1+e). \tag{2.4.15}$$

For orbits around the earth, which is considered to be located at the prime focus F, the periapsis is called *perigee*; r_p is the perigee distance from the focus. For orbits around the sun, the periapsis is called *perihelion*.

If $\theta = 180°$ then for point B in Figure 2.4.2 we have

$$r_a = p/(1-e). \tag{2.4.16}$$

Point B is called the *apoapsis* and is the point on the ellipse with the maximum distance from the focus located at F; r_a denotes the apoapsis radius vector. The apoapsis of an elliptic orbit in the solar system is called *aphelion*. The apoapsis of an earth-orbiting spacecraft is called its *apogee*. In this textbook, the "apogee" and "perigee" terminology for the apoapsis and periapsis will be used exclusively.

From Eq. 2.4.15 and Eq. 2.4.16, $r_a/r_p = (1+e)/(1-e)$, from which it follows that

$$e = \frac{r_a - r_p}{r_a + r_p}. \tag{2.4.17}$$

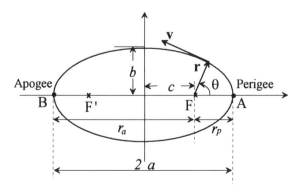

Figure 2.4.2 Geometric definitions of the elliptic orbit.

On the ellipse, the major axis is equal to $2a = r_a + r_p = 2p/(1-e^2)$, hence

$$p = a(1-e^2) = h^2/\mu; \qquad (2.4.18)$$

a will be called the *semimajor axis*.

From Eq. 2.1.11 and Eq. 2.4.12, the total energy of a body with unit mass in the orbit will be:

$$E = \frac{v^2}{2} - \frac{\mu}{r} = \frac{(e^2-1)\mu^2}{2h^2} = \frac{(e^2-1)\mu}{2p} = \frac{(e^2-1)\mu}{2a(1-e^2)} = -\frac{\mu}{2a}, \qquad (2.4.19)$$

from which it follows that

$$\frac{v^2}{2} = \frac{\mu}{r} - \frac{\mu}{2a}. \qquad (2.4.20)$$

The condition for an orbit to be elliptic becomes

$$\frac{v^2}{2} < \frac{\mu}{r}. \qquad (2.4.21)$$

From Eq. 2.4.19 and Eq. 2.4.12, the energy is equal to $-\mu/2a$, which is called the *energy constant*.

For an ellipse, we know that $c = ae$. In Eq. 2.4.18 we found that $p = a(1-e^2)$. It is also easily shown that

$$b = \sqrt{a^2 - c^2} = \sqrt{a^2 - a^2 e^2} = a\sqrt{1-e^2}.$$

Yet $a = p/(1-e^2)$, so it follows that

$$b = \frac{p\sqrt{1-e^2}}{1-e^2} = \frac{p}{\sqrt{1-e^2}}; \qquad (2.4.22)$$

b is called the *semiminor axis* of the elliptic orbit. Moreover,

$$c = \frac{pe}{1-e^2}, \qquad (2.4.23)$$

where c is the distance between the prime focus F and the geometrical center of the ellipse (see Figure 2.4.2).

Parabolic Orbits

Parabolic orbits are of no practical importance. Their special feature is that $E = 0$ and $e = 1$, from which it follows that

$$r = \frac{p}{1 + \cos(\theta)}, \quad r_p = \frac{p}{2}, \quad \text{and} \quad r = \frac{2r_p}{1 + \cos(\theta)}.$$

From the equality $E = 0$ it follows that $a \to \infty$ and also that

$$v^2 = 2\mu/r. \qquad (2.4.24)$$

The velocity in Eq. 2.4.24 is the *escape velocity* necessary to leave the parabolic orbit around the central body located at F in Figure 2.4.2, which also means that the s/c can approach new attracting central bodies, such as the moon. It is interesting also to notice that this escape velocity is larger by a factor of only $\sqrt{2}$ than the velocity of a circular orbit at the same distance r from F.

2.4 / Equation of Motion of a Particle in a Force Field

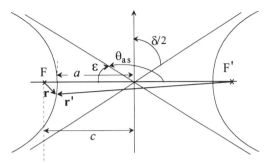

Figure 2.4.3 Geometry of the hyperbolic orbit.

Hyperbolic Orbits

For this important class of orbits, the total energy E is positive, $E > 0$. This means that the kinetic energy of the satellite is larger than its potential energy, so that the s/c is able to leave the gravitational attraction field of the central body. A satellite moving on a hyperbolic orbit does not revolve about the central body.

Because $E > 0$, it follows from Eq. 2.4.19 that $a = -\mu/(2E) < 0$. If we want to stay with $a > 0$, then Eq. 2.4.18 must be changed slightly to $p = a(e^2 - 1)$, so that $p > 0$ and

$$r = \frac{a(e^2 - 1)}{1 + e\cos(\theta)} = \frac{p}{1 + e\cos(\theta)}. \tag{2.4.25}$$

However, as r increases to infinity, $1 + e\cos(\theta)$ must decrease to zero since p is constant for a given orbit. In this situation (see Figure 2.4.3), the equation of the asymptotes becomes

$$\cos(\theta_\infty) = \cos(\theta_{as}) = -1/e. \tag{2.4.26}$$

Because $\epsilon = \pi - \theta_\infty = \pi/2 - \delta/2$ implies $\pi/2 + \delta/2 = \theta_\infty$, we have

$$\cos\left(\frac{\pi}{2} + \frac{\delta}{2}\right) = \cos(\theta_\infty) = -\frac{1}{e} = -\sin\left(\frac{\pi}{2}\right)\sin\left(\frac{\delta}{2}\right)$$

and, finally,

$$\sin\left(\frac{\delta}{2}\right) = \frac{1}{e}. \tag{2.4.27}$$

Hyperbolic orbits are useful for transplanetary spacecraft voyages. Their behavior at $r \to \infty$ is of special interest (Kaplan 1976); see Figure 2.4.4. Consider the moment of momentum at a point **x** very far away from the focus F. We have $h = Vr\sin(\alpha) = V\Delta = V_\infty\Delta$. The total energy at infinity is

$$E = \frac{V_\infty^2}{2} - \frac{\mu}{r_\infty} = -\frac{\mu}{2a},$$

so that

$$a = -\frac{\mu}{V_\infty^2}. \tag{2.4.28}$$

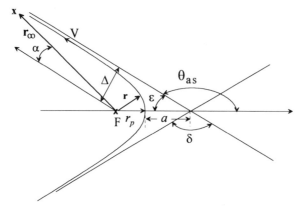

Figure 2.4.4 Geometry of the hyperbolic orbit for transplanetary voyages; adapted from Kaplan (1976) by permission of John Wiley & Sons.

But also $p = a(e^2 - 1)$, so

$$p = \frac{\mu}{V_\infty^2}(e^2 - 1) = \frac{V_\infty^2 \Delta^2}{\mu} = \frac{h^2}{\mu}$$

and

$$\Delta V_\infty = \frac{\mu}{V_\infty}\sqrt{e^2 - 1}. \tag{2.4.29}$$

If Δ and V_∞ are known, we can find a, e, and also δ:

$$e^2 = 1 + \frac{V_\infty^4 \Delta^2}{\mu^2}. \tag{2.4.30}$$

The practical meaning of these equations is as follows. A s/c navigating toward a far planet has a known velocity V_∞. Also, knowledge of its direction of motion allows us to calculate Δ. The parameters of the orbit that the s/c will follow can be obtained from Eq. 2.4.28 and Eq. 2.4.30. In a similar way, the s/c can be forced to approach the planet in a desired orbit path by appropriately manipulating Δ and V_∞. For more about interplanetary transfers, the reader is referred to Kaplan (1976) and Battin (1990).

2.5 Time and Keplerian Orbits

2.5.1 *True and Eccentric Anomalies*

The location of a body in any orbit can be described either in terms of its angular deviation from the major axis or by the time elapsed from its passage at the perigee. We use Figure 2.5.1 to help define the *true* and the *eccentric anomalies* of an ellipse. The true anomaly θ is defined as the angle between (i) the major axis pointing to the perigee and (ii) the radius vector from the prime focus F to the moving body. To define the eccentric anomaly, we draw an auxiliary circle with radius a centered at the middle of the major axis. The eccentric anomaly ψ is then defined as shown in Figure 2.5.1.

2.5 / Time and Keplerian Orbits

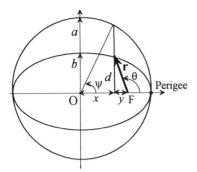

Figure 2.5.1 Geometry for finding the relationship between θ and ψ.

In this section we shall find some important relationships between the true and the eccentric anomalies. Referring to Figure 2.5.1, we have: $x + y = ae = c$, $x = a\cos(\psi)$, and $y = r\cos(180 - \theta) = -r\cos(\theta)$; hence $x + y = a\cos(\psi) - r\cos(\theta) = ae$. Using Eq. 2.4.13:

$$a\cos(\psi) = ae + \frac{a(1-e^2)\cos(\theta)}{1+e\cos(\theta)} = \frac{ae + a\cos(\theta)}{1+e\cos(\theta)} \quad \text{and}$$

$$\cos(\psi) = \frac{e + \cos(\theta)}{1 + e\cos(\theta)}, \quad \sin(\psi) = \frac{\sin(\theta)\sqrt{1-e^2}}{1+e\cos(\theta)}, \tag{2.5.1}$$

$$\cos(\theta) = \frac{\cos(\Psi) - e}{1 - e\cos(\Psi)}, \quad \sin(\theta) = \frac{\sin(\psi)\sqrt{1-e^2}}{1-e\cos(\psi)}. \tag{2.5.2}$$

Also (cf. Deutsch 1963),

$$\tan\left(\frac{\theta}{2}\right) = \sqrt{\frac{1+e}{1-e}} \tan\left(\frac{\psi}{2}\right). \tag{2.5.3}$$

The importance of these equations will become apparent in the following sections. Another important relation may be derived from Figure 2.5.1: substitute $\cos(\theta)$ from Eq. 2.5.2 into Eq. 2.4.13 to obtain

$$r = a[1 - e\cos(\psi)]. \tag{2.5.4}$$

2.5.2 Kepler's Second Law (Law of Areas) and Third Law

In Figure 2.5.2, the radius vector **r** sweeps in a differential period of time the differential area $\Delta A = (\Delta\theta rr)/2$, from which we can derive

$$\frac{dA}{dt} = \frac{1}{2}\left(r^2 \frac{d\theta}{dt}\right) = \frac{1}{2}h = \text{const}; \tag{2.5.5a}$$

in other words, the time rate of change in area is constant, which is Kepler's second law. After integration of this equation, the area swept amounts to

$$A = \tfrac{1}{2}ht. \tag{2.5.5b}$$

Because the area of an ellipse is $A = \pi ab$, if the time period of the orbit is $t = T$ then, according to Eq. 2.5.5b,

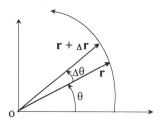

Figure 2.5.2 Geometry for deriving the law of areas.

$$T = \frac{2A}{h} = \frac{2\pi ab}{\sqrt{p\mu}} = \frac{2\pi ab}{\sqrt{a(1-e^2)\mu}} = \frac{2\pi a^2\sqrt{1-e^2}}{\sqrt{a(1-e^2)\mu}} = 2\pi\sqrt{\frac{a^3}{\mu}} = \frac{2\pi}{n}. \quad (2.5.6)$$

Equation 2.5.6 is the third law of Kepler, stating that the orbital period is proportional to $a^{3/2}$. The term n denotes *mean motion,* and $M = n(t - t_p)$ is called the *mean anomaly;* M enables calculation of the time elapsed from the perigee passage at time t_p, given the true anomaly θ.

2.5.3 Kepler's Time Equation

Knowledge of the true anomaly θ allows us to find the value of the eccentric anomaly ψ. Knowing ψ, the elapsed time from the periapsis passage can be computed. For that purpose, the law of areas is used in the following way. In Figure 2.5.3, we observe that $t_M = (T/\pi ab)S(LOM)$, where S stands for the area of planar surface and $t_M = t - t_p$. Hence

$$S(LOM) = \frac{b}{a}S(LOM') \quad \text{and}$$

$$S(LOM') = \frac{\psi}{2\pi}(\pi a^2) - S(OQM') = \frac{\psi}{2\pi}\pi a^2 - \frac{1}{2}ac\sin(\psi)$$

$$= \frac{\psi}{2\pi}(\pi a^2) - \frac{1}{2}a^2 e\sin(\psi) = \frac{\psi}{2}a^2 - \frac{1}{2}a^2 e\sin(\psi),$$

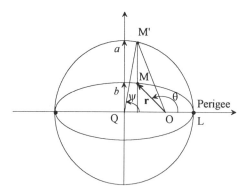

Figure 2.5.3 Derivation of Kepler's time equation.

2.5 / Time and Keplerian Orbits

so it follows that

$$t_M = \frac{T}{\pi ab} \frac{b}{a} \frac{a^2}{2} [\psi - e\sin(\psi)] = \frac{T}{2\pi}[\psi - e\sin(\psi)]; \qquad (2.5.7)$$

recall that $t_M = t - t_p$ and t_p is the passage time from the perigee. We can also write Eq. 2.5.7 in the more compact form

$$t_M \frac{2\pi}{T} = (t - t_p)\frac{2\pi}{T} = (t - t_p)n = M = \psi - e\sin(\psi). \qquad (2.5.8)$$

Knowing the eccentricity e and the true anomaly θ, we can find the eccentric anomaly ψ by Eq. 2.5.1 and the elapsed time $t - t_p$ from the perigee passage using Eq. 2.5.7.

A different problem is to find the position of a satellite θ at any time, given the orbital parameters. The mean anomaly is found from Eq. 2.5.8, but unfortunately this equation is not solvable in closed form; it must be solved numerically. After solving for ψ, θ can be found from Eq. 2.5.2.

EXAMPLE 2.5.1 Let us take as an example a *Molniya*-type satellite orbit, with an apogee altitude of 40,000 km and a perigee altitude of 500 km. We need to compute the fraction of time that the satellite remains in its operational range of ±30° from the apogee.

For this orbit, $e = 0.7417$ (Eq. 2.4.17). The orbital period is equal to $T = 43,243.3$ sec (Eq. 2.5.6). The mean motion $n = 1/6,882.3882$ rad/sec (Eq. 2.5.6). The operational range is ±30° from the apogee. This means that the satellite is outside the operational range for $|\theta| < 150°$ from the perigee. Next, we shall compute the time for which the satellite is not in the operational range, $M = n(t - t_p)$. If $\theta = 150°$ then $\psi = 110.34°$ (Eq. 2.5.1). Use Eq. 2.5.8 to find $M = 1.2295$, from which we also find $t - t_p = 8,461.87$ sec. Finally, the operative time period is $T - 2(t - t_p) = 26,319.645$ sec, which is more than 60% of the orbital time period.

A solution of Eq. 2.5.8 in the form of a trigonometric series was developed by Lagrange (see Pritchard and Sciulli 1986):

$$\psi = M + 2\sum_{n=1}^{\infty} \frac{1}{n} J_n(ne)\sin(nM), \qquad (2.5.9)$$

where J_n is a Bessel function of the first kind of order n. The true anomaly can also be expressed directly in the series form:

$$\theta = M + \left[2e - \frac{e^3}{4}\right]\sin(M) + \frac{5}{4}e^2\sin(2M) + \frac{13}{12}e^3\sin(3M). \qquad (2.5.10)$$

Equation 2.5.7 is valid for motion in an elliptic orbit. For a hyperbolic orbit (in which $e > 1$), we must define $\sinh(\Phi) = [\sqrt{e^2-1}\sin(\theta)]/[1 + e\cos(\theta)]$, and Kepler's time equation takes the form

$$e\sinh(\Phi) - \Phi = n_{\text{hyp}}(t - t_p) = M, \qquad (2.5.11)$$

with $n_{\text{hyp}} = \sqrt{\mu/a^3}$. The subscript "hyp" stands for *hyperbolic*.

As in the previous case of an elliptic orbit, given the eccentricity e and true anomaly θ, we can find the mean anomaly M. In fact, by integrating Eq. 2.4.2 with time, we can obtain equations that relate the time $t - t_p$ directly with the true anomaly θ for both elliptic and hyperbolic orbits. From Eq. 2.4.2,

$$\frac{p^2 d\theta}{[1+e\cos(\theta)]^2} = dt\sqrt{p\mu} = dt\sqrt{a(1-e^2)\mu}$$

(see Thomson 1986). Integrating the left side of the equation with θ and the right side of the equation with time, we can obtain the following final equations for elliptic and hyperbolic orbits:

$$t_{\text{ell}} = \frac{a^{3/2}}{\sqrt{\mu}} \left\{ 2 \arctan\left[\sqrt{\frac{1-e}{1+e}} \tan\left(\frac{\theta}{2}\right)\right] - \frac{e\sqrt{1-e^2}\sin(\theta)}{1+e\cos(\theta)} \right\} \quad (2.5.12)$$

for $e < 1$, and

$$t_{\text{hyp}} = \frac{a^{3/2}}{\sqrt{\mu}} \left\{ \frac{e\sqrt{e^2-1}\sin(\theta)}{1+e\cos(\theta)} - \ln \frac{\sqrt{e+1} + \sqrt{e-1}\tan(\theta/2)}{\sqrt{e+1} - \sqrt{e-1}\tan(\theta/2)} \right\} \quad (2.5.13)$$

for $e > 1$.

This completes our analytical treatment of Keplerian planar orbits. It is important to summarize the parameters that represent the orbital motion in a plane. The physical parameters sufficient to define an orbit are its total energy E and its momentum h. The geometric parameters sufficient to define the orbit are the semimajor axis a and the eccentricity e. The mean anomaly M enables finding the location of the moving body in the orbit with time. In the next section we shall find that three more parameters are necessary to define an orbit in space.

2.6 Keplerian Orbits in Space

2.6.1 *Definition of Parameters*

For earth-orbiting spacecraft, it is common to define an inertial coordinate system with the center of mass of the earth as its origin (a geocentric system) and whose direction in space is fixed relative to the solar system. Astronomical measurements have shown that this system can be a suitable inertial system for practical purposes. The earth moves in an almost circular orbit around the sun with a long period (a whole year), so its motion is practically unaccelerated for our purposes, and the reference system can be accepted as being inertial or Galilean.

The **Z** axis is the axis of rotation of the earth in a positive direction, which intersects the celestial sphere at the *celestial pole*. The **X-Y** plane of this coordinate system is taken as the equatorial plane of the earth, which is perpendicular to the earth's axis of rotation.

The direction of the axis of rotation of the earth relative to the inertial star system is not constant, since it is perturbed by forces due to the sun and the moon. The consequences are a precessional motion due to the sun (with a period of 25,800 years and an amplitude of 23.5°), together with a superimposed periodic nutational motion due to the moon (with a period of 18.6 years and an amplitude of 9″21).

Next, we shall define the inertial **X** axis of the geocentric inertial system. As is well known, the earth's equatorial plane is inclined to the *ecliptic plane,* which is the

2.6 / Keplerian Orbits in Space

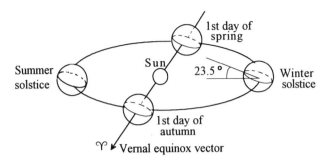

Figure 2.6.1 Vernal equinox inertial direction.

plane of the earth orbit around the sun, by an angle of 23.5° (see Figure 2.6.1). Both planes intersect along a line that is quasi-inertial in space with respect to the stars. The **X** axis of our inertial system coincides with this line, which is called the *vernal equinox* vector (or direction) and which intersects the celestial sphere at a point named the *first point of the Aries* ♈, or the *vernal point*. The third axis **Y** completes an orthogonal right-handed system.

Unfortunately, both the equatorial plane and the ecliptic plane move slowly with respect to the true celestial inertial coordinate system, centered in the center of mass of the solar system. The planets affect the orientation of the ecliptic plane in the slow rotational motion of *planetary precession*. As the **Z** axis precesses, so does the equatorial plane, which is perpendicular to it. The conclusion is that the geocentric inertial system moves slowly relative to the stars, and it is necessary to define the system with reference to a certain date. For instance, it was once common to classify the stars relative to the geocenter inertial system as defined on January 1, 1950. Today, it is common to use as reference the geocentric system for the year 2000.

Having defined the geocentric coordinate system, we can now discuss the three additional parameters necessary to position an orbit in space. In Figure 2.6.2 (see also Bate, Mueller, and White 1971 or Bernard 1983), the plane of the orbit is inclined

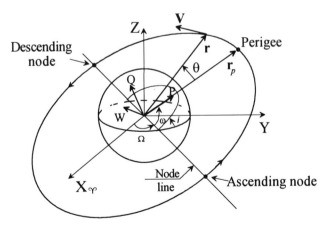

Figure 2.6.2 Parameters that define location of orbits in space.

to the **X-Y** plane – which is also the equatorial plane of the earth – by an angle i, the inclination of the orbit. The orbital plane and the equatorial plane intersect at the *node line*. The angle in the equator plane that separates the node line from the $\mathbf{X}_\Upsilon$ axis is called the *right ascension*, Ω. In the orbit plane, **r** is the radius vector to the moving body; $\mathbf{r}_p$ is the radius vector to the perigee of the orbit. The angle between $\mathbf{r}_p$ and the node line is ω, the *argument of perigee*. These three parameters, together with the three parameters (a, e, and M) of the orbit in the plane, complete a system of six parameters that suffices to define the location in space of a body moving in any Keplerian orbit. These parameters are known as the *classical orbit parameters*, which are redefined as follows:

(1) a, the semimajor axis;
(2) e, the eccentricity;
(3) i, the inclination;
(4) Ω, the right ascension of the ascending node;
(5) ω, the argument of perigee; and
(6) $M = n(t - t_0)$, the mean anomaly (where n is the mean motion).

It is convenient to define the vector $[\boldsymbol{\alpha}] = [a \ e \ i \ \Omega \ \omega \ M]^T$.

Although the classical parameters completely define an orbit in space, some of them (e.g. Ω) are poorly defined if the inclination angle i is very small, as with geostationary orbits (Section 2.9.1). In such cases, a variation of the six listed parameters will be preferred. Orbits with very small inclinations are called *equatorial* orbits.

The six parameters are convenient for defining an orbit in space in the inertial system defined by its three axes $\mathbf{X}_\Upsilon$, **Y**, and **Z**, as shown in Figure 2.6.2. However, it can also be useful to express the location of a moving body in other parameters, such as Cartesian or polar.

2.6.2 *Transformation between Cartesian Coordinate Systems*

The following basic information will be useful through the entire textbook in many different variations. In Figure 2.6.3, **R** is the radius vector of a point relative to the origin of both Cartesian systems $[\mathbf{I}, \mathbf{J}]$ and $[\mathbf{i}, \mathbf{j}]$. System $[\mathbf{i}, \mathbf{j}]$ is rotated by an angle Ω with respect to system $[\mathbf{I}, \mathbf{J}]$. The components of **R** are respectively X, Y and x, y in both systems. See also Appendix A. Here, **I**, **J** are unit vectors in $[X, Y]$ and **i**, **j** are unit vectors in $[x, y]$. For a transformation in the plane,

$$\mathbf{R} = X\mathbf{I} + Y\mathbf{J} = x\mathbf{i} + y\mathbf{j}. \tag{2.6.1}$$

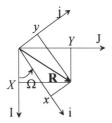

Figure 2.6.3 Two-dimensional coordinate transformation.

2.6 / Keplerian Orbits in Space

By taking the scalar product of the vector **R** and the unit vector **i**, we have

$$X\mathbf{I}\cdot\mathbf{i} + Y\mathbf{J}\cdot\mathbf{i} = x\mathbf{i}\cdot\mathbf{i} + y\mathbf{j}\cdot\mathbf{i}.$$

Since $\mathbf{I}\cdot\mathbf{i} = \cos(\Omega)$, $\mathbf{J}\cdot\mathbf{i} = \sin(\Omega)$, $\mathbf{i}\cdot\mathbf{i} = 1$, and $\mathbf{i}\cdot\mathbf{j} = 0$, it follows that $x = X\cos(\Omega) + Y\sin(\Omega)$. Similarly, by taking the scalar product of **R** and the unit vector **j**, we obtain

$$X\mathbf{I}\cdot\mathbf{j} + Y\mathbf{J}\cdot\mathbf{j} = x\mathbf{i}\cdot\mathbf{j} + y\mathbf{j}\cdot\mathbf{j}.$$

Since $\mathbf{I}\cdot\mathbf{j} = -\sin(\Omega)$, $\mathbf{J}\cdot\mathbf{j} = \cos(\Omega)$, $\mathbf{i}\cdot\mathbf{j} = 0$, and $\mathbf{j}\cdot\mathbf{j} = 1$, we have $y = X[-\sin(\Omega)] + Y\cos(\Omega)$.

The rotation is about an axis **K**, perpendicular to both the **I** and **J** axes. Looking on that transformation as a three-dimensional transformation in space with Z along the **K** axis and z along the **k** axis, we have $Z = z$. Finally, we get

$$\begin{bmatrix} x \\ y \\ z \end{bmatrix} = \begin{bmatrix} \cos\Omega & \sin\Omega & 0 \\ -\sin\Omega & \cos\Omega & 0 \\ 0 & 0 & 1 \end{bmatrix} \begin{bmatrix} X \\ Y \\ Z \end{bmatrix}; \qquad (2.6.2)$$

see also Appendix A. This three-dimensional transformation can be performed as many times as necessary to achieve a desired overall transformation in space around different axes. For simplicity, let us define this transformation by the angle Ω as $[\mathbf{r}] = [\mathbf{A}(\Omega)][\mathbf{R}]$.

For example, suppose we wish to perform three subsequent transformations from inertial coordinates **X**, **Y**, and **Z** to orbit coordinates **P**, **Q**, and **W** as defined in Figure 2.6.2. (**P** is a unit vector directed from the center of the orbit to the perigee, **W** is a unit vector along the momentum axis of the orbit, $\mathbf{h} = \mathbf{r} \times \mathbf{v}$, and $\mathbf{Q} = \mathbf{P} \times \mathbf{W}$.) The transformation proceeds as follows:

$$\begin{bmatrix} P \\ Q \\ W \end{bmatrix} = [\mathbf{A}_z(\omega)][\mathbf{A}_x(i)][\mathbf{A}_z(\Omega)] \begin{bmatrix} X \\ Y \\ Z \end{bmatrix}$$

$$= \begin{bmatrix} c\omega & s\omega & 0 \\ -s\omega & c\omega & 0 \\ 0 & 0 & 1 \end{bmatrix} \begin{bmatrix} 1 & 0 & 0 \\ 0 & ci & si \\ 0 & -si & ci \end{bmatrix} \begin{bmatrix} c\Omega & s\Omega & 0 \\ -s\Omega & c\Omega & 0 \\ 0 & 0 & 1 \end{bmatrix} \begin{bmatrix} X \\ Y \\ Z \end{bmatrix}. \qquad (2.6.3)$$

In this equation, c_- and s_- stand for $\cos_-$ and $\sin_-$, and ω, i, and Ω are as defined in Figure 2.6.2; $[\mathbf{A}_d(\_)]$ stands for a transformation matrix about an axis d by an angle $\_$.

In the following sections, it will be necessary to define what is known as a *local coordinate system*. In this system, **R** is a unit vector along the radius vector **r**, **W** is a unit vector along the momentum vector **h**, and **S** completes a right-handed orthogonal coordinate system. In this system, the third rotation is by an angle $\theta + \omega$, the argument of the location of the body from the ascending node. The overall transformation is found to be

$$\begin{bmatrix} R \\ S \\ W \end{bmatrix} = [\mathbf{A}_z(\omega+\theta)][\mathbf{A}_x(i)][\mathbf{A}_z(\Omega)] \begin{bmatrix} X \\ Y \\ Z \end{bmatrix} =$$

$$= \begin{bmatrix} c(\omega+\theta)c\Omega - cis(\omega+\theta)s\Omega & c(\omega+\theta)s\Omega + s(\omega+\theta)cic\Omega & s(\omega+\theta)si \\ -s(\omega+\theta)c\Omega - cis\Omega c(\omega+\theta) & -s(\omega+\theta)s\Omega + c(\omega+\theta)cic\Omega & c(\omega+\theta)si \\ sis\Omega & -sic\Omega & ci \end{bmatrix} \begin{bmatrix} X \\ Y \\ Z \end{bmatrix}.$$

(2.6.4)

2.6.3 Transformation from $\alpha = [a \, e \, i \, \Omega \, \omega \, M]^T$ to $[\mathbf{v}, \mathbf{r}]$

Here our problem is to find the Cartesian coordinates of a satellite in the inertial frame defined by $\mathbf{X}_\Upsilon$, $\mathbf{Y}$, and $\mathbf{Z}$ (shown in Figure 2.6.2) given the six classical orbit parameters a, e, i, Ω, ω, and M. Since a Keplerian orbit is in a plane, we can define a coordinate system $\mathbf{x}, \mathbf{y}$ in a plane with $z = 0$; see Figure 2.6.4. In this figure,

$$x = a\cos(\psi) - c = a\cos(\psi) - ae \quad \text{and}$$
$$y = [b\sin(\psi)a]/a = a\sin(\psi)\sqrt{1-e^2}, \tag{2.6.5}$$

where $\mathbf{r} = \hat{\mathbf{i}}x + \hat{\mathbf{j}}y$. Also, for a plane orbit, $z = 0$. The term Ψ can be found by using the Kepler equation for a given $M = nt$, since $\Psi - e\sin(\Psi) = M = nt$. To find Ψ, a simple numeric procedure (based on successive approximations) may be used if $e < 1$:

$$\Psi_0 = M = n(t - t_0) \quad (t_0 = \text{time of passage at the perigee});$$
$$\Psi_1 = M + e\sin(\Psi_0); \; \Psi_2 = M + e\sin(\Psi_1); \; \ldots; \; \Psi_{n+1} = M + e\sin(\Psi_n).$$

This sequence converges for $e < 1$. For $e > 1$, a different procedure is necessary for convergence (see e.g. Battin 1990, Chobotov 1991). Given Ψ, from Eq. 2.6.5 we can solve for components x and y of the satellite in the orbit coordinate system shown in Figure 2.6.4.

To find $\mathbf{r}$ and $d\mathbf{r}/dt$ (see Figures 2.6.2 and 2.6.4), let us put $\hat{\mathbf{i}} \equiv \mathbf{P}$ and $\hat{\mathbf{j}} \equiv \mathbf{Q}$; then

$$\mathbf{r} = a[\cos(\psi) - e]\mathbf{P} + a\sqrt{1-e^2}\sin(\psi)\mathbf{Q} = x\hat{\mathbf{i}} + y\hat{\mathbf{j}} = x\mathbf{P} + y\mathbf{Q} \tag{2.6.6}$$

(Bate et al. 1971, Bernard 1983). The inverse transformation of Eq. 2.6.3, with x, y, and z known, takes the following form:

$$\begin{bmatrix} X \\ Y \\ Z \end{bmatrix} = [\mathbf{A}_z(\Omega)]^{-1}[\mathbf{A}_x(i)]^{-1}[\mathbf{A}_z(\omega)]^{-1} \begin{bmatrix} x \\ y \\ 0 \end{bmatrix}; \tag{2.6.7}$$

X, Y, and Z are the inertial coordinates of the moving satellite.

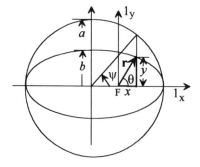

Figure 2.6.4 Transformation from orbit parameters to Cartesian inertial frame coordinates.

2.6 / Keplerian Orbits in Space

To calculate the velocity vector,

$$\mathbf{v} = \frac{d\mathbf{r}}{dt} = \frac{d\mathbf{r}}{d\Psi}\frac{d\Psi}{dt},$$

we must first find $d\Psi/dt$ from Kepler's time equation:

$$\frac{dM}{dt} = n = \frac{d\Psi}{dt} - e\cos(\Psi)\frac{d\Psi}{dt},$$

from which, together with Eq. 2.5.4, we derive

$$\frac{d\Psi}{dt} = \frac{n}{1 - e\cos(\Psi)} = \frac{an}{r}. \tag{2.6.8}$$

Differentiating Eq. 2.6.6, we use Eq. 2.6.8 to obtain

$$\frac{d\mathbf{r}}{dt} = \mathbf{v} = \frac{a^2 n}{r}[-\sin(\Psi)\mathbf{P} + \sqrt{1-e^2}\cos(\Psi)\mathbf{Q}] \tag{2.6.9}$$

(see also Deutsch 1963, Balmino 1980). We have thus found $\mathbf{v} = d\mathbf{r}/dt$ also. Knowing the components of $\mathbf{v}$ on the $\mathbf{P}$ and $\mathbf{Q}$ axes, we can use once more the transformation in Eq. 2.6.3 to find the velocity components in the inertial coordinate system. The inverse transformation will be similar to that of Eq. 2.6.7.

2.6.4 Transformation from $[\mathbf{v}, \mathbf{r}]$ to $\alpha = [a\ e\ i\ \Omega\ \omega\ M]^T$

If $\mathbf{r}$ and $\mathbf{v}$ are known in their Cartesian coordinates X, Y, Z, V_x, V_y, and V_z, then the classical orbit parameters can be easily computed. From Eq. 2.4.20 we find

$$a = \frac{\mu}{2\left[\frac{\mu}{r} - \frac{v^2}{2}\right]} \tag{2.6.10}$$

(Bate et al. 1971, Chobotov 1991), and from $\mathbf{h} = \mathbf{r} \times \mathbf{v} = |\mathbf{h}|\mathbf{W}$ we can find i and Ω. To find i we use the relation

$$\cos(i) = h_z/|\mathbf{h}|. \tag{2.6.11}$$

To find Ω with the correct sign, we use the relations

$$\sin(\Omega) = \frac{h_x}{\sqrt{h_x^2 + h_y^2}} \quad \text{and} \quad \cos(\Omega) = \frac{-h_y}{\sqrt{h_x^2 + h_y^2}}; \tag{2.6.12}$$

i and Ω are then known. The terms h_x, h_y, and h_z are the components of $\mathbf{h}$, calculated from the Cartesian coordinates of $\mathbf{v}$ and $\mathbf{r}$ by calculating the vector product $\mathbf{r} \times \mathbf{v}$.

Since $p = h^2/\mu = a(1-e^2)$, it follows that

$$e^2 = 1 - h^2/\mu a. \tag{2.6.13}$$

Next, we must calculate ψ. Having found a and e, and knowing $|\mathbf{r}|$ from the coordinates of $\mathbf{r}$, we can calculate $\cos(\psi)$ from Eq. 2.5.4:

$$\psi = \cos^{-1}\left[\frac{a - |\mathbf{r}|}{ae}\right]. \tag{2.6.14}$$

From Eq. 2.5.4, Eq. 2.6.6, and Eq. 2.6.9, we easily find that

$$\sin(\psi) = \frac{\mathbf{r} \cdot \mathbf{v}}{a^2 n e} = \frac{\mathbf{r} \cdot \mathbf{v}}{e\sqrt{\mu a}}. \qquad (2.6.15)$$

From Eq. 2.6.14 and Eq. 2.6.15 we can now calculate ψ with its correct sign. We can find the true anomaly θ from Eq. 2.5.2 and the mean anomaly M from Kepler's time equation (Eq. 2.5.8).

We are left with one more unknown parameter, the argument of perigee ω. To find it, we use the transformation of Eq. 2.6.7, in which x and y are the components of $\mathbf{r}$ in the plane of the orbit:

$$x = r\cos(\theta); \qquad y = r\sin(\theta). \qquad (2.6.16)$$

The components of $\mathbf{r}$ in space are known: X, Y, and Z. Having already found the inclination angle i and the right ascension of the ascending node Ω, we can find the argument of perigee ω from the transformation in Eq. 2.6.7. Inserting x, y, X, Y, and Z in the equation yields the following equalities:

$$\sin(\omega + \theta) = \frac{Z}{r \sin(i)}; \qquad \cos(\omega + \theta) = \frac{X\cos(\Omega) + Y\sin(\Omega)}{r} \qquad (2.6.17)$$

(Bernard 1983). From the equalities in Eqs. 2.6.17 we can finally calculate ω, since θ can be found from Eq. 2.5.2.

2.7 Perturbed Orbits: Non-Keplerian Orbits

2.7.1 *Introduction*

In previous sections, ideal Keplerian orbits were treated under the basic assumptions that the motion of a body in these orbits is a result of the gravitational attraction between two bodies. This ideal situation does not exist in practice. In the solar system, it is true that the mass of the sun is 1,047 times larger than that of the largest planet (Jupiter), but still there exist eight additional planets perturbing the motion of each individual planet in its particular orbit. In fact, the two-body problem of motion is an idealization, and additional forces acting on any moving body must be taken into account. The perturbing forces caused by the additional bodies are conservative field forces, already encountered in Sections 2.1.2 and 2.2. In the case of high-orbit satellites (e.g. geostationary satellites), the effects of the conservative perturbing forces of the sun and the moon on the motion of the satellite cannot be ignored, as we shall see in Section 2.8.3. Their major contribution is to change the inclination of the geostationary orbit.

There are also nonconservative perturbing forces, such as the solar pressure. In the case of geostationary orbits, solar pressure tends to change the eccentricity of the orbit. Another nonconservative force that disturbs the motion of a satellite is the atmospheric force (also called atmospheric drag), which is pertinent to low-altitude orbits circling the earth. Such forces tend to decrease the major axis of the orbit, eventually causing a satellite to fall back to the earth's surface. There are many other perturbing forces that cause ideal Keplerian orbits to acquire strange properties. These forces may strongly affect the orbital motion of spacecraft, and will be treated in the remaining part of this chapter.

2.7.2 The Perturbed Equation of Motion

Equation 2.2.3 is the basic dynamical equation of motion for a Keplerian orbit. It can be rewritten in the following form:

$$\frac{d^2\mathbf{r}}{dt^2} = -\mu \frac{\mathbf{r}}{r^3} = \gamma_K \tag{2.7.1}$$

with initial conditions $\mathbf{r}(0), \mathbf{v}(0)$. For Keplerian orbits,

$$\frac{da}{dt} = \frac{de}{dt} = \frac{d\Omega}{dt} = \frac{d\omega}{dt} = \frac{di}{dt} = 0; \quad \frac{dM}{dt} = n. \tag{2.7.2}$$

For the general case, including perturbing forces of any kind, the equation of motion of the satellite becomes:

$$\frac{d^2\mathbf{r}}{dt^2} = \gamma_K + \gamma_p, \tag{2.7.3}$$

with initial conditions $\mathbf{r}(t_0) = \mathbf{r}_0$ and $\mathbf{v}(t_0) = \mathbf{v}_0$. Here γ_K and γ_p stand (respectively) for the Keplerian and perturbing accelerations caused by the Keplerian and perturbing forces.

Equation 2.7.3 is the general equation for the motion of a body in any orbit. In the following analysis, the perturbation acceleration (force) γ_p is to be appreciably smaller than the Keplerian acceleration (force) γ_K. According to Section 2.6.4, knowing the radius vector $\mathbf{r}$ and the velocity vector $\mathbf{v}$ at any time allows us to find the orbit parameters a, e, i, Ω, ω, and M, which together define the vector $\boldsymbol{\alpha}$. Suppose that, at any time t_0, the perturbing acceleration γ_p is removed. Since we know $\mathbf{r}(t_0)$ and $\mathbf{v}(t_0)$, we can find the evolving Keplerian orbit, called the *osculating orbit;* its parameters are $\boldsymbol{\alpha} = [a \ e \ i \ \Omega \ \omega \ M]^T$. See Figure 2.7.1. In fact, the orbit parameters are dependent also on time, since the perturbing acceleration is dependent on the radius vector $\mathbf{r}$, the velocity vector $\mathbf{v}$, and the time: $\gamma_p = \gamma_p(\mathbf{r}, \mathbf{v}, t)$. For example, the moon's perturbing acceleration on the s/c depends on the moon's position in its orbit relative to the earth.

The equations of motion become

$$\frac{d\mathbf{v}}{dt} = \mathbf{F}(\mathbf{r}, \mathbf{v}, t) \quad \text{and} \quad \frac{d\mathbf{r}}{dt} = \mathbf{v}. \tag{2.7.4}$$

According to Section 2.6.3, the vectors $\mathbf{r}(t)$ and $\mathbf{v}(t)$ can be expressed in terms of the vector $\boldsymbol{\alpha}$ defined previously, whose elements are the six classical orbit parameters. We have $\mathbf{r} = \mathbf{r}(\alpha_i, t)$ and $\mathbf{v} = \mathbf{v}(\alpha_i, t)$. Accordingly, the vector equations of Eq. 2.7.4 can be decomposed into their Cartesian coordinates in terms of α_i as follows:

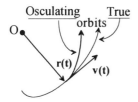

Figure 2.7.1 Definition of true and osculating orbits.

$$\sum_i \frac{\partial v_x}{\partial \alpha_i} \frac{d\alpha_i}{dt} = F_x(\alpha_i, t),$$

$$\sum_i \frac{\partial v_y}{\partial \alpha_i} \frac{d\alpha_i}{dt} = F_y(\alpha_i, t),$$

$$\sum_i \frac{\partial v_z}{\partial \alpha_i} \frac{d\alpha_i}{dt} = F_z(\alpha_i, t),$$

$$\sum_i \frac{\partial r_x}{\partial \alpha_i} \frac{d\alpha_i}{dt} = v_x(\alpha_i),$$

$$\sum_i \frac{\partial r_y}{\partial \alpha_i} \frac{d\alpha_i}{dt} = v_y(\alpha_i),$$

$$\sum_i \frac{\partial r_z}{\partial \alpha_i} \frac{d\alpha_i}{dt} = v_z(\alpha_i).$$

(2.7.5)

The first three of Eqs. 2.7.5 describe the Cartesian components of the body acceleration $d\mathbf{v}/dt$; the last three describe the components of the velocity vector $\mathbf{v} = d\mathbf{r}/dt$.

Finally, an inverse to the set of Eqs. 2.7.5 can be obtained:

$$\frac{da}{dt} = F_a(a, e, i, \Omega, \omega, M, t),$$

$$\frac{de}{dt} = F_e(a, e, i, \Omega, \omega, M, t),$$

$$\frac{di}{dt} = F_i(a, e, i, \Omega, \omega, M, t),$$

$$\frac{d\Omega}{dt} = F_\Omega(a, e, i, \Omega, \omega, M, t),$$

$$\frac{d\omega}{dt} = F_\omega(a, e, i, \Omega, \omega, M, t),$$

$$\frac{dM}{dt} = F_M(a, e, i, \Omega, \omega, M, t).$$

(2.7.6)

In Eqs. 2.7.6, for Keplerian orbits with $\gamma_p = 0$, the first five expressions are equal to zero; the last one is constant, $dM/dt = n$.

In the balance of this chapter, we shall look for the solution of Eqs. 2.7.6 for non-Keplerian orbits with different kinds of perturbing forces.

2.7.3 The Gauss Planetary Equations

Before we can solve Eqs. 2.7.6, we must find a general formulation for the right sides of these equations expressing arbitrary perturbing forces (accelerations). When these formal expressions are included, Eqs. 2.7.6 are known as the *Gauss equations*.

To find the solutions of Eqs. 2.7.6, we shall decompose the perturbing force (acceleration) along the axes of a moving Cartesian frame defined in the following way (see Figure 2.7.2): **R** - along the radius vector **r**; **S** - in the local plane of the osculating

2.7 / Perturbed Orbits: Non-Keplerian Orbits

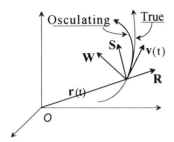

Figure 2.7.2 Definition of an orthogonal axis frame for the perturbing forces; adapted from Bernard (1983) by permission of Cépaduès-Éditions.

orbit, perpendicular to **R**, and in the direction of the satellite motion; **W** – perpendicular to both **R** and **S**, in the direction of the momentum vector **R**×**S**. Any perturbing acceleration (force) can then be expressed as

$$\gamma_p = R\mathbf{R} + S\mathbf{S} + W\mathbf{W} \tag{2.7.7}$$

(Bernard 1983).

We shall exemplify the derivation of the Gauss equation for da/dt. Remember (cf. Eq. 2.1.6) that $E = \int \mathbf{F} \cdot d\mathbf{r} = \int \mathbf{F} \cdot \mathbf{v}\, dt$, so that

$$dE = \mathbf{F} \cdot \mathbf{v}\, dt \Rightarrow \frac{dE}{dt} = \mathbf{F} \cdot \mathbf{v}.$$

Since (according to Eq. 2.4.19) $E = -\mu/2a$, we find that

$$\frac{dE}{dt} = \frac{\mu}{2a^2}\frac{da}{dt} = \mathbf{v} \cdot (\gamma_p + \gamma_K).$$

However, in a Keplerian orbit, E is constant, so the last equation reduces to

$$\frac{dE}{dt} = \mathbf{v} \cdot \gamma_p = \frac{\mu}{2a^2}\frac{da}{dt}. \tag{2.7.8}$$

The velocity vector **v** can be decomposed along the unit vectors **R** and **S**. If β is the angle between **v** and the unit vector **S**, then $\mathbf{v} = v\sin(\beta)\mathbf{R} + v\cos(\beta)\mathbf{S}$, leading to

$$\frac{dE}{dt} = Rv\sin(\beta) + Sv\cos(\beta). \tag{2.7.9}$$

To find β, consider Figure 2.3.2. In this figure, the radial velocity is along the radius vector **r**, and $\sin(\beta) = (1/v)(dr/dt)$. However,

$$\frac{dr}{dt} = \frac{dr}{d\theta}\frac{d\theta}{dt} = \frac{dr}{d\theta}\frac{h}{r^2}.$$

Since $h = r^2(d\theta/dt)$, from Eq. 2.3.4 and Eq. 2.4.13 we have

$$\frac{dr}{dt} = \frac{dr}{d\theta}\frac{\sqrt{p\mu}}{r^2} = \frac{pe\sin(\theta)}{[1+e\cos(\theta)]^2}\frac{\sqrt{p\mu}}{r^2} = \frac{\sqrt{p\mu}\,e\sin(\theta)}{p} \quad \text{and}$$

$$\sin(\beta) = \frac{e\sin(\theta)}{v}\sqrt{\frac{\mu}{p}}.$$

We would like to express $\sin(\beta)$ and $\cos(\beta)$ in terms of the orbital parameters. Use of Eq. 2.4.20 leads to the following expressions for $\sin(\beta)$ and $\cos(\beta)$:

$$\sin(\beta) = \frac{e\sin(\theta)}{\sqrt{1+2e\cos(\theta)+e^2}}; \quad \cos(\beta) = \frac{1+e\cos(\theta)}{\sqrt{1+2e\cos(\theta)+e^2}}. \quad (2.7.10)$$

Substituting Eq. 2.7.10 into Eq. 2.7.9 and using Eq. 2.7.8, we easily obtain the final result:

$$\frac{da}{dt} = \frac{2}{n\sqrt{1-e^2}}\{e\sin(\theta)R + [1+e\cos(\theta)]S\}. \quad (2.7.11)$$

Equation 2.7.11 is the Gauss equation for da/dt.

In a similar way, the remaining expressions for Eqs. 2.7.6 can be developed (see e.g. Escobal 1965, Bernard 1983). The resulting Gauss equations are:

$$\frac{de}{dt} = \frac{\sqrt{1-e^2}}{na}\{\sin(\theta)R + [\cos(\psi)+\cos(\theta)]S\}, \quad (2.7.12)$$

$$\frac{di}{dt} = \frac{1}{na\sqrt{1-e^2}}\frac{r}{a}\cos(\theta+\omega)W, \quad (2.7.13)$$

$$\frac{d\Omega}{dt} = \frac{1}{na\sqrt{1-e^2}}\frac{r}{a}\frac{\sin(\theta+\omega)}{\sin(i)}W, \quad (2.7.14)$$

$$\frac{d\omega}{dt} = \frac{\sqrt{1-e^2}}{nae}\left\{-R\cos(\theta) + \left[1+\frac{1}{1+e\cos(\theta)}\right]\sin(\theta)S - \frac{d\Omega}{dt}\cos(i)\right\}, \quad (2.7.15)$$

$$\frac{dM}{dt} = n + \frac{1-e^2}{nae}\left\{\left[\frac{-2e}{1+e\cos(\theta)} + \cos(\theta)\right]R - \left[1+\frac{1}{1+e\cos(\theta)}\right]\sin(\theta)S\right\}. \quad (2.7.16)$$

Equations 2.7.11–2.7.16 show that if the perturbing force vector is known then the differential changes of all six orbit parameters can be calculated analytically. The force vector can be conservative or nonconservative. For a known perturbing vector γ_p and initial conditions of the vector $\alpha(t_0)$, the Gauss equations can be continuously integrated to calculate the evolution with time of the classical orbit parameters. As an example, let us calculate the influence of aerodynamic forces acting on a satellite's major axis (da/dt).

EXAMPLE 2.7.1 Differential Change of a Due to Perturbing Aerodynamic Forces
For low-orbit satellites, the air density is high enough to produce a perturbing force, which in turn tends to decrease orbit altitude. Suppose that the orbit is circular, and that the density ρ of the air is constant for the entire orbit period. The perturbing force F_p due to atmospheric drag will be $F_p = \frac{1}{2}\rho v^2 C_d S$, where v is the satellite velocity, C_d is the drag coefficient of the satellite, and S is the equivalent satellite surface in the direction of motion of the satellite. The perturbing acceleration will be $\gamma_p = \mathbf{F}_p/m_S$, with m_S the mass of the satellite. In a circular orbit, $\mathbf{F}_p$ is opposite to the direction of motion of the satellite, so that

$$\frac{\mathbf{v}}{|\mathbf{v}|} = \frac{-\gamma_p}{|\gamma_p|} \quad \text{and} \quad \mathbf{v}\cdot\gamma_p = \frac{-v(\rho v^2 C_d S)}{2m_S}.$$

The term $C_d S/(2m_S)$ is known as the *ballistic coefficient*. Using also Eq. 2.7.8 and Eq. 2.4.14, we find that

$$\frac{da}{dt} = -\rho\sqrt{a\mu}\,\frac{C_d S}{m_S}. \tag{2.7.17}$$

This equation illustrates that the rate of decrease of the semimajor axis a is linearly proportional to both the air density ρ (which depends on the altitude of the satellite and on the atmospheric temperature) and the geometric properties of the satellite, but is inversely proportional to its mass. To maintain a constant altitude, the satellite must provide a force to balance the perturbing force $\mathbf{F}_p$, with an inevitable fuel consumption. The necessary fuel mass will be discussed in Section 3.3.3.

2.7.4 Lagrange's Planetary Equations

For cases where the perturbing force is *conservative* – that is, derived from a scalar function $\mathbf{F}_p = -\operatorname{grad} U(\mathbf{r})$ (see Section 2.1.2) – the Gauss equations can be simplified. The development of the Lagrange equations for the differential change of the six orbit parameters can be found in Escobal (1965) and Balmino (1980). The results are summarized as follows:

$$\frac{da}{dt} = \frac{2}{na}\frac{\partial U}{\partial M}, \tag{2.7.18}$$

$$\frac{de}{dt} = \frac{1-e^2}{na^2 e}\frac{\partial U}{\partial M} - \frac{\sqrt{1-e^2}}{na^2 e}\frac{\partial U}{\partial \omega}, \tag{2.7.19}$$

$$\frac{di}{dt} = \frac{-1}{na^2\sqrt{1-e^2}\sin(i)}\left[\frac{\partial U}{\partial \Omega} + \cos(i)\frac{\partial U}{\partial \omega}\right], \tag{2.7.20}$$

$$\frac{d\Omega}{dt} = \frac{1}{na^2\sqrt{1-e^2}\sin(i)}\frac{\partial U}{\partial i}, \tag{2.7.21}$$

$$\frac{d\omega}{dt} = \frac{\sqrt{1-e^2}}{na^2 e}\frac{\partial U}{\partial e} - \frac{\cos(i)}{na^2\sqrt{1-e^2}\sin(i)}\frac{\partial U}{\partial i}, \tag{2.7.22}$$

$$\frac{dM}{dt} = n - \frac{2}{na}\frac{\partial U}{\partial a} - \frac{1-e^2}{na^2 e}\frac{\partial U}{\partial e}. \tag{2.7.23}$$

Equations 2.7.18–2.7.23 are the classical form of Lagrange's planetary equations. Since numerous perturbing forces are conservative, the Lagrange equations are frequently used in the analysis of non-Keplerian orbits. The Lagrange equations can also be integrated with time to calculate the evolutions of the classical orbit parameters.

2.8 Perturbing Forces and Their Influence on the Orbit

2.8.1 *Definition of Basic Perturbing Forces*

Before we can use the Gauss and the Lagrange equations for evaluating perturbations of Keplerian orbits, we must clearly define the perturbing accelerations (forces) acting on the satellite.

One of the most important perturbing forces on earth-orbiting satellites arises from the nonhomogeneity of the earth. The earth globe is not a perfect sphere, and neither is its mass distribution homogeneous. These physical facts produce perturbing accelerations on the moving body. The consequences of these accelerations are variations of the orbital parameters of earth-orbiting satellites. Analysis of these perturbing forces will be performed in Section 2.8.2.

As already mentioned, a true Keplerian orbit is obtained for a two-body system. The existence of additional celestial bodies produces perturbing forces with the heavy consequence that a three- (or more) body problem must be solved. For such problems, a closed-form analytical solution might not exist. Moreover, we shall see in Section 2.8.3 that the gravitational perturbing forces of the sun and the moon cause serious complications in high-altitude geostationary orbits.

The solar pressure exerted by the sun on large satellites can be ignored in low-altitude orbits, where aerodynamic perturbing forces predominate. For high-altitude orbits, where aerodynamic forces are negligible, the perturbing solar pressure forces cannot be ignored; this will be analyzed in Section 2.8.4. On the other hand, for interplanetary voyages, the solar pressure may be used to obtain accelerating forces on the satellite. This is the "solar sail" mode of interplanetary voyages, which will not be treated in this book.

2.8.2 The Nonhomogeneity and Oblateness of the Earth

Because the force exerted by the earth on a body outside its sphere is a conservative force, it can be derived from a gradient of a scalar potential function $U(r) = -\mu/r$. This would be completely true if the earth were modeled as a mass concentrated in a single point, or as a homogeneous sphere. Unfortunately, this is not the case: the earth is an oblate body, and its mass distribution is not homogeneous. Correction factors must therefore be added to the scalar potential function.

It is convenient to express the corrected potential of the earth in the following form:

$$U(r, \phi, \lambda) = -\frac{\mu}{r} + B(r, \phi, \lambda), \qquad (2.8.1)$$

where $B(r, \phi, \lambda)$ is the appropriate spherical harmonic expansion used to correct the gravitational potential for the earth's nonsymmetric mass distribution; see Figure 2.8.1. If we define R_e as the mean radius of the earth at the equator, then

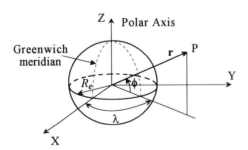

Figure 2.8.1 Coordinates for the derivation of the earth's external gravitational potential.

2.8 / Perturbing Forces and Their Influence on the Orbit

$$B(r, \phi, \lambda) = \frac{\mu}{r} \left\{ \sum_{n=2}^{\infty} \left[\left(\frac{R_e}{r}\right)^n J_n P_n(s\phi) \right. \right.$$
$$\left. \left. + \sum_{m=1}^{n} \left(\frac{R_e}{r}\right)^n (C_{nm} \cos m\lambda + S_{nm} \sin m\lambda) P_{nm}(s\phi) \right] \right\}. \quad (2.8.2)$$

Equation 2.8.2 is the infinite series of the geopotential function at any point P outside the earth's sphere when r, ϕ, and λ are its spherical coordinates (Kaplan 1976, Wertz 1978, Chobotov 1991). The parameters are defined as follows:

- r – geocentric distance of point P;
- ϕ – geocentric latitude;
- λ – geographical longitude;
- R_e – mean equatorial radius of the earth;
- $s\phi = \sin(\phi)$;
- $\cos m\lambda$ and $\sin m\lambda$ – harmonics in λ;
- J_{nm} – zonal harmonic coefficients;
- J_n – zonal harmonic coefficients of order 0;
- P_{nm} – associated Legendre polynomial of degree n and order m;
- P_n – Legendre polynomial degree n and order 0;
- C_{nm} – tesseral harmonic coefficients for $n \neq m$; and
- S_{nm} – sectoral harmonic coefficients for $n = m$.

From 2.8.2 we see that the zonal harmonics depend on the latitude only. These coefficients are a consequence of the earth's oblateness. The tesseral harmonics represent longitudinal variations of the earth's shape. Values of the listed coefficients are obtained from satellite observations and appropriate measurements; these values are time-dependent. Some values for the WGS (World Geodetic Survey) model for the year 1984 are:

$$J_2 = 1{,}082.6 \times 10^{-6}, \quad J_3 = -2.53 \times 10^{-6}, \quad J_4 = -1.61 \times 10^{-6},$$
$$C_{21} = S_{21} = 0, \quad C_{22} = 1.57 \times 10^{-6};$$
$$S_{22} = -0.9 \times 10^{-6}, \quad C_{31} = 2.19 \times 10^{-6}, \quad S_{31} = 0.27 \times 10^{-6},$$
$$C_{32} = 0.31 \times 10^{-6}, \quad S_{32} = -0.21 \times 10^{-6}.$$

It is important to realize that the successive coefficients C_{nm} and S_{nm} do not necessarily decrease; however, the factor $(R_e/r)^n$ tends to diminish each term of the series in Eq. 2.8.2. (Additional series expansion models of the geopotential function can be found, e.g., in Campan 1983.) Comparison of these coefficients shows that the magnitude of J_2 is at least 400 times larger than the other J_n coefficients, which can be disregarded for many engineering purposes. Given this, the potential function $U(r, \phi, \lambda)$ of Eq. 2.8.1 takes the simplified form

$$U \approx -\frac{\mu}{r}\left[1 - \sum_{n=2}^{\infty}\left(\frac{R_e}{r}\right)^n J_n P_n(s\phi)\right] = \frac{\mu}{r}[U_0 + U_{J2} + U_{J3} + \cdots]. \quad (2.8.3)$$

Writing P_2 and P_3 explicitly, for the simplified potential function we have

$$U_0 = -1; \quad U_{J2} = \left(\frac{R_e}{r}\right)^2 J_2 \frac{1}{2}(3\sin^2\phi - 1); \quad U_{J3} = \left(\frac{R_e}{r}\right)^3 J_3 \frac{1}{2}(5\sin^3\phi - 3\sin\phi).$$

Approximating U in the Lagrange equations with U_{J2} yields some important results for the perturbing J_2 coefficient (Borderies 1980).

Before U_{J2} can be used to derive Eqs. 2.7.18–2.7.23 explicitly, $\sin^2\phi$ must first be expressed in terms of orbital parameters. At any point on the orbit in Figure 2.6.2, ϕ is identified as the geographic latitude. In the right spherical triangle formed by the angles i, ϕ, and $\omega+\theta$, the following equality holds: $\sin(\phi) = \sin(i)\sin(\omega+\theta)$. Using this equality in the expression for U_{J2}, some lengthy trigonometric manipulations result in the final expression

$$U_{J2} = \frac{\mu}{a}\left(\frac{R_e}{a}\right)^2 J_2\left[\frac{1}{2} - \frac{3}{4}\sin^2(i)\right](1-e^2)^{-3/2}. \tag{2.8.4}$$

With this result, the explicit calculation of Eqs. 2.7.18–2.7.23 is easily performed, yielding

$$\frac{da}{dt} = \frac{de}{dt} = \frac{di}{dt} = 0. \tag{2.8.5}$$

This means that the average change of the parameters a, e, and i per orbit is null (see Example 2.8.3). We also have

$$\frac{d\Omega}{dt} = -\frac{3}{2}\frac{nJ_2\cos(i)}{(1-e^2)^2}\left(\frac{R_e}{a}\right)^2 = n_\Omega, \tag{2.8.6}$$

$$\frac{d\omega}{dt} = -\frac{3nJ_2[1-5\cos^2(i)]}{4(1-e^2)^2}\left(\frac{R_e}{a}\right)^2 = n_\omega, \tag{2.8.7}$$

$$\frac{dM}{dt} = n + \frac{3nJ_2[3\cos^2(i)-1]}{4(1-e^2)^{3/2}}\left(\frac{R_e}{a}\right)^2 = n_M. \tag{2.8.8}$$

With known initial conditions, Eqs. 2.8.5–2.8.8 can be written as

$$\begin{aligned}
a &= a_0, \\
e &= e_0, \\
i &= i_0, \\
\Omega &= \Omega_0 + n_\Omega(t-t_0), \\
\omega &= \omega_0 + n_\omega(t-t_0), \\
M &= M_0 + n_M(t-t_0)
\end{aligned} \tag{2.8.9}$$

(Escobal 1965). In Eqs. 2.8.9, some of the orbital parameters change with time. The following examples illustrate the significance of these changes.

EXAMPLE 2.8.1 Basic Parameters for Sun-Synchronous Orbits For those low-orbit nadir-pointing satellites (such as the French *Spot* satellite) that carry earth-scanning optical instrumentation, it is imperative that during the surveying stage the sun be behind the satellite, so that the best sun–satellite–target conditions are achieved. With heliosynchronous orbits, this optimal condition is achieved constantly. The idea is to obtain an orbit with the secular rate of the right ascension Ω of the ascending node (n_Ω) equal to the right ascension rate of the mean sun; namely, $d\Omega/dt = 360°/\text{yr} = 0.986°/\text{day} = n_\Omega$ and $\Omega = \Omega_0 + [0.986°/\text{day}](t-t_0)$.

2.8 / Perturbing Forces and Their Influence on the Orbit

To satisfy Eq. 2.8.6, the three parameters a, e, and i must be chosen judiciously. In order to survey most of the globe, including the polar regions, the inclination of the orbit must be as high as possible. The altitude of the orbit near the surveyed region must be low enough to obtain good optical resolution. Suppose $h_a = 1,000$ km and $h_p = 500$ km; then $e = 0.03057$. Using Eq. 2.8.6 we find that the inclination should be $i = 98.368°$. In this special case of a heliosynchronous orbit, the perturbing acceleration caused by the J_2 term helps to obtain a "useful perturbed" orbit.

EXAMPLE 2.8.2 Constant Perigee Argument In telecommunications satellite systems based on elliptic orbits (e.g., the *Molniya*-type satellites), it is important that the perigee remain constant relative to the line of nodes, so that the apogee remains above the region of communication. This condition is achieved by setting $n_\omega = 0$. In Eq. 2.8.7, this corresponds to $i = 63.43°$, which is called the *critical inclination*.

EXAMPLE 2.8.3 Write the perturbed equations of motion of a satellite, taking into account the J_2 zonal harmonic coefficient only.

Solution According to Eq. 2.8.3, the gravitational potential function is approximated by

$$U = \frac{\mu}{r}[U_0 + U_{J2}] = \frac{\mu}{r}\left\{-1 + \left(\frac{R_e}{r}\right)^2 J_2 \frac{1}{2}[3\sin^2(\phi) - 1]\right\}.$$

In this equation, the perturbing potential function is dependent only on the elevation of the satellite above the earth's equatorial plane. In the inertial coordinate system,

$$\sin(\phi) = \frac{z}{r} = \frac{z}{\sqrt{x^2 + y^2 + z^2}}.$$

The gravitational forces acting on the satellite are obtained from the relation $\mathbf{F} = -\text{grad}\, U(x, y, z)$. Namely, we find that

$$F_x = -\frac{\partial U}{\partial x} = \mu\left[-\frac{x}{r^3} + A_{J2}\left(15\frac{xz^2}{r^7} - 3\frac{x}{r^5}\right)\right],$$

$$F_y = -\frac{\partial U}{\partial y} = \mu\left[-\frac{y}{r^3} + A_{J2}\left(15\frac{yz^2}{r^7} - 3\frac{y}{r^5}\right)\right],$$

$$F_z = -\frac{\partial U}{\partial z} = \mu\left[-\frac{z}{r^3} + A_{J2}\left(15\frac{z^3}{r^7} - 9\frac{z}{r^5}\right)\right],$$

where $A_{J2} = \frac{1}{2}J_2 R_e^2$ and R_e is the mean radius of the earth at the equator. Integration of these equations in the Cartesian inertial axis frame for known initial conditions of the vectors $\mathbf{r}(0)$ and $\mathbf{v}(0)$ will give us the motion in the disturbed Keplerian orbit, in which the right ascension Ω and the argument of perigee ω will not be constant but instead will evolve according to Eq. 2.8.6 and Eq. 2.8.7.

To exemplify the results, let us simulate the motion of a satellite in an orbit having the following classical orbit parameters: $a = 12,000$ km, $e = 0.3$, $i = 20°$, $\Omega = 40°$, $\omega = 60°$, and $\theta = 80°$. To integrate the equations of motion (Eqs. 2.7.4), we first need to find the initial conditions of $\mathbf{r}(0)$ and $\mathbf{v}(0)$ in Cartesian coordinates. This is easily

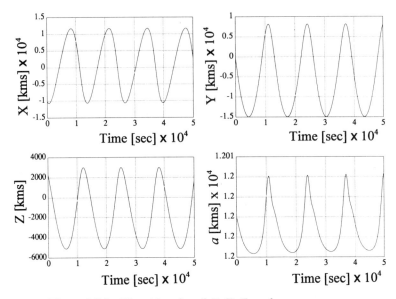

Figure 2.8.2 Time histories of X, Y, Z, and a.

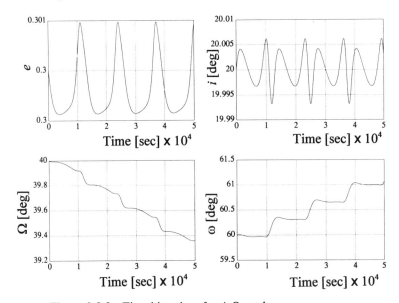

Figure 2.8.3 Time histories of e, i, Ω, and ω.

done by using the procedure of Section 2.6.3. We find that $X(0) = -10,121.0$ km, $Y(0) = -308.219$ km, $Z(0) = 2,281.8$ km, $V_x(0) = -1.929$ km/sec, $V_y(0) = -6.184$ km/sec, and $V_z(0) = -1.727$ km/sec. The time histories are given in Figure 2.8.2 and Figure 2.8.3.

Some important conclusions can be derived from this example. First, although the classical orbit parameters a, e, and i do change slightly with time, their average

2.8 / Perturbing Forces and Their Influence on the Orbit

change per orbit is null as expected from the equalities in Eq. 2.8.5. On the other hand, the average values of Ω and ω shown in Figure 2.8.3 do have a net change with time, exactly as predicted by Eq. 2.8.6 and Eq. 2.8.7. In Figure 2.8.3,

$$n_\Omega = d\Omega/dt = -13.17 \times 10^{-6} \text{ deg/sec} = -0.2297 \times 10^{-6} \text{ rad/sec} \quad \text{and}$$

$$n_\omega = d\omega/dt = +25.16 \times 10^{-6} \text{ deg/sec} = 0.438 \times 10^{-6} \text{ rad/sec},$$

as expected.

2.8.3 A Third-Body Perturbing Force

A third body, like the sun or the moon, creates a perturbing force with respect to an earth-orbiting satellite that can change appreciably the parameters of its nominal Keplerian orbit. The two-body problem treated in the beginning of this chapter can be generalized to the much more difficult n-body problem in the following way.

In a system consisting of n bodies, the sum of the forces acting on the ith body will be

$$\mathbf{F}_i = G \sum_{j=1}^{j=n} \frac{m_i m_j}{r_{ij}^3} (\mathbf{r}_j - \mathbf{r}_i), \quad i \neq j. \tag{2.8.10}$$

According to Newton's second law of motion (Eq. 2.1.1), for constant masses $\mathbf{F}_i = m_i(d^2\mathbf{r}_i/dt^2)$, from which it follows that

$$\frac{d^2\mathbf{r}_i}{dt^2} = G \sum_{j=1}^{j=n} \frac{m_j}{r_{ij}^3} (\mathbf{r}_j - \mathbf{r}_i), \quad i \neq j. \tag{2.8.11}$$

In Figure 2.8.4, m_1 stands for the earth and m_2 for the satellite. Extracting these two masses from the summation in Eq. 2.8.11, the accelerations for m_1 and m_2 become

$$\frac{d^2\mathbf{r}_1}{dt^2} = G\frac{m_2}{r_{12}^3}(\mathbf{r}_2 - \mathbf{r}_1) + G\sum_{j=3}^{j=n} \frac{m_j}{r_{1j}^3}(\mathbf{r}_j - \mathbf{r}_1), \tag{2.8.12}$$

$$\frac{d^2\mathbf{r}_2}{dt^2} = G\frac{m_1}{r_{21}^3}(\mathbf{r}_1 - \mathbf{r}_2) + G\sum_{j=3}^{j=n} \frac{m_j}{r_{2j}^3}(\mathbf{r}_j - \mathbf{r}_2) \tag{2.8.13}$$

(Battin 1990). These are the equations of motion with respect to the inertial coordinate axes. As in Section 2.2, we define $\mathbf{r} = \mathbf{r}_2 - \mathbf{r}_1 = \mathbf{r}_{12}$; also, $\mathbf{r}_{2j} = \boldsymbol{\rho}_j$ and $\mathbf{r}_{1j} = \mathbf{r}_{pj}$. If we choose $m_1 = M_e$ to be the mass of the earth, $m_2 = m_S$ to be the mass of the satellite, and $m_j = m_{pj}$ to be the mass of the j perturbing body, and if we locate the origin

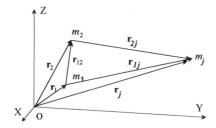

Figure 2.8.4 Simplified model for the n-body dynamics perturbing function.

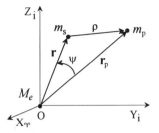

Figure 2.8.5 Addition of a perturbing third body; reproduced from Campan (1983) by permission of Cépaduès-Éditions.

of the inertial frame at the center of the earth ($\mathbf{r}_1 = 0$), then subtraction of Eq. 2.8.12 from Eq. 2.8.13 leads to the final result:

$$\frac{d^2\mathbf{r}}{dt^2} + G\frac{\mathbf{r}}{r^3}(M_e + m_S) = G\sum_{j=3}^{j=n} m_{pj}\left[\frac{\boldsymbol{\rho}_j}{\rho_j^3} - \frac{\mathbf{r}_{pj}}{r_{pj}^3}\right]. \tag{2.8.14}$$

This equation is identical to Eq. 2.2.3 if no third body exists. The perturbing acceleration due to the $n-2$ perturbing bodies becomes

$$\boldsymbol{\gamma}_p = \sum_{j=3}^{j=n} \mu_{pj}\left[\frac{\boldsymbol{\rho}_j}{\rho_j^3} - \frac{\mathbf{r}_{pj}}{r_{pj}^3}\right], \tag{2.8.15}$$

where $\mu_{pj} = Gm_{pj}$.

Figure 2.8.4 is adapted in Figure 2.8.5 to the special case of the three-body problem. The earth is at the origin O, ψ is the angle between the radius vectors to the satellite m_S and to the perturbing body m_p, and $\boldsymbol{\rho}$ is the vector from the satellite to the perturbing body. The perturbing acceleration becomes

$$\boldsymbol{\gamma}_p = \left(\frac{\boldsymbol{\rho}}{\rho^3} - \frac{\mathbf{r}_p}{r_p^3}\right)\mu_p, \tag{2.8.16}$$

where $\mu_p = Gm_p$, the gravity constant of the jth perturbing body. Equation 2.2.3 is replaced with the following equation:

$$\ddot{\mathbf{r}} + G(M_e + m_S)\frac{\mathbf{r}}{r^3} = \boldsymbol{\gamma}_p. \tag{2.8.17}$$

With some "relieving" assumptions, Eq. 2.8.17 can be analytically solved for the three-body system.

It can be shown that the perturbing acceleration satisfies the equality $\boldsymbol{\gamma}_p = -\partial U_p/\partial \mathbf{r}$, where U_p has the form

$$U_p = \mu_p\left(\frac{1}{\rho} - \frac{1}{r_p^3}\mathbf{r}\cdot\mathbf{r}_p\right) \tag{2.8.18}$$

(Battin 1990). According to Figure 2.8.5,

$$\mathbf{r}\cdot\mathbf{r}_p = (rr_p)\cos(\Psi) \quad \text{and} \quad \frac{\mathbf{r}\cdot\mathbf{r}_p}{r_p^3} = \frac{r}{r_p^2}\cos(\psi).$$

Also

$$\rho^2 = r^2 + r_p^2 - 2rr_p\cos(\Psi) = r_p^2\left[1 + \left(\frac{r}{r_p}\right)^2 - 2\frac{r}{r_p}\cos(\Psi)\right] \tag{2.8.19}$$

2.8 / Perturbing Forces and Their Influence on the Orbit

and

$$\frac{1}{\rho} = \frac{1}{r_p} \left[1 - 2\frac{r}{r_p} \cos(\Psi) + \frac{r^2}{r_p^2} \right]^{-1/2}.$$

Since $r/r_p \ll 1$, we find that

$$\frac{1}{\rho} \approx \frac{1}{r_p} \left[1 + \frac{r}{r_p} \cos(\Psi) - \frac{1}{2}\frac{r^2}{r_p^2} + \frac{3}{2}\frac{r^2}{r_p^2} \cos^2(\Psi) \right]. \tag{2.8.20}$$

Substitute $1/\rho$ and $\mathbf{r} \cdot \mathbf{r}_p / r_p^3$ into Eq. 2.8.18 to obtain

$$U_p = \frac{\mu_p}{r_p} \left[1 - \frac{1}{2}\left(\frac{r}{r_p}\right)^2 + \frac{3}{2}\left(\frac{r}{r_p}\right)^2 \cos^2(\Psi) \right]. \tag{2.8.21}$$

The complete expression in terms of Legendre polynomials is

$$U_p = \frac{\mu_p}{r_p} \sum_{n=2}^{\infty} \left(\frac{r}{r_p}\right)^n P_n[\cos(\Psi)]. \tag{2.8.22}$$

It is interesting to mention that

$$\left.\frac{\mu_p}{r_p^3}\right|_{\text{moon}} = 8.6 \times 10^{-14} \text{ sec}^{-2} \quad \text{and} \quad \left.\frac{\mu_p}{r_p^3}\right|_{\text{sun}} = 3.96 \times 10^{-14} \text{ sec}^{-2}.$$

The values established here will be used in Section 2.9 to derive the inclination drift for perturbed high-altitude geostationary satellites.

2.8.4 Solar Radiation and Solar Wind

Solar radiation comprises all the electromagnetic waves radiated by the sun with wavelengths from X-rays to radio waves. The *solar wind* consists mainly of ionized nuclei and electrons. Both kinds of radiation may produce a physical pressure when acting on any surface of a body. This pressure is proportional to the *momentum flux* (momentum per unit area per unit time) of the radiation. The solar radiation momentum flux is greater than that of the solar wind, by a factor of 100 to 1,000, so solar wind pressure is of secondary importance.

The mean solar energy flux of the solar radiation is proportional to the inverse square of the distance from the sun. The mean integrated energy flux at the earth's position is given by

$$F_e = \frac{1{,}358}{1.0004 + 0.0334\cos(D)} \text{ W/m}^2.$$

Here D is the "phase" of the year, which is calculated as starting on July 4, the day of earth aphelion (Wertz 1978). This is equivalent to a mean momentum flux of $P = F_e/c = 4.5 \times 10^{-6}$ kg-m^{-1}-sec^{-2}, where c is the velocity of light.

The *solar radiation pressure* $|\mathbf{F}_R|$ is proportional to P, to the cross-sectional area A of the satellite perpendicular to the sun line, and to a coefficient C_P that is dependent on the absorption characteristic of the spacecraft:

$$|\mathbf{F}_R| = PAC_P. \tag{2.8.23}$$

The value of C_P lies between 0 and 2; $C_P = 1$ for a black body, a perfectly absorbing material, whereas $C_P = 2$ for a body reflecting all light back toward the sun.

2.9 Perturbed Geostationary Orbits

2.9.1 *Redefinition of the Orbit Parameters*

Geosynchronous orbits are orbits with orbital period equal to the period of revolution of the earth about its own axis of rotation. The motion of a satellite on such orbits is synchronized with the rotational motion of the earth. For these orbits, there is no restriction on their inclination and eccentricity. With reference to an inertial frame, the period of rotation of the earth is the sidereal day, equal to $T_s = 86,400/(1+1/365.25) = 86,164.1$ sec, from which follows that the semimajor axis of a geosynchronous satellite is $a = 42,164.157$ km.

The geostationary satellite orbit is a geosynchronous orbit that has some special characteristics. Namely, its inclination is null or close to null (of the order of zero to several degrees), and the eccentricity too is very small, of the order of 10^{-4}. The geostationary orbit characterizes a mission where it is important to keep the satellite fixed in apparent position relative to the earth. For such equatorial orbits, the right ascension Ω is ill-defined; hence, redefinition of the six classical orbit parameters is necessary.

The orbit in Figure 2.9.1 is close to being equatorial, so Ω, ω, and θ can be assimilated in a single argument, $\alpha = \Omega + \omega + \theta$, which is the sidereal angle of the satellite.

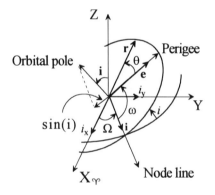

Figure 2.9.1 Orbital parameters in space for a geostationary orbit.

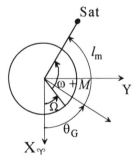

Figure 2.9.2 Orbital parameters of a GEO spacecraft in the equatorial plane.

2.9 / Perturbed Geostationary Orbits

To define this argument, the equatorial plane of the earth and the inertial X_Υ-Y coordinate system are shown in Figure 2.9.2, where θ_G is the angle between the vernal equinox X_Υ and the Greenwich longitudinal meridian; this is the sidereal angle of the Greenwich meridian. Instead of the mean anomaly (see Section 2.6.1) we shall use the *mean longitude*, defined as $l_m = \Omega + \omega + M - \theta_G(t)$. The inclination i and the eccentricity e will be redefined as vectors in the X_Υ-Y plane.

In Figure 2.9.1, with a very small inclination i, we can suppose that the parameters Ω, ω, and θ are virtually in the same plane. There are two different definitions of the inclination vector **i**. Some authors (Legendre 1980, Alby 1983) define the inclination vector as a vector aligned with the node line and whose norm is equal to the value i of the inclination. (The second definition will be discussed later.) With the first definition, the inclination vector has components i_x and i_y that are proportional to the right ascension Ω; see Eq. 2.9.3 and Eq. 2.9.4.

In geostationary orbits, the eccentricity is also very small, so that the locations of the perigee and the apogee become rather doubtful. The eccentricity e is also redefined as a vector **e**, aligned along the radius vector pointing from the location of the center of the attracting body to the perigee; **e** is dependent on ω and Ω. The components of **e** are e_x and e_y, aligned (respectively) along the X_Υ and Y inertial axes. With these definitions (see also Figures 2.9.1 and 2.9.2), the six orbit parameters take the following form:

$$a, \tag{2.9.0}$$

$$e_x = e\cos(\omega + \Omega), \tag{2.9.1}$$

$$e_y = e\sin(\omega + \Omega), \tag{2.9.2}$$

$$i_x = i\cos(\Omega), \tag{2.9.3}$$

$$i_y = i\sin(\Omega), \tag{2.9.4}$$

$$l_m = M + \omega + \Omega - \theta_G(t), \tag{2.9.5}$$

where $\theta_G(t)$ is the Greenwich (meridian) sidereal time and l_m is the angular location of the satellite relative to the Greenwich meridian. This parameter is a geographical longitude expressed in terms of the classical orbit parameters and the Greenwich sidereal time.

The second definition for the inclination is based on the *orbital pole*, which is a unit vector perpendicular to the orbit plane (Soop 1988); here, $\sin(i)$ is the component of the orbital pole on the equatorial plane. For small inclinations, $\sin(i) \approx i$. With this approximation, the inclination vector is expressed as

$$i_x = i\sin(\Omega), \tag{2.9.6}$$

$$i_y = -i\cos(\Omega). \tag{2.9.7}$$

The components i_x, i_y, e_x, and e_y will be put in vector forms:

$$\mathbf{i} = \begin{bmatrix} i_x \\ i_y \end{bmatrix}; \quad \mathbf{e} = \begin{bmatrix} e_x \\ e_y \end{bmatrix}.$$

2.9.2 Introduction to Evolution of the Inclination Vector

In this section we shall develop the equations for the inclination vector evolution caused by perturbing forces. The primary causes of this evolution are the sun's

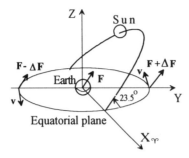

Figure 2.9.3 Solar attraction that causes inclination perturbation; adapted from Soop (1983) by permission of European Space Agency.

and the moon's attracting forces, and also the J_2 coefficient of the gravitational potential function of the earth (Pocha 1987). Before proceeding with the analytical details, we should like to have some physical insight into the problem at hand.

Figure 2.9.1 showed a geostationary orbit located in the equatorial plane of the earth. In winter, the sun will be located to the south side of the orbit; in summer, to the north side, as in Figure 2.9.3. Exactly in midspring and in midautumn, the sun will be in the orbital plane. The last case is illustrated in Figure 2.9.4.

At midwinter and midsummer, the sun is in the **Y-Z** plane of the inertial system, above or below the geostationary orbit plane. In Figure 2.9.3, the sun exerts a mean gravity force **F** on the satellite directed toward its center of mass. During half the day, when the sun is to the north of the orbit plane, a positive average attracting force $\mathbf{F}+\Delta\mathbf{F}$ is exerted on the satellite. During the second half of the day, an average force $\mathbf{F}-\Delta\mathbf{F}$ is exerted on the satellite, since the satellite is now more distant from the sun and so less attracted to it (see Soop 1983). The net effect is that the remaining average moment on the orbit about the $\mathbf{X}_\gamma$ axis will lead to a movement of the orbital pole about the **Y** axis, toward the positive $\mathbf{X}_\gamma$ axis. This is equivalent to the increase of the inclination i. In the other half of the year, when the sun is to the south of the orbit plane, the net moment applied on the orbit has the same direction as in the first part of the year, and an additional increase of the inclination follows. Exactly in midspring and midautumn, the sun is in the equatorial plane, so that no moment is applied on the orbit and the change in the inclination is null. Between the seasons, smaller changes of the inclination are induced, but the gravitational force of the sun exerted on the satellite causes an overall net increase of the inclination.

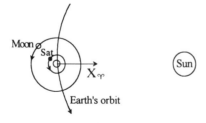

Figure 2.9.4 The sun and the moon as attracting bodies in spring; adapted from Soop (1983) by permission of European Space Agency.

2.9 / Perturbed Geostationary Orbits

The same effect is produced by the moon. The gravitational force exerted by the moon on the satellite has the similar effect of increasing the inclination. Although the moon is much smaller than the sun, its effect on the satellite is stronger because it is so much closer to the earth (see Section 2.8.3). The perturbation effects of both the sun and the moon are added vectorially.

The moon's orbit about the earth is inclined about 5.3° to the ecliptic plane. The orbital period of the moon about the earth is approximately 28 days. The difference in orbital periods of the moon and the sun about the earth makes analytical computation of the inclination evolution extremely complicated. As can be seen in Figure 2.9.4, if both the sun and the moon are to the right of the earth then the gravitational forces they exert on the s/c have the same direction. When the moon is on the left of the earth, as is shown in the figure, the forces applied by the moon and the sun tend to cancel each other. Hence, different geometric relations induce different rates of inclination drift.

2.9.3 Analytical Computation of Evolution of the Inclination Vector

Our analytical computation follows closely the presentation in Alby (1983). The model of the inclination vector **i** to be adopted is that of the definitions given by Eq. 2.9.3 and Eq. 2.9.4, which means that **i** lies on the line of nodes. Since a third-body attraction force is a conservative force, the Lagrange equations are easier to use. The potential function U_p is independent of the perigee argument, but does depend strongly on the right ascension Ω of the satellite orbit. Only the out-of-plane component of the perturbing force can produce a moment on the orbit that will force the orbital pole to move toward the $\mathbf{X}_\Upsilon$ axis, thus increasing the inclination. The value of this out-of-plane component depends on Ω. Using the Lagrange equations (Eq. 2.7.20 and Eq. 2.7.21) for di/dt and $d\Omega/dt$ and with U_p standing for the perturbing potential function, we can write the following approximated relations for orbits with very small inclination and eccentricity:

$$\frac{di}{dt} = \frac{-1}{ina^2} \frac{\partial U_p}{\partial \Omega}, \tag{2.9.8}$$

$$\frac{d\Omega}{dt} = \frac{1}{ina^2} \frac{\partial U_p}{\partial i}. \tag{2.9.9}$$

Differentiating Eq. 2.9.3 and Eq. 2.9.4 yields

$$\frac{di_x}{dt} = \frac{di}{dt} \cos(\Omega) - \frac{d\Omega}{dt} i \sin(\Omega), \tag{2.9.10}$$

$$\frac{di_y}{dt} = \frac{di}{dt} \sin(\Omega) + \frac{d\Omega}{dt} i \cos(\Omega). \tag{2.9.11}$$

Together with Eq. 2.9.8 and Eq. 2.9.9, we have

$$\frac{di_x}{dt} = -\frac{\cos(\Omega)}{ina^2} \frac{\partial U_p}{\partial \Omega} - \frac{\sin(\Omega)}{na^2} \frac{\partial U_p}{\partial i}, \tag{2.9.12}$$

$$\frac{di_y}{dt} = \frac{-\sin(\Omega)}{ina^2} \frac{\partial U_p}{\partial \Omega} + \frac{\cos(\Omega)}{na^2} \frac{\partial U_p}{\partial i}. \tag{2.9.13}$$

Since U_p is a function of the inclination, the following relation holds:

$$\partial U_p = \frac{\partial U_p}{\partial i_x} di_x + \frac{\partial U_p}{\partial i_y} di_y. \tag{2.9.14}$$

Because i_x and i_y are related through i and Ω, it follows (together with Eq. 2.9.3 and Eq. 2.9.4) that

$$\begin{aligned}\frac{\partial U_p}{\partial \Omega} &= \frac{\partial U_p}{\partial i_x}\frac{\partial i_x}{\partial \Omega} + \frac{\partial U_p}{\partial i_y}\frac{\partial i_y}{\partial \Omega} = \frac{-\partial U_p}{\partial i_x}i\sin(\Omega) + \frac{\partial U_p}{\partial i_y}i\cos(\Omega),\\ \frac{\partial U_p}{\partial i} &= \frac{\partial U_p}{\partial i_x}\frac{\partial i_x}{\partial i} + \frac{\partial U_p}{\partial i_y}\frac{\partial i_y}{\partial i} = \frac{\partial U_p}{\partial i_x}\cos(\Omega) + \frac{\partial U_p}{\partial i_y}\sin(\Omega). \end{aligned} \tag{2.9.15}$$

If we substitute Eqs. 2.9.15 into Eq. 2.9.12 and Eq. 2.9.13, the final equations follow:

$$\begin{aligned}\frac{di_x}{dt} &= \frac{-1}{na^2}\frac{\partial U_p}{\partial i_y},\\ \frac{di_y}{dt} &= \frac{1}{na^2}\frac{\partial U_p}{\partial i_x}, \end{aligned} \tag{2.9.16}$$

where a is the semimajor axis of the orbit, n the mean motion, and U_p the perturbing potential function. In this case, the perturbing function is due to the gravity forces applied by the moon and the sun on the satellite. The perturbing function caused by a third-body perturbing force was analyzed in Section 2.8.3 (see Figure 2.8.5 and Eq. 2.8.21). In order to solve Eqs. 2.9.16, we must first expand Eq. 2.8.21 and find the needed partial derivatives. The solution of the equations follows.

In Figure 2.8.5, we define r_x, r_y, r_z, r_{px}, r_{py}, and r_{pz} as the vector components of $\mathbf{r}$ and $\mathbf{r}_p$. The satellite vector components can be found by the inverse of Eq. 2.6.4. If these components are normalized to the norm of the respective vector then they are called the *direction cosines* (see Appendix A). The direction cosine components a_x, a_y, a_z, a_{px}, a_{py}, and a_{pz} are the components of the unit vectors $\mathbf{r}/|\mathbf{r}|$ and $\mathbf{r}_p/|\mathbf{r}_p|$. They are necessary in order to find the value of $\cos(\psi)$ in Eq. 2.8.21. To begin with, they can be found by taking the inverse of Eq. 2.6.4, in which $\mathbf{R}$ is a unit vector in the direction of $\mathbf{r}$. For a nearly circular orbit, $\theta + \omega$ can be written as

$$\omega + \theta = \omega + M + \Omega - \Omega = \alpha - \Omega.$$

With this definition, $\alpha = \omega + M + \Omega$ is the argument of the location of the satellite relative to the $\mathbf{X}_{\Upsilon}$ axis, the vernal equinox (see Figure 2.6.2). Inversion of Eq. 2.6.4 will provide the direction cosines of the vector $\mathbf{r}$ in the inertial reference frame. The resulting values are:

$$\begin{aligned}a_x &= \cos(\omega+\theta)\cos(\Omega) - \sin(\omega+\theta)\sin(\Omega)\cos(i)\\ &= \cos(\alpha-\Omega)\cos(\Omega) - \sin(\alpha-\Omega)\sin(\Omega)\cos(i),\\ a_y &= \cos(\omega+\theta)\sin(\Omega) + \sin(\omega+\theta)\cos(\Omega)\cos(i)\\ &= \cos(\alpha-\Omega)\sin(\Omega) + \sin(\alpha-\Omega)\cos(\Omega)\cos(i),\\ a_z &= \sin(\omega+\theta)\sin(i) = \sin(\alpha-\Omega)\sin(i). \end{aligned} \tag{2.9.17}$$

For a geostationary orbit, the inclination is close to null, so $\sin(i) \approx i$. Thus, using the definitions in Eq. 2.9.3 and Eq. 2.9.4 for i_x and i_y, we have

2.9 / Perturbed Geostationary Orbits

$$a_x = \cos(\alpha),$$
$$a_y = \sin(\alpha), \quad (2.9.18)$$
$$a_z = i_x \sin(\alpha) - i_y \cos(\alpha).$$

But $\cos(\Psi) = a_x a_{xp} + a_y a_{yp} + a_z a_{zp}$, so

$$\cos^2(\Psi) = \{a_{xp}\cos(\alpha) + a_{yp}\sin(\alpha) + a_{zp}[i_x\sin(\alpha) - i_y\cos(\alpha)]\}^2. \quad (2.9.19)$$

To find di_x/dt, we must first derive

$$\frac{\partial U_p}{\partial i_y} = \frac{\partial U_p}{\partial \cos^2(\Psi)} \frac{\partial \cos^2(\Psi)}{\partial i_y}.$$

From Eq. 2.8.21:

$$\frac{\partial U_p}{\partial \cos^2(\Psi)} = \frac{3}{2}\frac{\mu_p r^2}{r_p^3}, \quad (2.9.20)$$

where μ_p is the gravity coefficient of the perturbing body (the sun or the moon).

Next, from Eq. 2.9.19 and assuming a very small inclination ($i \approx 0$):

$$\frac{\partial \cos^2(\Psi)}{\partial i_y} = -[a_{xp}a_{zp} + a_{xp}a_{zp}\cos(2\alpha) + a_{yp}a_{zp}\sin(2\alpha)]. \quad (2.9.21)$$

The final results, as obtained by Alby (1983), become:

$$\frac{di_x}{dt} = \frac{3}{2}\frac{\mu_p r^2}{na^2 r_p^3}[a_{xp}a_{zp} + a_{xp}a_{zp}\cos(2\alpha) + a_{yp}a_{zp}\sin(2\alpha)]. \quad (2.9.22)$$

Equation 2.9.22 is the result for di_x/dt. In a similar procedure, we can also find

$$\frac{di_y}{dt} = \frac{3}{2}\frac{\mu_p r^2}{na^2 r_p^3}[a_{yp}a_{zp} - a_{yp}a_{zp}\cos(2\alpha) + a_{xp}a_{zp}\sin(2\alpha)]. \quad (2.9.23)$$

In the last two equations, $a = r$ for a geostationary circular orbit. Let us define

$$K_p = \frac{\mu_p r^2}{n r_p^3 a^2} = \frac{\mu_p}{n r_p^3}.$$

We find that

$$K_{\text{moon}} = 5.844 \times 10^{-3} \text{ deg/day},$$
$$K_{\text{sun}} = 2.69 \times 10^{-3} \text{ deg/day}.$$

Finally, we must interpret the meaning of the last two equations. For this we adapt Figures 2.9.1 and 2.9.3 to the geostationary satellite case, including both the sun and the moon orbits; see Figures 2.9.5 and 2.9.6.

Relative Motion between Earth, Sun, and Moon

The eccentricity of the apparent sun orbit about the earth is $e_s = 0.016726$, and that of the moon is $e_m = 0.0549$. As a convenient aid to obtaining meaningful results, we shall assume that the orbits of the perturbing bodies are circular orbits, $e_p = 0$, with constant inclinations i_p.

In Figures 2.9.5 and 2.9.6, the inclination of the apparent orbit of the sun about the earth is $i_s = 23.45°$, and the average inclination of the moon orbit to the ecliptic

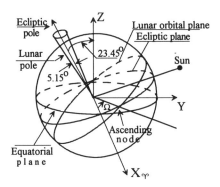

Figure 2.9.5 Apparent motion of the sun and the moon; reproduced from Agrawal (1986) by permission of Prentice-Hall.

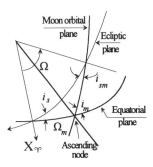

Figure 2.9.6 Orbital parameters of the moon orbit; reproduced from Agrawal (1986) by permission of Prentice-Hall.

orbit is 5°8' (this value ranges from $i_m = 4°59'$ to $5°18'$). The moon orbits the earth with a period of 27.3 days, and the earth orbits the sun with a period of 365.25 days. The ecliptic and the lunar poles are directional vectors perpendicular to the ecliptic and lunar orbit planes. The lunar pole precesses about the ecliptic pole with a period of 18.6 years and an average cone angle of 5.133°. The earth axis precesses about the ecliptic pole with an angle of 23.45° and a period of 25,800 years.

Equations 2.9.18, when referring to the perturbing bodies, require knowledge of the inclination i_p of their orbits and the right ascension Ω_p of the ascending node. The right ascension of the ecliptic orbit is null, by definition. The right ascension Ω_m of the moon orbit is analytically easier to express through the right ascension angle Ω of the moon orbit in the ecliptic orbit from the vernal equinox axis; see Figure 2.9.6. The parameter Ω has an appropriate analytical expression (Agrawal 1986):

$$\Omega[\deg] = 178.78 - 0.05295t, \qquad (2.9.24)$$

where t is the number of days since January 1, 1960.

Using spherical trigonometry for the spherical triangle in Figure 2.9.6, we find that

$$\cos(i_m) = \cos(i_s)\cos(i_{sm}) - \sin(i_s)\sin(i_{sm})\cos(\Omega),$$
$$\sin(\Omega_m) = [\sin(i_{sm})\sin(\Omega)]/\sin(i_m). \qquad (2.9.25)$$

2.9 / Perturbed Geostationary Orbits

In these equations, i_m is the inclination of the moon orbit to the equatorial plane, i_{sm} is the inclination of the moon orbit relative to the ecliptic orbit plane, and Ω is known from Eq. 2.9.24. The value of i_m for calendar dates can be found in celestial almanacs; it may also be approximated by the following equation, which is accurate enough for our demands:

$$i_m \text{ [deg]} = 23.736 + 5.133 \sin[\omega_m(t - 2{,}001.7433)], \tag{2.9.26}$$

where ω_m is the frequency of the motion of the lunar pole with a time period of 18.6 years, and t is the calendar time expressed in years.

Computation of the Inclination Derivatives

With the preceding definitions of celestial motion of the moon about the earth, and given the apparent motion of the sun about the earth, we can return to our initial problem of computing the inclination derivatives of Eq. 2.9.22 and Eq. 2.9.23. In these two equations, $\sin(2\alpha)$ and $\cos(2\alpha)$ are fast harmonic motions with period of half a solar day and a null average. Since we are interested only in the secular inclination derivatives, harmonic terms in 2α shall be ignored. With this assumption, Eq. 2.9.22 and Eq. 2.9.23 become (respectively)

$$\frac{di_x}{dt} = \frac{3}{2} K_p a_{xp} a_{zp}, \tag{2.9.27}$$

$$\frac{di_y}{dt} = \frac{3}{2} K_p a_{yp} a_{zp}, \tag{2.9.28}$$

where a_{xp}, a_{zp}, and a_{yp} are as given in Eqs. 2.9.17 for the perturbing bodies' orbits. In order to simplify the analytical treatment, we suppose that the orbits of the perturbing bodies are circular. In this case, $\omega_p + \theta_p = \omega_p + M_p$ in Eq. 2.9.17 and Eq. 2.9.18. Set $\lambda_p = \omega_p + M_p$. This, together with Eqs. 2.9.17, transforms Eq. 2.9.27 and Eq. 2.9.28 as follows:

$$\left.\frac{di_x}{dt}\right|_p = \frac{3}{8} K_p [-\sin(\Omega_p) \sin(2i_p) + 2 \sin(2\lambda_p) \sin(i_p) \cos(\Omega_p) \\ + \cos(2\lambda_p) \sin(\Omega_p) \sin(2i_p)], \tag{2.9.29}$$

$$\left.\frac{di_y}{dt}\right|_p = \frac{3}{8} K_p [\cos(\Omega_p) \sin(2i_p) + 2 \sin(2\lambda_p) \sin(i_p) \sin(\Omega_p) \\ + \cos(2\lambda_p) \cos(\Omega_p) \sin(2i_p)]. \tag{2.9.30}$$

The results in Eq. 2.9.29 and Eq. 2.9.30 are taken from Alby (1983). These are the final equations we sought. They contain almost constant elements like i_p and Ω_p as well as harmonic terms in $2\lambda_p$.

We are especially interested in the secular term of the inclination derivative, which means that we can ignore harmonic terms in λ_p. In this case, the equations become

$$\frac{di_x}{dt} = H = -\frac{3}{8} K_{\text{moon}} \sin(\Omega_m) \sin(2i_m) + 0, \tag{2.9.31}$$

$$\frac{di_y}{dt} = K = \frac{3}{8} K_{\text{moon}} \cos(\Omega_m) \sin(2i_m) + \frac{3}{8} K_{\text{sun}} \sin(2i_s). \tag{2.9.32}$$

In these equations, i_s is constant. The evolutions of Ω_m and i_m are given by Eq. 2.9.25 and Eq. 2.9.26.

Table 2.9.1 *Evolution of inclination derivatives*

year	Ω	i_m	Ω_m	H	K	N	Ω_d
1990	319.1	27.5	-7.3	0.08	0.92	0.92	84.8
1991	299.7	26.2	-10.2	0.11	0.89	0.89	82.8
1992	280.3	24.5	-12.2	0.13	0.86	0.87	81.5
1993	261.1	22.8	-13.2	0.13	0.83	0.84	81.1
1994	241.7	21.2	-12.6	0.12	0.79	0.80	81.6
1995	222.3	19.8	-10.2	0.09	0.77	0.78	83.3
1996	203.1	18.9	-6.2	0.05	0.76	0.76	86.1
1997	183.7	18.6	-1.0	0.01	0.75	0.75	89.3
1998	164.4	18.8	4.3	-0.04	0.76	0.76	92.7
1999	145.1	19.6	8.8	-0.08	0.77	0.77	95.7
2000	125.7	20.9	11.8	-0.11	0.79	0.80	97.8
2001	106.4	22.5	12.9	-0.13	0.82	0.83	98.8
2002	87.1	24.2	12.6	-0.13	0.85	0.86	98.7
2003	67.7	25.8	10.9	-0.12	0.89	0.89	97.6
2004	48.4	27.3	8.4	-0.11	0.91	0.92	95.9
2005	29.1	28.3	5.3	-0.06	0.93	0.94	93.7
2006	9.8	28.8	1.8	-0.02	0.94	0.95	91.3
2007	350.4	28.8	-1.8	0.02	0.94	0.94	88.7
2008	331.1	28.1	-5.3	0.06	0.93	0.93	86.2
2009	311.8	27.0	-8.5	0.10	0.91	0.91	84.3
2010	292.4	25.5	-11.1	0.12	0.88	0.89	82.3

The angle between H and K is of importance during the stage in which the inclination of the satellite orbit is to be controlled (see Chapter 3). We define

$$\Omega_d = \arctan(K/H). \qquad (2.9.33)$$

The norm of the terms of the inclination derivative is also important:

$$\frac{di}{dt} = N = \sqrt{H^2 + K^2}. \qquad (2.9.34)$$

The values of Ω, Ω_m, i_m, H, K, N, and Ω_d are given in Table 2.9.1, and graphed in Figures 2.9.7. For the perturbing moon, the period of the harmonic term of the derivative is 13.66 days with an amplitude of 0.0035°; for the perturbing sun, the period will be of 182.65 days with an amplitude of 0.023°.

In order to find the inclination vector evolution, Eq. 2.9.29 and Eq. 2.9.30 must be integrated with time. The results for a period of 400 days (beginning January 1990) are shown in Figure 2.9.8. The short-period second harmonics of the sun's and moon's motions are clearly seen in these figures.

2.9.4 *Evolution of the Eccentricity Vector*

As in the previous section's treatment of the inclination vector's evolution, here also it is instructive first to describe the physical mechanism of the eccentricity vector's evolution (see Pocha 1987).

2.9 / Perturbed Geostationary Orbits

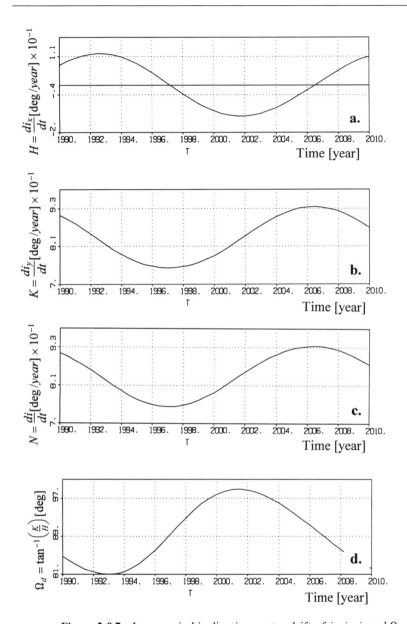

Figure 2.9.7 Long-period inclination vector drift of i_x, i_y, i, and Ω_d.

The sun applies pressure on the satellite that results in an acceleration of the s/c in the sun–satellite direction. This solar pressure on the geostationary satellite produces a long-period cyclic perturbation in the orbit eccentricity, with no change in the semi-major axis. Suppose that the orbit is initially circular. The effect of the solar pressure on the satellite, integrated over the lower part of the orbit, may be approximated by a differential increment $\Delta \mathbf{V}$ at point 1 added to the nominal circular velocity $\mathbf{V}_c$. The solar pressure exerted on the satellite during the opposite part of the orbit produces the same increment in velocity and in the same direction, but located now at point 2,

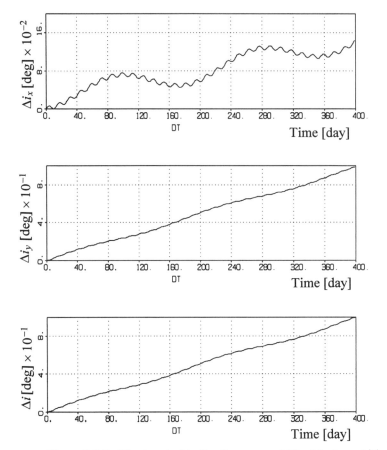

Figure 2.9.8 Short-period inclination vector drift of i_x, i_y, and $|\mathbf{i}|$.

so that $\Delta \mathbf{V}$ is now opposite to $\mathbf{V}_c$. The increment of velocity at point 1 tends to increase the altitude of the orbit at point 2, which is similar to creating an apogee at point 2. On the other hand, the decrease in velocity at point 2 tends to reduce the altitude of the orbit at point 1; this decrease of altitude has the effect of creating a perigee. The net result is that a small eccentricity vector has been created, with its direction perpendicular to the solar radiation direction; see Figure 2.9.9.

Let us assume that the area/mass ratio and reflectance of the satellite are constant over a long annual period, that the sun is in the equatorial plane, and that the distance of the sun to the satellite is constant over the year. With these assumptions, the eccentricity vector will increase by constant increments perpendicular to the sun pressure direction over constant time periods. Because the sun has an apparent circular motion around the earth, the tip of the eccentricity vector will describe a circle, called the *eccentricity circle;* see Figure 2.9.10. As we shall see in this section, the radius of this circle depends on the solar pressure P, and also on the physical properties of the reflectance C_P, the value of which lies between 0 and 2 (see Section 2.8.4).

The analytical development would be similar to that followed for the inclination vector derivatives in Section 2.9.2. However, only the final results for the eccentricity evolution (as developed by Alby 1983) are given here:

2.9 / Perturbed Geostationary Orbits

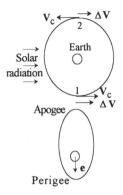

Figure 2.9.9 Solar pressure perturbation and the eccentricity evolution; reproduced from Pocha (1987) by permission of D. Reidel Publishing Co.

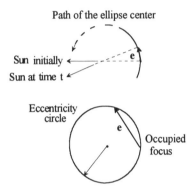

Figure 2.9.10 Evolution of the eccentricity vector; reproduced from Pocha (1987) by permission of D. Reidel Publishing Co.

$$\frac{de_x}{dt} = -\frac{1}{na^2}\frac{\partial U_p}{\partial e_y},$$
$$\frac{de_y}{dt} = \frac{1}{na^2}\frac{\partial U_p}{\partial e_x}. \quad (2.9.35)$$

(The geometry of the evolution of the eccentricity vector is shown in Figure 2.9.12.) To solve these equations, we need an analytical expression for the perturbing function U_p.

Solar Radiation Perturbing Function

The perturbing acceleration will be $\gamma_p = C_P(A_S/M_S)P\mathbf{S}$, where $\mathbf{S}$ is a unit vector in the sun-satellite direction and where A, M, and C_P denote (respectively) area, mass, and the coefficient of specular reflection. We now define the factor σ, which depends on the radiation pressure and on the physical properties of the satellite: $\sigma = C_P(A_S/M_S)P$. With this definition, the perturbing function becomes

$$U_p = -\sigma r \cos(\delta), \quad (2.9.36)$$

where r is the earth–satellite distance. We denote by $\mathbf{r}$ the vector from the center of mass (cm) of the earth to the satellite, $\mathbf{r} = [r_x \ r_y \ r_z]^T$. For a geostationary satellite the inclination is null ($i = 0$), and the elements of $\mathbf{r}$ can be approximated as

$$r_x = r\cos(\omega + \Omega + \theta),$$
$$r_y = r\sin(\omega + \Omega + \theta), \qquad (2.9.37)$$
$$r_z = 0.$$

According to Eq. 2.5.4, $r = a[1 - e\cos(\psi)]$. For low-eccentricity orbits,

$$\theta \approx M + 2e\sin(M) + \tfrac{5}{4}e^2 \sin(2M) + \cdots$$

(see Wertz 1978). Retaining the first two members of the series and defining $\alpha = \omega + \Omega + M$, we have

$$r_x = a[\cos(\alpha) - 1.5e_x + 0.5e_x \cos(2\alpha) + 0.5e_y \sin(\alpha)],$$
$$r_y = a[\sin(\alpha) - 1.5e_y + 0.5e_x \sin(2\alpha) - 0.5e_y \cos(2\alpha)], \qquad (2.9.38)$$
$$r_z = 0.$$

Let us define the sun–earth vector $\mathbf{S}$ as $\mathbf{S} = [s_x \ s_y \ s_z]^T$. In this case, in the geostationary orbit plane with the sun in the orbital plane, $\cos(\delta) = r_x s_x + r_y s_y$. Together with Eqs. 2.9.35 and Eq. 2.9.36, we have

$$\frac{de_x}{dt} = \frac{\sigma}{na}\left[\frac{1}{2}s_x \sin(2\alpha) - \frac{1}{2}s_y \cos(2\alpha) - \frac{3}{2}s_y\right],$$
$$\frac{de_y}{dt} = \frac{\sigma}{na}\left[-\frac{1}{2}s_y \sin(2\alpha) - \frac{1}{2}s_x \cos(2\alpha) + \frac{3}{2}s_x\right]. \qquad (2.9.39)$$

Because we have assumed the apparent sun orbit to be circular and in the equatorial plane of the earth (see Figure 2.9.11), it follows that $s_x = \cos(\lambda)$ and $s_y = \sin(\lambda)$ with $d\lambda/dt = 0.9856°/\text{day} = \omega_s$. It also follows that Eqs. 2.9.39 contain short-term periods of one day and a long-term period of one year. If we neglect the one-day short-term periods, the equations become

$$\frac{de_x}{dt} = -\frac{3}{2}\frac{\sigma}{na}\sin(\lambda),$$
$$\frac{de_y}{dt} = \frac{3}{2}\frac{\sigma}{na}\cos(\lambda). \qquad (2.9.40)$$

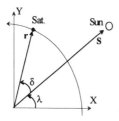

Figure 2.9.11 Relative position of the sun and the earth in the equatorial plane; reproduced from Alby (1983) by permission of Cépaduès-Éditions.

2.9 / Perturbed Geostationary Orbits

These equations show that the eccentricity derivatives are proportional to σ, which depends on the mass of the satellite as well as on the surface reflectance and the effective area directed toward the sun. This area is not constant, and depends on the geometry of the satellite. In order to obtain some reasonable and simplified results, we assume that C_P and A_S are constant, or that they have some average value during the one-year time period.

Eccentricity Vector Evolution

With the foregoing assumptions, time integration of Eqs. 2.9.40 yields the following equations for the eccentricity vector:

$$e_x(t) = e_x(t_0) + R_{ec}\{\cos[\lambda(t)] - \cos[\lambda(t_0)]\},$$
$$e_y(t) = e_y(t_0) + R_{ec}\{\sin[\lambda(t)] - \sin[\lambda(t_0)]\}, \qquad (2.9.41)$$

where the *natural eccentricity radius* $R_{ec} = \frac{3}{2}(\sigma/na\omega_s)$. For the geostationary orbit, $R_{ec} = 1.115 \times 10^{-2} C_P(A_S/M_S)$. During a one-year period, the eccentricity radius describes a circle with radius R_{ec} whose center is located at

$$c_x = e_x(t_0) - R_{ec}\cos[\lambda(t_0)],$$
$$c_y = e(t_0) - R_{ec}\sin[\lambda(t_0)]; \qquad (2.9.42)$$

see Figure 2.9.12.

To get an idea of practical values for R_{ec}, suppose $C_P = 1.5$, $A_S = 10 \text{ m}^2$, and $M_S = 1{,}500$ kg. Then $R_{ec} = 1.115 \times 10^{-4}$. The value of the eccentricity vector depends also on c_x, c_y, and the initial value of $\mathbf{e}_0$. In Chapter 3, we shall see that for a geostationary satellite the need to actively control the eccentricity will depend on R_{ec}.

This completes our simplified discussion of the evolution of the eccentricity vector. However, note that additional moon–sun perturbing forces (creating only very small perturbing effects) have been ignored. If taken into account, the intermediate moon gravity perturbation would create moon period waves superposed on the eccentricity circle of Figure 2.9.12. Figure 2.9.13 shows the evolution of the norm of the eccentricity vector for a period of 725 days, beginning January 1995. For this special case, $C_P(A_S/M_S) = 0.037$, and the additional moon–sun and earth gravitation perturbations have been included.

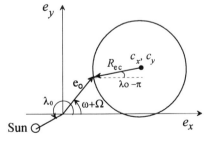

Figure 2.9.12 Evolution of the eccentricity vector; reproduced from Alby (1983) by permission of Cépaduès-Éditions.

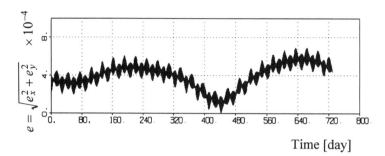

Figure 2.9.13 Evolution of the norm of the eccentricity vector beginning January 1995.

2.9.5 *Longitudinal Acceleration Due to Oblateness of the Earth*

In Section 2.9.1 we defined the mean longitude l_m of a satellite relative to some reference geographic longitudinal meridian (e.g., the Greenwich meridian). In an ideal Keplerian geostationary orbit, the satellite should remain indefinitely at its initial geographic position. Unfortunately, different perturbing forces cause the satellite to deviate from its initial location relative to the earth. The primary perturbing force is due to the ellipticity of the earth equatorial plane. The tesseral terms in the harmonic expansion of the earth's gravitational potential function are responsible for that ellipticity (see Section 2.8.2 and Eq. 2.8.2). This effect is illustrated in Figure 2.9.14.

In the equatorial plane, the earth's gravitational acceleration can be decomposed into a Keplerian term γ_K and into a tangential term γ_T, which is in fact the perturbing acceleration that tends to accelerate the satellite along the nominal orbit path. The value and the direction of γ_T depend on the earth's geographic longitude corresponding to the location of the satellite. In Figure 2.9.14, points 1, 2, 3, and 4 are equilibrium points on the orbit, since at these points γ_T is null. However, only points 1 and 3 are stable equilibrium points, as will be shown later in this section. (The analytical derivation of the longitudinal perturbing acceleration can be found in the literature, and will not be derived here.)

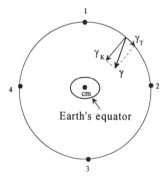

Figure 2.9.14 Perturbing longitudinal acceleration.

2.10 / Euler–Hill Equations

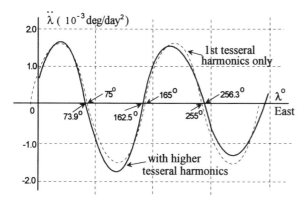

Figure 2.9.15 Longitudinal drift acceleration due to earth's ellipticity; adapted from Agrawal (1986) by permission of Prentice-Hall.

It is common in this context to designate the longitude of the satellite by λ. The first-order approximation, taking into account only the first tesseral harmonic term, leads to the simple differential equation

$$\frac{d^2\lambda}{dt} = -0.00168 \sin 2(\lambda - \lambda_s) \text{ deg}/\text{day}^2. \quad (2.9.43)$$

The two stable longitude points 1 and 3 in Figure 2.9.14 are located at $\lambda_s = 75°$ and 255° E. Taking into consideration higher tesseral harmonic terms, Eq. 2.9.43 is somewhat altered, and the stable points move to $\lambda_s = 73.9°$ and 256.3° E; see Figure 2.9.15.

To show the stability of points 1 and 3 in Figure 2.9.14, suppose that the satellite is located at 75° E and a small drift of the satellite is sensed toward the east direction. In this case, a negative longitudinal acceleration will force the satellite to move back toward the west direction. Should the small drift of the satellite be to the left of the stable point, a positive longitudinal acceleration would force the satellite to move toward the stable point to the east. The same arguments hold for a location close to the stable longitude point at 255° E.

To show the instability of points 2 and 4 in Figure 2.9.14, suppose that the satellite is drawn to the right of the 165° E equilibrium point in Figure 2.9.15. In this case, a positive longitude acceleration will tend to drive the satellite more to the east, and the positive acceleration will increase until the satellite reaches the stable longitude position at 255° E.

2.10 Euler–Hill Equations

2.10.1 Introduction

In the previous sections of this chapter, the motion of a satellite in any orbit was expressed in the inertial coordinate frame with its Cartesian coordinates. The distance between two satellites moving on two different orbits could be calculated by use of those coordinates. However, there are special situations in which two satellites move in almost identical orbits and are very near to each other. In these cases, it is easier to analyze the evolution of the distance between the two satellites by defining a coordinate system that is not inertial. Rather, this system's origin is fixed at the center of mass of one of the satellites and is moving with it. The coordinates of the other satellite are calculated in this moving coordinate frame. These equations describe the relative motion of satellites in neighboring orbits. With some constraints (to be stated in Section 2.10.2), the equations of motion of the second satellite are developed with respect to the moving frame, and are called the *Euler-Hill equations* or simply the *Hill equations*.

These equations can be used in many special problems to obtain elegant analytical solutions. Some examples of problems easily solved by use of the Hill equations are: (1) the influence of small perturbations on satellites, such as drag forces; (2) the relative motion of two neighboring satellites moving in two almost identical orbits but having slightly different velocity vectors at some epoch; and (3) the very important *rendezvous* problem between two spacecraft.

2.10.2 Derivation

To obtain the Hill equations, we must first define the moving frame in which the equations of relative motion of a neighboring spacecraft are to be developed. See Figure 2.10.1.

If $\mathbf{r}_1$ and $\mathbf{r}_2$ are (respectively) the distances of the reference and the second satellite from the central body M in Figure 2.10.1, then the relative distance between the satellites is $\boldsymbol{\rho} = \mathbf{r}_2 - \mathbf{r}_1$. Since $\ddot{\mathbf{r}}_1 = -\mu \mathbf{r}_1/r_1^3$ and by definition $\ddot{\mathbf{r}}_2 = -\mu \ddot{\mathbf{r}}_2/r_2^3 + \mathbf{f}$, where $\mathbf{r}_2$ can be expressed in terms of $\mathbf{r}_1$ and of $\boldsymbol{\rho}$, for small ρ the relative acceleration, seen in the rotating frame, becomes

$$\ddot{\boldsymbol{\rho}} = \frac{\mu}{r_1^3}\left[-\boldsymbol{\rho} + 3\left(\frac{\mathbf{r}_1}{r_1} \cdot \boldsymbol{\rho}\right)\frac{\mathbf{r}_1}{r_1}\right] + \mathbf{f} + O(r^2). \tag{2.10.1}$$

If we assume an almost circular orbit (a very small eccentricity e), and if we neglect terms of order e^2 and ρ^2 and products of ρ and e, then Eq. 2.10.1 becomes, in terms of components of relative motion,

$$\ddot{x} - 2n\dot{y} - 3n^2 x = f_x, \tag{2.10.2}$$

$$\ddot{y} + 2n\dot{x} = f_y, \tag{2.10.3}$$

$$\ddot{z} + n^2 z = f_z \tag{2.10.4}$$

(Breakwell and Roberson 1970; Kaplan 1976). Here x and y, as defined in Figure 2.10.1, are the plane distances between the two neighboring satellites, while z is the

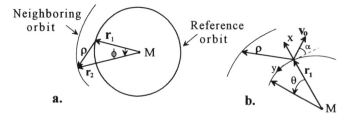

Figure 2.10.1 Relative positions of two satellites in two neighboring orbits.

2.10 / Euler-Hill Equations

off-plane distance; n is the almost constant angular velocity of the reference orbit, which we have assumed to be nearly circular. Equations 2.10.2–2.10.4 can be Laplace-transformed.

The solution of the third expression, Eq. 2.10.4, is independent from the first two. Hence, its solution is simply

$$z(s) = \frac{sz_0 + \dot{z}_0}{s^2 + n^2} + \frac{f_z(s)}{s^2 + n^2}. \tag{2.10.5}$$

If we assume that for practical purposes f_z is an impulse function of strength f_z, then

$$z(t) = z_0 \cos(nt) + \frac{\dot{z}_0}{n} \sin(nt) + \frac{f_z}{n} \sin(nt). \tag{2.10.6}$$

Equation 2.10.2 and Eq. 2.10.3 must be solved simultaneously:

$$\begin{bmatrix} x(s) \\ y(s) \end{bmatrix} = \frac{1}{s^2(s^2 + n^2)} \begin{bmatrix} s^2 & 2ns \\ -2ns & s^2 - 3n^2 \end{bmatrix} \begin{bmatrix} sx_0 + \dot{x}_0 - 2ny_0 + f_x(s) \\ sy_0 + \dot{y}_0 + 2nx_0 + f_y(s) \end{bmatrix}. \tag{2.10.7}$$

For practical purposes we assumed f_x and f_y to be impulsive forces, so the time-domain solution of Eq. 2.10.7 becomes

$$x(t) = [\dot{x}_0 + f_x] \frac{\sin(nt)}{n}$$
$$- \left[3x_0 + \frac{2(\dot{y}_0 + f_y)}{n} \right] \cos(nt) + \frac{2}{n}(\dot{y}_0 + f_y) + 4x_0, \tag{2.10.8}$$

$$y(t) = [4(\dot{y}_0 + f_y) + 6nx_0] \frac{\sin(nt)}{n}$$
$$+ \frac{2}{n}(\dot{x}_0 + f_x) \cos(nt) + y_0 - \frac{2}{n}(\dot{x}_0 + f_x) - 3(\dot{y}_0 + f_y + 2nx_0)t. \tag{2.10.9}$$

In these last two equations, $\dot{x}_0, f_x$ and $\dot{y}_0, f_y$ always appear in pairs, indicating that the initial velocity conditions and the impulsive forces have equivalent influence on satellite motion. In terms of the initial relative velocity vector $\mathbf{V}_0$ (see Figure 2.10.1.b), Eqs. 2.10.8–2.10.9 can be rewritten as

$$x(t) = \frac{V_0}{n} \{\sin(\alpha) \sin(nt) + 2 \cos(\alpha)[\cos(nt) - 1]\}, \tag{2.10.10}$$

$$y(t) = \frac{V_0}{n} \{2 \sin(\alpha)[\cos(nt) - 1] - 4 \cos(\alpha) \sin(nt) + 3(nt) \cos(\alpha)\}. \tag{2.10.11}$$

(Notice that $\dot{y}_0 < 0$ in Figure 2.10.1.b.)

When using these equations we have to remember that $x(t)$ and $z(t)$ must be assumed to be small; this is not true of $y(t)$, which does not appear explicitly in the Hill equations. Moreover, $y(t)$ is in the direction of motion of the satellite, so that y is in fact a part of an arc on the orbit, of length $r_1\theta$, where θ is shown in Figure 2.10.1.b. These equations can be used in different ways, including application to the rendezvous problem. The following example treats a related case, and will clarify the use of the Hill equations.

EXAMPLE 2.10.1 Suppose that a satellite has been put on its final orbit by an apogee boost motor (ABM). The satellite and the ABM are then separated by giving the ABM a small velocity opposite to the initial direction of motion of both bodies, $\dot{y}_0 = -V_0 \cos(\alpha)$, with $\alpha = 0$ nominally, so that the ABM will be removed to a safe distance from the satellite. The velocity imparted to the ABM body may contain a parasitic component in the x direction, $\dot{x}_0 = V_0 \sin(\alpha)$. Could this parasitic velocity lead to a crash meeting between the two bodies at any future time?

Solution Assume that both bodies remain in the same plane after separation, which means that $\dot{z}(0) \equiv \dot{z}_0 = 0$. Also, $x_0 = y_0 = z_0 = 0$. However, $\dot{x}(0) \equiv \dot{x}_0 \neq 0$ and $\dot{y}(0) \equiv \dot{y}_0 \neq 0$. In a first check, suppose that an initial velocity in only the x direction has been imparted to the ABM: $\alpha = 90°$, $V_0 = 0.01$ m/sec. Then Eq. 2.10.10 and Eq. 2.10.11 become

$$x(t) = V_0 \frac{\sin(nt)}{n} \quad \text{and} \quad y(t) = \frac{2V_0}{n}[\cos(nt) - 1].$$

We must check if $\rho(t)$ in Figure 2.10.1.b can ever become null. To satisfy $x(t) = y(t) = 0$ simultaneously, it is necessary that $nt = 2m\pi$, $m = 1, 2, 3, \ldots$. Observe the development of $\rho(t)$ in Figure 2.10.2. On the x-y phase plane, it is clear that the motion of the ABM body about the satellite is harmonic; at the end of each orbital period, the two bodies collide.

Next, let us suppose that $\alpha = 0°$. In this case $x(t) = (V_0/n)2[\cos(nt) - 1]$, so $x(t)$ is null at $nt = 2\pi m$ for $m = 1, 2, 3, \ldots$; $y(t) = (V_0/n)[-4\sin(nt) + 3(nt)]$, so $y(t)$ cannot be null at $nt = 2\pi m$ because of the $3(nt)$ term (except at $nt = 0$, which is a trivial solution). Hence no collision is to be expected.

The minimum distance between the satellite and the separated ABM depends on the velocity vector initial conditions, V_0 in Figure 2.10.1. Figure 2.10.3 shows the

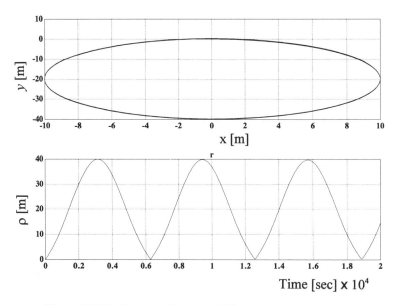

Figure 2.10.2 Relative distance $\rho(t)$ between the two separated bodies with initial conditions $V_0 = 0.01$ m/sec and $\alpha = 90°$.

2.10 / Euler-Hill Equations

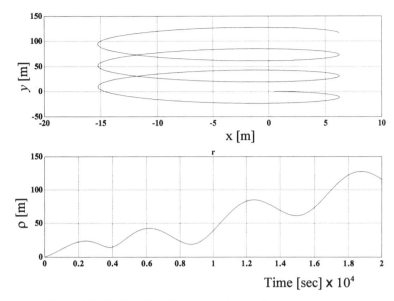

Figure 2.10.3 Relative distances between the two separated bodies with initial conditions $V_0 = 0.01$ m/sec and $\alpha = 77°$.

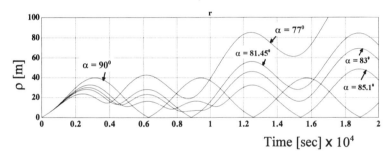

Figure 2.10.4 Relative distances between the two separated bodies with initial conditions $V_0 = 0.01$ m/sec and $\alpha = 77°$, 81.456°, 83°, 85.08°, and 90°.

evolution of the relative distance components with initial conditions $V_0 = 0.01$ m/sec and $\alpha = 77°$. The minimal distance is 15 m. In this example, $n = 0.001$ rad/sec, which is equivalent to a circular orbit with altitude of $h = 981.39$ km. The minimal distance can be augmented also by increasing V_0.

In the general case, there might be values of $\alpha \neq 90°$ for which both $x(t)$ and $y(t)$ are null simultaneously, thus leading to collision. By numerical computation, it was found that a value of either $\alpha = 81.456°$ or $\alpha = 85.08°$ for the direction of the initial relative vector velocity will null $x(t)$ and $y(t)$ simultaneously in Eq. 2.10.10 and Eq. 2.10.11. This happens at $nt = 15.3$. The relative distances $\rho(t)$ are shown in Figure 2.10.4 for $\alpha = 77, 81.456, 83, 85.08$, and 90 degrees.

Example 2.10.1 suggests that when the separation stage between the two bodies is designed, care must be taken to prevent a catastrophic collision. However, since the two separated bodies normally have different aerodynamic profiles, it is natural that they will be acted upon by different aerodynamic drag forces, thus increasing the distance between them.

2.11 Summary

Chapter 2 dealt with the dynamics of spacecraft orbits. First, the classical Keplerian orbit was treated and analyzed. A second purpose of the chapter was to state the different forces that tend to perturb the ideal Keplerian orbits. The influence of these forces on the classical parameters of practical orbits was exemplified with a detailed treatment of geostationary orbits. The effects of these perturbing forces on the orbit can be annoying (as when a geostationary orbit must be maintained), but are sometimes welcomed by the spacecraft engineer (as in the case of achieving a heliosynchronous orbit). The Hill equations were also presented, and their use exemplified by calculating the collision hazard between two artificially separated bodies.

References

Agrawal, B. N. (1986), *Design of Geosynchronous Spacecraft*. Englewood Cliffs, NJ: Prentice-Hall.

Alby, F. (1983), "Les Perturbations de l'Orbite Géostationaire," in *The Motion of the Satellite, Lectures and Exercises on Space Mechanics*. Toulouse, France: Cépaduès-Éditions, pp. 569–611.

Balmino, G. (1980), "Le Mouvement Elliptique Perturbé," in *Le Mouvement du Véhicule Spatial en Orbite*. Toulouse, France: Centre National d'Études Spatiales, pp. 59–84.

Bate, R. R., Mueller, D. D., and White, J. E. (1971), *Fundamentals of Astrodynamics*. New York: Dover.

Battin, R. H. (1990), *An Introduction to the Mathematics and Methods of Astrodynamics*. Washington, DC: AIAA.

Bernard, J. (1983), "Mouvement Keplerien et Perturbé," in *The Motion of the Satellite, Lectures and Exercises on Space Mechanics*. Toulouse, France: Cépaduès-Éditions, pp. 25–86.

Borderies, N. (1980), "Les Perturbation d'Orbites: les Effets," in *Le Mouvement du Véhicule Spatial en Orbite*. Toulouse, France: Centre National d'Études Spatiales, pp. 131–51.

Breakwell, J., and Roberson, R. (1970), "Orbital and Attitude Dynamics," Lecture notes, Stanford University.

Campan, G. (1983), "Les Perturbations d'Orbite," in *The Motion of the Satellite, Lectures and Exercises on Space Mechanics*. Toulouse, France: Cépaduès-Éditions, pp. 129–57.

Chobotov, V. A. (1991), *Orbital Mechanics*. Washington, DC: AIAA.

Deutsch, R. (1963), *Orbital Dynamics of Space Vehicles*. Englewood Cliffs, NJ: Prentice-Hall.

Escobal, P. R. (1965), *Methods of Orbit Determination*. Malabar, FL: Krieger.

Kaplan, M. H. (1976), *Modern Spacecraft Dynamics and Control*. New York: Wiley.

Legendre, P. (1980), "Mantien a Poste des Satellites Géostationaires: Stratégie des Corrections d'Orbite," in *Le Mouvement du Véhicule Spatial en Orbite*. Toulouse, France: Centre National d'Études Spatiales, pp. 583–609.

References

Pocha, J. J. (1987), *An Introduction to Mission Design for Geostationary Satellites.* Dordrecht: Reidel.

Pritchard, L. W., and Sciulli, A. J. (1986), *Satellite Communication Systems Engineering.* Englewood Cliffs, NJ: Prentice-Hall.

Soop, E. M. (1983), *Introduction to Geostationary Orbits.* Paris: European Space Agency.

Thomson, W. T. (1986), *Introduction to Space Dynamics.* New York: Dover.

Wertz, J. R. (1978), *Spacecraft Attitude Determination and Control.* Dordrecht: Reidel.

CHAPTER 3

Orbital Maneuvers

3.1 Introduction

From the moment that a satellite is launched into its initial orbit, commonly called a *transfer orbit,* multiple orbital changes must be performed. Changes of the transfer orbit are necessary in order to obtain the desired final orbit because a launch vehicle usually cannot put the satellite in its final orbit. Even if it were possible, a satellite launch that placed the s/c in its final orbit would not be optimal from the point of view of fuel consumption (see Duret and Frouard 1980).

Fuel consumption is a crucial factor in orbital maneuvers. Any orbital change is accompanied by a velocity change of the satellite, which necessitates a certain quantity of fuel consumption. As we shall see in this chapter, minimization of fuel consumption is essential because the weight of the useful payload that can be carried to the desired orbit depends on this minimization.

This chapter will also consider different kinds of orbit maneuvers and changes. For instance, orbits may be adjusted by single or multiple thrust impulses. It will be seen that with a single thrust impulse, very limited kinds of orbit changes can be achieved, whereas multiple thrust impulses can effect any desired orbit change. It is comparatively easy to analyze the change in orbit parameters due to an impulsive thrust; the analytical treatment usually ends with a closed-form solution. Unfortunately, an impulsive thrust is an idealization that cannot be met in practice.

Thus, another way of viewing orbital maneuvers concerns the duration of the thrust. During an orbit change, thrust is applied for a length of time (sometimes hours) that depends on the thrust magnitude (see Redding 1984). What we call an *impulsive thrust* depends very much on the thrust duration relative to the orbit's natural period. The principal drawback in applying a *non*impulsive thrust is that the application of a finite-time thrust is accompanied by a nonconstant thrust direction during burns. The nonconstant thrust direction results in a velocity loss because only a component of the thrust acceleration, rather than its entire value, acts along the mean thrust direction (see Robins 1966).

The classical analysis of orbit dynamics, treated in Chapter 2, is sufficient for the purpose of introducing basic notions of orbit maneuvering and adjustments, with the practical approximation that the applied thrust is impulsive. Once in its operational orbit, the satellite is subject to different disturbance forces. In order for the satellite to perform its mission successfully, the orbit must be corrected accordingly. Such corrections are also treated in this chapter. The propulsion equation (Eq. C.2.5) developed in Appendix C is important to this chapter because it shows the mass of consumed fuel necessary to change the velocity of a satellite during any orbit maneuver. Appendix C also details the hardware of propulsion systems.

3.2 / Single-Impulse Orbit Adjustment

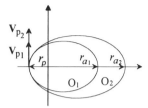

Figure 3.2.1 Change of the apogee radius vector.

3.2 Single-Impulse Orbit Adjustment

Single-impulse orbit adjustment is a very restricted class of orbital maneuvers. Nevertheless, it is commonly used to circularize elliptic orbits or to change the eccentricity of an orbit, the altitudes of the perigee or the apogee, or the argument of perigee measured from the line of apsides of the initial orbit. These cases will be described in this section.

3.2.1 Changing the Altitude of Perigee or Apogee

In this case, the radius vector of the apogee will be changed from r_{a1} to r_{a2} (see Figure 3.2.1). For known orbits O_1 and O_2, the velocities V_{p1} and V_{p2} at the perigee – which is common to both orbits – can be computed, since the initial r_{a1} and the desired r_{a2} are also known. Using Eq. 2.4.20, we can write

$$V_{p1} = \sqrt{2\left(\frac{\mu}{r_p} - \frac{\mu}{r_p + r_{a1}}\right)} \quad \text{and} \quad V_{p2} = \sqrt{2\left(\frac{\mu}{r_p} - \frac{\mu}{r_p + r_{a2}}\right)}. \tag{3.2.1}$$

In Eqs. 3.21, since $r_{a2} > r_{a1}$ it follows that $V_{p2} > V_{p1}$, and $\Delta V = V_{p2} - V_{p1}$ is the velocity impulse to be added to the satellite at the perigee that is common to both orbits in order to increase the apogee radius vector to r_{a2}.

3.2.2 Changing the Semimajor Axis a_1 and Eccentricity e_1 to a_2 and e_2

In Figure 3.2.2, the velocity vector at the apogee of O_1, which is perpendicular to r_{a1}, can be expressed as

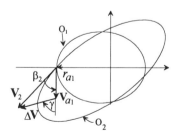

Figure 3.2.2 Change of a and e of a Keplerian orbit.

$$V_{a1}^2 = 2\mu\left[\frac{1}{r_{a1}} - \frac{1}{2a_1}\right] = 2\mu\left[\frac{1}{a_1(1+e_1)} - \frac{1}{2a_1}\right] = \mu\frac{[1-e_1]}{a_1[1+e_1]}. \quad (3.2.2)$$

For the final orbit O_2, $r_2 = r_{a1}$ at the apogee of O_1. To find the velocity vector $\mathbf{V}_2$, we write

$$V_2^2 = 2\mu\left[\frac{1}{r_2} - \frac{1}{2a_2}\right] = 2\mu\left[\frac{1}{r_{a1}} - \frac{1}{2a_2}\right] = 2\mu\left[\frac{1}{a_1(1+e_1)} - \frac{1}{2a_2}\right]. \quad (3.2.3)$$

Next, we must find the direction of $\mathbf{V}_2$. For this we use Eq. 2.4.18 and the definition in Section 2.3 of the momentum $h = rv\cos(\beta_2)$. We obtain $p = a_2(1 - e_2^2) = [r_{a1}v_2\cos(\beta_2)]^2/\mu$, from which it follows that

$$\cos^2(\beta_2) = \frac{\mu a_2(1-e_2^2)}{v_2^2 a_1^2(1+e_1)^2}. \quad (3.2.4)$$

Knowing the velocity vector $\mathbf{V}_2$, we can calculate the velocity vector change $\Delta\mathbf{V}$ to be added to $\mathbf{V}_{a1}$ so that $\mathbf{V}_2$ is achieved. In Figure 3.2.2, v_2, v_{a1}, and β_2 are known, from which it is easily found that

$$\Delta V = \sqrt{[v_2\cos(\beta_2) - v_{a1}]^2 + v_2^2\sin^2(\beta_2)}; \quad \sin(\gamma) = \sin(\beta_2)v_2/\Delta V. \quad (3.2.5)$$

Equation 3.2.5 is the desired result, because it gives the velocity vector change $\Delta\mathbf{V}$ to be added at the apogee of the initial orbit O_1.

EXAMPLE 3.2.1 The initial orbit is defined as: $a_1 = 5R_e$; $e_1 = 0.7$. The final orbit is defined as: $a_2 = 10R_e$; $e_2 = 0.3$. Here $R_e = 6{,}378.6$ km, the mean radius of the earth at the equator. The initial mass of the satellite is 1,000 kg. Given $I_{sp} = 200$ sec, what is the required propellant mass for the orbit change? (I_{sp} denotes "specific impulse"; see Appendix C.)

Solution We find that $v_{a1} = 1.485$ km/sec, $v_2 = 2.9078$ km/sec, and $\beta_2 = 15.23°$, from which it follows that $\Delta V = 1.526$ km/sec and $\gamma = 30.87°$. Using Eq. C.2.5, we calculate that $m_{\text{prop}} = 540.1$ kg. This is the minimum mass of propellant that the satellite needs in order to achieve the desired orbit change.

In this example, the desired change in the parameters a_1 and e_1 could be achieved. However, not all specifications on a_2 and e_2 are attainable. In the general case of Example 3.2.1, given a_1 and e_1, there exist limits on the values of a_2 and e_2 that can be achieved with a single-impulse thrust adjustment.

In Figure 3.2.3, the orbit change is performed anywhere on the original orbit O_1, say, at $\mathbf{r}_1$. From Eq. 2.4.20 we can find the velocity v_2, whose square must be a positive value:

$$\frac{v_2^2}{2} = \frac{\mu}{r_1} - \frac{\mu}{2a_2} \geq 0 \Rightarrow 2a_2 \geq r_1. \quad (3.2.6)$$

Moreover, using Eq. 2.4.18 for the target orbit O_2, we can write

$$h_2^2 = \mu a_2(1-e_2^2) = [v_2 r_1 \cos(\beta_2)]^2,$$

from which it follows that

3.2 / Single-Impulse Orbit Adjustment

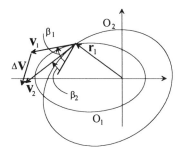

Figure 3.2.3 The general case of changing a and e.

$$\cos^2(\beta_2) = \frac{\mu a_2(1-e_2^2)}{r_1^2 v_2^2} = \frac{\mu a_2(1-e_2^2)}{r_1^2\left(\frac{2\mu}{r_1} - \frac{\mu}{a_2}\right)} \leq 1 \quad (3.2.7)$$

or (alternatively) $a_2(1-e_2^2) \leq 2r_1 - r_1^2/a_2$. We must solve the quadratic inequality

$$r_1^2 - 2a_2 r_1 + a_2^2(1-e_2^2) \leq 0. \quad (3.2.8)$$

The roots of r_1 are

$$r_{11}, r_{12} = \tfrac{1}{2}[2a_2 \pm \sqrt{4a_2^2 - 4a_2^2(1-e_2^2)}] = a_2 \pm a_2 e_2. \quad (3.2.9)$$

The inequality in Eq. 3.2.8 can be written as

$$[r_1 - r_{11}][r_1 - r_{12}] = [r_1 - a_2(1+e_2)][r_1 - a_2(1-e_2)] \leq 0; \quad (3.2.10)$$

this inequality holds if

$$r_{p2} = a_2(1-e_2) < r_1 < a_2(1+e_2) = r_{a2}. \quad (3.2.11)$$

Equation 3.2.11 is described graphically in Figure 3.2.4, in which r_{a1} and r_{p1} are the apogee and perigee radius vectors of the initial orbit O_1; $r_{a1} = a_1(1+e_1)$ and $r_{p1} = a_1(1-e_1)$. The screened region in the figure shows the range of achievable parameters a_2 and e_2 for O_2, related to the given parameters a_1 and e_1 of O_1. This region is bounded by the two inequalities of Eq. 3.2.11 and the two extreme values of r_1, which are r_{a1} and r_{p1}.

The selection of the point in the orbit at which the impulsive velocity change is to be performed depends on different operational constraints. However, Eq. 3.2.6

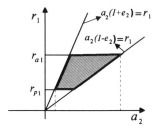

Figure 3.2.4 Limits on achievable a_2 and e_2 for given a_1 and e_1.

clearly shows that v_2 will be minimal for the maximum possible value of r_1. After $\mathbf{r}_1$ is chosen, we can compute v_1, v_2, β_1, and β_2, from which ΔV is found:

$$\Delta V^2 = v_1^2 + v_2^2 - 2v_1 v_2 \cos(\beta_1 - \beta_2). \qquad (3.2.12)$$

EXAMPLE 3.2.2 The orbit parameters of $a_1 = 3R_e$ and $e_1 = 0.1$ are to be changed to $a_2 = 10R_e$ and $e_2 = 0.5$.

Solution As before, R_e is the mean radius of the earth. Using the inequality in Eq. 3.2.11 yields $r_1 = a_2(1-e_2) < r_{a1} = 3R_e(1+e_1)$; hence, $a_2 < 3R_e(1.1)/0.5 = 6.6R_e$. In the same way we find that $a_2 > 3R_e(0.9)/1.5 = 1.8R_e$. Hence $6.6R_e > a_2 > 1.8R_e$ and so $a_2 = 10R_e$ cannot be achieved. If a_2 is the more important parameter to be achieved, then e_2 must be redefined in order for the inequality in Eq. 3.2.11 to be satisfied. In this case, the minimum value of e_2 would be 0.67.

3.2.3 Changing the Argument of Perigee

In Section 2.6.1 (see also Figure 2.6.2), the angle ω was defined as the *argument of perigee*. There is a variety of communications satellite systems using elliptical orbits (e.g., the *Molniya* system; see Pritchard and Sciulli 1986) for which the apogee footprint on the earth surface must remain fixed at a certain geographical latitude. This situation occurs for satellite orbits having an inclination angle of 63.435° (see Example 2.8.2). Moreover, depending on the initial satellite launch parameters, the initial perigee argument might not be located at the correct geographical latitude. Alignment and adjustments of the argument of perigee then become imperative.

Suppose we wish to change the perigee argument by an amount α without altering the remaining two planar parameters of the orbit, the semimajor axis a and the eccentricity e. The orbit maneuver is performed in-plane. The two orbits have a common point at P, and $r_1 = r_2 = r$. Since the shapes of the two orbits remain unchanged, their angular momentums h are equal and hence $p_1 = p_2 = a(1-e^2) = h^2/\mu$. This means that, according to Eq. 2.3.4, $\beta_1 = \beta_2 = \beta$ in Figure 3.2.5.b.

The argument between the two apsides is α, which is also the desired change in the perigee argument. As the shapes of the two orbits are identical and $r_1 = r_2$, according to Eq. 2.4.13 we have $\delta_1 = \delta_2 = \delta$ in Figure 3.2.5.b, $\alpha = 2\delta$, and $V_2 = V_1 = V$.

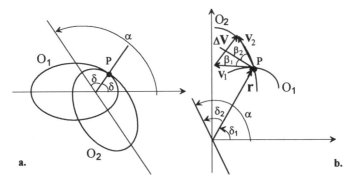

Figure 3.2.5 Geometry for changing ω, the argument of perigee.

3.2 / Single-Impulse Orbit Adjustment

In order to find the required change in the velocity vector ΔV we must find V and β, since $\Delta V = 2V\sin(\beta)$ in Figure 3.2.5.b.

The equations needed to solve this problem are:

$$p = a(1-e^2) = \frac{v^2 r^2 \cos^2(\beta)}{\mu},$$

$$r = \frac{p}{1+e\cos(\delta)},$$

$$v^2 = \frac{2\mu}{r} - \frac{\mu}{a}.$$

These equations yield

$$v^2 = \frac{2\mu}{r} - \frac{\mu}{a} = \frac{2\mu}{p}[1+e\cos(\delta)] - \frac{\mu}{a} = \frac{\mu}{a}\frac{[1+e^2+2e\cos(\delta)]}{1-e^2}, \quad (3.2.13)$$

$$\cos^2(\beta) = \frac{\mu a(1-e^2)}{v^2 r^2} = \frac{[1+e\cos(\delta)]^2}{1+e^2+2e\cos(\delta)}. \quad (3.2.14)$$

From Eq. 3.2.14 we find that

$$\sin^2(\beta) = 1 - \cos^2(\beta) = \frac{e^2 \sin^2(\delta)}{1+e^2+2e\cos(\delta)}. \quad (3.2.15)$$

Finally, from Eq. 3.2.13 and Eq. 3.2.15 we obtain

$$\Delta V = 2v\sin(\beta) = 2\sqrt{\frac{\mu}{a(1-e^2)}}\, e \sin\left(\frac{\alpha}{2}\right). \quad (3.2.16)$$

Equation 3.2.15 and Eq. 3.2.16 determine the vector ΔV that must be added to the velocity vector V at point P in order to change the perigee argument by an amount α.

3.2.4 Restrictions on Orbit Changes with a Single Impulsive ΔV

As exemplified in Section 3.2.3, applying a single impulsive thrust is not sufficient to intentionally change all the parameters of the planar orbit; certain restrictions apply. Table 3.2.1 summarizes the restrictions on coplanar transfer orbit changes.

Table 3.2.1 *Coplanar transfer orbit restrictions*
(reproduced from Deutsch 1963 by permission of Prentice-Hall)

Element changed	Fixed elements	Restrictions
a	e, ω	Impossible
a, e, ω	None	$1 + D > \frac{a_1}{a_2} > 1 - D$
e, ω	a	$\left(\frac{e_1}{e_2}\right)^2 + 1 - 2\left(\frac{e_1}{e_2}\right)\cos(\Delta\omega) > 0$
a, ω	e	$1 + D' > \frac{a_1}{a_2} > 1 - D'$
a, e	ω	$1 \pm \left(\frac{a_1}{a_2}e_1 - e_2\right) \geq \frac{a_1}{a_2} \geq 1 \pm \left(e_2 - \frac{a_1}{a_2}e_1\right)$
e	a, ω	None
ω	a, e	None

In Table 3.2.1,

$$D^2 = \left(\frac{a_1}{a_2}\right)^2 e_1^2 + e_2^2 - 2\frac{a_1}{a_2} e_1 e_2 \cos(\Delta\omega) \quad \text{and}$$

$$D' = e\left[\left(\frac{a_1}{a_2}\right)^2 + 1 - 2\left(\frac{a_1}{a_2}\right)\cos(\Delta\omega)\right]^{1/2}.$$

(Deutsch 1963 gives a table summarizing restrictions on changes for noncoplanar orbits also.)

3.3 Multiple-Impulse Orbit Adjustment

A principal characteristic of – and a major restriction on – single-impulse orbital adjustment is that the initial and the final orbits will always have at least one common point. The physical reason for this is that, after application of the $\Delta \mathbf{V}$ impulse at some location in the orbit, the altered orbit will repeat itself and pass over the same point at which the $\Delta \mathbf{V}$ impulse was applied. The only way to achieve a final orbit that does not intersect the initial orbit is to apply multiple thrust impulses. Such multiple adjustments also eliminate the drawback of the restrictions treated in Sections 3.2.2 and 3.2.4.

3.3.1 Hohmann Transfers

The Hohmann transfer between orbits was originally intended to allow a transfer between two circular orbits with the minimum consumption of fuel, which is equivalent to a transfer with the minimum total ΔV. From this point of view, the Hohmann transfer is optimal, as long as the ratio of the large to the small radius of the two circular orbits is less than 11.8 (see Kaplan 1976, Prussing 1991). Once more, it is assumed that the applied thrust is impulsive, which means that the added velocity changes are instantaneous.

According to the Hohmann orbital transfer principle, changing a circular orbit O_1 of radius r_1 to a coplanar and concentric circular orbit O_2 of radius r_2 requires an initial transfer of the first orbit to an intermediate transfer orbit (TO). The TO's perigee radius vector must equal the radius of O_1 ($r_p = r_1$), and its apogee radius vector must equal the radius of O_2 ($r_a = r_2$); see Figure 3.3.1.

At any point in the orbit O_1, an impulsive thrust is applied such that the additional velocity $\Delta \mathbf{V}_1$ will raise the energy of the initial orbit to that of an elliptic orbit with

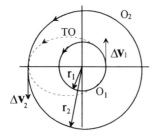

Figure 3.3.1 Hohmann transfer between two circular orbits.

3.3 / Multiple-Impulse Orbit Adjustment

$r_p = r_1$ and $r_a = r_2$. The velocity of the circle orbit O_1 is $v_1 = \sqrt{\mu/r_1}$ (see Eq. 2.4.14). Knowing r_a and r_p of the transfer orbit TO, we can use Eq. 2.4.20 to find the velocity at its perigee:

$$v_p = \sqrt{2\mu\left(\frac{1}{r_1} - \frac{1}{2a}\right)} = \sqrt{2\mu\left(\frac{1}{r_1} - \frac{1}{r_1+r_2}\right)} = \sqrt{\frac{2\mu r_2}{r_1(r_1+r_2)}}. \tag{3.3.1}$$

It follows that the velocity to be added at the perigee in the direction of motion of the orbit O_1 is

$$\Delta v_1 = v_p - v_1 = \sqrt{\frac{\mu}{r_1}}\left[\sqrt{\frac{2r_2}{r_1+r_2}} - 1\right]. \tag{3.3.2}$$

At the apogee of the TO, an additional change Δv_2 is added so that the velocity of the satellite will be increased to that of the circular orbit O_2 with velocity $v_2 = \sqrt{\mu/r_2}$. We will calculate that change as follows. First, we must find the velocity of the satellite at the apogee of the TO:

$$v_a = \sqrt{\frac{2\mu r_p}{r_a(r_a+r_p)}} = \sqrt{\frac{2\mu r_1}{r_2(r_1+r_2)}}. \tag{3.3.3}$$

The velocity to be added at the apogee (in the direction of motion of the satellite) will then be

$$\Delta v_2 = v_2 - v_a = \sqrt{\frac{\mu}{r_2}}\left[1 - \sqrt{\frac{2r_1}{r_1+r_2}}\right]. \tag{3.3.4}$$

Finally, the total change in velocity to be given to the satellite is

$$\Delta V = \Delta v_1 + \Delta v_2. \tag{3.3.5}$$

Kaplan (1976) contains a proof that this calculated ΔV is optimal in the sense of minimizing ΔV, which is the same as minimizing the fuel consumption. In the case where the transfer is from O_2 to O_1, the velocities will be added in the direction opposite to the motion of the satellite.

3.3.2 Transfer between Two Coplanar and Coaxial Elliptic Orbits

Exactly as in the previous section, the transfer is performed in two stages via a transfer orbit; see Figure 3.3.2. The minimum-energy transfer between the two orbits consists of Δv_1 at the perigee of the O_1 orbit and Δv_2 at the apogee of the elliptic transfer orbit. With known r_{a1}, r_{p1}, r_{a2}, and r_{p2}, the semimajor axis of the TO

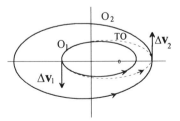

Figure 3.3.2 Transfer between two elliptic orbits.

will be $a_{TO} = \frac{1}{2}(r_{p1} + r_{a2})$, with $r_{aTO} = r_{a2}$ and $r_{pTO} = r_{p1}$. Following the same derivation as in Section 3.3.1, we obtain

$$\Delta v_1 = \sqrt{\frac{2\mu}{r_{p1}} - \frac{\mu}{a_{TO}}} - \sqrt{\frac{2\mu}{r_{p1}} - \frac{\mu}{a_1}} \text{ and } \Delta v_2 = \sqrt{\frac{2\mu}{r_{a2}} - \frac{\mu}{a_2}} - \sqrt{\frac{2\mu}{r_{aTO}} - \frac{\mu}{a_{TO}}};$$
$$\Delta V = \Delta v_1 + \Delta v_2.$$
(3.3.6)

In Sections 3.3.1 and 3.3.2, only two thrust impulses are used to obtain the desired changes in orbit parameters. In the present section, the initial and final orbits remain coaxial. If the major axis of the final orbit is to be rotated relative to the initial major axis, then an additional thrust can be applied (see Section 3.2.3). Using multiple thrust impulses enables an infinite number of orbit adjustments, including noncoplanar transfers.

Depending on the characteristics of the orbital change, the associated fuel consumption can be quite high. In the preliminary design stage, a minimum quantity of fuel must be calculated for the satellite to fulfill its mission. Orbit changes related to geostationary orbits will be dealt with in Section 3.4.

3.3.3 Maintaining the Altitude of Low-Orbit Satellites

As pointed out in Example 2.7.1, the semimajor axis a tends to decrease owing to atmospheric drag. The drag depends on atmospheric density conditions and on the satellite size, weight, and altitude, and can easily be estimated in advance. In order to keep the altitude of a circular orbit within prescribed limits, $\Delta \mathbf{V}$ maneuvers in the direction of the satellite motion are executed when the altitude reaches the specified limits. In the preliminary stages of satellite technical definition and planning, it is important to estimate the mass of fuel necessary for these maneuvers. The estimation is very simple.

Suppose that an average disturbing drag force F_d is expected. Using Eq. C.2.4, F_d for very small ΔV changes can be expressed as $m_f = m_i(1 - \Delta V/gI_{sp})$, where m_f (resp. m_i) is the final (initial) mass of the satellite after (before) fuel expenditure. This equation can be written in the alternative form

$$(m_i - m_f)gI_{sp} = \Delta m_{prop} gI_{sp} = m_i \Delta V,$$
(3.3.7)

where Δm_{prop} is the consumed propellant mass. For a finite increment ΔV, Newton's second law can be written as $\mathbf{F} = m(\Delta V/\Delta t)$. In our case, $\mathbf{F} = \mathbf{F}_d$. Together with Eq. 3.3.7 this yields

$$F_d \Delta t = \Delta m_{prop} gI_{sp}.$$
(3.3.8)

In Eq. 3.3.8, Δm_{prop} is the consumed propellant mass for the time interval Δt with an assumed atmospheric drag of F_d.

As an example, suppose that the expected disturbance acting on a satellite located in a circular orbit of $h = 450$ km is $F_d = 10^{-3}$ N. If $I_{sp} = 250$ sec, how much fuel mass will be expended in one year? The answer is:

$$\Delta m_{prop} = \frac{F_d \times 3{,}600 \times 24 \times 365}{9.81 \times 250} = 12.8 \text{ kg/yr}.$$

3.4 Geostationary Orbits

3.4.1 *Introduction*

There are two major stages in the life of a geostationary satellite. The first one is the transfer of the satellite from the *geosynchronous transfer orbit* (GTO) to the final *geostationary orbit* (GEO) in which the satellite is intended to perform its mission. This critical stage may last between one and four weeks. The second stage is the mission stage, which is generally expected to last more than ten years.

In the first stage, the satellite is put into the GTO by the launch vehicle. The quantity of fuel needed to transfer the satellite from the GTO to the GEO usually approximates the dry weight of the satellite. The quantity of fuel needed to keep the satellite in its mission orbit during the next ten years amounts to 10%–20% of its dry weight, or about 2% per year. This means that wasting 2% of the fuel during a GTO-to-GEO transfer is equivalent to the loss of one year of communications service. This is a tremendous monetary loss, which is why the design of geostationary communications missions has drawn so much attention over the last decade. See Duret and Frouard (1980), Belon (1983), Soop (1983), and Pocha (1987). The minimization of fuel consumption during orbital changes is also closely connected to the accuracy with which the $\Delta \mathbf{V}$ is delivered to the satellite; this is discussed in Section 6.2.

3.4.2 *GTO-to-GEO Transfers*

As mentioned in the previous chapter, GEOs are circular orbits located in the earth's equatorial plane. In order to launch the satellite directly into this plane, the site of the launch vehicle must also be located in the equatorial plane. Unfortunately, no such launch sites exist and a price in fuel must be paid for that drawback.

Figure 3.4.1 shows a hypothetical launch site. Launch sites are located at different geographic latitudes. Different launch vehicles use slightly different launch profiles, but they all put the geostationary satellite in an orbit with quite similar parameters: a perigee altitude of 180–200 km and an apogee located in the equatorial plane at the geostationary altitude of $h_a = 35{,}786.2$ km above earth.

The *Delta* vehicle launch site is located at Cape Canaveral, Florida, at a latitude of 28.5°, so the minimal achievable GTO inclination is 28.5°. The *Ariane* vehicle launch site is located at Kourou, French Guyana, at a latitude of 5.2°; from this

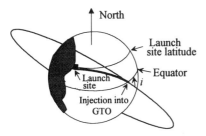

Figure 3.4.1 Launch trajectory; adapted from Pocha (1987) by permission of D. Reidel Publishing Co.

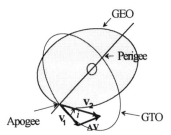

Figure 3.4.2 Transfer from GTO to GEO.

launch site, owing to different mission constraints, a GTO with inclination of 7° is usually achieved. The difference in inclinations has a pronounced effect on fuel consumption in a GTO-to-GEO transfer.

The GTO is designed so that the initial perigee and the apogee lie in the equatorial plane, or close to it, depending on physical and technical constraints. There are two tasks to be achieved: first, the elliptic orbit of the GTO must be circularized; second, the initial inclination must be zeroed. Each task can be performed individually and in any order, or a combined single ΔV maneuver can transfer the GTO to the final GEO. In Figure 3.4.2, the apogee of the GTO is in the plane of the GEO; V_1 is the velocity vector at the apogee of the GTO that lies in the plane of the GEO, and V_2 is the velocity vector of the circular GEO.

From a technical point of view, fuel consumption for the different approaches will ideally be the same. The problem is that, during the orbit change maneuver, the attitude control system of the satellite does not allow a ΔV application in the correct nominal direction (see Section 3.4.3). An average attitude error of as little as one or two degrees may cause intolerably excessive fuel consumption. This point will be clarified in the next section. Meanwhile, let us proceed to the inclination correction maneuver.

Zeroing the Inclination of the Initial GTO

As seen in Figure 3.4.2, the apogee of the GTO lies in the plane of the GEO. Maneuvering at the apogee is the cheapest location from the point of view of fuel consumption, because the velocity at this location is the lowest in the orbit. Dealing with the inclination only is equivalent to changing the spatial attitude of the orbit plane by an angle i, with no other orbit parameter alterations. This applies to any equatorial orbit, whether geostationary or not.

In Figure 3.4.3, the initial and final velocities V_i and V_f are equal. Only the direction of V_i needs to be changed, so that the inclination i is zeroed. Here $V_f = V_i = V_a$, where V_a denotes velocity at apogee, so

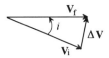

Figure 3.4.3 Inclination maneuver.

3.4 / Geostationary Orbits

$$\Delta V_1 = 2V_a \sin(\tfrac{1}{2}i). \tag{3.4.1}$$

To find the fuel consumption mass, use Eq. C.2.5.

Circularization of the GTO

Once the initial GTO orbit is brought to the GEO inclination, it must be circularized. The apogee is initially at the geostationary altitude, with $r_a = 42{,}164.2$ km. The perigee is at an approximate altitude of 200 km. To raise the perigee altitude to geostationary altitude, $\Delta V_2 = V_{\text{cir}} - V_a$ is to be applied at the apogee of the orbit, with a direction parallel to the velocity vector of the apogee. The term V_{cir} denotes the velocity of the GEO circular orbit, and V_a is the velocity at the apogee of the GTO. The r_a and r_p of the initial GTO are known: $r_a = h_a + R_e$ and $r_p = h_p + R_e$, with $2a = r_a + r_p$ and $r_{\text{cir}} = r_a$. Hence, using Eq. 3.3.4, we find that

$$\Delta V_2 = \sqrt{\frac{\mu}{r_{\text{cir}}}} \left(1 - \sqrt{\frac{2r_p}{r_{\text{cir}} + r_p}} \right). \tag{3.4.2}$$

Maneuvering from the GTO to the GEO via two individual orbital changes is not economical from the perspective of expended fuel, as in this case $\Delta V = \Delta V_1 + \Delta V_2$.

Combined GTO-to-GEO Maneuver

In order to perform the overall maneuver with minimum addition of velocity to the satellite, a combined maneuver is performed so that the in-plane maneuver (circularization of the GTO) and the out-of-plane maneuver (zeroing the inclination) are carried out in one $\Delta \mathbf{V}$ stage, as in Figure 3.4.4. In this case,

$$\Delta V^2 = V_{\text{GTO}}^2 + V_{\text{GEO}}^2 - 2V_{\text{GTO}} V_{\text{GEO}} \cos(i). \tag{3.4.3}$$

A numerical example will clarify the quantities of fuel involved in these maneuvers.

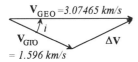

Figure 3.4.4 Combined GTO-to-GEO maneuver; velocity vector diagram at apogee burn.

EXAMPLE 3.4.1 Suppose first that a launch from Kourou is attempted, $i = 7°$. The mass of the satellite in the initial GTO orbit is 2,000 kg. The GTO has a perigee altitude of 200 km, and $I_{\text{sp}} = 300$ sec. In this case, according to Eq. 2.4.20 and Eq. 3.4.1, for the GTO inclination cancellation we have $\Delta V_1 = 194.97$ m/sec. For the GTO-to-GEO maneuver (Eq. 3.4.2), $\Delta V_2 = 1{,}477.76$ m/sec. The overall $\Delta V = 1{,}672.73$ m/sec. For a combined maneuver (Eq. 3.4.3), $\Delta V_{\text{com}} = 1{,}502.4$ m/sec.

Fuel consumption for the first case, using Eq. 3.4.1 and Eq. 3.4.2, amounts to $\Delta m = 867$ kg. Fuel consumption for the combined maneuver is only $\Delta m_{\text{com}} = 800$ kg, a difference of 67 kg. (For the Cape Canaveral launch, $\Delta V_{\text{com}} = 1{,}803.2$ km/sec and $\Delta m_{\text{com}} = 916.2$ kg.)

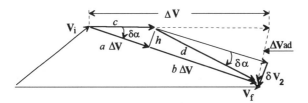

Figure 3.4.5 Division of ΔV for optimizing the orbit change under attitude error constraints.

3.4.3 Attitude Errors During GEO-to-GEO Transfer

In general, the GTO-to-GEO orbital maneuver is executed in more than one firing. The ΔV is added vectorially to the initial velocity vector at the apogee; however, because of attitude errors during the ΔV process, the obtained orbit will not be the desired one. The correction of this orbit might be very expensive in fuel if the ΔV is not correctly divided into two or more smaller ΔVs. Figure 3.4.5 shows the division of ΔV into $\Delta V_1 = a\Delta V$ and $\Delta V_2 = b\Delta V$.

In Figure 3.4.5, ΔV is to be added vectorially to V_i so that the nominal V_f can be achieved. However, if an average attitude error $\delta\alpha$ exists during the firing stage then the achieved velocity vector will not coincide with the nominal V_f. In this case, an additional correcting maneuver must be performed (with a corresponding expenditure of fuel). The ΔV_{ad} to be added is

$$\Delta V_{ad} = 2\Delta V \sin(\delta\alpha/2). \tag{3.4.4}$$

The problem is to divide the ΔV into two parts, $\Delta V = a\Delta V + b\Delta V$, such that the velocity loss is minimized. In Figure 3.4.5,

$$h = a\Delta V \delta\alpha, \qquad c = \sqrt{[a\Delta V]^2 + h^2} = a\Delta V \sqrt{1+\delta\alpha^2},$$
$$d = \sqrt{[b\Delta V]^2 + h^2} = \Delta V \sqrt{b^2 + [a\delta\alpha]^2}, \quad \text{and} \quad a+b = 1.$$

Hence
$$d = \Delta V\sqrt{[1-a]^2 + a^2\delta\alpha^2} = \Delta V\sqrt{a^2[1+\delta\alpha^2]+1-2a},$$
$$c+d = \Delta V[a\sqrt{1+\delta\alpha^2} + \sqrt{a^2(1+\delta\alpha^2)+1-2a}].$$

There are two velocity terms that define the loss in ΔV due to attitude error during the firing stage. The first one is obviously $\epsilon V_1 = (c+d-a-b)\Delta V$, and the second is $\epsilon V_2 = d\delta\alpha$. Because $a+b = 1$, we have

$$\epsilon V_1 = \Delta V[a\sqrt{1+\delta\alpha^2} + \sqrt{a^2(1+\delta\alpha^2)+1-2a} - 1]. \tag{3.4.5}$$

Moreover,

$$\epsilon V_2 = d\delta\alpha = \Delta V \delta\alpha \sqrt{a^2[1+\delta\alpha^2]+1-2a}, \tag{3.4.6}$$

$$\epsilon V = \epsilon V_1 + \epsilon V_2 = \Delta V[a\sqrt{1+\delta\alpha^2} + (1+\delta\alpha)\sqrt{a^2(1+\delta\alpha^2)+1-2a} - 1]. \tag{3.4.7}$$

Here ϵV is the total velocity loss due to attitude errors during the ΔV_1 and ΔV_2 firings. Only one parameter, a, can minimize ϵV:

3.4 / Geostationary Orbits

Table 3.4.1 *Division of ΔV for the purpose of minimizing fuel loss due to attitude errors δα*

δα [deg]	a	b	ε₁ %	ε₂ %	ε =[εV / ΔV] %
0.	1.	0.	0.	0.	0.
0.5	0.93	0.07	0.05	0.06	0.11
1.	0.91	0.09	0.15	0.17	0.31
1.5	0.89	0.11	0.26	0.31	0.57
2.	0.87	0.13	0.4	0.47	0.87
4.	0.81	0.19	1.04	1.37	2.4
6.	0.77	0.23	1.78	2.57	4.35
8.	0.73	0.27	2.57	4.03	6.6
10.	0.69	0.31	3.38	5.72	9.1

$$\frac{d\epsilon V}{da} = \sqrt{1+\delta\alpha^2}+(1+\delta\alpha)\frac{2a(1+\delta\alpha^2)-2}{2\sqrt{a^2(1+\delta\alpha^2)+1-2a}} = 0.$$

Define $A = 1+\delta\alpha^2$ and $B = [1+\delta\alpha]^2$. We must solve the quadratic equation

$$[1-B]A^2a^2 + 2A[B-1]a + A - B = 0, \qquad (3.4.8a)$$

from which the minimizing a can be solved:

$$a = \frac{1}{1+\delta\alpha^2}\left[1-\sqrt{\frac{\delta\alpha}{2+\delta\alpha}}\right]. \qquad (3.4.8b)$$

The results for some values of δα are given in Table 3.4.1 and in Figure 3.4.6.

In Table 3.4.1, we see that as the attitude error δα increases, a decreases. This means that a larger part of the entire ΔV must be postponed for the second apogee firing. The term ε in the last column is the ratio of the total loss εV to the total nominal ΔV; for δα = 4°, the εV loss will be 2.4% of ΔV. This is definitely not a negligible value. In medium-sized geostationary satellites, moving the satellite from the GTO

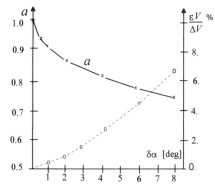

Figure 3.4.6 The ratio factor a for minimizing εV losses during GTO-to-GEO transfer.

to the GEO orbit is accompanied by fuel consumption of about 500 kg, 2.4% of which would amount to a loss of about 12 kg. Since the annual fuel consumption for such a satellite is about 10 kg/yr, this means that one year of mission life would be wasted by an average attitude error of 4° (see Section 3.5.3).

With one-shot firing in the GTO-to-GEO maneuver, the same attitude error would (according to Eq. 3.4.4) yield a fuel loss of $(2 \times 500) \sin(2°) = 34.9$ kg. This corresponds to a loss in mission life of about 3.5 years. A good attitude control design should aim to achieve errors of the order of 2° or less. Even with this accuracy and with two apogee firings, there will be a loss of $(0.869 \times 500)/100 = 4.34$ kg, which is not insignificant.

3.4.4 Station Keeping of Geostationary Satellites

As we have seen in Chapter 2, the basic parameters of the geostationary orbit change with time owing to perturbing forces, such as the moon and sun attraction forces, solar pressure, and lateral forces caused by the earth's nonhomogeneity. The primary parameters that change are the inclination, the eccentricity, and the longitude of the satellite relative to its nominal geographic longitude. Hence *station keeping* (SK) of geostationary orbits consists primarily of the following orbit adjustments:

(1) longitude (east–west) SK;
(2) inclination (north–south) SK (see Section 3.5.1); and
(3) eccentricity corrections, if necessary (see Section 3.5.2).

The remainder of this section will be devoted to a detailed account of longitude station keeping.

Figure 3.4.7 shows a satellite that is situated nominally at the longitude λ_0 but is allowed to move inside the limits $\pm \Delta\lambda$. (See also Agrawal 1986, Pocha 1987.) The perturbing longitudinal acceleration $\ddot{\lambda}$ is known from Figure 2.9.15 and Eq. 2.9.43. Because of this acceleration, the longitude of the satellite evolves in a parabolic trajectory. Suppose that the satellite is initially situated at $(\lambda_0 - \Delta\lambda, \dot{\lambda}_0)$. To keep it inside the permitted longitude limits, the satellite is given a linear velocity change ΔV when it reaches the western limit with an angular velocity $-\dot{\lambda}_0$ (see Figure 3.4.7), so that it is transferred once more to its initial location $(\lambda_0 - \Delta\lambda, \dot{\lambda}_0)$. Our task is to compute the drift velocity $\dot{\lambda}_0$ for a given $|\ddot{\lambda}|$ and required limit $\Delta\lambda$. We will then calculate

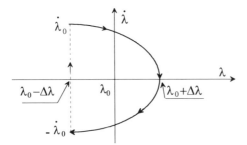

Figure 3.4.7 Phase trajectory for longitude station keeping (E–W SK).

3.4 / Geostationary Orbits

the time interval T between two consecutive E-W SK maneuvers, as well as the ΔV added during each maneuver.

Computation of $\dot{\lambda}_0$

In the following derivation, suppose that the longitudinal acceleration is negative:
$$\ddot{\lambda} = -|\ddot{\lambda}|.$$
Then
$$\dot{\lambda} = \ddot{\lambda}t + \dot{\lambda}_0 = -|\ddot{\lambda}|t + \dot{\lambda}_0, \quad (3.4.9)$$
from which it follows that
$$-|\ddot{\lambda}|\frac{t^2}{2} + \dot{\lambda}_0 t + \lambda_0 = \lambda. \quad (3.4.10)$$
At time t_1, $\dot{\lambda} = 0$, so that $-|\ddot{\lambda}|t_1 = -\dot{\lambda}_0$ and $t_1 = \dot{\lambda}_0/|\ddot{\lambda}|$. This is the time it takes for the longitude to evolve from $\lambda_0 - \Delta\lambda$ to $\lambda_0 + \Delta\lambda$. The complete cycle will take place in a period of
$$t_f = 2t_1 = 2\frac{\dot{\lambda}_0}{|\ddot{\lambda}|}. \quad (3.4.11)$$
In the period t_1, λ has changed from $\lambda_0 - \Delta\lambda$ to $\lambda_0 + \Delta\lambda$, so
$$-|\ddot{\lambda}|\frac{\dot{\lambda}_0^2}{2\ddot{\lambda}^2} + \frac{\dot{\lambda}_0^2}{|\ddot{\lambda}|} - \Delta\lambda = \Delta\lambda,$$
from which it follows that $\dot{\lambda}_0^2 = 2\Delta\lambda|\ddot{\lambda}|2$. Finally,
$$\dot{\lambda}_0 = 2\sqrt{\Delta\lambda|\ddot{\lambda}|}. \quad (3.4.12)$$

Computation of the Time Interval T between Two E-W SK Maneuvers

It is important to calculate the time interval between successive maneuvers. This interval T is the time required for the longitudinal velocity to change from $\dot{\lambda}_0$ to $-\dot{\lambda}_0$. From Eq. 3.4.9 we have $-\dot{\lambda}_0 = -|\ddot{\lambda}|T + \dot{\lambda}_0$, so
$$T = \frac{2\dot{\lambda}_0}{|\ddot{\lambda}|} = t_f. \quad (3.4.13)$$
Using Eq. 3.4.12, we conclude that
$$T = 4\sqrt{\frac{\Delta\lambda}{|\ddot{\lambda}|}}. \quad (3.4.14)$$
The s/c motion during T changes the drift rate from $\dot{\lambda}_0$ to $-\dot{\lambda}_0$. Next, we must find the required velocity change ΔV for a maneuver that will change the angular velocity from $-\dot{\lambda}_0$ to $\dot{\lambda}_0$.

Computation of the Velocity Change ΔV

The ΔV required is calculated by using Eq. 2.4.20 and Eq. 2.5.6. First, differentiating Eq. 2.4.20 while keeping r constant during the $\Delta \mathbf{V}$ process, we obtain

$$2V\Delta V = \mu \frac{\Delta a}{a^2}. \tag{3.4.15}$$

Next, we differentiate Eq. 2.5.6; this yields

$$\Delta n = -\frac{3}{2}\frac{\sqrt{\mu}}{a^{2.5}}\Delta a. \tag{3.4.16}$$

Substituting Δa from Eq. 3.4.15, we have

$$\Delta n = -3\frac{V}{\sqrt{a\mu}}\Delta V. \tag{3.4.17}$$

Finally,

$$\Delta V = -\frac{\sqrt{a\mu}}{3v}\Delta n = -\frac{a^2 n}{3v}\Delta n, \tag{3.4.18}$$

which is our desired result. Also, $\Delta n \equiv \Delta \dot{\lambda}$.

For the geostationary satellite, $a = 42{,}164.2$ km, $n = 2\pi/(23.9 \times 3{,}600)$ rad/sec, and $V = 3.0746$ km/sec. Using these values in Eq. 3.4.18, we obtain

$$\Delta V = 14{,}019.23 \Delta n \text{ km/sec}, \tag{3.4.19}$$

where Δn is proportional to the longitudinal drift rate change $\Delta \dot{\lambda}$. If Δn is defined in units of deg/day and ΔV in units of m/sec, then

$$\Delta V = \frac{14{,}019.24(1{,}000\Delta \dot{\lambda})}{(180/\pi)24(3{,}600)} = 2.83\Delta \dot{\lambda} \text{ m/sec}.$$

For one maneuver, $\Delta \dot{\lambda} = 2\dot{\lambda}_0$; hence,

$$\Delta V/\text{maneuver} = 5.66\dot{\lambda}_0 \text{ m-sec}^{-1}/\text{maneuver} = 11.32[|\ddot{\lambda}|\Delta\lambda]^{0.5} \text{ m-sec}^{-1}/\text{maneuver}.$$

Since we know the time interval T between maneuvers (Eq. 3.4.14), we can find the necessary ΔV per year: $\Delta V/\text{yr} = 5.66\dot{\lambda}_0(365/T)$ m-sec^{-1}/yr. After substitution of $\dot{\lambda}_0$ and T from Eq. 3.4.12 and Eq. 3.4.14, respectively, we obtain

$$\Delta V/\text{yr} = 1{,}032.95|\ddot{\lambda}| \text{ m-sec}^{-1}/\text{yr} = 1.74 \sin[2(\lambda - \lambda_s)] \text{ m-sec}^{-1}/\text{yr}. \tag{3.4.20}$$

This equation shows also that the ΔV necessary to keep the satellite inside the $\Delta \lambda$ limits is independent of the limit $\Delta \lambda$, and is dependent only on the geographic longitude of the satellite.

3.5 Geostationary Orbit Corrections

Figure 3.5.1 shows the possibility of correcting the parameters of the orbit by sequentially adding velocity components: (1) ΔV_R, along the radius vector to the satellite; (2) ΔV_N, normal to the orbit plane; and (3) ΔV_T, tangential to the velocity vector. We first calculate the change of the inclination vector $\mathbf{i}$ as a result of an impulsive change ΔV_N at the location of the satellite designated by the radius vector $\mathbf{R}$.

The orbital pole $\mathbf{I}$ is defined by the vector product $\mathbf{V}_T \times \mathbf{R}$, where $\mathbf{V}_T$ is the tangential velocity of the circular orbit. The vector change of the orbit momentum, as a result of the impulsive ΔV_N, will be $\Delta \mathbf{V}_N \times \mathbf{R}$. In a circular orbit, $\mathbf{V}_T$ is perpendicular to $\mathbf{R}$. Hence, for small inclination changes, $\Delta i = \Delta V_N / V_T$. Define $\alpha = \Omega + \omega + \theta$.

3.5 / Geostationary Orbit Corrections

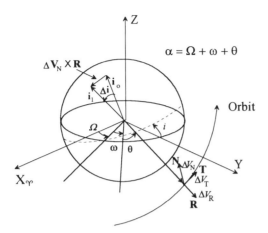

Figure 3.5.1 Geometry for achieving a change of the inclination vector **i**.

According to the definitions of the inclination vector given in Eq. 2.9.3 and Eq. 2.9.4, we can write the changes of the inclination vector as

$$\Delta i_x = \frac{\Delta V_N}{V_T} \cos(\alpha), \tag{3.5.1}$$

$$\Delta i_y = \frac{\Delta V_N}{V_T} \sin(\alpha) \tag{3.5.2}$$

(Robert and Foliard 1980, Alby 1983). In a similar way, we can find that

$$\Delta e_x = \frac{2\Delta V_T}{V_T} \cos(\alpha) + \frac{\Delta V_R}{V_T} \sin(\alpha), \tag{3.5.3}$$

$$\Delta e_y = \frac{2\Delta V_T}{V_T} \sin(\alpha) - \frac{\Delta V_R}{V_T} \cos(\alpha), \tag{3.5.4}$$

$$\Delta l_m = \frac{-2\Delta V_R}{V_T}, \tag{3.5.5}$$

where l_m is the mean longitude.

Note that $\Delta \mathbf{V_R}$ is seldom used to correct the eccentricity vector, because it is less than half as effective as $\Delta \mathbf{V_R}$ used for the same purpose. The meaning of the angle α in the foregoing equations will be clarified in the next section.

3.5.1 North-South (Inclination) Station Keeping

The angular location α of the satellite in its orbit has an important meaning: α is the *right ascension* of the location of the satellite with respect to the vernal equinox vector. Suppose that $\alpha = \Omega$. Then $\Delta V_N/V_T$ will increase proportionally the components of the inclination vector, Eq. 2.9.3 and Eq. 2.9.4, such that the $\Delta \mathbf{i}$ will be collinear with the inclination vector **i**. In some respects this is the optimal location for changing the inclination, and the only location for zeroing the inclination using a

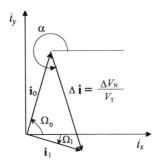

Figure 3.5.2 Correction of the initial inclination vector $\mathbf{i}_0$; adapted from Robert and Foliard (1980) by permission of Centre National d'Études Spatiales.

single impulsive change of velocity. But this is not always the case. If the inclination correction is to be performed at $\alpha \neq \Omega$, then the following considerations hold.

In Figure 3.5.2, an initial inclination vector $\mathbf{i}_0$ is shown (see also Robert and Foliard 1980; Slavinskas et al. 1988). As we shall see, it is sometimes necessary to change the initial inclination vector $\mathbf{i}_0$ to a desired inclination vector $\mathbf{i}_1$ using a single impulsive $\Delta \mathbf{V}$. This can be done at will by performing the $\Delta \mathbf{V}$ impulsive correction at the proper location – defined by the angle α – in the orbit, as shown in Figure 3.5.2. The final inclination vector $\mathbf{i}_1$ is determined according to the inclination correction strategy; this strategy is shown in Figure 3.5.3.

The mission-permitted inclination circle limits the maximum permitted drift of the inclination vector. For geostationary satellites, the required radius of this circle ranges from $0.05°$ to $0.1°$. For practical reasons, such as uncertainties in the determination of the inclination vector, the limits of the operational circle are somewhat reduced in order to compensate for the disabilities of the attitude control system. The strategy for keeping the inclination inside the permitted limits is as follows.

Suppose that, at some epoch, the inclination vector is located at point 1 in Figure 3.5.3. The evolution of the inclination vector (as explained in Section 2.9.3) will follow the path shown in the figure. When it reaches the operationally permitted

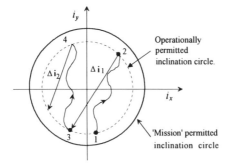

Figure 3.5.3 Strategy for keeping the inclination vector inside the permitted inclination circle; adapted from Belon (1983) by permission of Cépaduès-Éditions.

3.5 / Geostationary Orbit Corrections

(dotted) circle, the inclination vector should be corrected by $\Delta\mathbf{i}_1$ so that the final vector will reach point 3 on the same circle. The direction of the correcting vector $\Delta\mathbf{i}$ must take into account the average direction angle Ω_d of the natural evolution of the inclination (see Eq. 2.9.33) that can be predicted analytically by the analysis in Section 2.9.3, so that correction of the evolution of the inclination vector will be optimal with regard to both fuel consumption and maximizing the time between corrections. As seen in Figure 3.5.3, when the inclination vector reaches the permitted boundary, the correcting strategy is not designed to *null* the inclination. Rather, to perform the correct change of the inclination vector, the straightforward vector addition process shown in Figure 3.5.2 is used.

In Figure 3.5.3, the initial inclination vector is known and is heading toward point 2. We wish to redirect the vector toward point 3. Knowing the distance between points 2 and 3, we can calculate ΔV_N. Recall that V_T is the velocity component that is tangential to the circular geostationary orbit; $V_T = 3.07$ km/sec. Since the arguments Ω of $\mathbf{i}_0$ and $\mathbf{i}_1$ are known from Figure 3.5.2, the right ascension α at which the ΔV_N correction is to be applied can be easily computed. In practical situations, the time periods between corrections range from 5 to 15 days.

In practice, there are additional operational constraints that influence the SK strategy, such as the field of view of the sun sensors used for attitude control during station keeping. With this constraint, the N-S station keeping cannot always be performed at the optimal location in the orbit (the argument $\alpha = \Omega + \omega + \theta$ in Figures 3.5.1 and 3.5.2), thus precluding optimal control from the point of view of minimizing fuel consumption. A programmed N-S SK algorithm based on the PEPSOC package (Soop and Morley 1989) has been prepared for the *AMOS1* geostationary satellite by M. Regenstreif (MBT, Israel Aircraft Industries). Figure 3.5.4 shows a time history of the inclination vector $\mathbf{i}$ (beginning January 1, 1997) obtained by adapting the PEPSOC package.

We should note that it is possible to leave the inclination uncontrolled; in six years, the inclination would change (roughly) between -3 and $+3$ degrees. However, this would require that the ground station antenna track the satellite continuously.

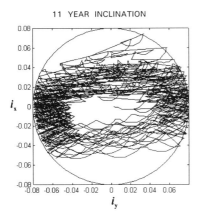

Figure 3.5.4 Time history of the controlled inclination vector $\mathbf{i}$ for an 11-year period.

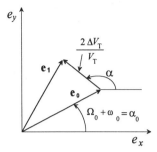

Figure 3.5.5 Eccentricity correction geometry; adapted from Legendre (1980) by permission of Centre National d'Études Spatiales.

3.5.2 *Eccentricity Corrections*

To correct the eccentricity, Eq. 3.5.3 and Eq. 3.5.4 are used. Figure 3.5.5 shows the geometry of the eccentricity correction. The calculation of the right ascension α at which ΔV_T is applied is performed by inspection of Figure 3.5.5. The same is true for finding the value of ΔV_T. However, the strategy for eccentricity corrections might be different from that of the inclination corrections.

As explained in Section 2.9.4, the value of the *natural eccentricity radius* depends on the geometrical and physical characteristics of the satellite. If the eccentricity radius happens to be smaller than the required eccentricity specification, all we need do is locate the center of the eccentricity circle at the origin. From here on, the evolution of the eccentricity will be a circle of radius e that is smaller than the demands of the operational requirements. If the eccentricity does evolve beyond the permitted bounds, it can be corrected in the same way as the initial eccentricity was corrected (see also Gantous 1986).

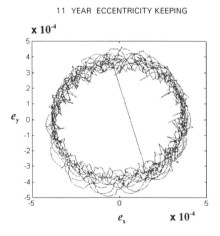

Figure 3.5.6 Time history of the controlled eccentricity vector **e** for an 11-year period.

3.5 / Geostationary Orbit Corrections

Based on the PEPSOC package, an E–W SK control algorithm has also been prepared for the *AMOS1* geostationary satellite by M. Regenstreif. A controlled eccentricity vector **e** time history for an 11-year period (beginning January 1, 1997) is shown in Figure 3.5.6.

3.5.3 Fuel Budget for Geostationary Satellites

The fuel budget for satellites with different missions is calculated in light of the two basic phases in the life of the satellite: (1) maneuvers from the transfer orbit to the final mission orbit; and (2) correcting adjustments in the mission orbit.

The primary factor that causes changes in the classical parameters of low orbits is atmospheric drag, especially at altitudes below 500 km. In order to keep the altitude of a satellite inside defined boundaries, the orbit control system must provide adequate velocity corrections (see Example 2.7.1 and Section 3.3.3). For higher-altitude satellites, solar pressure predominates. The exact mass of fuel to be reserved for keeping the satellite's orbit with predefined limits depends on the nominal orbit definitions, which vary widely from s/c to s/c. Consequently, different parameters will influence the estimation of fuel to be carried in the satellite for the necessary orbital corrections. With geostationary satellites, the perturbing forces influencing the fuel budget estimation are well defined, as we have seen in previous sections.

It is instructive to summarize the resulting fuel budget. It is more practical to express the budget in terms of ΔV, since the fuel mass to be expended depends on the initial mass of the satellite. We shall take into consideration only the most important factors for fuel consumption: GTO-to-GEO maneuvers, inclination correction (N–S SK), and longitude correction (E–W SK). There may be other reasons for orbit corrections (e.g., eccentricity evolution, station repositioning, de-orbit of the satellite at the end of its life, etc.), but their influence on the fuel budget is minor and so will not be taken into account here.

The necessary ΔV for inclination corrections depends on the year epoch (see Table 2.9.1). Hence, we shall use an average value for the years for which fuel consumption is to be estimated. The ΔV for E–W station keeping depends on the nominal geographic longitude of the satellite. The fuel mass will be calculated for a s/c with an initial mass of 2,000 kg and for $I_{sp} = 300$ sec. We also assume that the initial GTO has a perigee altitude of 200 km. The ΔV necessary to transfer this GTO to the final GEO depends on the inclination of the GTO. We shall assume the two most-used GTO inclinations, 7° and 28.5°.

Using Eq. 2.4.20, we have $V_{a7} = V_{a28.5} = 1.596$ km/sec and $V_{GEO} = 3.07466$ km/sec. Using Eq. 3.4.3, we find that $\Delta V_7 = 1.5024$ km/sec and $\Delta V_{28.5} = 1.837$ km/sec. For a satellite to be operational for 10 years (beginning 1995) the average annual inclination corrections will amount to 0.8115 deg/yr; see Table 2.9.1. Using Eq. 3.4.1 shows that $\Delta V = 43.54$ m-sec^{-1}/yr $= 435.4$ m-sec^{-1}/(10 yr).

For longitude station keeping, we shall assume that the satellite is located either at 120° E or at 300° E, where the lateral perturbing acceleration is at its maximum (see Eq. 3.4.20). With this assumption, $\Delta V = 1.74$ m-sec^{-1}/yr $= 17.4$ m-sec^{-1}/(10 yr). The approximate total ΔV and propellant mass for the two assumed 10-year orbits is thus

$\Delta V_{\text{TOT}7} = 1,955.8$ m/sec, $\quad m_{\text{prop}7} = 800.0$ kg;

$\Delta V_{\text{TOT}28.5} = 2,289.8$ m/sec, $\quad m_{\text{prop}28.5} = 1,081.4$ kg.

3.6 Summary

The present chapter dealt with classical orbital maneuvers and adjustments. It was shown that with multi-impulse corrections, any orbit change can be achieved. However, the strategy of corrections must be such as to achieve the orbit changes with minimum consumption of fuel. Note that the expected fuel consumption should be calculated in the early design stages of the attitude and orbit control system, because the additional mass of the fuel will have a strong influence on the design of the satellite structure.

References

Agrawal, B. N. (1986), *Design of Geosynchronous Spacecraft.* Englewood Cliffs, NJ: Prentice-Hall.

Alby, F. (1983), "Les Perturbations de l'Orbite Géostationaire," in *The Motion of the Satellite, Lectures and Exercises on Space Mechanics.* Toulouse, France: Cépaduès-Éditions, pp. 569–611.

Belon, B. (1983), "Stratégie de Maintien a Poste des Satellites Géostationaires," in *The Motion of the Satellite, Lectures and Exercises on Space Mechanics.* Toulouse, France: Cépaduès-Éditions, pp. 613–42.

Deutsch, R. (1963), *Orbital Dynamics of Space Vehicles.* Englewood Cliffs, NJ: Prentice-Hall.

Duret, F., and Frouard, J. P. (1980), *Conception Générale des Systèmes Spatiaux, Conception des Fusées Porteuses.* Toulouse, France: Ecole National de l'Aéronautique et de l'Espace.

Gantous, D. J. (1986), "Eccentricity Control Strategy for Geosynchronous Communication Satellites," in *Space Dynamics for Geostationary Satellites.* Toulouse, France: Cépaduès-Éditions, pp. 693–704.

Kaplan, M. H. (1976), *Modern Spacecraft Dynamics and Control.* New York: Wiley.

Legendre, P. (1980), "Maintien a Poste des Satellites Géostationaires: Stratégie des Corrections d'Orbite," in *Le Mouvement du Véhicule Spatial en Orbite.* Toulouse, France: Centre National d'Études Spatiales, pp. 583–609.

Pocha, J. J. (1987), *An Introduction to Mission Design for Geostationary Satellites.* Dordrecht: Reidel.

Pritchard, W., and Sciulli, J. (1986), *Satellite Communication Systems Engineering.* Englewood Cliffs, NJ: Prentice-Hall.

Prussing, J. (1991), "Simple Proof of the Global Optimality of the Hohmann Transfer," *Journal of Guidance, Control, and Dynamics* 15(4): 1037–8.

Redding, D. C. (1984), "Highly Efficient, Very-Low-Thrust Transfer to Geosynchronous Orbit: Exact and Approximate Solutions," *Journal of Guidance, Control, and Dynamics* 7(2): 141–7.

Robbins, H. M. (1966), "An Analytical Study of the Impulsive Approximation," *AIAA Journal* 4: 1417–23.

Robert, J. M., and Foliard, J. (1980), "Satellite Géostationaire," in *Le Mouvement du Véhicule Spatial en Orbite.* Toulouse, France: Centre National d'Études Spatiales, pp. 505–57.

References

Slavinskas, D., Dabbaghi, H., Bendent, W., and Johnson, G. (1988), "Efficient Inclination Control for Geostationary Satellites," *Journal of Guidance, Control, and Dynamics* 11(6): 584–9.

Soop, M. E. (1983), *Introduction to Geostationary Orbits*. Paris: European Space Agency.

Soop, M. E., and Morley, T. (1989), "A Portable Program Package for Geostationary Orbit Control (PEPSOC)," Spacecraft Trajectory Branch, Orbit Attitude Division, European Space Operations Center, Darmstadt, Germany.

CHAPTER 4

Attitude Dynamics and Kinematics

4.1 Introduction

The dynamics of spacecraft orbits was treated in Chapter 2. Understanding the natural motion of an orbiting s/c was necessary before we could deal with the control of orbits, the subject of Chapter 3. A similar progression will be followed in our study of first the dynamics and then control of the attitude motion of spacecraft.

Such basic physical notions as *angular kinetic energy, angular momentum,* and *moment about the mass center* will be stated and used in the derivation of the fundamental laws of *angular motion,* which are based on Euler's moment equations. In this chapter we state the angular dynamical equations of motion for spinning and nonspinning rigid bodies. Based on these equations, the attitude control of spacecraft, with different conceptual principles, will be treated in Chapters 5–9. In Chapter 10, structural and sloshing dynamics are appended to the equations of motion for a rigid body.

4.2 Angular Momentum and the Inertia Matrix

Let us suppose that a rigid body is moving in an inertial frame. This motion can be described by the translation motion of its center of mass (cm), together with a rotational motion of the body about some axis through its center of mass.

In the following analysis we shall use the well-known operator equation acting on a given vector $\mathbf{A}$,

$$\left.\frac{d}{dt}\mathbf{A}\right|_I = \left.\frac{d\mathbf{A}}{dt}\right|_B + \boldsymbol{\omega} \times \mathbf{A}, \qquad (4.2.1)$$

which simply states that the rate of change of the vector $\mathbf{A}$ as observed in the fixed coordinate system (I – "inertial" in our case) equals the rate of change of the vector $\mathbf{A}$ as observed in the rotating coordinate system (B – "body" in our case) with angular velocity $\boldsymbol{\omega}$, plus the vector product $\boldsymbol{\omega} \times \mathbf{A}$ (see Goldstein 1964, Thomson 1986). This differentiation law will be used in many applications and in several different contexts.

In Figure 4.2.1, suppose that an orthogonal triad axis frame has its origin O located at the center of mass of the body; $\mathbf{i}, \mathbf{j}, \mathbf{k}$ are the respective unit vectors along the body frame axes. For any particle m_i in the body B, $\mathbf{R}_i = \mathbf{R}_0 + \mathbf{r}_i$, so that

$$\dot{\mathbf{R}}_i = \dot{\mathbf{R}}_0 + \dot{\mathbf{r}}_i + \boldsymbol{\omega} \times \mathbf{r}_i = \mathbf{v}_0 + \mathbf{v}_i + \boldsymbol{\omega} \times \mathbf{r}_i, \qquad (4.2.2)$$

where $\boldsymbol{\omega}$ denotes the angular velocity vector of the body B with respect to the inertial frame. The moment of momentum of a body particle m_i is

$$\mathbf{h}_i = \mathbf{r}_i \times m_i \dot{\mathbf{R}}_i = \mathbf{r}_i \times m_i (\dot{\mathbf{R}}_0 + \dot{\mathbf{r}}_i + \boldsymbol{\omega} \times \mathbf{r}_i) \qquad (4.2.3)$$

(cf. the discussion in Section 2.3). However, by definition we have $\dot{\mathbf{r}}_i = 0$ in a rigid body, so in this case it follows that

4.2 / Angular Momentum and the Inertia Matrix

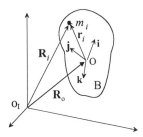

Figure 4.2.1 Angular motion of a rigid body.

$$\mathbf{h}_i = \mathbf{r}_i \times m_i(\dot{\mathbf{R}}_0 + \boldsymbol{\omega} \times \mathbf{r}_i) = -\mathbf{v}_0 \times m_i \mathbf{r}_i + \mathbf{r}_i \times m_i(\boldsymbol{\omega} \times \mathbf{r}_i). \tag{4.2.4}$$

To find the angular momentum of the entire body, we shall sum the momentum components of all the mass particles:

$$\mathbf{h} = \sum_{m_i} -\mathbf{v}_0 \times m_i \mathbf{r}_i + \sum_{m_i} \mathbf{r}_i \times (\boldsymbol{\omega} \times \mathbf{r}_i)m_i = -\mathbf{v}_0 \times \sum_{m_i} m_i \mathbf{r}_i + \sum_{m_i} \mathbf{r}_i \times (\boldsymbol{\omega} \times \mathbf{r}_i)m_i. \tag{4.2.5}$$

Since the angular motion is about the center of mass, $\sum_{m_i} m_i \mathbf{r}_i = 0$ holds. Finally:

$$\mathbf{h} = \sum_{m_i} \mathbf{r}_i \times (\boldsymbol{\omega} \times \mathbf{r}_i)m_i. \tag{4.2.6}$$

After performing the vector triple product, we get the following equations:

$$\mathbf{h} = \mathbf{i}\left[\omega_x \sum_{m_i}(y_i^2 + z_i^2)m_i - \omega_y \sum_{m_i} y_i x_i m_i - \omega_z \sum_{m_i} x_i z_i m_i\right]$$
$$+ \mathbf{j}\left[\omega_y \sum_{m_i}(x_i^2 + z_i^2)m_i - \omega_x \sum_{m_i} x_i y_i m_i - \omega_z \sum_{m_i} y_i z_i m_i\right]$$
$$+ \mathbf{k}\left[\omega_z \sum_{m_i}(x_i^2 + y_i^2)m_i - \omega_x \sum_{m_i} x_i z_i m_i - \omega_y \sum_{m_i} y_i z_i m_i\right]. \tag{4.2.7}$$

In Eq. 4.2.7, x_i, y_i, z_i are the coordinates of a particle i in the body axis frame and $\omega_x, \omega_y, \omega_z$ are the angular velocity components around the $\mathbf{i}, \mathbf{j}, \mathbf{k}$ body axes. The summations of the squared coordinate components are easily identified as the three moments of inertia of the body about its three orthogonal axes. The summations of the products of the coordinate components are identified as the products of inertia. With these definitions, Eq. 4.2.7 takes the following form:

$$\mathbf{h} = \mathbf{i}[\omega_x I_{xx} - \omega_y I_{xy} - \omega_z I_{xz}] + \mathbf{j}[\omega_y I_{yy} - \omega_x I_{yx} - \omega_z I_{yz}] + \mathbf{k}[\omega_z I_{zz} - \omega_x I_{zx} - \omega_y I_{zy}]$$
$$= \mathbf{i}h_x + \mathbf{j}h_y + \mathbf{k}h_z; \tag{4.2.8}$$

$\mathbf{h}$ is the *angular momentum* vector of the rigid body.

If we define the angular velocity vector as: $\boldsymbol{\omega} = [\omega_x \ \omega_y \ \omega_z]^T$, then Eq. 4.2.8 can be put in the matrix form

$$\mathbf{h} = \begin{bmatrix} I_{xx} & -I_{xy} & -I_{xz} \\ -I_{yx} & I_{yy} & -I_{yz} \\ -I_{zx} & -I_{zy} & I_{zz} \end{bmatrix} \begin{bmatrix} \omega_x \\ \omega_y \\ \omega_z \end{bmatrix} = [\mathbf{I}]\boldsymbol{\omega}. \tag{4.2.9}$$

From symmetry considerations, it is easily deduced that $I_{xy} = I_{yx}$, $I_{xz} = I_{zx}$, and $I_{yz} = I_{zy}$. $[\mathbf{I}]$ is the *inertia tensor*, or *inertia matrix*.

4.3 Rotational Kinetic Energy of a Rigid Body

Let us consider a rigid body moving in space together with a set of body axes whose origin coincides with the body's center of mass. To begin with, if T is the body's kinetic energy then $\Delta T = 0.5 v^2 (\Delta m)$. According to Figure 4.2.1 and using Eq. 4.2.1, $\mathbf{v}_i = \mathbf{v}_0 + \boldsymbol{\omega} \times \mathbf{r}_i$; hence

$$v_i^2 = \mathbf{v}_i \cdot \mathbf{v}_i = v_0^2 + 2\mathbf{v}_0 \cdot (\boldsymbol{\omega} \times \mathbf{r}_i) + (\boldsymbol{\omega} \times \mathbf{r}_i) \cdot (\boldsymbol{\omega} \times \mathbf{r}_i). \tag{4.3.1}$$

To find the kinetic energy,

$$\begin{aligned}
T &= \frac{1}{2}\int_M v^2 \, dm = \frac{1}{2}\int v_0^2 \, dm + \frac{2}{2}\int \mathbf{v}_0 \cdot (\boldsymbol{\omega} \times \mathbf{r}) \, dm + \frac{1}{2}\int (\boldsymbol{\omega} \times \mathbf{r}) \cdot (\boldsymbol{\omega} \times \mathbf{r}) \, dm \\
&= \frac{1}{2} v_0^2 M + \mathbf{v}_0 \cdot \boldsymbol{\omega} \times \int \mathbf{r} \, dm + \frac{1}{2}\int (\boldsymbol{\omega} \times \mathbf{r}) \cdot (\boldsymbol{\omega} \times \mathbf{r}) \, dm \\
&= T_{\text{transl}} + \mathbf{v}_0 \cdot \boldsymbol{\omega} \times 0 + \frac{1}{2}\int (\boldsymbol{\omega} \times \mathbf{r}) \cdot (\boldsymbol{\omega} \times \mathbf{r}) \, dm \\
&= T_{\text{transl}} + T_{\text{rot}}. \tag{4.3.2}
\end{aligned}$$

By definition, the angular motion is about the center of mass of the rigid body, so $\int \mathbf{r} \, dm = 0$. From Eq. 4.3.2, the rotational kinetic energy is

$$\begin{aligned}
T_{\text{rot}} &= \frac{1}{2}\int_M (\boldsymbol{\omega} \times \mathbf{r}) \cdot (\boldsymbol{\omega} \times \mathbf{r}) \, dm \\
&= \frac{1}{2}\int_M [(\omega_y z - \omega_z y)^2 + (\omega_z x - \omega_x z)^2 + (\omega_x y - \omega_y x)^2] \, dm.
\end{aligned}$$

After integrating over the mass of the body M and using the definitions of moments of inertia and products of inertia from the previous section, we find that

$$\begin{aligned}
T_{\text{rot}} &= \tfrac{1}{2}[\omega_x^2 I_x + \omega_y^2 I_y + \omega_z^2 I_z - 2\omega_y \omega_x I_{yx} - 2\omega_y \omega_z I_{zy} - 2\omega_z \omega_x I_{zx}] \\
&= \tfrac{1}{2}[\omega_x(\omega_x I_x - \omega_z I_{zx} - \omega_y I_{yx}) \\
&\quad + \omega_y(\omega_y I_y - \omega_z I_{yz} - \omega_x I_{yx}) + \omega_z(\omega_z I_z - \omega_y I_{zy} - \omega_x I_{zx})]. \tag{4.3.3}
\end{aligned}$$

This may be written in matrix form as

$$T_{\text{rot}} = \tfrac{1}{2} \boldsymbol{\omega}^T [\mathbf{I}] \boldsymbol{\omega}, \tag{4.3.4}$$

where the rotational kinetic energy is expressed in terms of the inertia matrix.

It is important to emphasize that together the angular momentum and the rotational kinetic energy completely define the rotational dynamic state of a rigid body.

4.4 Moment-of-Inertia Matrix in Selected Axis Frames

4.4.1 *Moment of Inertia about a Selected Axis in the Body Frame*

It is generally possible to compute the moment of inertia of the body about any axis ξ passing through the center of mass of the body. In this case, $T_{\text{rot}} = \tfrac{1}{2} I_\xi \omega^2$. Using also Eq. 4.3.3, we find that

4.4 / Moment-of-Inertia Matrix in Selected Axis Frames

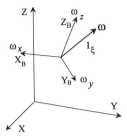

Figure 4.4.1 Coordinate axes of the X, Y, Z inertial system and of the $\mathbf{X}_B, \mathbf{Y}_B, \mathbf{Z}_B$ body frame.

$$I_\xi \omega^2 = \omega_x^2 I_x + \omega_y^2 I_y + \omega_z^2 I_z - 2\omega_y \omega_x I_{yx} - 2\omega_y \omega_z I_{zy} - 2\omega_z \omega_x I_{zx}. \tag{4.4.1}$$

Equation 4.4.1 can be put in a more convenient form as follows:

$$\begin{aligned} I_\xi &= \left(\frac{\omega_x}{\omega}\right)^2 I_x + \left(\frac{\omega_y}{\omega}\right)^2 I_y + \left(\frac{\omega_z}{\omega}\right)^2 I_z \\ &\quad - 2\left(\frac{\omega_y}{\omega}\right)\left(\frac{\omega_z}{\omega}\right) I_{zy} - 2\left(\frac{\omega_x}{\omega}\right)\left(\frac{\omega_z}{\omega}\right) I_{xz} - 2\left(\frac{\omega_x}{\omega}\right)\left(\frac{\omega_y}{\omega}\right) I_{yx}. \end{aligned} \tag{4.4.2}$$

The inertial and the body axis frames are shown in Figure 4.4.1. The $\mathbf{1}_\xi$ axis is the direction of the rotational axis of the body with angular velocity ω. The angular velocity vector ω has components $\omega_x, \omega_y, \omega_z$ along the $\mathbf{X}_B, \mathbf{Y}_B, \mathbf{Z}_B$ body axes. We can define $a_x = \omega_x/\omega$, $a_y = \omega_y/\omega$, and $a_z = \omega_z/\omega$, which are the direction cosines of the vector ω in the body frame (see Appendix A). Given these definitions, Eq. 4.4.2 can be written in the more compact form

$$I_\xi = a_x^2 I_x + a_y^2 I_y + a_z^2 I_z - 2a_y a_z I_{zy} - 2a_x a_z I_{xz} - 2a_x a_y I_{yx}. \tag{4.4.3}$$

Considering the complexity of these equations – which is due to the existence of the products of inertia – it is attractive to choose a body axis frame in which all products of inertia vanish. The control engineer generally prefers a satellite in which there are negligible products of inertia, as it is easier to design the attitude control systems for such satellites. The existence of parasitic products of inertia can then be treated as mere disturbances to be taken into account in the design stage of the attitude control systems. In the next section we show how to choose a body axis frame in which all products of inertia are practically eliminated. (There do exist special circumstances in which small products of inertia are advantageously used for some special control tasks; see Section 8.8.5.)

4.4.2 Principal Axes of Inertia

The problem at hand is to transform the general inertia matrix, Eq. 4.2.9, into a diagonal one. Transformation of a nondiagonal real square symmetric matrix into a diagonal one is a common procedure that is treated in linear algebra (see Hildebrand 1968). For our case of an orbiting satellite (see Kaplan 1976), we shall give the results but not the mathematical details of the transformation.

The components of the angular velocity ω in the initial body axis frame are labeled $\omega_x, \omega_y, \omega_z$. To obtain a new axis frame in which the inertia matrix will be diagonal, an axis rotation must be performed. In the new axis frame, the components of ω will be changed to ω' by the vector transformation $\omega = [\mathbf{A}]\omega'$. Returning to Eq. 4.3.4, we can write

$$2T_{\text{rot}} = ([\mathbf{A}]\omega')^T[\mathbf{I}][\mathbf{A}]\omega' = \omega'^T[\mathbf{A}]^T[\mathbf{I}][\mathbf{A}]\omega' = \omega'^T[\mathbf{I}']\omega'. \quad (4.4.4)$$

The matrix [**A**] is the transformation matrix from the old orthogonal axis frame to the new orthogonal axis frame, in which the inertia matrix will be diagonal.

The eigenvalues λ_i of the inertia matrix [**I**] are the principal moments of inertia, the elements of the diagonal matrix [**I**']. They can be found by evaluating $\det\{[\mathbf{I}] - \lambda[\mathbf{1}]\} = 0$, where [**1**] is the diagonal unit matrix. The eigenvectors $\mathbf{e}_1, \mathbf{e}_2, \mathbf{e}_3$ of [**I**] are the column vectors of the transformation matrix [**A**]:

$$[\mathbf{A}] = \begin{bmatrix} e_{1x} & e_{2x} & e_{3x} \\ e_{1y} & e_{2y} & e_{3y} \\ e_{1z} & e_{2z} & e_{3z} \end{bmatrix}. \quad (4.4.5)$$

To find the eigenvectors $\mathbf{e}_i$, we solve the set of i equations $\lambda_i \mathbf{e}_i = [\mathbf{I}]\mathbf{e}_i$, $i = 1, 2, 3$. To summarize, the diagonal terms of the diagonal intertia matrix are known as the *principal moments of inertia,* and the corresponding new axes are called *principal axes*. The three principal axes include the axes of maximum and minimum inertia, referred to as the *major* and *minor* axes, respectively.

EXAMPLE 4.4.1 Suppose that the inertia matrix is given by

$$[\mathbf{I}] = \begin{bmatrix} 20 & -10 & 0 \\ -10 & 30 & 0 \\ 0 & 0 & 40 \end{bmatrix} \text{N-m-sec}^2.$$

To find the eigenvalues λ_i, we solve for $\det\{[\mathbf{I}] - [\mathbf{1}]\lambda\} = 0$. We find that $\lambda_1 = 13.82$, $\lambda_2 = 36.18$, and $\lambda_3 = 40$, which are also the principal moments of inertia in the new axis frame.

To find the eigenvector $\mathbf{e}_1$, we write $[\mathbf{I}]\mathbf{e}_1 = \lambda_1 \mathbf{e}_1$ and obtain the following three equations:

$$(20 - 13.82)e_{1x} - 10e_{1y} + 0e_{1z} = 0,$$
$$-10e_{1x} + (30 - 13.82)e_{1y} + 0e_{1z} = 0,$$
$$0e_{1x} + 0e_{1y} + (40 - 13.82)e_{1z} = 0.$$

Clearly $e_{1z} = 0$. To find e_{1x} and e_{1y}, put $e_{1x} = c_1$, from which it follows that $e_{1y} = c_1(20 - 13.82)/10 = 0.618c_1$. Finally,

$$\mathbf{e}_1 = \begin{bmatrix} c_1 \\ 0.618c_1 \\ 0 \end{bmatrix} = \begin{bmatrix} 0.85066 \\ 0.527 \\ 0 \end{bmatrix}.$$

(The norm of $\mathbf{e}_1$ has been normalized to unity.)

In a similar way, $\mathbf{e}_2$ and $\mathbf{e}_3$ are found to be

$$\mathbf{e}_2 = [-0.52571 \ 0.85066 \ 0]^T \quad \text{and} \quad \mathbf{e}_3 = [0 \ 0 \ 1]^T,$$

and the matrix $[\mathbf{A}]$ is

$$[\mathbf{A}] = \begin{bmatrix} 0.85066 & -0.52571 & 0 \\ 0.527 & 0.85066 & 0 \\ 0 & 0 & 1 \end{bmatrix}.$$

From $\omega = [\mathbf{A}]\omega'$, we find after inversion of $[\mathbf{A}]$ that $\omega' = [\mathbf{A}]^{-1}\omega$. However, since $[\mathbf{A}]$ is orthogonal, we also have the more direct solution $\omega' = [\mathbf{A}]^T\omega$. It is also easy to check that

$$[\mathbf{I}'] = [\mathbf{A}]^T[\mathbf{I}][\mathbf{A}] = \begin{bmatrix} 13.82 & 0 & 0 \\ 0 & 36.18 & 0 \\ 0 & 0 & 40 \end{bmatrix},$$

which is diagonal as expected.

The columns of $[\mathbf{A}]$ are the direction cosines of the principal axes with respect to the initial body axes $\mathbf{X}_B, \mathbf{Y}_B, \mathbf{Z}_B$. The direction cosine matrix in Example 4.4.1 shows that, in order to obtain the principal axes, the original axes must be rotated about the $\mathbf{Z}_B$ axis only. The $\mathbf{Z}_B$-axis moment of inertia does not change. From now on, the body axis frame is assumed to be for principal axes, unless otherwise stated. With this assumption, Eq. 4.2.8, Eq. 4.3.3, and Eq. 4.4.2 are greatly simplified.

4.4.3 Ellipsoid of Inertia and the Rotational State of a Rotating Body

With the body axes chosen to be principal axes, Eq. 4.4.2 becomes:

$$I_\xi = \left(\frac{\omega_x}{\omega}\right)^2 I_x + \left(\frac{\omega_y}{\omega}\right)^2 I_y + \left(\frac{\omega_z}{\omega}\right)^2 I_z. \tag{4.4.6}$$

As depicted in Figure 4.4.2, the moment of inertia about any instantaneous axis of rotation $\mathbf{1}_\xi$ will be

$$I_\xi = a_x^2 I_x + a_y^2 I_y + a_z^2 I_z. \tag{4.4.7}$$

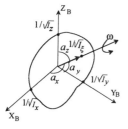

Figure 4.4.2 Ellipsoid of inertia.

Such values can be calculated for any direction of 1_ξ defined in the body axis frame by its direction cosines. With the resulting values drawn graphically onto the body axis frame, a closed surface will emerge, called an *ellipsoid of inertia* (Figure 4.4.2). Equation 4.4.7 can be put in the following form:

$$1 = \frac{a_x^2}{I_\xi}I_x + \frac{a_y^2}{I_\xi}I_y + \frac{a_z^2}{I_\xi}I_z = X^2 I_x + Y^2 I_y + Z^2 I_z, \quad (4.4.8)$$

which is the general equation of an ellipsoid with axes dimensions $1/\sqrt{I_x}$, $1/\sqrt{I_y}$, and $1/\sqrt{I_z}$.

The angular momentum for principal moments of inertia can be found from Eq. 4.2.8. The angular momentum, together with the rotational kinetic energy, describes the dynamic state of the rotating body:

$$h^2 = I_x^2 \omega_x^2 + I_y^2 \omega_y^2 + I_z^2 \omega_z^2. \quad (4.4.9)$$

With principal moments of inertia, Eq. 4.3.3 reduces to

$$2T_{\text{rot}} = I_x \omega_x^2 + I_y \omega_y^2 + I_z \omega_z^2. \quad (4.4.10)$$

From here on, for notational simplicity it will be understood that T stands for T_{rot}.

The preceding two equations can be written in normalized form as

$$\frac{\omega_x^2}{(h/I_x)^2} + \frac{\omega_y^2}{(h/I_y)^2} + \frac{\omega_z^2}{(h/I_z)^2} = 1, \quad (4.4.11)$$

$$\frac{\omega_x^2}{(\sqrt{2T/I_x})^2} + \frac{\omega_y^2}{(\sqrt{2T/I_y})^2} + \frac{\omega_z^2}{(\sqrt{2T/I_z})^2} = 1. \quad (4.4.12)$$

These equations describe ellipsoids of the angular momentum and of the rotational kinetic energy, with the components of the angular velocity vector ω as variables.

For any given free body with known rotational kinetic energy T and momentum h, all instantaneous values of moment of inertia and of angular velocity are defined by the two ellipsoids. Any ω that satisfies the momentum ellipsoid of Eq. 4.4.11 must also satisfy the kinetic energy ellipsoid of Eq. 4.4.12. Hence, both ellipsoids intersect along a curve that is the locus of all possible ωs satisfying both equations. This curve is called a *polhode*, and is shown in Figure 4.4.3. For different angular momentum and rotational kinetic energy, different polhodes will be generated.

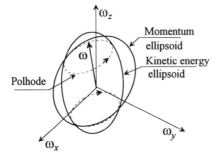

Figure 4.4.3 Satisfaction of both momentum and kinetic energy ellipsoids: the polhode.

4.5 Euler's Moment Equations

We found in Section 2.2 that a moment, acting on a body about its center of mass, equals the time rate of change of its angular momentum. In this section we will examine the rotational motion of a body caused by the applied moment.

Using once more the identity of Eq. 4.2.1 for the angular momentum vector **h**, we can write

$$\mathbf{M} = \dot{\mathbf{h}}_I = \dot{\mathbf{h}}_B + \boldsymbol{\omega} \times \mathbf{h}. \tag{4.5.1}$$

This is the well-known *Euler's moment equation*. In this equation, the subscript "I" indicates a derivative in the *inertial* frame, while the subscript "B" indicates a derivative in the rotating *body* frame.

Assuming that $\mathbf{X}_B, \mathbf{Y}_B, \mathbf{Z}_B$ are the principal axes of inertia and performing the vector product, we obtain the three scalar equations

$$\begin{aligned} M_x &= I_x \dot{\omega}_x + \omega_y \omega_z (I_z - I_y), \\ M_y &= I_y \dot{\omega}_y + \omega_x \omega_z (I_x - I_z), \\ M_z &= I_z \dot{\omega}_z + \omega_x \omega_y (I_y - I_x). \end{aligned} \tag{4.5.1'}$$

These equations are nonlinear, so they do not have an analytical closed-form solution. However, they can be solved under some relieving conditions as follows.

4.5.1 *Solution of the Homogeneous Equation*

The homogeneous equations will be solved under the assumption that the rotating body is axisymmetric, with, for instance, $I_x = I_y$. Suppose also that the body is spinning about the $\mathbf{Z}_B$ axis, which is the axis of symmetry, with a constant angular velocity $\omega_z = \text{const} = n$.

For this situation, the simplified Euler equations have an analytic solution, and we can find the time-domain angular motion of the body for any initial conditions. The equations become:

$$\begin{aligned} I_x \dot{\omega}_x + \omega_y n (I_z - I_y) &= 0, \\ I_y \dot{\omega}_y + \omega_x n (I_x - I_z) &= 0, \\ \dot{\omega}_z I_z &= 0. \end{aligned} \tag{4.5.2}$$

The third of Eqs. 4.5.2 simply means that $\omega_z = n = \text{const}$, which has already been stated. The first two equations are linear, and if we define $\lambda = n(I_z - I_x)/I_x$ then they can be put in the following form:

$$\begin{aligned} \dot{\omega}_x + \lambda \omega_y &= 0, \\ \dot{\omega}_y - \lambda \omega_x &= 0. \end{aligned} \tag{4.5.3}$$

If we multiply the first equation by ω_x, multiply the second by ω_y, and add the first to the second, we have $\omega_x \dot{\omega}_x + \omega_y \dot{\omega}_y = 0$, from which follows the important result

$$\omega_{xy}^2 = \omega_x^2 + \omega_y^2 = \text{const}_{xy}; \tag{4.5.4}$$

ω_{xy} is the component of angular velocity in the $\mathbf{X}_B$-$\mathbf{Y}_B$ plane of the rotating body.

Because ω_z is also constant, we find that the norm of the angular velocity remains constant also:
$$|\omega| = \sqrt{\omega_{xy}^2 + \omega_z^2} = \sqrt{\omega_x^2 + \omega_y^2 + \omega_z^2} = \text{const.}$$

From Eq. 4.5.3, we can find relations between the initial conditions of ω_x, ω_y, and their derivatives as follows:
$$\begin{aligned}\dot{\omega}_x(0) &= -\lambda \omega_y(0), \\ \dot{\omega}_y(0) &= +\lambda \omega_x(0).\end{aligned} \quad (4.5.5)$$

From this stage, the solution of the homogeneous equation is a standard process. Let us differentiate the first of Eqs. 4.5.3 with time; this yields
$$\ddot{\omega}_x + \lambda \dot{\omega}_y = \ddot{\omega}_x + \lambda^2 \omega_x = 0. \quad (4.5.6)$$

Taking the Laplace transform, together with Eqs. 4.5.5 we have
$$\omega_x(s) = \frac{\dot{\omega}_x(0) + s\omega_x(0)}{s^2 + \lambda^2}. \quad (4.5.7)$$

The time-response solution of Eq. 4.5.7 is
$$\omega_x(t) = \omega_x(0)\cos(\lambda t) + \frac{\dot{\omega}_x(0)}{\lambda}\sin(\lambda t). \quad (4.5.8)$$

In a similar way, integrating the second of Eqs. 4.5.3, we find that
$$\omega_y(t) = \frac{-\dot{\omega}_x(t)}{\lambda} = \omega_x(0)\sin(\lambda t) - \frac{\dot{\omega}_x(0)}{\lambda}\cos(\lambda t). \quad (4.5.9)$$

By adding ω_x and ω_y in quadrature in the $\mathbf{X}_B$-$\mathbf{Y}_B$ plane equations of angular motion of the rotating body, we can write
$$\begin{aligned}\omega_{xy}(t) &= \omega_x + j\omega_y = \omega_x(0)[\cos(\lambda t) + j\sin(\lambda t)] + \frac{\dot{\omega}_x(0)}{\lambda}[\sin(\lambda t) - j\cos(\lambda t)] \\ &= \left[\omega_x(0) - j\frac{\dot{\omega}_x(0)}{\lambda}\right][\cos(\lambda t) + j\sin(\lambda t)] \\ &= [\omega_x(0) + j\omega_y(0)]e^{j\lambda t} = \omega_{xy}(0)e^{j\lambda t},\end{aligned} \quad (4.5.10)$$

where $j = \sqrt{-1}$. It is important to remember that the final result in Eq. 4.5.10 has been derived under the assumption that the body axes were chosen to be principal axes – so that no products of inertia exist – and that the body is axisymmetric, $I_x = I_y$.

4.5.2 Stability of Rotation for Asymmetric Bodies about Principal Axes

In the previous discussion we assumed an axisymmetric body, with $I_x = I_y$. In the present section, this assumption is relieved. We wish to find conditions for stability about any principal axis with no external moments acting on the body. For that purpose we use Eqs. 4.5.1' with $M_x = M_y = M_z = 0$. Let us suppose that stability conditions for rotation about the $\mathbf{Z}_B$ body axis are sought. In this case let $\omega_z = n + \epsilon$, where ϵ is a small disturbance; $\dot{\omega}_z = \dot{n} + \dot{\epsilon} = \dot{\epsilon}$. If $\epsilon \to 0$ then Eqs. 4.5.1' become:

$$I_x\dot{\omega}_x + \omega_y n(I_z - I_y) = 0,$$
$$I_y\dot{\omega}_y + \omega_x n(I_x - I_z) = 0, \qquad (4.5.11)$$
$$I_z\dot{\epsilon} + \omega_x \omega_y (I_y - I_x) = 0.$$

The first two equations are linear, and a second differentiation of the first together with the second yields

$$\ddot{\omega}_x + n^2 \frac{I_z - I_y}{I_y} \frac{I_z - I_x}{I_x} \omega_x = 0. \qquad (4.5.12)$$

Taking the Laplace transform of Eq. 4.5.12, we have

$$\left[s^2 + n^2 \frac{I_z - I_y}{I_y} \frac{I_z - I_x}{I_x}\right]\omega_x(s) = 0 \Rightarrow s^2 + \beta^2 = 0 \quad \text{with}$$
$$s^2 = -n^2 \frac{I_z - I_y}{I_y} \frac{I_z - I_x}{I_x} = -\beta^2 \quad \text{and} \quad \beta = n\left[\left(1 - \frac{I_z}{I_x}\right)\left(1 - \frac{I_z}{I_y}\right)\right]^{1/2}. \qquad (4.5.13)$$

For stability we need β to be real, which means that the conditions for stability are

$$I_z > I_x, I_y \quad \text{or} \quad I_z < I_x, I_y. \qquad (4.5.14)$$

In words, if a body is spinning about its axis of minimum (or maximum) moment of inertia then the angular motion will be stable about these axes. But suppose that the body is spinning about the axis with the intermediate value of moment of inertia and so Eq. 4.5.14 is not satisfied. This means that

$$I_x > I_z > I_y \quad \text{or} \quad I_x < I_z < I_y. \qquad (4.5.15)$$

In this case β will be imaginary, so that one of the roots of the determinant of Eq. 4.5.12 will be positive and real (unstable), and the second will be negative and real (stable).

These conclusions hold for the trivial case where the momentum of the body is constant (no applied external moments on the body) and the rotational kinetic energy is constant. In practical situations, these assumptions do not generally hold, as will be explained in Section 4.6.2.

4.5.3 Solution of the Homogeneous Equation for Unequal Moments of Inertia

In Section 4.5.1, the homogeneous equations were solved for the case where two of the principal moments of inertia were equal. If these two moments of inertia are not equal but do have close values then the solution may still be valid from an engineering point of view, but care must be taken to avoid wrong conclusions.

An exact solution can be found for the case of distinct moments of inertia. In this case, the solution of Eqs. 4.5.1' will result in elliptic functions, which are not easy to use for practical purposes. The complete solution for distinct moments of inertia will not be pursued in this textbook; the reader is referred to Thomson (1986) for further reading on the subject.

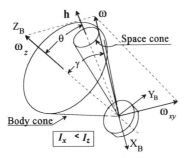

Figure 4.6.1 Body and space cones for $I_x < I_z$; adapted from Thomson (1986) by permission of Dover Publications, Inc.

4.6 Characteristics of Rotational Motion of a Spinning Body

4.6.1 *Nutation of a Spinning Body*

As we saw in Section 4.5.1, if the body is in rotational motion caused by initial conditions but not under the influence of external moments, then the norm of the angular velocity remains constant, $|\omega|$ = const (see also Eq. 4.5.4). Moreover, since there are no applied moments on the body, $\mathbf{M} = \dot{\mathbf{h}}_I = 0$, it follows from Eq. 4.5.1 that $\mathbf{h}_I$ = const; the momentum vector is constant in inertial space.

In this section the angular momentum and the angular velocity vectors will be resolved into two components, one in the $\mathbf{X}_B$-$\mathbf{Y}_B$ plane and another along the $\mathbf{Z}_B$ body axis (assume once more that the body is symmetric, $I_x = I_y$): $\omega = \omega_{xy} + \omega_z$ and $\mathbf{h} = I_x \omega_{xy} + I_z \omega_z$. Since ω and $\mathbf{h}$ have components in the same directions (ω_{xy} and ω_z), it is evident that $\mathbf{h}$, ω, and ω_z are coplanar. But since the momentum vector $\mathbf{h}$ is constant, its direction is fixed in space. The vector ω_{xy} rotates in the $\mathbf{X}_B$-$\mathbf{Y}_B$ body plane, so the angular velocity vector ω must also rotate about $\mathbf{h}$; see Figure 4.6.1.

We now define the angles θ and γ as follows:

$$\tan(\theta) = \frac{h_{xy}}{h_z} = \frac{I_x \omega_{xy}}{I_z \omega_z}, \quad (4.6.1)$$

$$\tan(\gamma) = \frac{\omega_{xy}}{\omega_z}; \quad (4.6.2)$$

θ is called the *nutation angle*. From Eq. 4.6.1 and Eq. 4.6.2 it follows that

$$\tan(\theta) = \frac{I_x}{I_z} \tan(\gamma). \quad (4.6.3)$$

From this we conclude that

$$\begin{aligned} \theta > \gamma & \quad \text{if } I_x > I_z, \\ \theta < \gamma & \quad \text{if } I_x < I_z. \end{aligned} \quad (4.6.4)$$

As is evident from Figures 4.6.1 and 4.6.2, the body cone "rolls" on the space cone, which is fixed in space. The $\mathbf{Z}_B$-ω plane rotates about the $\mathbf{h}$ vector, which is

4.6 / Characteristics of Rotational Motion of a Spinning Body

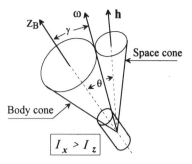

Figure 4.6.2 Body and space cones for $I_x > I_z$; adapted from Kaplan (1976) by permission of John Wiley & Sons.

also fixed in space. Whenever the body $\mathbf{Z}_B$ axis deviates from the momentum vector $\mathbf{h}$, the body is said to *nutate*. This nutation forces the spin axis to deviate from the nominal desired direction. Keeping the nutation angle small is one of the important tasks of attitude control systems, to be discussed in subsequent chapters.

4.6.2 Nutation Destabilization Caused by Energy Dissipation

As seen in Section 4.4, the angular kinetic energy about any momentary axis ξ can be written as

$$T = T_{\text{rot}} = 0.5 I_\xi \omega^2 = 0.5 \frac{h^2}{I_\xi}. \tag{4.6.5}$$

With no applied moments on the spinning body, the momentum $\mathbf{h}$ will remain constant. As already mentioned, the value of I_ξ depends on the direction of the axis of rotation in the body axes frame (Eq. 4.4.7). Since also h is constant:

$$T_{\max} = 0.5 \frac{h^2}{I_{\min}} \quad \text{(at the minor axis);} \tag{4.6.6}$$

$$T_{\min} = 0.5 \frac{h^2}{I_{\max}} \quad \text{(at the major axis).} \tag{4.6.7}$$

According to Eq. 4.6.6 with h constant, if the body spins about the minor axis and there is some internal energy dissipation that tends to decrease the rotational kinetic energy to its minimum, then the body will transfer the spin of rotation to the major axis in order to satisfy Eq. 4.6.7. (This phenomenon was identified in the satellite *Explorer I*, which became rotationally unstable; see Bracewell and Garriott 1958.)

From Eq. 4.2.8, for the symmetrical body case $I_x = I_y$ and with no products of inertia:

$$h^2 = [\omega_x^2 + \omega_y^2] I_x^2 + \omega_z^2 I_z^2. \tag{4.6.8}$$

From Eq. 4.3.3,

$$2T = [\omega_x^2 + \omega_y^2] I_x + \omega_z^2 I_z. \tag{4.6.9}$$

Multiply Eq. 4.6.9 by I_x and subtract from Eq. 4.6.8 to obtain

$$h^2 - 2TI_x = \omega_z^2 I_z (I_z - I_x). \qquad (4.6.10)$$

By definition, $\cos(\theta) = h_z/h = I_z\omega_z/h$; hence

$$2TI_x = h^2 - h^2 \cos^2(\theta)\frac{I_z - I_x}{I_z}. \qquad (4.6.11)$$

Differentiating this equation with time yields

$$2\dot{T}I_x = -h^2 2\cos(\theta)[-\sin(\theta)]\dot{\theta}\left(1 - \frac{I_x}{I_z}\right),$$

and finally we get

$$\dot{T} = \frac{h^2}{I_z}\cos(\theta)\sin(\theta)\left(\frac{I_z}{I_x} - 1\right)\dot{\theta}. \qquad (4.6.12)$$

Equation 4.6.12 is very important. It entails that:

(1) If $\dot{T} < 0$ and $I_z > I_x$ then $\dot{\theta} < 0$, which implies nutation stability since an initial nutation angle, under the influence of the energy dissipation inside the body, will decrease to zero under ideal conditions.
(2) If $\dot{T} < 0$ and $I_z < I_x$ then $\dot{\theta} > 0$, which implies nutation instability (!) since an initial nutation angle, under the influence of energy dissipation inside the body, will increase until the spin is transferred to the major axis.

These conclusions confirm our claim in Section 4.6.1 that, in the presence of energy dissipation, *a spinning body is in stable angular motion only if the spin is about the major axis.*

As we shall see in future chapters, inertial stabilization by spinning the satellite is common in many practical applications. The first *Pioneer* satellites were spin-stabilized; this is a cheap and efficient solution. However, with further development of space technology and sophistication of space instrumentation, the space community understood that more ingenious, and higher-quality, space tasks could be achieved by using "three-axis–stabilized" satellites. Roughly speaking, this means that the satellite is not spinning about one of its body axes for the achievement of attitude stabilization. In the following sections we shall develop the attitude kinematics and dynamics equations of motion of nonspinning three-axis–stabilized satellites.

4.7 Attitude Kinematics Equations of Motion for a Nonspinning Spacecraft

4.7.1 Introduction

It is useful to clarify some notions about *attitude kinematics* in space before proceeding (in the next section) to the derivation of the attitude dynamics equations of motion for three-axis–stabilized satellites.

The choice of an axis system is closely related to the satellite's tasks. It is not uncommon for a s/c to be transferred from one reference coordinate system to another system that is more appropriate for the particular tasks that are pertinent to a given period of the satellite's life. For instance, a spacecraft voyaging toward some planet in the solar system will begin its life in a reference coordinate system moving with the initial orbit that circles the earth. Next, on the way toward the planet, an inertial coordinate system will be used for its attitude definition in space. Finally, close to the planet, a third reference coordinate system will be chosen, similar to the first one but this time attached to the orbit circling the planet.

4.7.2 Basic Coordinate Systems

In this section we make use of the material presented in Appendix A concerning methods of attitude determination in space. We must first define the reference coordinate frame in which the satellite is to be three-axis attitude-stabilized. Different reference axis frames are chosen for different s/c tasks.

For planet-orbiting s/c, including the earth-orbiting craft, it is most convenient to define the *orbit reference frame* as follows. In Figure 4.7.1, the origin of this orbit reference frame moves with the cm (center of mass) of the satellite in the orbit. The Z_R axis points toward the cm of the earth (the subscript "R" stands for *reference*). The X_R axis is in the plane of the orbit, perpendicular to the Z_R axis, in the direction of the velocity of the spacecraft. The Y_R axis is normal to the local plane of the orbit, and completes a three-axis right-hand orthogonal system. The term ω_{RI} is the angular velocity vector of this frame, relative to the inertial axis frame defined in Section 4.7.3.

The inertial axis frame has its origin at the cm of the earth. The satellite's axis frame is defined by X_B, Y_B, and Z_B (where, as before, the subscript "B" denotes *body*). The satellite's attitude with respect to any reference frame is defined by a direction cosine matrix [**A**], by its quaternion vector **q**, or by the Euler angles (see Appendix A). In our treatment, the Euler angles are defined as the rotational angles about the body axes as follows: ϕ, about the X_B axis; θ, about the Y_B axis; and ψ, about the Z_B axis.

The direction cosine matrices can be expressed in terms of the Euler angles. According to the different order of rotation of the axes of the moving body with respect to the reference frame, there may be as many as 12 direction cosine matrices expressed in trigonometric functions of the Euler angles. Some of these matrices are detailed in Appendix A.

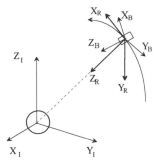

Figure 4.7.1 Definition of the orbit reference frame.

4.7.3 Angular Velocity Vector of a Rotating Frame

Two important factors in satellite kinematics are (1) the angular velocities of the body axis frame with respect to the reference axis frame and (2) the velocity of the body axis frame with respect to the inertial axis frame. In general, the angular velocity vector of the body frame relative to the reference frame is denoted by $\omega_{BR} = p\mathbf{i} + q\mathbf{j} + r\mathbf{k}$, and the angular velocity vector of the reference frame relative to the inertial frame is denoted as $\omega_{RI} = \omega_{RIx}\mathbf{1}_x + \omega_{RIy}\mathbf{1}_y + \omega_{RIz}\mathbf{1}_z$. When ω_{RI} is expressed in the body frame, it takes the form $\omega_{RIB} = \omega_{RIBx}\mathbf{i} + \omega_{RIBy}\mathbf{j} + \omega_{RIBz}\mathbf{k}$. With these definitions, the velocity vector of the body frame relative to the inertial frame becomes

$$\omega_{BI} = \omega_{BR} + \omega_{RIB}. \tag{4.7.1}$$

The angular velocity vector ω_{BR} is of importance, because it allows us to calculate the Euler angles of the moving body with respect to any defined reference frame in space. To begin with, suppose that the initial body axes were aligned with the reference axes $\mathbf{X}_R$, $\mathbf{Y}_R$, and $\mathbf{Z}_R$. We shall illustrate the procedure for calculating the Euler angles from the body angular velocity vector for some common and useful orders of attitude transformations.

Angular Velocities for the Transformation $\psi \to \theta \to \phi$

When we choose the order of axes transformation as $\psi \to \theta \to \phi$ (with axes order of rotation $3 \to 2 \to 1$), $[\mathbf{A}_\psi]$ is the first angular rotation about the $\mathbf{Z}_B$ body axis to be performed. The next rotation will be about the new $\mathbf{Y}_{B1}$ axis by an angle θ, and so on. Finally:

$$[\mathbf{A}_{\psi\theta\phi}] = [\mathbf{A}_\phi][\mathbf{A}_\theta][\mathbf{A}_\psi]. \tag{4.7.2}$$

Performing the matrix multiplications in Eq. 4.7.2, we find that

$$[\mathbf{A}_{321}] = [\mathbf{A}_{\psi\theta\phi}] = \begin{bmatrix} c\theta c\psi & c\theta s\psi & -s\theta \\ -c\phi s\psi + s\phi s\theta c\psi & c\phi c\psi + s\phi s\theta s\psi & s\phi c\theta \\ s\phi s\psi + c\phi s\theta c\psi & -s\phi c\psi + c\phi s\theta s\psi & c\phi c\theta \end{bmatrix} \tag{4.7.3}$$

(see Appendix A), where c and s denote cos and sin, respectively. Equation 4.7.3 is a direction cosine matrix expressed in the Euler angles, and it shows the rotation of the body axes relative to the reference axis frame.

In the process of angular rotation – for instance, rotation about the $\mathbf{Z}_B$ axis – a derivative of the ψ angle, $d\psi/dt$, is sensed about the same axis. We shall next find the relationships between the Euler angles and their derivatives as well as the angular velocity vector of the body in its rotational motion.

The first rotation, about the $\mathbf{Z}_B$ axis (axis 3), leads to the derivative $d\psi/dt$ about the $\mathbf{Z}_B$ axis; this derivative is subject to three successive angular transformations: ψ, θ, and finally one about $\mathbf{X}_{B2}$ by the angle ϕ. The second transformation, about the $\mathbf{Y}_{B1}$ axis (axis 2) by the Euler angle θ, produces the derivative $d\theta/dt$, which is subject to two angular rotations: first about $\mathbf{Y}_{B2}$ by an angle θ and then about $\mathbf{X}_{B3}$ by the angle ϕ. The last rotation, about the axis $\mathbf{X}_{B3}$ (axis 1), produces the derivative $d\phi/dt$. This derivative is subject to only one attitude transformation before final angular position of the body coordinate system is reached relative to the reference coordinate system.

4.7 / Attitude Kinematics Equations of Motion for Nonspinning Spacecraft

These derivatives are transformed to the body angular rates p, q, and r by the following equation:

$$\begin{bmatrix} p \\ q \\ r \end{bmatrix} = \omega_{BR} = [\mathbf{A}_\phi][\mathbf{A}_\theta][\mathbf{A}_\psi] \begin{bmatrix} 0 \\ 0 \\ \dot{\psi} \end{bmatrix} + [\mathbf{A}_\phi][\mathbf{A}_\theta] \begin{bmatrix} 0 \\ \dot{\theta} \\ 0 \end{bmatrix} + [\mathbf{A}_\phi] \begin{bmatrix} \dot{\phi} \\ 0 \\ 0 \end{bmatrix}. \qquad (4.7.4)$$

After performing the matrix multiplications we get

$$\begin{aligned} p &= \dot{\phi} - \dot{\psi}\sin(\theta), \\ q &= \dot{\theta}\cos(\phi) + \dot{\psi}\cos(\theta)\sin(\phi), \\ r &= \dot{\psi}\cos(\theta)\cos(\phi) - \dot{\theta}\sin(\phi). \end{aligned} \qquad (4.7.5)$$

These equations can be solved for $\dot{\phi}$, $\dot{\theta}$, and $\dot{\psi}$ as follows:

$$\begin{aligned} \dot{\phi} &= p + [q\sin(\phi) + r\cos(\phi)]\tan(\theta), \\ \dot{\theta} &= q\cos(\phi) - r\sin(\phi), \\ \dot{\psi} &= [q\sin(\phi) + r\cos(\phi)]\sec(\theta). \end{aligned} \qquad (4.7.6)$$

The first and last of Eqs. 4.7.6 show a singularity at $\theta = 90°$. This is the reason why, in certain engineering situations, a special order of rotation might be preferred. In some gyroscopic inertial systems, for instance, this singularity might cause the phenomenon called *gimbal lock*. It will be shown in Section 4.7.5 that such singularities do not occur in quarternion terminology.

Measurements of the body axes angular rates p, q, and r relative to the reference frame, together with knowledge of the initial conditions of the Euler angles, allow us to integrate with time the set of Eqs. 4.7.6. We thereby obtain the Euler angles ψ, θ, and ϕ by which the body frame is rotated relative to the reference frame. It is also important to note that p, q, and r are not necessarily inertial angular velocities. This fact will be treated in future sections. Results similar to Eqs. 4.7.5 and Eqs. 4.7.6 can be found for different orders of rotation about the body axes, as presented in Appendix A. Two such results will be described next.

Angular Velocities for the Transformation $\theta \to \phi \to \psi$

The results for the transformation with the order of rotations $\theta \to \phi \to \psi$ about the axes $2 \to 1 \to 3$ are as follows:

$$\begin{aligned} p &= \dot{\theta}\cos(\phi)\sin(\psi) + \dot{\phi}\cos(\psi), \\ q &= \dot{\theta}\cos(\phi)\cos(\psi) - \dot{\phi}\sin(\psi), \\ r &= -\dot{\theta}\sin(\phi) + \dot{\psi}. \end{aligned} \qquad (4.7.7)$$

From Eqs. 4.7.7 we also find that

$$\begin{aligned} \dot{\theta} &= [p\sin(\psi) + q\cos(\psi)]/\cos(\phi), \\ \dot{\phi} &= p\cos(\psi) - q\sin(\psi), \\ \dot{\psi} &= [p\sin(\psi) + q\cos(\psi)]\tan(\phi) + r. \end{aligned} \qquad (4.7.8)$$

Angular Velocities for the Transformation $\phi \to \theta \to \psi$

Using the methods described previously generates the following results for the angular velocities of the transformation order $1 \to 2 \to 3$:

$$p = \cos(\psi)\cos(\theta)\dot{\phi} + \sin(\psi)\dot{\theta},$$
$$q = -\sin(\psi)\cos(\theta)\dot{\phi} + \cos(\psi)\dot{\theta}, \quad (4.7.9)$$
$$r = \sin(\theta)\dot{\phi} + \dot{\psi}.$$

After some simple algebraic operations, we have also

$$\dot{\theta} = p\sin(\psi) + q\cos(\psi),$$
$$\dot{\phi} = [p\cos(\psi) - q\sin(\psi)]/\cos(\theta), \quad (4.7.10)$$
$$\dot{\psi} = r - [p\cos(\psi) - q\sin(\psi)]\tan(\theta).$$

We should note that, for very small Euler angles, Eqs. 4.7.6, Eqs. 4.7.8, and Eqs. 4.7.10 show that $p \approx \dot{\phi}$, $q \approx \dot{\theta}$, and $r \approx \dot{\psi}$ (see Appendix A). These approximations will be used in the derivation of the linearized attitude dynamic equations of the satellite.

4.7.4 Time Derivation of the Direction Cosine Matrix

Knowledge of direction cosine matrix elements is equivalent to knowing the attitude of the s/c relative to the reference frame in which the transformation matrix [**A**] is defined. In general, for a rotating body, the elements of this matrix change with time. In Wertz (1978) it is shown that:

$$\frac{d}{dt}[\mathbf{A}] = [\mathbf{\Omega}][\mathbf{A}], \quad (4.7.11)$$

with

$$[\mathbf{\Omega}] = \begin{bmatrix} 0 & \omega_z & -\omega_y \\ -\omega_z & 0 & \omega_x \\ \omega_y & -\omega_x & 0 \end{bmatrix}. \quad (4.7.12)$$

The terms $\omega_x, \omega_y, \omega_z$ are angular velocities about the body coordinate axes. The preceding equations are used when the angular velocity vector of the body can be measured (by inertial measurement instrumentation) to find the evolving direction cosine matrix. Numerical integration of Eq. 4.7.11 requires knowledge of the initial conditions of [**A**(0)]. Integration of this equation system is excessively time-consuming and hence seldom carried out. The integration of the quaternion vector (introduced in Appendix A) is much more efficient and thus much more common.

4.7.5 Time Derivation of the Quaternion Vector

As with the previous case, a differential vector equation for **q** can be written if the angular velocity vector **ω** of the body frame is known with respect to another reference frame. The differential equations of the quaternion system become

$$\frac{d}{dt}\mathbf{q} = \frac{1}{2}[\mathbf{\Omega}']\mathbf{q}, \quad (4.7.13)$$

4.7 / Attitude Kinematics Equations of Motion for Nonspinning Spacecraft

with $[\Omega']$ defined as

$$[\Omega'] = \begin{bmatrix} 0 & \omega_z & -\omega_y & \omega_x \\ -\omega_z & 0 & \omega_x & \omega_y \\ \omega_y & -\omega_x & 0 & \omega_z \\ -\omega_x & -\omega_y & -\omega_z & 0 \end{bmatrix}.$$

In this case also, knowing the initial condition of $\mathbf{q}(0)$, the numerical integration of the system of Eq. 4.7.13 will provide us with the time evolution of the quaternion vector $\mathbf{q}$.

4.7.6 Derivation of the Velocity Vector ω_{RI}

In deriving the attitude dynamics equations of a s/c, it is important to know the inertial velocity vector ω_{BI} of the satellite, which is evaluated by inertial measuring instrumentation such as precision rate integrating gyros. In order to find the evolution of the Euler angles of the satellite in the orbit reference frame, we must know ω_{BR} from the equality in Eq. 4.7.1: $\omega_{BR} = \omega_{BI} - \omega_{RIB}$. In this section we derive the angular velocity vector ω_{RI} of the orbit reference coordinate frame with respect to the inertial coordinate frame, as defined in Section 4.7.3 and in Figure 4.7.1 and Figure 4.7.2; we shall then express this vector in the body coordinate frame as ω_{RIB}.

In Figure 4.7.2, the orbit reference frame is defined by the triad of unit vectors $\mathbf{i}, \mathbf{j}, \mathbf{k}$. The definition of these unit vectors, based on the radius vector $\mathbf{r}$ and the velocity vector $\mathbf{v}$ of the orbit, is as follows:

$$\mathbf{k} = -\frac{\mathbf{r}}{|\mathbf{r}|},$$
$$\mathbf{j} = \frac{\mathbf{v} \times \mathbf{r}}{|\mathbf{v} \times \mathbf{r}|}, \quad (4.7.14)$$
$$\mathbf{i} = \mathbf{j} \times \mathbf{k}.$$

Next, we determine the angular velocities about the $\mathbf{i}, \mathbf{j}$, and $\mathbf{k}$ axes. As in Section 2.4, for a positive clockwise rotation δ about $\mathbf{j}$ we have

$$\frac{d\mathbf{i}}{dt} = \frac{d\mathbf{i}}{d\delta}\frac{d\delta}{dt} = -\mathbf{k}\omega_j,$$

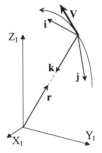

Figure 4.7.2 Derivation of ω_{RI}.

since obviously $d\mathbf{i}/d\delta = -\mathbf{k}$. Finally,
$$\omega_j = -\frac{d\mathbf{i}}{dt} \cdot \mathbf{k}.$$

Repeating the last procedure, we find also that
$$\omega_i = \frac{d\mathbf{j}}{dt} \cdot \mathbf{k} = -\frac{d\mathbf{k}}{dt} \cdot \mathbf{j},$$
$$\omega_j = -\frac{d\mathbf{i}}{dt} \cdot \mathbf{k} = \frac{d\mathbf{k}}{dt} \cdot \mathbf{i}, \qquad (4.7.15)$$
$$\omega_k = \frac{d\mathbf{i}}{dt} \cdot \mathbf{j} = -\frac{d\mathbf{j}}{dt} \cdot \mathbf{i}.$$

Together with Eqs. 4.7.14, we have
$$\omega_i = -\frac{d\mathbf{k}}{dt} \cdot \mathbf{j} = \frac{1}{|\mathbf{r}|} \frac{d\mathbf{r}}{dt} \cdot \frac{\mathbf{v} \times \mathbf{r}}{|\mathbf{v} \times \mathbf{r}|}$$
$$= \frac{1}{|\mathbf{r}||\mathbf{v} \times \mathbf{r}|} \mathbf{v} \cdot (\mathbf{v} \times \mathbf{r}) = \frac{1}{|\mathbf{r}||\mathbf{v} \times \mathbf{r}|} \mathbf{v} \times \mathbf{v} \cdot \mathbf{r} = 0 \qquad (4.7.16)$$

(see Kaplan 1953 for relationships between vector and scalar products). In a similar way, we can find
$$\omega_j = \frac{d\mathbf{k}}{dt} \cdot \mathbf{i} = -\frac{1}{|\mathbf{r}|} \mathbf{v} \cdot \frac{(\mathbf{v} \times \mathbf{r}) \times (-\mathbf{r})}{|\mathbf{v} \times \mathbf{r}||\mathbf{r}|} = -\frac{1}{|\mathbf{r}|} \mathbf{v} \cdot \frac{[(\mathbf{r} \cdot \mathbf{v})\mathbf{r} - (\mathbf{r} \cdot \mathbf{r})\mathbf{v}]}{|\mathbf{r}||\mathbf{v} \times \mathbf{r}|}$$
$$= \frac{-1}{r^2|\mathbf{v} \times \mathbf{r}|} \mathbf{v} \cdot [(\mathbf{r} \cdot \mathbf{v})\mathbf{r} - (r^2 \mathbf{v})]. \qquad (4.7.17)$$

For a circular orbit, $\mathbf{v}$ is perpendicular to $\mathbf{r}$; hence $|\mathbf{v} \times \mathbf{r}| = vr$ and $\mathbf{r} \cdot \mathbf{v} = 0$. Finally,
$$\omega_j = \frac{v^2 r^2}{r^3 v} = \frac{v}{r} = \omega_o, \qquad (4.7.18)$$

where ω_o is the angular orbital velocity of the s/c (see also Eq. 2.5.6).

In order to find ω_k, we calculate as follows:
$$\omega_k = -\frac{d\mathbf{j}}{dt} \cdot \mathbf{i} = -\frac{1}{|\mathbf{v} \times \mathbf{r}|} \cdot \frac{d}{dt}(\mathbf{v} \times \mathbf{r}) \cdot \mathbf{i} = -\frac{1}{|\mathbf{v} \times \mathbf{r}|}[\dot{\mathbf{v}} \times \mathbf{r} + \mathbf{v} \times \dot{\mathbf{r}}] \cdot \mathbf{i}$$
$$= -\frac{1}{|\mathbf{v} \times \mathbf{r}|}[\dot{\mathbf{v}} \times \mathbf{r} + 0] \cdot \mathbf{i} = -\frac{1}{|\mathbf{v} \times \mathbf{r}|}[\dot{\mathbf{v}} \times \mathbf{r}] \cdot \frac{(\mathbf{v} \times \mathbf{r}) \times (-\mathbf{r})}{|\mathbf{r}||\mathbf{v} \times \mathbf{r}|}$$
$$= \frac{1}{|\mathbf{v} \times \mathbf{r}|^2 |\mathbf{r}|}[\ddot{\mathbf{r}} \times \mathbf{r}] \cdot [(\mathbf{v} \times \mathbf{r}) \times \mathbf{r}]. \qquad (4.7.19)$$

In Keplerian orbits, there are no out-of-plane accelerations and so $\omega_k = 0$. Finally, for a circular orbit,
$$\omega_{\text{RI}} = [0 \ -\mathbf{j}\omega_o \ 0]^T. \qquad (4.7.20)$$

4.8 Attitude Dynamics Equations of Motion for a Nonspinning Satellite

4.8.1 *Introduction*

The attitude dynamics equations will be obtained from Euler's moment equation (Eq. 4.5.1). In Section 4.5, the body was presumed to be rigid, with no moving elements inside it. However, in the present derivations we will allow for the existence of rotating elements inside the satellite – known as *momentum exchange devices* – and for other kinds of *gyroscopic* devices. The most common momentum exchange devices are the *reaction wheel*, the *momentum wheel*, and the *control moment gyro*. In this textbook, only the first two will be used for attitude control. (These devices will be described in Chapter 7.) Meanwhile, we must remember that these rotating elements have their own angular momentum, which becomes a part of the momentum of the entire system in Euler's moment equations.

Returning to Eq. 4.5.1, **M** is the total external moment acting on the body, which is equal to the inertial momentum change of the system. External inertial moments are products of aerodynamic, solar, or gravity gradient forces, or of magnetic torques or reaction torques produced by particles expelled from the body, such as hydrazine gas or ion particles (see Appendix C).

4.8.2 *Equations of Motion for Spacecraft Attitude*

Equation 4.5.1 may be rewritten as follows:

$$\mathbf{T} = \dot{\mathbf{h}}_I = \dot{\mathbf{h}} + \boldsymbol{\omega} \times \mathbf{h}. \tag{4.8.1}$$

For practical reasons, **T** was substituted for **M**. It is also understood that in this equation $\dot{\mathbf{h}}$ denotes differentiation of **h** in the body frame. We shall break down the external torque **T** into two principal parts: $\mathbf{T}_c$, the control moments to be used for controlling the attitude motion of the satellite; and $\mathbf{T}_d$, those moments due to different disturbing environmental phenomena. The total torque vector is thus $\mathbf{T} = \mathbf{T}_c + \mathbf{T}_d$.

The momentum of the entire system will be divided between the momentum of the rigid body $\mathbf{h}_B = [h_x \ h_y \ h_z]^T$ and the momentum of the moment exchange devices $\mathbf{h}_w = [h_{wx} \ h_{wy} \ h_{wz}]^T$. Finally, $\mathbf{h} = \mathbf{h}_B + \mathbf{h}_w$. With these definitions, the general equations of motion become

$$\begin{aligned}\mathbf{T} = \mathbf{T}_c + \mathbf{T}_d &= [\dot{h}_x + \dot{h}_{wx} + (\omega_y h_z - \omega_z h_y) + (\omega_y h_{wz} - \omega_z h_{wy})]\mathbf{i} \\ &+ [\dot{h}_y + \dot{h}_{wy} + (\omega_z h_x - \omega_x h_z) + (\omega_z h_{wx} - \omega_x h_{wz})]\mathbf{j} \\ &+ [\dot{h}_z + \dot{h}_{wz} + (\omega_x h_y - \omega_y h_x) + (\omega_x h_{wy} - \omega_y h_{wx})]\mathbf{k}.\end{aligned} \tag{4.8.2}$$

Here, **i, j, k** are the unit direction vectors of the body axis frame. In Eq. 4.8.2, the body momentum components h_x, h_y, h_z are defined as in Eq. 4.2.8, and contain all the moments and products of inertia. The terms h_{wx}, h_{wy}, h_{wz} are the vector components of the sum of the angular momentum of all the momentum exchange devices (see Section 7.3). Equation 4.8.2 summarizes the full attitude dynamics that must be implemented in the complete six-degrees-of-freedom (6-DOF) simulation necessary for analyzing the attitude control system. Care must be taken in deriving the vector ω, since it must be expressed in the correct coordinate frame.

4.8.3 *Linearized Attitude Dynamics Equations of Motion*

In the first phase of the design stage, it is important to transform Eq. 4.8.2 into a more easily treatable form. If the design problem at hand allows working with principal axes, then the products of inertia may be eliminated from the dynamic equations, thus simplifying them considerably. Moreover, the angular motion can be approximated by infinitesimal angular motion, which means small Euler angles and angle derivatives. With these assumptions, the dynamics equations can be Laplace-transformed, thus gaining the important advantage of using linear control theory.

Gravity Gradient Moments

Before we can write the linearized attitude dynamic equations of motion, we must state and analyze one important external moment, that is, the *gravitational moment*. This moment is inherent in low-orbit satellites, and cannot be neglected when dealing with passively attitude-controlled satellites. A short description of the dynamics of these moments follows.

An asymmetric body subject to a gravitational field will experience a torque tending to align the axis of least inertia with the field direction. A full development of the gravitational moments equations can be found in Greensite (1970). Only an outline of the development, with the final results, is given here. For the following discussion, we suppose that the moving satellite is at a distance R_0 from the center of mass of the earth.

The orbit reference axis frame is defined as in Section 4.7.2 and Figure 4.7.2. In Figure 4.8.1, $\mathbf{i}_R, \mathbf{j}_R, \mathbf{k}_R$ are the unit vectors of the reference axis frame. The origin of the reference frame is located in the cm of the body. The attracting gravity force is aligned along the $\mathbf{k}_R$ axis; ρ is the distance between the cm of the body and any mass element dm in the body; and $\mathbf{i}_B, \mathbf{j}_B, \mathbf{k}_B$ are the unit vectors of the body coordinates axis frame (not shown in the figure).

We can find the components of the vector $\mathbf{R} = -R_0 \mathbf{k}_R$ in the body axes by using any one of the Euler angle transformations of Section A.3 - for instance, the trans-

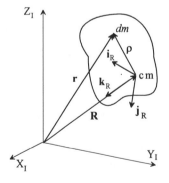

Figure 4.8.1 Gravitational moments on an asymmetric spacecraft.

4.8 / Attitude Dynamics Equations of Motion for a Nonspinning Satellite

formation $[\mathbf{A}_{\psi\theta\phi}]$ of Eq. A.3.6. The components of the vector $\mathbf{R}$ in the body axes will be labeled R_x, R_y, and R_z. We have

$$\begin{bmatrix} R_x \\ R_y \\ R_z \end{bmatrix} = [\mathbf{A}_{\psi\theta\phi}] \begin{bmatrix} 0 \\ 0 \\ -R_0 \end{bmatrix}. \tag{4.8.3}$$

It follows that

$$\begin{aligned} R_x &= R_0 \sin(\theta) &= a_{13}(-R_0), \\ R_y &= -R_0 \sin(\phi)\cos(\theta) &= a_{23}(-R_0), \\ R_z &= -R_0 \cos(\phi)\cos(\theta) &= a_{33}(-R_0). \end{aligned} \tag{4.8.4}$$

Define the gravity gradient vector as $\mathbf{G} = [G_x\ G_y\ G_z]^T$. The force exerted on a mass element due to gravity is $d\mathbf{F} = -[(\mu\, dm)/|\mathbf{r}|^3]\mathbf{r}$, where $\mathbf{r} = \mathbf{R}+\boldsymbol{\rho}$ is the distance from the earth's cm to the mass dm. Since $\rho \ll R_0$, the moment about the cm of the body becomes

$$d\mathbf{G} = \boldsymbol{\rho} \times d\mathbf{F} = -\frac{\mu\, dm}{|\mathbf{r}|^3} \boldsymbol{\rho} \times \mathbf{r}, \tag{4.8.5}$$

where $\boldsymbol{\rho}$ is the radius vector from the body center of mass to a generic mass element dm. With $\rho \ll R_0$, $1/r^3$ can be approximated as

$$\frac{1}{r^3} \approx \frac{1}{R_0^3}\left[1 - \frac{3\mathbf{R}\cdot\boldsymbol{\rho}}{R_0^2}\right]. \tag{4.8.6}$$

Integration of Eq. 4.8.5 over the entire body mass, together with Eq. 4.8.6, leads to

$$\mathbf{G} = \frac{3\mu}{R_0^5} \int_M [\mathbf{R}\cdot\boldsymbol{\rho}][\boldsymbol{\rho}\times\mathbf{R}]\, dm. \tag{4.8.7}$$

After calculating the scalar and vector products, a procedure similar to that of Section 4.2 is used to obtain the final results:

$$\begin{aligned} G_x &= \frac{3\mu}{2R_0^3}(I_z - I_y)\sin(2\phi)\cos^2(\theta) = \frac{3\mu}{R_0^3}(I_z - I_y)a_{23}a_{33}, \\ G_y &= \frac{3\mu}{2R_0^3}(I_z - I_x)\sin(2\theta)\cos(\phi) = \frac{-3\mu}{R_0^3}(I_z - I_x)a_{13}a_{33}, \\ G_z &= \frac{3\mu}{2R_0^3}(I_x - I_y)\sin(2\theta)\sin(\phi) = \frac{-3\mu}{R_0^3}(I_x - I_y)a_{13}a_{23}. \end{aligned} \tag{4.8.8}$$

These are the gravity gradient moment components of $\mathbf{G}$. As is easily seen, the gravity moment vector $\mathbf{G}$ can be expressed not only in terms of Euler angles but also in terms of elements of the direction cosine matrix, which is the transformation of the attitude angles of the body axis frame to those of the reference axis frame.

Equations 4.8.8 can be simplified by linearization for a body in a circular orbit using small-angle approximations for ϕ and θ. We have previously found that the lateral velocity of a body in a circular orbit of radius R_0 is $v = \sqrt{\mu/R_0}$ (see Section 2.4.2); thus the angular orbital velocity of the body (also called orbital *rate* or *frequency*) becomes $\omega_o = v/R_0 = \sqrt{\mu/R_0^3}$. Hence Eqs. 4.8.8 take the approximate form

$$G_x = 3\omega_o^2(I_z - I_y)\phi,$$
$$G_y = 3\omega_o^2(I_z - I_x)\theta, \qquad (4.8.9)$$
$$G_z = 0.$$

The linear components of the moment vector **G** in Eqs. 4.8.9 are used for the derivation of the linearized attitude dynamics equations.

Linearized Attitude Dynamics Equations of Motion

In Eq. 4.7.1, ω_{BI} is the inertial angular velocity of the body. In order to simplify the notation, set $\omega = \omega_{BI}$. To find ω we use Eq. 4.7.1 and Eqs. 4.7.7. However, ω_{RI} is to be expressed in the body frame and so will be renamed ω_{RIB}. For small Euler angles, the following relation exists:

$$\begin{bmatrix} \omega_{RIBx} \\ \omega_{RIBy} \\ \omega_{RIBz} \end{bmatrix} = \begin{bmatrix} 1 & \psi & -\theta \\ -\psi & 1 & \phi \\ \theta & -\phi & 1 \end{bmatrix} \begin{bmatrix} 0 \\ -\omega_o \\ 0 \end{bmatrix} = \begin{bmatrix} -\psi\omega_o \\ -\omega_o \\ \phi\omega_o \end{bmatrix}. \qquad (4.8.10)$$

With known ω_{RIB} (Eq. 4.8.10), $\omega = \omega_{BI} = \omega_{BR} + \omega_{RIB}$ becomes

$$\begin{bmatrix} \omega_x \\ \omega_y \\ \omega_z \end{bmatrix} = \begin{bmatrix} p \\ q \\ r \end{bmatrix} + \begin{bmatrix} -\psi\omega_o \\ -\omega_o \\ \phi\omega_o \end{bmatrix}. \qquad (4.8.11)$$

According to Eqs. 4.7.5, Eqs. 4.7.7, or Eqs. 4.7.9, p, q, and r can for small Euler angles be approximated as $p \approx \dot{\phi}$, $q \approx \dot{\theta}$, and $r \approx \dot{\psi}$. With these approximations, Eq. 4.8.11 becomes

$$\begin{bmatrix} \omega_x \\ \omega_y \\ \omega_z \end{bmatrix} = \begin{bmatrix} \dot{\phi} - \psi\omega_o \\ \dot{\theta} - \omega_o \\ \dot{\psi} + \phi\omega_o \end{bmatrix}. \qquad (4.8.12)$$

From Eq. 4.8.12 it easily follows that

$$\dot{\omega}_x = \ddot{\phi} - \omega_o \dot{\psi},$$
$$\dot{\omega}_y = \ddot{\theta}, \qquad (4.8.13)$$
$$\dot{\omega}_z = \ddot{\psi} + \omega_o \dot{\phi}.$$

In future chapters, we shall also be concerned with momentum-biased satellites. In this type of satellite, a constant momentum bias h_{wyo} is applied along the $\mathbf{Y_B}$ axis to give inertial angular stability about the $\mathbf{Y_B}$ axis of the s/c (see Chapter 8). With this assumption, Eq. 4.8.2 - together with Eqs. 4.8.9, Eq. 4.8.12, and Eqs. 4.8.13 - become the desired linearized attitude dynamics equations of motion:

$$\begin{aligned} T_{dx} + T_{cx} &= I_x\ddot{\phi} + 4\omega_o^2(I_y - I_z)\phi + \omega_o(I_y - I_z - I_x)\dot{\psi} + \dot{h}_{wx} - \omega_o h_{wz} \\ &\quad - \dot{\psi}h_{wyo} - \phi\omega_o h_{wyo} - I_{xy}\ddot{\theta} - I_{xz}\ddot{\psi} - I_{xz}\omega_o^2\psi + 2I_{yz}\omega_o\dot{\theta}, \\ T_{dy} + T_{cy} &= I_y\ddot{\theta} + 3\omega_o^2(I_x - I_z)\theta + \dot{h}_{wy} \\ &\quad - I_{xy}(\ddot{\phi} - 2\omega_o\dot{\psi} - \omega_o^2\phi) + I_{yz}(-\ddot{\psi} - 2\omega_o\dot{\phi} + \omega_o^2\psi), \\ T_{dz} + T_{cz} &= I_z\ddot{\psi} + \omega_o(I_z + I_x - I_y)\dot{\phi} + \omega_o^2(I_y - I_x)\psi + \dot{h}_{wz} + \omega_o h_{wx} \\ &\quad + \dot{\phi}h_{wyo} - \psi\omega_o h_{wyo} - I_{yz}\ddot{\theta} - I_{xz}\ddot{\phi} - 2\omega_o I_{xy}\dot{\theta} - \omega_o^2 I_{xz}\phi. \end{aligned} \qquad (4.8.14)$$

In Eqs. 4.8.14, h_{wx}, h_{wy}, h_{wz} are the momentum components of the wheels with axes of rotation along the X_B, Y_B, and Z_B body axes of the satellite; $h_{wx} = I_{wx}\omega_{wx}, h_{wy} = I_{wy}\omega_{wy} + h_{wyo}$, and $h_{wz} = I_{wz}\omega_{wz}$, where I_{wx}, I_{wy}, I_{wz} are the moments of inertia of the individual wheels and $\omega_{wx}, \omega_{wy}, \omega_{wz}$ are the angular velocities of the wheels. The terms $\dot{h}_{wx}, \dot{h}_{wy}, \dot{h}_{wz}$ are the angular moments that the wheels exert on the s/c along the body axes. If $\dot{\omega}_{wx}$ is the angular acceleration of the X_B axis wheel, then $\dot{h}_{wx} = I_{wx}\dot{\omega}_{wx}$ is the negative of the angular moment that the X_B wheel exerts on the satellite about its X_B axis. The same applies for the Y_B and Z_B axes wheel components. Attitude control of a s/c can be achieved by controlling these angular accelerations, which are internal torques exerted on the satellite. If, in addition, external (inertial) torques such as magnetic or reaction torques are applied to the satellite, they are incorporated in T_c, the vector of control torques.

In general, as we shall see in later chapters, the rotation axes of the wheels are not necessarily aligned along the satellite's body axes. Moreover, there may be more (or less) than three wheels in the satellite. In such cases, the momentum and the angular acceleration of the wheels will be transformed to the body axes, so that they comply with Eqs. 4.8.14. If the body coordinate axes are principal axes then the products of inertia are canceled, and Eqs. 4.8.14 are reduced to the minimum possible number of terms.

4.9 Summary

The purpose of this chapter was to state and analyze the attitude dynamics equations of spinning and nonspinning satellites. The equations were first written and used in their nonlinear form, and then manipulated to their linear form in order to simplify the analysis and design of the attitude control system in the following chapters.

Satellites can be either three-axis attitude-stabilized or, by taking advantage of the spin stabilizing effect, single-axis stabilized. Both cases will be analyzed in Chapters 5–8.

References

Bracewell, R. N., and Garriott, O. K. (1958), "Rotation of Artificial Earth Satellites," *Nature* 82: 760-2.
Goldstein, H. (1964), *Classical Mechanics*. Reading, MA: Addison-Wesley.
Greensite, A. (1970), *Analysis and Design of Space Vehicles Flight Control Systems*. New York: Spartan Books.
Hildebrand, F. (1968), *Methods of Applied Mathematics*. New Delhi: Prentice-Hall.
Kaplan, M. (1976), *Modern Spacecraft Dynamics and Control*. New York: Wiley.
Kaplan, W. (1953), *Advanced Calculus*. Reading, MA: Addison-Wesley.
Thomson, W. T. (1986), *Introduction to Space Dynamics*. New York: Dover.
Wertz, J. R. (1978), *Spacecraft Attitude Determination and Control*. Dordrecht: Reidel.

CHAPTER 5

Gravity Gradient Stabilization

5.1 Introduction

The present and remaining chapters deal with *attitude control of spacecraft;* this section serves as an introduction to all of them. The expression *attitude control* has the general meaning of controlling the attitude of the satellite. In practice, there exist a multitude of variations to this simple and apparently straightforward expression. The following are some examples of primary control tasks for which the attitude control system is responsible.

(1) In orbital maneuvering and adjustments, the attitude of the satellite must be pointed and held in the desired $\Delta \mathbf{V}$ direction.
(2) A spin-stabilized satellite may be designed to keep the spin axis of its body pointed at some particular direction in space.
(3) A nadir-pointing three-axis–stabilized satellite must keep its three Euler angles close to null relative to the orbit reference frame; this is true of most communications satellites.
(4) In earth-surveying satellites, the attitude control system is designed to allow the operative payload to track defined targets on the earth's surface.
(5) A scientific satellite observing the sky must maneuver its optical instruments toward different star targets on the celestial sphere in some prescribed pattern of angular motion.

The few examples listed and the many others not mentioned suggest a multitude of different tasks and missions to be performed by the attitude control system. However, we shall see that some features are common to all such systems.

An important distinction for attitude control concepts is between *passive* and *active* attitude control. Passive attitude control is attractive because the hardware required is less complicated and relatively inexpensive. Natural physical properties of the satellite and its environment are used to control the s/c attitude. However, the achievable accuracies with passive attitude control are generally much lower than those that are possible with active attitude control, which uses sophisticated (and much more expensive) control instrumentation.

Another important distinction is between *attitude-maneuvering* and *nadir-pointing* (earth-pointing) *stabilized* satellites. The attitude control hardware and the appropriate design concepts used in these two classes are quite different. In fact, there are many possible classifications of satellite control schemes. Our approach will be to present the material following the basic line of advancing from the simpler to the more elaborate attitude control schemes.

The attitude and orbit control of a satellite is performed with the aid of hardware that can be classified as either *attitude determination* or *control* hardware. The attitude determination hardware enables direct measurement (or estimation) of the s/c attitude with respect to some reference coordinate system in space. Examples of attitude determination hardware are earth sensors, sun and star sensors, and integrating gyros (see Appendix B). Control hardware provides translational and angular accelerations so that the orbit and the location of the satellite within that orbit, as well as its angular attitude, can be varied at will. Control hardware includes reaction and momentum wheels, control moment gyros, reaction thrusters, and magnetic torquers (also called *torqrods*); see Appendix C.

5.2 The Basic Attitude Control Equation

Generally speaking, the satellite attitude dynamics equations (see Section 4.5) are three second-order nonlinear equations. Automatic control theory does not provide exact analytical solutions and design procedures for such dynamic plants, so linearization of these equations is necessary if the satellite control engineer wishes to use standard automatic control techniques. Linearization for small Euler angles was presented in Section 4.8.3. The attitude dynamics equations need not be based on the Euler angles only, as in Eqs. 4.8.14. Suppose that one of the body axes of the satellite is to be aligned with the sun's direction; in this case, the projection of the sun vector into two correctly defined perpendicular planes in the satellite body are the attitude errors to be controlled. The appropriate equations for this control task can also be linearized.

In Eqs. 4.8.14 it is clearly shown that (in automatic control terminology) the "plant," with respect to a single satellite body axis, consists of two integrators. To control such a system, "control torques" are necessary. These torques can be produced actively with control hardware instruments, or by natural effects such as gravity gradient moments. The control torques to be activated are always a function of the attitude errors. Since we are dealing with second-order systems, some damping control must also be provided for improved stability. This means that the control torques will have to include a term that is dependent on the attitude rates to be measured or estimated. If, in addition, the steady-state error is to be nulled, then an integral of the attitude error can be added to the control torque equation.

Control torque equations can be written in the following form:

$$Tc_i = Kp_i(\text{er}) + Kd_i \frac{d}{dt}(\text{er}) + Ki_i \int (\text{er}) \, dt, \quad i = 1, 2, 3, \tag{5.2.1}$$

for each of the three body axes. These are the well-known PID controller equations (for control gains that are Kp proportional, Ki integral, and Kd derivative). The goal of controlling a satellite's attitude is met by optimally defining and mechanizing the physical errors for the different attitude control tasks.

In practice, the attitude dynamics equations of the satellite are more complicated than those shown in Eqs. 4.8.14. There may exist side effects such as structural dynamics of the body or of the appended solar panels, sloshing effects in the fuel tanks, and sensor noise. Although the basic form of Eq. 5.2.1 remains unaffected, the control equations will need "filters" to handle these complicating effects.

5.3 Gravity Gradient Attitude Control

5.3.1 *Purely Passive Control*

We shall first derive the linearized angular equations of motion and the stability conditions for purely passive gravity gradient (GG) attitude control. The dynamics of motion for GG can be derived from Eqs. 4.8.14; since the system is passively controlled, T_{cx}, T_{cy}, T_{cz}, h_{wx}, h_{wy}, and h_{wz} do not exist. The equations thus reduce to:

$$T_{dx} = I_x \ddot{\phi} + 4\omega_o^2(I_y - I_z)\phi - \omega_o(I_x + I_z - I_y)\dot{\psi},$$
$$T_{dz} = I_z \ddot{\psi} + \omega_o^2(I_y - I_x)\psi + \omega_o(I_z + I_x - I_y)\dot{\phi}, \qquad (5.3.1)$$
$$T_{dy} = I_y \ddot{\theta} + 3\omega_o^2(I_x - I_z)\theta.$$

In Eqs. 5.3.1, angular motion can be activated only by disturbing torques and initial angles of the Euler angles and their derivatives. For ease of notation, we define

$$\sigma_y = (I_x - I_z)/I_y, \quad \sigma_x = (I_y - I_z)/I_x, \quad \sigma_z = (I_y - I_x)/I_z. \qquad (5.3.2)$$

Stability about the Y_B Body Axis

By Laplace-transforming the third of Eqs. 5.3.1, the characteristic equation for the motion about the Y_B axis becomes

$$s^2 + 3\omega_o^2(I_x - I_z)/I_y = 0. \qquad (5.3.3)$$

Equation 5.3.3 has one unstable root if $I_x < I_z$, so the condition for stability becomes

$$I_x > I_z. \qquad (5.3.4)$$

There is no damping factor in this second-order equation, so it follows that for any initial condition or nonzero disturbance T_{dy}, the satellite will oscillate in a stable motion about the Y_B axis with an amplitude proportional to the initial condition $\theta(0)$ and the level of the disturbance T_{dy} about this axis. Note that the third of Eqs. 5.3.1 is independent of the first two.

Stability about the X_B and Z_B Body Axes

Using the definitions in Eq. 5.3.2, the first two of Eqs. 5.3.1 become

$$\ddot{\phi} + 4\omega_o^2 \sigma_x \phi - \omega_o(1 - \sigma_x)\dot{\psi} = \frac{T_{dx}}{I_x},$$
$$\ddot{\psi} + \omega_o^2 \sigma_z \psi + \omega_o(1 - \sigma_z)\dot{\phi} = \frac{T_{dz}}{I_z}. \qquad (5.3.5)$$

It is important to keep in mind that the values of σ_x and σ_z are limited and that they are smaller than unity. To show this we make use of the definition of moments of inertia:

$$\sigma_x = \frac{I_y - I_z}{I_x} = \frac{\int(x^2 + z^2)\,dm - \int(x^2 + y^2)\,dm}{\int(y^2 + z^2)\,dm} = \frac{\int(z^2 - y^2)\,dm}{\int(z^2 + y^2)\,dm} < 1. \qquad (5.3.6a)$$

In the same way, it may be shown that

$$\sigma_z < 1 \quad \text{and} \quad \sigma_y < 1. \qquad (5.3.6b)$$

5.3 / Gravity Gradient Attitude Control

After taking the Laplace transform of Eqs. 5.3.5, their determinant is found to be

$$s^4 + \omega_o^2[3\sigma_x + \sigma_x\sigma_z + 1]s^2 + 4\omega_o^4\sigma_x\sigma_z = 0. \tag{5.3.7}$$

We shall next derive the stability conditions for Eq. 5.3.7. Its solution in terms of s^2 is

$$\frac{s^2}{\omega_o^2} = \frac{-(3\sigma_x + \sigma_x\sigma_z + 1) \pm \sqrt{(3\sigma_x + \sigma_x\sigma_z + 1)^2 - 16\sigma_x\sigma_z}}{2}. \tag{5.3.8}$$

If s_1 is a root of Eq. 5.3.7 then so is $-s_1$. For s_1 to be a root with no positive real part, it is necessary that s_1 be imaginary ($s^2 < 0$) and also that the term under the radical be positive. This means that the following three conditions must be fulfilled:

$$\begin{aligned} 3\sigma_x + \sigma_x\sigma_z + 1 &> 4\sqrt{\sigma_x\sigma_z}, \\ \sigma_x\sigma_z &> 0, \\ 3\sigma_x + \sigma_x\sigma_z + 1 &> 0. \end{aligned} \tag{5.3.9}$$

Remember also the first condition: $I_x > I_z$. From Eq. 5.3.6a it also follows that

$$I_y < I_x + I_z. \tag{5.3.10}$$

The inequalities of Eqs. 5.3.9 may be translated to the σ_x-σ_z plane as in Figure 5.3.1, where the regions for stability (and instability) about the X_B and Z_B axes are displayed. To find these regions, we first multiply both sides of the inequality of Eq. 5.3.10 by $(I_x - I_z)$. By simple algebraic manipulation, and remembering the condition $I_x > I_z$ for stability of the Y_B (pitch) axis, we find that $I_x^2 - I_z^2 > I_xI_y - I_yI_z$ or

$$I_z(I_y - I_z) = I_yI_z - I_z^2 > I_xI_y - I_x^2 = I_x(I_y - I_x),$$

from which it follows that

$$\sigma_x = (I_y - I_z)/I_x > (I_y - I_x)/I_z = \sigma_z$$

or, finally,

$$\sigma_x > \sigma_z. \tag{5.3.11}$$

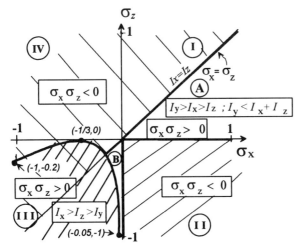

Figure 5.3.1 Stability regions for GG-stabilized satellites; adapted from Kaplan (1986) by permission of John Wiley & Sons.

Next, we find the regions on the σ_x-σ_z plane where the satellite is in stable attitude conditions. The condition of Eq. 5.3.11 is the region below the line $\sigma_x = \sigma_z$ in Figure 5.3.1, wherein the four quadrants are labeled I to IV. The second inequality in Eqs. 5.3.9 excludes regions II and IV as stability regions, because in these regions $\sigma_z \sigma_x < 0$. We are left with half of the regions I and III, which lie below the line $\sigma_x = \sigma_z$. However, in region III there is an additional forbidden area due to the first inequality in Eqs. 5.3.9. Squaring both sides of this inequality yields

$$\sigma_x^2(9 + \sigma_z^2 + 6\sigma_z) + \sigma_x(6 - 14\sigma_z) + 1 > 0. \tag{5.3.12}$$

The plot of the solution to this inequality is located in region III. Since $|\sigma_x| < 1$ and also $|\sigma_z| < 1$, we shall look for the values of this function on the boundaries $\sigma_x = -1$ and $\sigma_z = -1$, and also on the σ_x axis, for which $\sigma_z = 0$. The results are:

$(\sigma_x, \sigma_z) = (-1/3, 0),$

$(\sigma_x, \sigma_z) = (-0.0505, -1),$

$(\sigma_x, \sigma_z) = (-1, -0.202).$

These three points of the inequality function are also shown in Figure 5.3.1. The region below this function pertains to the nonstable solution of the passive gravity gradient attitude control. The remaining subregion B in III is permitted from the point of view of stability, but is seldom used owing to practical structural difficulties. Subregion A in region I is a stable region normally used in practical designs of GG-stabilized spacecraft.

Constraints on the Moments of Inertia in Subregion A   It is important to translate the stable σ_x-σ_z regions into constraints on the moments of inertia of the satellite. One of these constraints has already been stated: $I_x > I_z$. In region I, $\sigma_x > 0$ and $\sigma_z > 0$. In subregion A, $\sigma_x > \sigma_z$. According to the definitions in Eq. 5.3.2 and the inequality in Eq. 5.3.4, it follows that

$$I_y > I_x > I_z. \tag{5.3.13}$$

Remember that the inequality of Eq. 5.3.10 must also hold in the GG-stable region A.

Equation 5.3.10 and the inequalities of Eq. 5.3.13 are constraints on the moments of inertia for which gravity gradient attitude stability can exist. At first it appears that any arbitrary choice of moments of inertia satisfying these inequalities will result in an attitude-stable system. Unfortunately, the inequality $I_y < I_x + I_z$ of Eq. 5.3.10, which applies to the stability subregion A in Figure 5.3.1, is a difficult constraint to accommodate from a structural standpoint (see Runavot 1980, William and Osborn 1987). The reason is as follows. Suppose that an angular motion with a small amplitude is permitted about the $\mathbf{Y}_B$ body axis, despite the existence of external disturbances. In this case, $I_x - I_z$ must be as large as possible (see Eq. 5.3.17). Let $I_x = 100$ kg-m^2 and $I_z = 10$ kg-m^2. In this case, the constraint on I_y will be $110 > I_y > 100$, which may be difficult to realize structurally.

Constraints on the Moments of Inertia in Subregion B   In region III, $\sigma_x < 0$ and $\sigma_z < 0$. According to the definitions in Eq. 5.3.2 and the inequality of Eq. 5.3.4, it follows that

$$I_x > I_z > I_y. \tag{5.3.14}$$

5.3 / Gravity Gradient Attitude Control

In subregion B we also have the inequality $I_x < I_y + I_z$. In the following section we analyze stability conditions for some singular cases.

5.3.2 Time-Domain Behavior of a Purely Passive GG-Stabilized Satellite

To find the attitude time response of a passive GG-stabilized satellite, we shall use the linearized Eqs. 5.3.1. With this relieving condition, the time response is easily found with the aid of Laplace transforms.

Time Response about the Y_B Pitch Axis

The motion of the satellite about the pitch axis depends on the initial conditions of the pitch angle θ and its derivative, and also on the external disturbances T_{dy}. In terms of the Laplace variable "s" we have

$$\theta(s) = \frac{T_{dy}}{I_y s(s^2 + 3\omega_o^2 \sigma_y)} + \frac{s\theta(0) + \dot{\theta}(0)}{s^2 + 3\omega_o^2 \sigma_y}. \tag{5.3.15}$$

The value of σ_y depends on I_x and I_z. We shall examine several cases.

Case A: For $I_x < I_z$, σ_y will be negative and one of the roots of Eq. 5.3.15 will be unstable. The pitch angle will diverge exponentially with time.

Case B: For $I_x = I_z$, which is the neutral case of stability, σ_y will be zero and the time response becomes

$$\theta(t) = \theta(0) + \dot{\theta}(0)t + \frac{T_{dy}t^2}{2I_y}. \tag{5.3.16}$$

Case C: For $I_x > I_z$, σ_y will be positive, and an oscillatory motion is to be expected for an external disturbance T_{dy}:

$$\theta(t) = \frac{T_{dy}}{I_y 3\omega_o^2 \sigma_y}[1 - \cos(\omega_o \sqrt{3\sigma_y} t)] = C[1 - \cos(\omega_0 \sqrt{3\sigma_y} t)], \tag{5.3.17}$$

where $C = T_{dy}/[3\omega_o^2(I_x - I_z)]$. The time behavior is a biased harmonic motion, with a constant average level of amplitude C. The frequency of oscillation depends on the relative values of I_x, I_y, and I_z, and also on the orbital rate ω_o. The amplitude of oscillation depends on the external disturbance T_{dy}, and is inversely proportional to the difference $I_x - I_z$. This means that the only way to limit the amplitude of oscillation is by choosing appropriate values for the satellite's moments of inertia. The coefficient of the s^1 term is zero in the determinant of Eq. 5.3.15, so the harmonic motion will be undamped. In the design of a GG-stabilized satellite, it will be necessary to add some passive or active damping.

Time Response in the X_B-Z_B Plane

To solve Eqs. 5.3.5 in the time domain, we take the Laplace transforms with initial conditions. This leads to the following equations:

$$(s^2 + 4\omega_o^2 \sigma_x)\phi - s\omega_o(1 - \sigma_x)\psi = \frac{T_{dx}}{I_x} + s\phi_0 + \dot{\phi}_0 - \omega_o(1 - \sigma_x)\psi_0,$$

$$(s^2 + \omega_o^2 \sigma_z)\psi + s\omega_o(1 - \sigma_z)\phi = \frac{T_{dz}}{I_z} + \omega_o(1 - \sigma_z)\phi_0 + s\psi_0 + \dot{\psi}_0, \tag{5.3.18}$$

where $\phi_0, \dot{\phi}_0, \psi_0, \dot{\psi}_0$ are the initial conditions. The equations can be rewritten in matrix form as

$$\begin{bmatrix} s^2+4\omega_o^2\sigma_x & -s\omega_o(1-\sigma_x) \\ s\omega_o(1-\sigma_z) & s^2+\omega_o^2\sigma_z \end{bmatrix} \begin{bmatrix} \phi(s) \\ \psi(s) \end{bmatrix} = \begin{bmatrix} T_{dx}/I_x + s\phi_0 + \dot{\phi}_0 - \omega_o(1-\sigma_x)\psi_0 \\ T_{dz}/I_z + \omega_o(1-\sigma_z)\phi_0 + s\psi_0 + \dot{\psi}_0 \end{bmatrix}. \quad (5.3.19)$$

The solutions of $\phi(s)$ and $\psi(s)$ are

$$\begin{bmatrix} \phi(s) \\ \psi(s) \end{bmatrix} = \frac{1}{\Delta(s)} \begin{bmatrix} s^2+\omega_o^2\sigma_z & s\omega_o(1-\sigma_x) \\ -s\omega_o(1-\sigma_z) & s^2+4\omega_o^2\sigma_x \end{bmatrix}$$
$$\times \begin{bmatrix} T_{dx}/I_x + s\phi_0 + \dot{\phi}_0 - \omega_o(1-\sigma_x)\psi_0 \\ T_{dz}/I_z + \omega_o(1-\sigma_z)\phi_0 + s\psi_0 + \dot{\psi}_0 \end{bmatrix}, \quad (5.3.20)$$

where $\Delta(s) = s^4 + s^2\omega_o^2[3\sigma_x + 1 + \sigma_x\sigma_z] + 4\omega_o^4\sigma_x\sigma_z$.

Time-domain analysis in the X_B-Z_B plane is more complicated because the determinant in Eq. 5.3.20 is of the fourth order. Hence we shall first analyze some singular and simpler cases for stability; later we will attempt to solve the general case in the time domain.

Symmetrical Case: $I_x = I_y$ In this special case $\sigma_z = 0$, and the determinant in Eq. 5.3.20 becomes simpler: $\Delta'(s) = s^2[s^2 + \omega_o^2(3\sigma_x + 1)]$. The two integrators outside the brackets indicate neutral stability. However, for stability of the remaining part of the determinant, the roots of the second-order term must be imaginary, which means that $3\sigma_x + 1 = 3(I_y - I_z)/I_x + 1 > 0$. In terms of the moments of inertia (and since, by definition, $I_x = I_y$), the condition for stability of the second term in brackets becomes

$$\frac{I_z}{I_x} = \frac{I_z}{I_y} < \frac{4}{3}. \quad (5.3.21)$$

If the inputs are disturbances of constant magnitude, then

$$\phi(s) = \frac{T_{dx}(s^2+\omega_o^2\sigma_z)}{I_x s^3[s^2+(3\sigma_x+1)\omega_o^2]} + \frac{T_{dz}(1-\sigma_x)\omega_o}{I_z s^2[s^2+(3\sigma_x+1)\omega_o^2]},$$
$$\psi(s) = \frac{-\omega_o T_{dx}(1-\sigma_z)}{I_x s^2[s^2+(3\sigma_x+1)\omega_o^2]} + \frac{T_{dz}(s^2+4\omega_o^2\sigma_x)}{I_z s^3[s^2+(3\sigma_x+1)\omega_o^2]}. \quad (5.3.22)$$

Both $\phi(t)$ and $\psi(t)$ will have divergent and oscillatory terms in their time response. Hence, the symmetrical satellite with $I_x = I_y$ cannot be passively stabilized.

Symmetrical Axis in the Direction of Satellite Motion: $I_y = I_z$ In this singular case, $\sigma_x = 0$ and $\Delta(s) = s^2[s^2 + \omega_o^2]$. As in the previous case, for constant-value disturbances about the X_B and Z_B axes, the yaw and the roll Euler angles will have divergent and oscillatory motions in their time responses.

General Case: Time Behavior of a GG Stable Spacecraft in Region I or III
As stated in Section 5.3.1 and as shown in Figure 5.3.1, both σ_x and σ_z are positive in region I. Because there are no damping factors in the determinant of Eq. 5.3.20 (the coefficients of s^1 and s^3 are null) and since the inequalities of Eq. 5.3.13 or Eq. 5.3.14 must be satisfied for stable gravity gradient attitude control, the determinant will consist of two oscillatory roots:

5.3 / Gravity Gradient Attitude Control

$$\Delta(s) = (s^2 + \omega_1^2)(s^2 + \omega_2^2) = s^4 + (\omega_1^2 + \omega_2^2)s^2 + \omega_1^2\omega_2^2; \tag{5.3.23}$$

ω_1 and ω_2 can be found by equating terms in the determinant of Eq. 5.3.20.

For constant disturbances T_{dx} and T_{dz} and in accordance with Eq. 5.3.20, the time responses of $\phi(t)$ and $\psi(t)$ can be derived from the inverse Laplace transforms of

$$\phi(s) = \frac{T_{dx}(s^2 + \omega_o^2\sigma_z)}{I_x s(s^2 + \omega_1^2)(s^2 + \omega_2^2)} + \frac{T_{dz}\omega_o(1-\sigma_x)}{I_z(s^2 + \omega_1^2)(s^2 + \omega_2^2)} \quad \text{and}$$

$$\psi(s) = \frac{-T_{dx}\omega_o(1-\sigma_x)}{I_x(s^2 + \omega_1^2)(s^2 + \omega_2^2)} + \frac{T_{dz}(s^2 + 4\omega_o^2\sigma_x)}{I_z s(s^2 + \omega_1^2)(s^2 + \omega_2^2)}. \tag{5.3.24}$$

The time responses of both $\psi(t)$ and $\phi(t)$ will consist of two harmonic terms with natural frequencies ω_1 and ω_2 superimposed on a bias of constant magnitude. The value of this bias will be calculated in the next section.

Equations 5.3.24 hold for small disturbances and initial attitude conditions, because they emerge from the linearized Eqs. 4.8.9. In practice, for the exact physical model, the time-domain simulation will use a set of nonlinear equations for which, in general, a closed-form solution does not exist.

EXAMPLE 5.3.1 In this example, a small satellite is assumed with moments of inertia $I_x = 6$, $I_y = 8$, $I_z = 4$ kg-m^2. The s/c is moving in a circular orbit of altitude $h = 800$ km. In order to achieve small attitude errors despite external disturbances about the $\mathbf{Y_B}$ axis - which is the primary disturbance acting on the satellite owing to aerodynamic forces in its direction of motion (see Example 2.7.1) - a mechanical boom has been extended along the $\mathbf{Z_B}$ axis, so that the moments of inertia about the $\mathbf{X_B}$ and $\mathbf{Y_B}$ axes are increased to $I_x = 80$, $I_y = 82$ kg-m^2. The time response for an initial condition of $\psi(0) = 5°$ is shown in Figure 5.3.2. For this orbit, $\omega_o = 0.00104$ rad/sec.

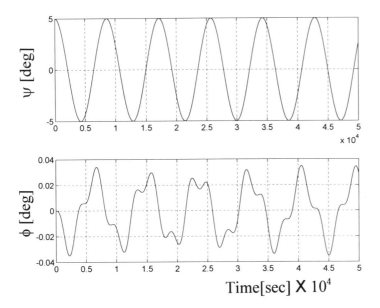

Figure 5.3.2 Time behavior for the initial condition $\psi(0) = 5°$.

From Eqs. 5.3.2 we have $\sigma_x = (82-4)/80 = 0.975$ and $\sigma_z = (82-80)/4 = 0.5$. We can compute ω_1 and ω_2 from Eq. 5.3.20 and Eq. 5.3.23 to find that: $\omega_1 = 1.9784\omega_o = 0.002054$ rad/sec, with a time period of $T_1 = 3,059$ sec; $\omega_2 = 0.7058\omega_o = 0.0007327$ rad/sec with $T_2 = 8,575$ sec. From Eq. 5.3.20 with $T_{dx} = T_{dz} = 0$, for an initial condition of $\psi(0)$ only we have

$$\phi(s) = -\omega_o^2 \sigma_z(1-\sigma_x)\psi(0)/\Delta(s),$$
$$\psi(s) = s[s^2 + \omega_o^2(1+\sigma_x\sigma_z + 3\sigma_x - \sigma_z)]\psi(0)/\Delta(s). \quad (5.3.25)$$

As we will see in the next section, achieving a low sensitivity to disturbances about the Y_B axis requires that we choose $I_x \gg I_z$. In Eqs. 5.3.24, we immediately perceive that the sensitivity of ψ to disturbances about the Z_B axis is greater than its sensitivity to disturbances about the X_B axis, by a factor of I_x/I_z (about 10 to 20). (Because of the inequality of Eq. 5.3.10, this is not true for the sensitivity of ϕ.)

Another important fact concerning GG stability about the Z_B axis can be deduced from Eqs. 4.8.8 and the linearized Eqs. 4.8.9: the gravity gradient moments about the Z_B axis are almost null for a nadir-pointing satellite ($\theta, \phi \approx 0$). This means that it is very difficult to GG-stabilize the Z_B axis against initial conditions in $\psi(0)$ and $\dot{\psi}(0)$ without active damping.

Returning to Eqs. 5.3.25 with the data of our present example, we find:

$$\phi(s) = 0.00366\omega_o \left[\frac{1}{s^2 + (1.9784\omega_o)^2} - \frac{1}{s^2 + (0.7058\omega_o)^2} \right]\psi(0),$$

$$\psi(s) = \frac{s[s^2 + 3.9125\omega_o^2]\psi(0)}{[s^2 + (1.9784\omega_o)^2][s^2 + (0.7058\omega_o)^2]} \approx \frac{s\psi(0)}{s^2 + (0.7058\omega_o)^2}.$$

The time responses become

$$\phi(t) = [0.001849\sin(1.978\omega_o t) - 0.005186\sin(0.7058\omega_o t)]\psi(0),$$
$$\psi(t) \approx [\cos(0.7058\omega_o t)]\psi(0). \quad (5.3.26)$$

The time-domain behavior of the system expressed by Eqs. 5.3.1 is shown in Figure 5.3.2. They agree fairly well with Eqs. 5.3.26.

A complete 6-DOF simulation has been carried out to show the time responses of the Euler angles due to constant disturbances T_{dx} or T_{dz}. The results are shown in Figures 5.3.3 and 5.3.4, respectively. According to Eqs. 5.3.24, with $T_{dz} = 0$ we have

$$\phi(s) = \frac{T_{dx}(s^2 + 0.5\omega_o^2)}{I_x s(s^2 + 1.9784^2\omega_o^2)(s^2 + 0.7058^2\omega_o^2)} \approx \frac{T_{dx}}{I_x s(s^2 + 1.9784^2\omega_o^2)},$$

$$\psi(s) = \frac{-T_{dx}\omega_o 0.5}{I_x(s^2 + 1.9784^2\omega_o^2)(s^2 + 0.7058^2\omega_o^2)}.$$

Taking the inverse Laplace transform for $T_{dx} = 10^{-5}$ N-m gives the following time responses:

$$\phi(t) = 1.699[1-\cos(1.9784\omega_o t)] \text{ deg},$$
$$\psi(t) = 0.4917\sin(1.9784\omega_o t) - 1.378\sin(0.7058\omega_o t) \text{ deg}.$$

The time responses for the complete 6-DOF simulation are shown in Figure 5.3.3.

5.3 / Gravity Gradient Attitude Control

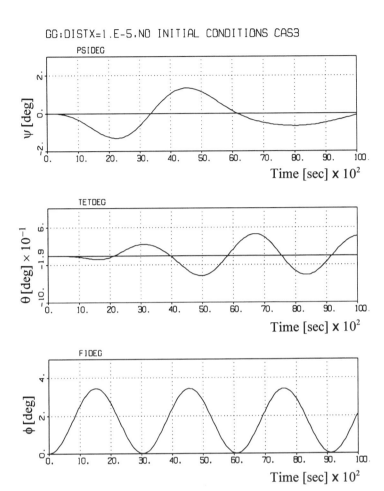

Figure 5.3.3 Time behavior of $\phi(t)$, $\theta(t)$, and $\psi(t)$ for T_{dx}.

For disturbances T_{dz} about the $\mathbf{Z}_B$ axis (and with $T_{dx} = 0$), we find:

$$\phi(s) = \frac{T_{dz}\omega_o}{I_x(s^2+1.9784^2\omega_o^2)(s^2+0.7058^2\omega_o^2)} \frac{(80-82+4)}{4},$$

$$\psi(s) = \frac{T_{dz}(s^2+4\omega_o^2 0.975)}{I_z s(s^2+1.9784^2\omega_o^2)(s^2+0.7058^2\omega_o^2)} \approx \frac{T_{dz}}{I_z s(s^2+0.7058^2\omega_o^2)}.$$

The time response of $\psi(t)$ is of importance in this case, because I_z is comparatively much smaller than the other moments of inertia. Hence, a much larger sensitivity to disturbances is to be expected. It is easily found that, for a disturbance of $T_{dz} = 10^{-5}$ N-m, the time response will be

$$\psi(t) = 267[1-\cos(0.7058\omega_o t)] \text{ deg}.$$

We conclude that GG stabilization does not provide enough immunity against disturbances about the $\mathbf{Z}_B$ axis. Moreover, a large change of the Euler yaw angle leads

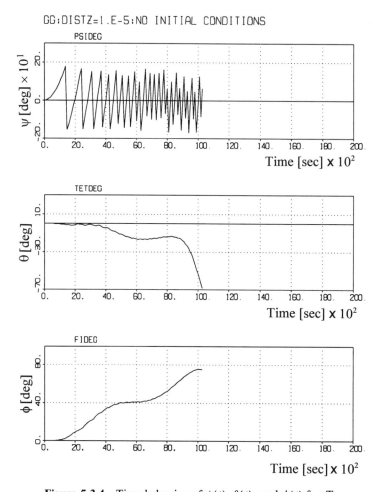

Figure 5.3.4 Time behavior of $\phi(t)$, $\theta(t)$, and $\psi(t)$ for T_{dz}.

to a slow interchange of the I_x and I_y between them: I_x becomes larger than I_y, and instability of the GG system follows. This phenomenon is seen in Figure 5.3.4, which shows the time responses for this example. Fortunately, disturbance torques about the $\mathbf{Z}_B$ axis are generally smaller than those about the other two body axes.

5.3.3 *Gravity Gradient Stabilization with Passive Damping*

As we have seen in previous sections, the gravity gradient stabilizes the s/c in the sense that there remains an amplitude-bounded harmonic angular motion about an average bias value. In order to obtain a useful system, the harmonic oscillatory angular motion must be damped and reduced to a minimum. The existence of external disturbances initiates the oscillatory motion and, with passive damping, the time to appreciably decrease the oscillatory motion might be very long. However,

5.3 / Gravity Gradient Attitude Control

since passive damping equipment is comparatively inexpensive, it is still occasionally used (Clopp and Osborn 1987, Hughes 1988).

Passive Dampers

Several kinds of passive dampers are mentioned in the literature (see Fleeter and Warner 1989, Hughes 1988). A short description of each kind follows.

Point-mass damper: The principle of this very simple damper is to use a mass-spring/dashpot damper mounted inside the satellite. The energy dissipated in the damper helps to damp the oscillatory motion of the satellite. For a complete analytical treatment of this damping system, see Kaplan (1976) and Hughes (1988).

Dampers mounted on external spring's boom: The extended boom that increases the moments of inertia in order to achieve desired GG stabilization can be mounted on a large external spring. On the tip of the spring, which has a helical form, a fluid damper can be mounted to increase the internal energy dissipation inside the satellite so that a damping effect is produced. The analytical treatment of this kind of damper is quite complicated because of the nonlumped character of the spring.

Magnetic hysteresis rod damper: If a rod of magnetically permeable material is located inside the satellite, then the angular motion of the satellite with respect to the earth's magnetic field will induce magnetic hysteresis losses. Magnetic hysteresis rods have been used on many satellites. Unfortunately, the strength of the earth's magnetic field is proportional to $1/R^3$, so this mode of damping is effective only for rather low orbits (see Fleeter and Warner 1989).

Damping by boom articulation: Many GG designs are based on a suitable mechanism at the hinges that join the augmenting booms to the satellite's body (Fleeter and Warner 1989). This mechanism usually has two degrees of freedom and so provides damping about two axes; this enables damping of both the pitch and the roll angles. Damping of the yaw axis is much more difficult to achieve.

Wheel damper: A wheel, immersed in a container holding a viscous fluid, can be effective in damping the angular motion of a satellite. If we align the axis of rotation of the free wheel with (say) the $\mathbf{Y}_B$ axis then the pitch oscillatory motion can be damped. Likewise, for a wheel whose axis of rotation is aligned with the $\mathbf{X}_B$ axis, the roll oscillatory motion can be damped.

Design of a GG-Stabilized Satellite Based on a Wheel Damper

In order to understand the effectiveness of such passive dampers, a complete analytical treatment of the wheel damper follows.

Dynamic Equations of the Damper First, we shall write the dynamic equations for a satellite with a GG control system that uses a passive wheel for damping purposes (see Bryson 1983). The third of Eqs. 5.3.1 can be augmented with the expression for the rotational dynamics of the wheel damper (Eq. 4.8.14b) as follows:

$$\ddot{\theta} I_y + 3\omega_o^2 (I_x - I_z)\theta + \dot{\omega}_w I_w = T_{dy},$$
$$I_w \dot{\omega}_w = D(\dot{\theta} - \omega_w) = D\Omega_w, \tag{5.3.27}$$

where D is the damping coefficient of the fluid in which the wheel is immersed, I_w is the moment of inertia of the wheel, and Ω_w is the angular velocity of the wheel relative to the body of the satellite. The system of equations can be written in matrix form as

$$\begin{bmatrix} s^2 + 3\omega_o^2 \sigma_y & s(I_w/I_y) \\ -Ds & sI_w + D \end{bmatrix} \begin{bmatrix} \theta(s) \\ \omega_w(s) \end{bmatrix} = \begin{bmatrix} T_{dy}/I_y + s\theta(0) + \dot{\theta}(0) \\ -D\theta(0) \end{bmatrix}. \qquad (5.3.28)$$

The characteristic equation is

$$\Delta(s) = I_w s^3 + D\left(\frac{I_w}{I_y} + 1\right)s^2 + 3I_w \omega_o^2 \sigma_y s + 3D\omega_o^2 \sigma_y, \qquad (5.3.29)$$

and the solution for a step disturbance T_{dy} is

$$\begin{bmatrix} \theta(s) \\ \omega_w(s) \end{bmatrix} = \frac{1}{\Delta(s)} \begin{bmatrix} sI_w + D & -s(I_w/I_y) \\ Ds & s^2 + 3\omega_o^2 \sigma_y \end{bmatrix} \begin{bmatrix} T_{dy}/(I_y s) + s\theta(0) + \dot{\theta}(0) \\ -D\theta(0) \end{bmatrix}. \qquad (5.3.30)$$

The solution for the pitch angle becomes

$$\theta(s) = \frac{(sI_w + D)[T_{dy}/(I_y s) + s\theta(0) + \dot{\theta}(0)] + s(I_w/I_y)D\theta(0)}{\Delta(s)}, \qquad (5.3.31)$$

and the steady-state error for the constant disturbance T_{dy} is

$$\theta_{ss} = \lim_{s \to 0} \frac{sT_{dy}(sI_w + D)}{I_y s \Delta(s)} = \frac{T_{dy}}{3\omega_o^2(I_x - I_z)}. \qquad (5.3.32)$$

Equation 5.3.32 shows that the steady-state error in the pitch angle is inversely proportional to the difference $I_x - I_z$. For known maximum external disturbances, given a specified acceptable error in θ_{ss} and an orbital rate ω_o, the needed difference in moments of inertia is easily calculated. Satellites are generally built with homogeneously distributed masses, which means that the three principal moments of inertia are of the same order of magnitude and differ by a factor of only two or three. In order to achieve a large difference $I_x - I_z$ in a GG-stabilized s/c, I_x must be much larger than I_z. Consequently, a moment of inertia–augmenting boom must be added to the satellite.

EXAMPLE 5.3.2 A basic satellite has the following moments of inertia: $I_x = 6$, $I_y = 8$, $I_z = 4$ kg-m². For a circular 800-km-altitude orbit, $\omega_o = 0.001038$ rad/sec. According to Eq. 5.3.32, for an expected constant disturbance of 10^{-5} N-m the error in pitch will be $\theta_{ss} = 88.67°$, which of course is unacceptable. The only way to decrease the steady-state error in pitch is to increase I_x. If a boom augmenting I_x (and also, inevitably, I_y) by $\Delta I = 74$ kg-m² is appended to the satellite, aligned with the $\mathbf{Z}_B$ axis, then I_x will increase to $I_x = 80$ kg-m² ($= 20I_z$). The steady-state error will decrease to $\theta_{ss} = 2.33°$, which is an acceptable error.

In Example 5.3.2, without the damping wheel discussed previously, an initial harmonic motion (due to initial conditions and transients) will oscillate constantly about the steady-state pitch error.

Design of the Damper The characteristic equation of the system, Eq. 5.3.29, can be arranged as follows using the root-locus design method (see Bryson 1983):

5.3 / Gravity Gradient Attitude Control

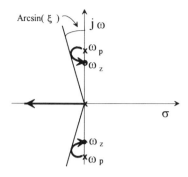

Figure 5.3.5 Root-locus representation of the damper and satellite dynamics.

$$1 + \frac{D\left(\frac{1}{I_w} + \frac{1}{I_y}\right)\left(s^2 + \frac{3\omega_o^2 \sigma_y}{1 + I_w/I_y}\right)}{s(s^2 + 3\omega_o^2 \sigma_y)} = 1 + \frac{D\left(\frac{1}{I_w} + \frac{1}{I_y}\right)(s^2 + s_z^2)}{s(s^2 + s_p^2)} = 1 + L(s), \quad (5.3.33)$$

where $s_p = j\omega_o\sqrt{3\sigma_y}$ and $s_z = s_p/\sqrt{1 + I_w/I_y}$. It follows immediately that $s_p > s_z$. In Eq. 5.3.33, the root-locus gain is $K = D(1/I_w + 1/I_y)$. The complete feedback system consists of one integrator, two imaginary zeros, and two imaginary poles, all on the imaginary axis. The root-locus plot is shown in Figure 5.3.5.

Even for a very large moment of inertia of the wheel (compared to the moments of inertia of the satellite body only) of $I_w \approx 1$ kg-m^2, and with $I_y = 82$ kg-m^2 for the example at hand, the imaginary zeros and poles are located quite close to each other. This means (see Figure 5.3.5) that the maximum achievable damping factor ξ will be very low, with no dependence on D. The value of D must be chosen such that, for the given moments of inertia I_w and I_y, the maximum damping coefficient may be obtained. A fast "cut-and-try" procedure using the root-locus method allows finding the damping factor D that will maximize the damping coefficient of the GG attitude control system. The intermediate results are shown in Table 5.3.1, with the selected $I_y = 82$ kg-m^2, for which $\omega_p = 1.91 \times 10^{-3}$ rad/sec.

Table 5.3.1 shows that, with a moment of inertia of the wheel $I_w = 1$ kg-m^2, the maximum obtained damping coefficient is $\xi \approx 0.0031$ with $D = 0.002$. This is a very low damping factor. Let us designate as $T_{0.1}$ the time needed to damp the initial oscillation to 1/10 of its value; we can write $e^{-\xi \omega_p T_{0.1}} = 0.1$ and $\xi \omega_p T_{0.1} = \xi \omega_o \sqrt{3\sigma_y} T_{0.1} = 2.3$. In the case of Example 5.3.2, $\sigma_y = (80 - 4)/82 = 0.9268$. Since $T_{orb} = 2\pi/\omega_o$, we have $\xi 2\pi 1.677 (T_{0.1}/T_{orb}) = 2.3$, so that $T_{0.1}/T_{orb} = 70.4$. In other words, about 70 orbit periods are necessary to damp initial oscillatory motion to 1/10 of its value. A 6-DOF simulation of this design example is shown in Figure 5.3.6.

Table 5.3.1 *Design stages of the passive damper*

$10^3 \, D$	2.0	1.0	0.75	1.0	2.0	3.0
I_w [kg-m^2]	0.4	0.4	0.4	1.0	1.0	1.0
$10^3 \, K$	5.02	2.51	1.88	1.01	2.02	3.04
$10^3 \, \xi$	0.75	1.1	1.2	2.6	3.1	2.6

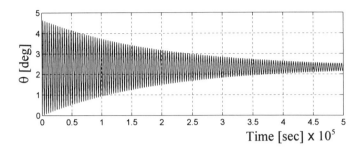

Figure 5.3.6 Damping of the oscillatory angular motion about the Y_B axis.

5.3.4 Gravity Gradient Stabilization with Active Damping

We have seen in the last section that stabilization about the Z_B axis is quite difficult owing to the absence of adequate gravity gradient moments about this axis (see also Figure 5.3.4). Some sophisticated boom structures have been proposed to magnify the GG moments about the Z_B axis, but they did not fulfill expectations. A more logical way to obtain a better stabilization of the Z_B axis against external disturbances is to use magnetic torqrods, which – by interacting with the earth's magnetic field – can produce the needed moments to counteract the disturbances. If **B** is the earth's magnetic field vector and if **M** is the magnetic dipole value produced by energizing the current-carrying coils inside the satellite, then the mechanical moment exerted on the body will be

$$\mathbf{T} = \mathbf{M} \times \mathbf{B} = \begin{bmatrix} \mathbf{i} & \mathbf{j} & \mathbf{k} \\ M_x & M_y & M_z \\ B_x & B_y & B_z \end{bmatrix}. \quad (5.3.34)$$

Suppose we wish to achieve a control torque vector $\mathbf{T}_c$. The vector product of Eq. 5.3.34 can be put in the form

$$\begin{bmatrix} T_{cx} \\ T_{cy} \\ T_{cz} \end{bmatrix} = \begin{bmatrix} 0 & B_z & -B_y \\ -B_z & 0 & B_x \\ B_y & -B_x & 0 \end{bmatrix} \begin{bmatrix} M_x \\ M_y \\ M_z \end{bmatrix}. \quad (5.3.35)$$

The value of $\mathbf{T}_c$ is computed according to Eq. 5.2.1, assuming that the attitude errors can be measured. The problem we are left with is finding the magnetic control dipole to be produced in the satellite with the three magnetic torqrods M_x, M_y, M_z. We need to find the inverse of Eq. 5.3.35, but the inverse does not exist because the matrix is singular. This means that it is impossible to achieve control of the three body axes by means of the created magnetic dipole vector in the satellite. (See also Section 7.4.) The physical reason is that Eq. 5.3.34 is a vector product, so no moment can be achieved about the direction of **B**. Control about two body axes should, however, be possible. In our case, the Y_B axis will be controlled by the gravity gradient moments about the Y_B axis, and the satellite can be stabilized about the X_B and Z_B axes

5.3 / Gravity Gradient Attitude Control

with the aid of two magnetic torqrods aligned along the X_B and Z_B body axes. In this terminology, $T_{cx} = -B_y M_z$ and $T_{cz} = B_y M_x$. Finally,

$$M_z = -T_{cx}/B_y \quad \text{and} \quad M_x = T_{cz}/B_y. \tag{5.3.36}$$

Using Eqs. 5.3.36, we can actively control the X_B and Z_B axes while the Y_B axis is purely GG-stabilized. Unfortunately, calculation of T_{cx} and T_{cz} necessitates attitude measurements or estimation of the roll (ϕ) and the yaw (ψ) Euler angles. For this simple control law, the earth's magnetic field must be known or measured with the aid of a magnetometer. As discussed in Martel, Pal, and Psiaki (1988), it is possible to estimate the three Euler angles with the aid of a three-axis magnetometer, and this equipment is quite simple and comparatively cheap.

EXAMPLE 5.3.3 Using the same satellite as described in Example 5.3.1, a gravity-gradient stabilization of the satellite is ameliorated by using two magnetic torqrods. Figure 5.3.7 shows the attitude errors. The s/c is well stabilized about the Z_B axis,

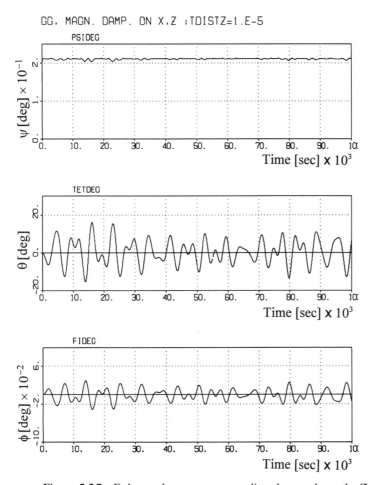

Figure 5.3.7 Euler angle responses to a disturbance about the Z_B axis.

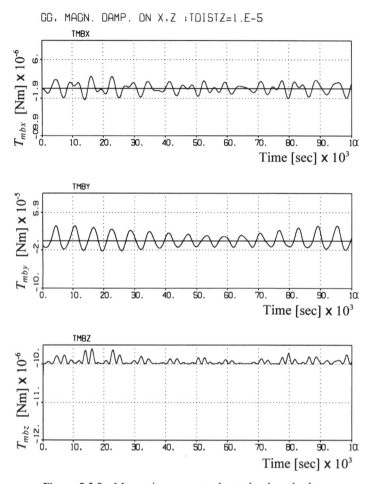

Figure 5.3.8 Magnetic moments about the three body axes induced by T_{dz}.

despite the high disturbance about this axis of $T_{dz} = 10^{-5}$ N-m. Figure 5.3.8 shows the moments produced with the magnetic torqrods. However, a part of the T_{dz} has been transferred to the Y_B axis (see Eq. 5.3.35). The magnetically induced moment is $T_{mby} = -B_z M_x + B_x M_z$. In steady state, $M_x = T_{dz}/B_y$ and $M_z = T_{dx}/B_y$. Hence, $T_{mby} = -B_z T_{dz}/B_y + B_x T_{dx}/B_y$. This is an additional disturbance that must be taken into consideration when designing the Y_B axis damper and calculating the needed boom moment of inertia. In spite of the disturbance about the Z_B axis, the yaw and roll Euler angles are now well controlled. However, the Y_B axis, which is only passively damped, responds to this produced moment as in Figure 5.3.7. The additional moment T_{mby} is shown in Figure 5.3.8. Because the earth's magnetic field components are harmonic with the orbital frequency (see Section 7.4.2), the induced disturbance about the Y_B axis is not constant and so cannot be damped very efficiently with a passive damper: the disturbance periodically induces oscillations about the Y_B axis, and there is not enough time to damp them between orbital periods. Martel et al. (1988) have suggested a novel control technique to alleviate this problem; this technique is described in the next section.

5.3.5 GG-Stabilized Satellite with Three-Axis Magnetic Active Damping

There is a possibility, despite the theoretical difficulties treated in the previous section, of damping all three body axes by active control with three magnetic torqrods aligned along the principal body axes. In this scenario, the *principle of perpendicularity* is used (see also Section 7.5.2).

As already mentioned, Eq. 5.3.35 has no inverse, and M_x, M_y, M_z cannot be found exactly. Let us return to the basic magnetic moment equation (Eq. 5.3.34), and multiply (using vector product) both sides by **B**. Using the scalar identity for the vector product, we have

$$\mathbf{B} \times \mathbf{T}_c = \mathbf{B} \times (\mathbf{M} \times \mathbf{B}) = (\mathbf{B} \cdot \mathbf{B})\mathbf{M} - (\mathbf{B} \cdot \mathbf{M})\mathbf{B}. \tag{5.3.37}$$

Suppose that the magnetic dipole vector **M**, applied inside the satellite, it always perpendicular to the earth's magnetic vector **B** (this is not actually true). If this condition is satisfied, then the scalar dot product **B·M** equals zero and Eq. 5.3.37 reduces to

$$\mathbf{M} = (\mathbf{B} \times \mathbf{T}_c)/B^2, \tag{5.3.38}$$

where B is the norm of **B**. The basic stability is achieved by the gravity gradient attributes of the extended boom, with damping provided by the active damping scheme described here. Since this control law is not exact, any potential design must be carefully checked by numerous simulations under different disturbance conditions. Moreover, since the earth's magnetic field characteristics for variously inclined orbits might be quite different, the design of the active magnetic damper should reflect these differences. Accuracies of the order of 2° are achievable with this technique.

EXAMPLE 5.3.4 In this example, the satellite has the following moments of inertia: $I_x = 50$, $I_y = 52$, $I_z = 3$ kg-m². The satellite must be capable of withstanding external disturbances of the order of $T_{dx} = T_{dz} = 10^{-6}$ N-m and $T_{dy} = 10^{-5}$ N-m. A simple PD (proportional–derivative) controller is used for all axes to achieve for each a basically second-order closed loop with a natural frequency $\omega_n = 0.05$ rad/sec and a damping factor $\xi = 0.9$. The time responses of the attitude errors are shown in Figure 5.3.9, and those of the activated magnetic dipole vector **M** inside the satellite in Figure 5.3.10.

5.4 Summary

In this chapter we explored gravity gradient stabilization. The attitude accuracy that can be achieved is quite poor with purely passive damping, on the order of 10°–30°, but can be improved to the order of 2°–10° by adding active damping. Gravity gradient stabilization is feasible for relatively low-orbit satellites.

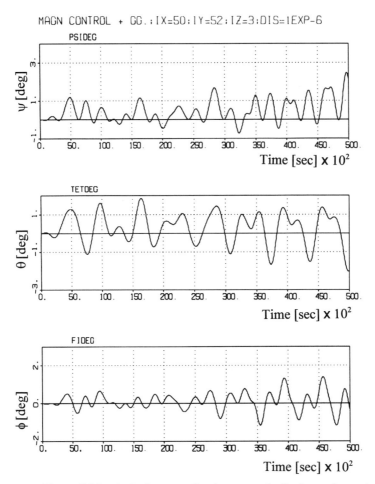

Figure 5.3.9 Attitude errors for the magnetically damped gravity gradient spacecraft.

References

Bryson, A. (1983), "Stabilization and Control of Spacecraft," Microfiche supplement to the *Proceedings of the Annual AAS Rocky Mountain Guidance and Control Conference* (5–9 February, Keystone, CO). San Diego, CA: American Astronautical Society.

Clopp, W., and Osborn, L. (1987), "Satellite Bus Design for the Multiple Satellite System," AIAA/DARPA Meeting on Lightweight Satellite Systems (4–6 August, Monterey, CA). Washington, DC: AIAA, pp. 105–15.

Fleeter, R., and Warner, R. (1989), "Guidance and Control of Miniature Satellites," Automatic Control in Aerospace, IFAC (Tsukuba, Japan). Oxford, UK: Pergamon, pp. 243–8.

Hughes, P. C. (1988), *Spacecraft Attitude Dynamics*. New York: Wiley.

Kaplan, M. (1976), *Modern Spacecraft Dynamics and Control*. New York: Wiley.

Martel, F., Pal, P., and Psiaki, M. (1988), "Active Magnetic Control System for Gravity Gradient Stabilized Spacecraft," Second Annual AIAA/USU Conference on Small Satellites (26–28 September, Logan, UT). Washington, DC: AIAA.

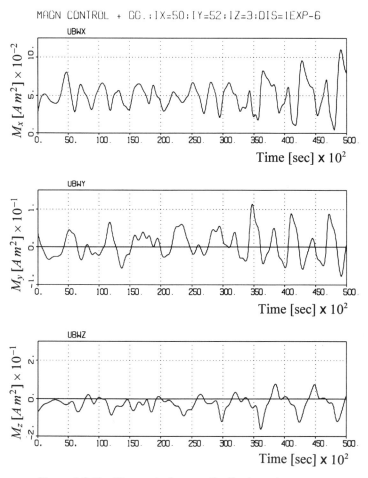

Figure 5.3.10 The control magnetic dipole vector.

Runavot, J. J. (1980), "Le Système de Commande d'Attitude et d'Orbite: Généralités et Définition du Système," in *Le Mouvement du Véhicule Spatial en Orbite*. Toulouse, France: Centre National d'Études Spatiales, pp. 909-28.

William, H. C., and Osborn, L. (1987), "Satellite Bus and Design for the Multiple Satellite System," AIAA/DARPA Meeting on Lightweight Satellite Systems (4-6 August, Monterey, CA). Washington, DC: AIAA.

CHAPTER 6

Single- and Dual-Spin Stabilization

6.1 Introduction

As discussed in Chapter 4, a body spinning about its major or minor axis will keep the direction of its spinning axis fixed with respect to the inertial space. This direction, according to Euler's moment equations of angular motion, will change only if external moments are applied about its center of mass and perpendicularly to the spin axis.

Almost all spacecraft employ the spin effect during part or all of their lifetime in space. Many satellites are spin-stabilized during the orbital maneuvering stage (e.g., in the transfer from the initial orbit to the final mission orbit) so that parasitic torque disturbances, produced by the high thrust of the apogee boost motor, do not appreciably change the nominal direction of the additional vector velocity ΔV imparted to the satellite. The dynamic attributes of spinning bodies are used also to stabilize satellites' attitude within the final mission orbit. Spin stabilization was used in the first communications satellites in the early sixties, and in a large number of modern satellites (see e.g. Fagg and MacLauchlan 1981, Fox 1986).

Single-spin attitude stabilization is a very simple concept from the perspective of attitude control, but it has some crucial drawbacks with respect to communication efficiency. Dual-spin three-axis attitude stabilization, which is an extension of the single-spin stabilization principle, alleviates the communication deficiency. We shall find conditions under which passive nutation damping is feasible for dual-spin stabilized spacecraft.

6.2 Attitude Spin Stabilization during the ΔV Stage

As already mentioned, a common way to stabilize the attitude of a s/c during orbital maneuvering (the ΔV stage) is to spin-stabilize the axis along which the propulsion thruster is aligned. Unfortunately, the direction of the applied thrust does not pass exactly through the center of mass (cm) of the satellite, and a parasitic torque results that tends to change the nominal attitude of the spin axis. Since the thruster is fixed to the satellite's body, change of the attitude of the satellite will induce an error in the application of ΔV, with the result that the required new orbit parameters will not be achieved (see Section 3.4.3). Moreover, the existence of a disturbing torque on the spinning body will induce a nutational motion about the spin axis, so that the added vector ΔV will have a time average that is different from the required nominal change in the velocity vector.

Suppose that the thrust **F** in Figure 6.2.1 is not aligned exactly with the nominal Z_B axis, so that an alignment error β remains between the thrust vector and the Z_B axis during the firing. Consequently, a transverse thrust component will induce a

6.2 / Attitude Spin Stabilization during the ΔV Stage

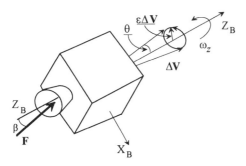

Figure 6.2.1 ΔV direction dispersion due to torque disturbances.

parasitic moment about the cm of the satellite, depending also on the distance d_{cm} between the cm and the intersection of the thrust vector with the Z_B axis. The level of the disturbing torque will be

$$T_d = F d_{cm} \sin(\beta). \tag{6.2.1}$$

Because of this torque, a nutational motion will be excited, and an average component $\epsilon \Delta V$ will be produced in a direction normal to the motion of the satellite. This $\epsilon \Delta V$ component will be lost for the linear increase in the velocity of the satellite in the desired nominal direction. The loss in velocity (with a consequent loss of fuel) will amount to:

$$\Delta V_{loss} = \Delta V [1 - \cos(\theta_{av})]. \tag{6.2.2}$$

The average nutation angle θ_{av} depends on the disturbing torque T_d (Eq. 6.2.1) and also on the initial angular momentum $I_z \omega_z = h_z$ imparted to the satellite. To compute θ_{av} we use Euler's moment equations of motion, Eqs. 4.5.1'. To linearize these equations, some practical assumptions are in order. With no moments about the Z_B axis, ω_z may be assumed to be constant during the entire ΔV period. To simplify the analysis, suppose also that the disturbing torque T_d acts only about the X_B axis and that the satellite is axisymmetric about the Z_B axis, $I_x = I_y$. With these assumptions, the third of Eqs. 4.5.1' leads to $\omega_z = n = \omega_z(0) = \text{const}$, and the solution of the first two of Eqs. 4.5.1' results in

$$I_x \omega_x(t) = h_x(t) = \frac{T_d}{\lambda} \sin(\lambda t), \tag{6.2.3}$$

$$I_y \omega_y(t) = I_x \omega_x(t) = h_y(t) = \frac{T_d}{\lambda} [1 - \cos(\lambda t)], \tag{6.2.4}$$

where $\lambda = \omega_z(0)(I_z - I_x)/I_x$.

The amplitude of the momentum component in the lateral X_B-Y_B plane of the satellite is

$$h_{xy}(t) = \sqrt{h_x^2 + h_y^2} = \frac{\sqrt{2} T_d}{\lambda} \sqrt{1 - \cos(\lambda t)} = \frac{\sqrt{2} T_d}{\omega_z (I_z/I_x - 1)} \sqrt{1 - \cos(\lambda t)}. \tag{6.2.5}$$

To find the nutation angle θ we use Eq. 4.6.1, which becomes

$$\tan(\theta) = \frac{h_{xy}}{h_z} = \frac{\sqrt{2}T_d}{\omega_z^2 I_z(I_z/I_x - 1)}\sqrt{1-\cos(\lambda t)} = \frac{2T_d}{\omega_z^2 I_z(I_z/I_x - 1)}\left|\sin\left(\frac{\lambda t}{2}\right)\right|. \quad (6.2.6)$$

We can also find the average value of the nutation angle. For small values of θ – which is a logical assumption, since we are interested in keeping a small nutation error in the ΔV stage – the nutation angle can be approximated as

$$\theta \approx \frac{2T_d}{\omega_z^2 I_z(I_z/I_x - 1)}\left|\sin\left(\frac{\lambda t}{2}\right)\right|. \quad (6.2.7)$$

From Eq. 6.2.5, θ is always positive because of the square-root radical; see also the nutation angle in Figures 6.2.2 and 6.2.3. After averaging $|\sin(\lambda t/2)|$ over one cycle, we obtain the final result for the averaged value of the nutation angle:

$$\theta_{av} = \frac{4}{\pi}\frac{T_d}{\omega_z^2 I_z(I_z/I_x - 1)}. \quad (6.2.8)$$

From Eqs. 6.2.6–6.2.8 it is apparent that θ is inversely proportional to ω_z^2, an important result that is demonstrated in the following example.

EXAMPLE 6.2.1 A satellite has the following moments of inertia: $I_x = I_y = 100$ kg-m^2; $I_z = 40$ kg-m^2. Suppose that a disturbance torque of 10 N-m acts on the satellite about the $\mathbf{X}_B$ body axis. Using Eq. 6.2.8, we find that if $\omega_z = \omega_z(0) = 10$ rad/sec then $\theta_{av} = 0.304°$, which is in good agreement with the time-simulation results shown in Figure 6.2.2.

If we decrease the spin velocity to $\omega_z = \omega_z(0) = 4$ rad/sec, Eq. 6.2.8 predicts an average nutation angle of $\theta_{av} = 1.9°$, also in good agreement with the simulation results in Figure 6.2.3.

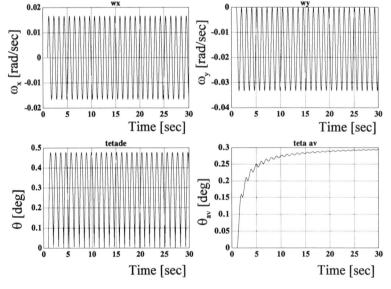

Figure 6.2.2 Time histories of $\omega_x(t)$, $\omega_y(t)$, $\omega_z(t)$, and $\theta_{av}(t)$ for $\omega_z(0) = 10$ rad/sec.

6.3 / Active Nutation Control

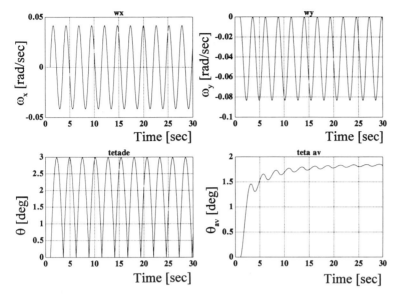

Figure 6.2.3 Time histories of $\omega_x(t)$, $\omega_y(t)$, $\omega_z(t)$, and $\theta_{av}(t)$ for $\omega_z(0) = 4$ rad/sec.

EXAMPLE 6.2.2 The linear velocity of a spin-stabilized satellite has been augmented with an ABM capable of $T = 400$ N. The thrust may deviate by $\beta = 0.8°$ from the nominal $\mathbf{Z}_B$ axis, thus causing a transverse torque disturbance of $T_d = 400 d_{cm} \sin(0.8°)$. If $d_{cm} = 1$ m, then $T_d = 5.58$ N-m. The moments of inertia of the satellite are $I_x = I_y = 500$ kg-m^2 and $I_z = 200$ kg-m^2. The loss in the increase of velocity due to the torque disturbances must be less than 0.5% of the nominal ΔV. What is the spin angular velocity ω_z to be imparted to the satellite in order to prevent the loss from being greater than 0.5%?

Solution From Eq. 6.2.2, limiting losses to less than 0.5% of ΔV requires maintaining a $\theta_{av} < 5.7°$ during the $\Delta \mathbf{V}$ stage. Using Eq. 6.2.8, we find that $\omega_z > 0.766$ rad/sec is necessary.

The angular velocity ω_z cannot be increased arbitrarily, because providing the satellite with a large angular momentum is accompanied by a larger quantity of fuel expenditure. Moreover, if the satellite is to be three-axis attitude-stabilized in its mission phase, the same angular momentum must be removed at the end of the $\Delta \mathbf{V}$ phase, with an additional quantity of fuel to be provided for that task. As we shall see in the following section, removing the initial angular momentum could consume a lot of fuel.

6.3 Active Nutation Control

Owing to geometrical constraints, a satellite in the transfer orbit stage spins about its minor axis (the axis with a minimum moment of inertia). As discussed in

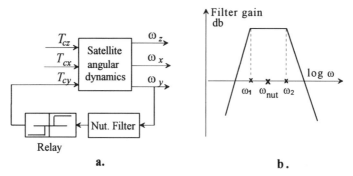

Figure 6.3.1 ANC system: (a) basic control system; (b) bandpass nutation filter.

Section 4.5.2, a body spinning about its maximum or minimum moment of inertia is dynamically stable in the absence of internal energy dissipation. However, if such dissipation does exist then a body spinning about its minor axis becomes dynamically unstable (Section 4.6.2). Active nutation control (ANC) is necessary in order to ensure attitude stability of the spinning satellite during the velocity vector change phase ($\Delta \mathbf{V}$).

Assuming that the body is spinning about its $\mathbf{Z}_B$ axis, a very simple and effective control logic for accomplishing ANC is to measure the components of the nutational angular motion and then to apply torques about the transverse axis $\mathbf{X}_B$ (or $\mathbf{Y}_B$, or both) proportional to the signs of ω_x or ω_y as shown in Figure 6.3.1.a. A filter is generally used to extract, from the measured angular rate components ω_x and/or ω_y, that part belonging to the nutation frequency: $\omega_{\text{nut}x}$ and/or $\omega_{\text{nut}y}$. The basic characteristics of such a filter are shown in Figure 6.3.1.b.

The ANC bandpass filter must be designed so that, despite the changes inherent in the nutation frequency ω_{nut}, the filter remains effective during the entire $\Delta \mathbf{V}$ stage. The nutation frequency may change appreciably during the apogee boost stage, because fuel consumption causes the moments of inertia of the satellite to decrease and hence the nutation frequency changes also. At least one differentiation is imperative in this filter in order to prevent constant biases, like those of the rate gyros measuring the angular velocity of the satellite, from causing augmentation of the nutation angle. The control law becomes:

$$T_{cx} = -F\Delta \,\text{sign}(\omega_{\text{nut}x}),$$
$$T_{cy} = -F\Delta \,\text{sign}(\omega_{\text{nut}y}). \quad (6.3.1)$$

In this equation, F is the level of the reaction thrust and Δ is the torque arm of the thruster.

Moreover, existing time delays and nonlinearities (such as dead zones) of the engineering hardware complicate the dynamics of the ANC system, and must be carefully taken into account in the final design stage. See Grasshoff (1968), Devey, Field, and Flook (1977), and Webster (1985).

EXAMPLE 6.3.1 A satellite has the following moments of inertia: $I_x = 100$, $I_y = 140$, $I_z = 40$ kg-m^2. The s/c spins about its minor axis with a frequency $\omega_z = 10$ rad/sec.

6.4 / Estimation of Fuel Consumed during Active Nutation Control 137

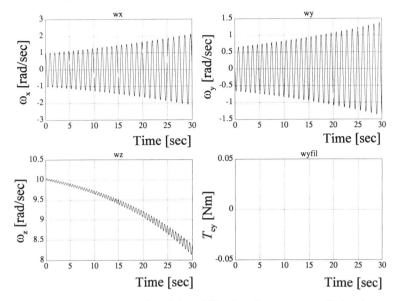

Figure 6.3.2 Nutational destabilization due to energy dissipation.

Because of energy dissipation (or for whatever physical reason, as explained in Section 4.6.2), the system is nutationally unstable. Equations 4.5.1′ have been adapted to exhibit energy dissipation dynamics due to fuel sloshing (see Grumer et al. 1992). The simulated unstable nutational motion is shown in Figure 6.3.2. An ANC system must be designed in order to prevent nutational destabilization of the spinning satellite. Rate gyros are used to measure the angular rates of the satellite. The thrust level is $F = 12$ N, with a torque arm of $\Delta = 0.5$ m.

Solution First, the nutation frequency must be computed. According to Eq. 4.5.13, $\omega_{\text{nut}} = 6.445$ rad/sec, so a bandpass filter around this frequency will be incorporated into the ANC system. Using the control law of Eq. 6.3.1 together with the bandpass filter, a simplified ANC system can prevent nutational destabilization. In order to have an effective filter despite uncertainties in ω_{nut}, we select $\omega_1 = 3$ rad/sec and $\omega_2 = 10$ rad/sec, as defined in Figure 6.3.1.b. Hysteresis is included in the relay to prevent unnecessary activation of the reaction pulses when the nutation level reaches an acceptable amplitude level.

In this example only the T_{cy} control was implemented, in order to emphasize the fact that nutational angular motion can be controlled by activating control torques about only one of the lateral axes of nutation. The time histories of the actively controlled spin motion are shown in Figure 6.3.3 (overleaf). The ANC also damps the initial nutational motion of $\omega_x(0)$ and $\omega_y(0)$.

6.4 Estimation of Fuel Consumed during Active Nutation Control

One of the important causes of energy dissipation during the apogee boosting stage with liquid fuel is the sloshing of the liquid. The energy dissipation level depends on the angular spin velocity of the satellite and on many other geometrical

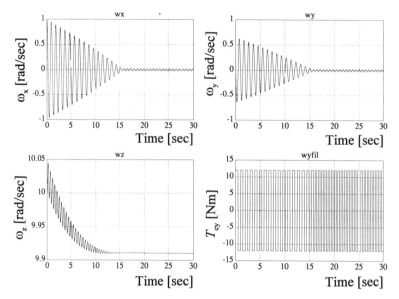

Figure 6.3.3 Effects of ANC system for preventing nutational destabilization and decreasing the initial nutational motion.

parameters of the spinning system; see Garg, Farimoto, and Vanyo (1986). In the following analysis, we assume that the energy dissipation rate $\dot{T}$ has already been estimated.

We use the *energy sink* equation presented in Section 4.6.2:

$$\dot{\theta} = \frac{2I_T I_S \dot{T}}{\sin(2\theta)(I_S - I_T)h^2}, \quad (6.4.1)$$

where I_S and I_T are the spin and the transverse (I_x, I_y) moments of inertia, respectively, assuming a symmetrical body in which $I_x = I_y = I_T$ and $I_S = I_z$. The term θ denotes the nutation angle, and h is the amplitude of the angular momentum of the spinning satellite:

$$h^2 = \omega_S^2 I_S^2 + (\omega_x^2 + \omega_y^2)I_T^2 = h_S^2 + \tan^2(\theta)h_S^2 = h_S^2[1 + \tan^2(\theta)]. \quad (6.4.2)$$

However, $\sin(\theta) = h_T/h$, so it follows that $\dot{\theta}\cos(\theta) = \dot{h}_T/h$. With the assumption that h is constant, we have

$$\dot{\theta} = \frac{\dot{h}_T}{h\cos(\theta)}. \quad (6.4.3)$$

Inserting this equality into Eq. 6.4.1 leads to

$$\frac{\dot{h}_T}{h\cos(\theta)} = \frac{2I_T I_S \dot{T}}{\sin(2\theta)(I_S - I_T)h^2}. \quad (6.4.4)$$

In order to counteract the produced angular moment $\dot{h}_T$, we need to apply a thrust that will produce a torque $F\Delta$, where F is the force and Δ is the torque arm. According to Appendix C and Eq. C.2.3,

$$\dot{h}_T = F\Delta = \dot{m}gI_{\text{sp}}\Delta. \quad (6.4.5)$$

Equation 6.4.5 expresses the transverse torque needed to control the nutation angle. Inserting Eq. 6.4.5 into Eq. 6.4.4, for small nutation angles θ we obtain

$$\dot{m} \approx \frac{I_T I_S \dot{T}}{(I_S - I_T) \sin(\theta) g I_{sp} h \Delta}, \tag{6.4.6}$$

where $h^2 = h_S^2 [1 + \tan^2(\theta)]$ with $h_S = I_S \omega_S$. With the last equality, Eq. 6.4.6 becomes

$$\dot{m} = \frac{I_T I_S \dot{T}}{(I_S - I_T) \sin(\theta) g I_{sp} \Delta (I_S \omega_S) \sqrt{1 + \tan^2(\theta)}} = \frac{1}{[I_S/I_T - 1] g I_{sp} \Delta} \frac{\dot{T}}{\tan(\theta) \omega_S}. \tag{6.4.7}$$

Equation 6.4.7 is the final equation sought. For a given s/c with known moments of inertia, spin velocity, torque arm, and (most important) a roughly estimated internal energy dissipation rate, this equation enables us to compute the amount of fuel necessary to keep the nutation angle below some predetermined maximum level. Equation 6.4.7 clearly shows that the rate of fuel consumption is linearly proportional to the rate of energy dissipation, which in turn is a complex function of the nutation angle θ and the spinning frequency ω_S (see Garg et al. 1986).

EXAMPLE 6.4.1 A satellite has the following physical characteristics:

$$I_T = 50 \text{ kg-m}^2, \quad I_S = 100 \text{ kg-m}^2, \quad I_{sp} = 200 \text{ sec},$$

$$\Delta = 0.5 \text{ m}, \quad dT/dt = 10^{-2} \text{ W}, \quad \omega_S = 2 \text{ rad/sec}.$$

What is the rate of fuel consumption required to ensure that the ANC will maintain a nutation angle smaller than 1°?

Solution Inserting the listed data into Eq. 6.4.7, we find that

$$\dot{m} = 1.0194 \times 10^{-3} \frac{10^{-2}}{2 \tan(1°)} = 2.91 \times 10^{-4} \text{ kg/sec}.$$

In these conditions, one hour of ANC control will cost 1.046 kg of fuel, which is not negligible.

In practice, evaluation of dm/dt with Eq. 6.4.7 is more complicated, because dT/dt is a function of ω_S, θ, and other geometric parameters of the fuel tank. The time constant τ of nutational stability is expressed as $1/\tau = \dot{\theta}/\theta$. For small nutation angles, Eq. 6.4.1 leads to the following approximate value:

$$\tau \approx I_S \left(\frac{I_S}{I_T} - 1\right) \omega_S^2 \frac{\theta^2}{\dot{T}}. \tag{6.4.8}$$

6.5 Despinning and Denutation of a Satellite

A significant number of today's satellites are three-axis attitude-stabilized in their final mission orbit. The primary benefit in spinning the satellite is to achieve an efficient orbit transfer in the presence of parasitic external disturbances during the velocity augmentation phase. However, once in the final mission orbit, the satellite

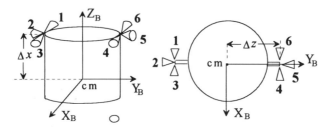

Figure 6.5.1 Simple reaction control system.

must be "despun" in order to become three-axis-stabilized. In fact, a spinning satellite always has some nutational angular motion resulting from initial attitude conditions or external disturbances produced by misalignment of the thruster's axis. In this section we develop a control law for despinning and denutating a satellite. Although not optimal from the perspective of fuel consumption, this technique minimizes the time of denutation.

Figure 6.5.1 depicts a simple but complete reaction control system that can provide both positive and negative moments about the three body axes. Thrusters 2 and 5 can provide negative and positive torques respectively about the X_B axis. Thrusters 1 and 6 provide positive torque about the Y_B axis, while 3 and 4 provide negative torques about the same axis. Thrusters 1 and 4 provide a positive torque about the Z_B axis, while 3 and 6 provide the necessary negative torque. A combination of thrusters 1, 3, 4, and 6 can provide the necessary simultaneous torques about both Y_B and Z_B axes.

We shall again make use of Euler's equations. For a satellite that is symmetric about the Z_B axis, the third of Eqs. 4.5.1' can be separated from the first two. In this way, the denutation and the despinning processes can be analyzed individually.

6.5.1 Despinning

The third of Eqs. 4.5.1' can be written as

$$\dot{h}_z = -2F\Delta z, \qquad (6.5.1)$$

where F is the thrust level and Δz is the torque arm of the spin axis. The solution is simply

$$h_z(t) = h_z(0) - 2F\Delta z t. \qquad (6.5.2)$$

The thrusters are stopped when the angular momentum decreases below some predetermined level – the "dead zone" (DZ in Figure 6.5.2). It is assumed that the angular rates of the satellite are measured or estimated with some adequate sensors. The spin velocity about the Z_B axis is maintained within the desired (or practically achievable) dead zone by activating short positive or negative pulses from the thrusters, depending on the direction of the residual angular momentum about the spin axis.

The time required to remove the initial angular spin momentum is simply

$$t_{\text{DESP}} = \frac{h_z(0)}{N_i F \Delta z}, \qquad (6.5.3)$$

6.5 / Despinning and Denutation of a Satellite

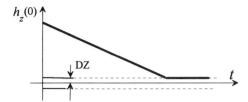

Figure 6.5.2 Time domain history for despinning of a satellite.

where N_i denotes the number of operational thrusters about the Z_B axis. In Figure 6.5.1, $N_i = 2$. The time for despinning the body is also inversely proportional to F and Δz. Figure 6.5.2 displays the time behavior according to Eq. 6.5.2.

To compute the fuel mass expended during the despinning phase, remember that $F = \dot{m} g I_{sp} = (dm/dt) g I_{sp}$. If N_i thrusters are used in the despinning process, then $dm = (N_i F\, dt)/g I_{sp}$. For small changes of the overall mass of the satellite system, including the fuel, we find that

$$m_{\text{FUEL}} \approx \frac{N_i F t_{\text{DESP}}}{g I_{sp}} = \frac{N_i F h_z(0)}{N_i F \Delta z\, g I_{sp}} = \frac{h_z(0)}{g I_{sp} \Delta z} \tag{6.5.4}$$

(cf. Eq. 3.3.8).

EXAMPLE 6.5.1 Assume a satellite with $I_z = 100$ kg-m^2, $\omega_z(0) = 10$ rad/sec, $I_{sp} = 200$ sec, and $\Delta z = 1$ m. Let $F = 5$ N and $N_i = 2$. In this case, $t_{\text{DESP}} = (100 \times 10)/(2 \times 5 \times 1) = 100$ sec. The consumed mass of fuel becomes

$m_{\text{FUEL}} = (100 \times 10)/(9.81 \times 200 \times 1) = 0.51$ kg.

6.5.2 Denutation

The denutation control will be based on the *bang-bang* principle, which in this case will not be optimal for conserving fuel. To simplify the analysis, we suppose that the satellite is axisymmetrical, $I_x = I_y$; according to Section 4.5.1, there is a phase delay of 90° between the X_B and Y_B axes' nutation rates. See also Figure 6.5.3.

In this simple control system, the angular rates are first measured; then torque commands that are proportional to the signs of ω_x and ω_y are applied about the X_B and Y_B axes, respectively:

$$\begin{aligned} T_{cx} &= -\text{sign}(\omega_x) F \Delta x, \\ T_{cy} &= -\text{sign}(\omega_y) F \Delta y. \end{aligned} \tag{6.5.5}$$

As in our treatment of the despinning phase, we shall compute the time to cancel the initial nutation of the spinning system as well as the mass of fuel consumed.

To compute the denutation time, let us first define the four regions designated in Figure 6.5.3. In region I, both ω_x and ω_y are positive; in region II, ω_x is positive and ω_y is negative; and so on. Consequently, in region I we need negative torques about

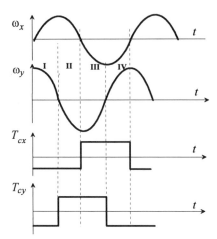

Figure 6.5.3 Denutation process and control command.

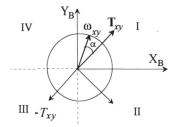

Figure 6.5.4 Torque control in the lateral X_B-Y_B plane of the satellite.

both the X_B and the Y_B axes. Since these torques are perpendicular to each other (see Figure 6.5.4), the net torque, which is the vector addition of both, has the value of $T_{xy} = \sqrt{T_x^2 + T_y^2}$. Also, the direction of T_{xy} is inclined 45° to both the X_B and Y_B axes. In Figure 6.5.4, ω_{xy} is the nutation angular velocity vector revolving in the X_B-Y_B plane (see also Section 4.6.1). The denutation process is 100% effective only when T_{xy} and ω_{xy} are collinear, which happens for only an infinitesimal period of time. If, for simplicity, we let $T_x = T_y$, then $T_{xy} = \sqrt{2} T_x$. The denutation efficiency depends on $\cos(\alpha)$, where α is the angle between T_{xy} and ω_{xy}. We can speak of an average torque in any one of the four regions in Figure 6.5.4.

For instance, in region I we have

$$T_{\text{av}} = \frac{1}{\pi/2} \int_{-\pi/4}^{\pi/4} \cos(\alpha) T_{xy}\, d\alpha = \frac{2}{\pi} \int_{-\pi/4}^{\pi/4} \cos(\alpha) \sqrt{2} F \Delta x\, d\alpha. \tag{6.5.6}$$

After integration,

$$T_{\text{av}} = \frac{2}{\pi} \sqrt{2} F \Delta x \sin(\alpha) \Big|_{-\pi/4}^{\pi/4} = \frac{4}{\pi} F \Delta x. \tag{6.5.7}$$

6.5 / Despinning and Denutation of a Satellite

The same results can be obtained in any one of the remaining three regions of Figure 6.5.3 and Figure 6.5.4. The meaning of Eq. 6.5.7 is that by simultaneously activating two thrusters – one producing a torque about the $\mathbf{X}_B$ axis and the other about the $\mathbf{Y}_B$ axis – we achieve a torque not of $2(F\Delta x)$ but only of $(4/\pi)(F\Delta x)$. This is the loss in torque due to the way in which denutation control is realized. To minimize fuel consumption, the denutating torques must be activated only when $\alpha \approx 0$, but then the denutation time period will increase accordingly.

The denutation can also be performed with one thruster only, albeit with the detrimental effect of increasing the denutation time. If N_i denotes the number of thrusters taking part in the denutation phase, then

$$t_{\text{DENU}} = \frac{h_{xy}(0)}{T_{\text{av}}} = \frac{h_{xy}(0)}{(4/\pi)F\Delta x}\left(\frac{2}{N_i}\right), \tag{6.5.8}$$

where $h_{xy}(0)$ is the initial momentum in the lateral plane of the satellite.

To compute the mass of fuel needed to denutate the satellite, once again we use $N_iF = N_i(dm/dt)gI_{\text{sp}}$. Since $dt = t_{\text{DENU}}$ and $h_{xy}(0)$ are already known, the expended fuel amounts to

$$N_iF\,dt = N_i\,dm\,gI_{\text{sp}} \Rightarrow N_iFt_{\text{DENU}} = N_im_{\text{DENU}}gI_{\text{sp}} = N_iF\frac{\pi h_{xy}(0)}{4F\Delta x}\left(\frac{2}{N_i}\right),$$

from which it follows that

$$m_{\text{DENUTOT}} = m_{\text{DENU}}N_i = \frac{\pi h_{xy}(0)}{2gI_{\text{sp}}\Delta x}. \tag{6.5.9}$$

In Eq. 6.5.9, m_{DENU} is the mass of fuel expelled by one thruster and m_{DENUTOT} is the total fuel expelled while denutating the satellite.

EXAMPLE 6.5.2 Suppose that, for the satellite in Example 6.5.1, $\omega_x(0) = 2$ rad/sec and $I_x = 200$ kg-m^2. Evaluate the time of denutation and the mass of fuel necessary to denutate the satellite. Two thrusters are used, one each for the $\mathbf{X}_B$ and $\mathbf{Y}_B$ body axes.

Solution Equation 6.5.8 gives $t_{\text{DENU}} = (\pi \times 200 \times 2)/(4 \times 5 \times 1) = 62.83$ sec. Equation 6.5.9 yields $m_{\text{DENUTOT}} = (200 \times 2 \times \pi)/(4 \times 9.81 \times 200 \times 1) = 0.16$ kg.

A complete simulation is shown in Figure 6.5.5 (overleaf). This is an idealized situation, in which different time delays in the control loop have been ignored. Moreover, rate measurements in realistic applications are accompanied by noise, which must be taken into consideration; a noise filter will cause additional delays in the control loop. A good *open-loop transfer function* can be achieved by use of standard classical control techniques in the frequency domain. As in the ANC system of Figure 6.2.1, a dead zone is added to prevent "chatter" and unnecessary fuel consumption. For the simulation in Figure 6.5.5, $\omega_z(0) = 5$ rad/sec and $\omega_x(0) = 0.5$ rad/sec; only the ω_x and ω_z rates are measured. The remaining angular rates after denutation and despinning are 0.02 rad/sec, as expected in light of incorporating a dead zone of 0.02 rad/sec.

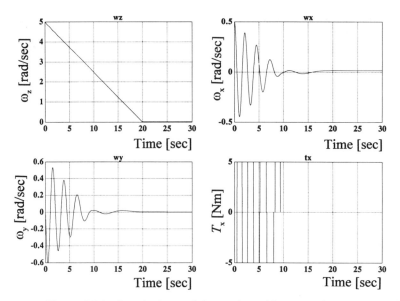

Figure 6.5.5 Despinning and denutation with an angular rate control dead zone of 0.02 rad/sec.

6.6 Single-Spin Stabilization

In the previous sections, spin was used to attitude-stabilize the satellite during its linear velocity augmentation in the transfer between orbits. In this section, the satellite will be spin-stabilized in its final, lifetime "mission" orbit. In Section 6.5, the satellite spin axis was in the direction of motion. If the same spin is to be utilized in the mission orbit, the spin axis must be first precessed until it becomes perpendicular to the orbit plane. With no external disturbances acting on the satellite, the satellite spin axis will be inertially stable. Any external disturbances acting perpendicularly to the spin axis will initiate an angular nutation motion, which may be damped in a number of different ways. As with the passive gravity gradient control of satellite attitude, we can use point-mass dampers, magnetic hysteresis rod dampers, boom articulation, and so on (see Section 5.3.3).

6.6.1 *Passive Wheel Nutation Damping*

In this section, we demonstrate one passive way to damp the nutation of a single-spin–stabilized satellite. Figure 6.6.1 shows a s/c spinning about its Z_B body axis. A nutation angle θ is to be damped by use of a wheel damper (immersed in viscous liquid) whose axis of revolution is aligned with the Y_B axis (see Bryson 1983).

In the configuration of the system in Figure 6.6.1, there are two free bodies that can be in angular motion. The total angular momentum is composed of the individual momentum of the wheel and that of the spinning body: $\mathbf{h} = \mathbf{h}_B + \mathbf{h}_w$. The angular rates of the body about the three orthogonal axes X_B, Y_B, Z_B are (respectively) $\omega_x, \omega_y, \omega_z$. The angular velocity of the wheel about its spin axis, and relative to the body, is Ω. As there is only one wheel in the satellite's body, $\mathbf{h}_w = \hat{\mathbf{j}}(\omega_y + \Omega) I_w$, where

6.6 / Single-Spin Stabilization

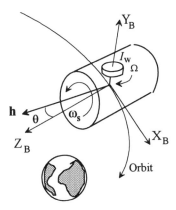

Figure 6.6.1 Wheel damper for a spin-stabilized satellite.

I_w is the moment of inertia of the wheel and $\hat{\mathbf{j}}$ is the unit direction vector along the $\mathbf{Y}_B$ body axis. The total angular momentum of the system becomes

$$\mathbf{h} = \hat{\mathbf{i}}\omega_x I_x + \hat{\mathbf{j}}[\omega_y(I_y + I_w) + I_w \Omega] + \hat{\mathbf{k}}\omega_z I_z. \tag{6.6.1}$$

If we assume that $I_y \gg I_w$, $I_x = I_y = I_T$, and $I_z = I_S$, then

$$\mathbf{h} \approx \hat{\mathbf{i}}\omega_x I_x + \hat{\mathbf{j}}[\omega_y I_y + I_w \Omega] + \hat{\mathbf{k}}\omega_z I_z. \tag{6.6.2}$$

Next, we use Euler's equation (Eq. 4.5.1) and obtain

$$\begin{aligned}
\dot{h}_{1x} &= 0 = \dot{\omega}_x I_T + \omega_y \omega_z (I_S - I_T) - \omega_z I_w \Omega, \\
\dot{h}_{1y} &= 0 = \dot{\omega}_y I_T + \dot{\Omega} I_w + \omega_z \omega_x (I_T - I_S), \\
\dot{h}_{1z} &= 0 = \dot{\omega}_z I_S + \omega_x \omega_y (I_y - I_x) + \omega_x I_w \Omega.
\end{aligned} \tag{6.6.3}$$

Since $I_x = I_y$ and $\omega_z = \omega_S$, Eqs. 6.6.3 become

$$\begin{aligned}
\dot{\omega}_x + \omega_y \omega_z \frac{I_S - I_T}{I_T} - \omega_z \frac{I_w}{I_T} \Omega &= 0, \\
\dot{\omega}_y + \omega_x \omega_z \frac{I_T - I_S}{I_T} + \dot{\Omega} \frac{I_w}{I_T} &= 0, \\
\dot{\omega}_z + \omega_x \frac{I_w}{I_S} \Omega &= 0.
\end{aligned} \tag{6.6.4}$$

From Eqs. 6.6.4 we conclude that $\omega_z = \omega_S =$ const, since $\omega_x \Omega (I_w / I_S)$ is negligible. Then we write the angular equation of motion of the wheel:

$$T_w = (\dot{\omega}_y + \dot{\Omega}) I_w + \Omega D = 0, \tag{6.6.5}$$

where D is the damping coefficient of the liquid in which the wheel is immersed. Because no external torque is applied on the wheel, $T_w = 0$. The first two of Eqs. 6.6.4, together with Eq. 6.6.5, can be put in the matrix form

$$\begin{bmatrix} s & -\omega_S[(I_T - I_S)/I_T] & -\omega_S(I_w/I_T) \\ \omega_S[(I_T - I_S)/I_T] & s & s(I_w/I_T) \\ 0 & s & s + D/I_w \end{bmatrix} \begin{bmatrix} \omega_x \\ \omega_y \\ \Omega \end{bmatrix} = 0. \tag{6.6.6}$$

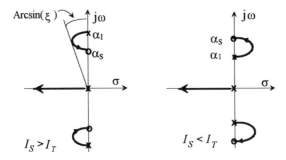

Figure 6.6.2 Stability of the damped system from root-locus considerations.

Define $\alpha_S = [(I_T - I_S)/I_T]\omega_S$ as the nutation frequency in the body frame, and set $\epsilon = I_w/I_T$ and $\sigma = D/I_w$. With these definitions, the matrix in Eq. 6.6.6 takes the form:

$$\begin{bmatrix} s & -\alpha_S & -\epsilon\omega_S \\ \alpha_S & s & \epsilon s \\ 0 & s & s+\sigma \end{bmatrix} \begin{bmatrix} \omega_x \\ \omega_y \\ \Omega \end{bmatrix} = 0. \tag{6.6.6'}$$

To assure stability, we must check the determinant $\Delta(s) = 0$. We find that

$$\Delta(s) = s^3(1-\epsilon) + s(\alpha_S^2 - \alpha_S\epsilon\omega_S) + \sigma s^2 + \alpha_S^2 \sigma = 0, \quad \text{or}$$

$$s(1-\epsilon)\left[s^2 + \frac{\alpha_S^2}{1-\epsilon}\left(1 - \frac{\epsilon\omega_S}{\alpha_S}\right)\right] + (s^2 + \alpha_S^2)\sigma = 0.$$

This characteristic equation can be arranged in root-locus form (Bryson 1983) as follows:

$$1 + \frac{\sigma}{1-\epsilon} \frac{s^2 + \alpha_S^2}{s\left\{s^2 + \alpha_S^2\left[1 + \frac{I_S I_w}{(I_S - I_T)(I_T - I_w)}\right]\right\}} = 1 + \frac{\sigma}{1-\epsilon} \frac{s^2 + \alpha_S^2}{s(s^2 + \alpha_1^2)}, \tag{6.6.7}$$

where

$$\alpha_1 = \alpha_S \sqrt{1 + \frac{I_S I_w}{(I_S - I_T)(I_T - I_w)}}.$$

It follows easily that

$$I_S > I_T \Rightarrow \alpha_1 > \alpha_S,$$
$$I_S < I_T \Rightarrow \alpha_1 < \alpha_S.$$

The stability conditions may be analyzed in the root-locus plane of the configuration in Figure 6.6.2. From these root-locus considerations it is easily seen that, for $I_S > I_T$, the roots are in the stable region with $\sigma < 0$. The system becomes nutationally unstable for $I_S < I_T$. The conclusion is that passive damping with a damper wheel is possible only if the spin axis is the major axis. See also Section 4.6.2.

6.6.2 *Active Wheel Nutation Damping*

It is also possible to damp a spinning spacecraft about its minor axis with an actively commanded wheel, where the damper wheel is torqued by an electrical

6.6 / Single-Spin Stabilization

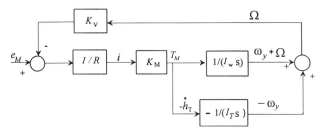

Figure 6.6.3 Dynamics of a DC electrical motor, in conjunction with the satellite.

motor. For such a system, we must first write a linear model of the electrical motor-satellite system.

In Figure 6.6.3, a direct current (DC) motor is used; R is the armature resistance of the motor on whose rotor the flywheel is aligned, K_M is the torque coefficient, and I_w is the total moment of inertia of the rotor. The i term is the current in the motor, and $T_M = iK_M$, where T_M is the torque produced by the motor on the rotor axis. The same torque, but opposite in sign, is applied on the satellite's body. The angular velocity of the damper wheel relative to the body is Ω.

The dynamic equations of the electrical motor are

$$T_M = K_M i = I_w(\dot{\omega}_y + \dot{\Omega}),$$
$$e_M = iR + K_V \Omega, \tag{6.6.8}$$

where e_M is the input voltage to the electrical motor. Equations 6.6.8 can be combined to yield ($N = K_V = K_M$ in MKS units)

$$I_w(\dot{\omega}_y + \dot{\Omega}) + \frac{N^2 \Omega}{R} = \frac{N}{R} e_M. \tag{6.6.9}$$

This equation is now added to the first two of Eqs. 6.6.4 to yield the final system:

$$\dot{\omega}_x + \omega_y \omega_S \frac{I_S - I_T}{I_T} - \omega_S \frac{I_w}{I_T} \Omega = 0,$$
$$\dot{\omega}_y + \omega_x \omega_S \frac{I_T - I_S}{I_T} + \dot{\Omega} \frac{I_w}{I_T} = 0, \tag{6.6.10}$$
$$\dot{\Omega} + \dot{\omega}_y + \frac{N^2}{RI_w} \Omega = \frac{N}{RI_w} e_M.$$

Define $\gamma = N^2/RI_w$ and $\beta = N/RI_w$; as before, $\alpha_S = [(I_T - I_S)/I_T]\omega_S$ and $\epsilon = I_w/I_T$. With these definitions, Eqs. 6.6.4 can be put in matrix form as follows:

$$\begin{bmatrix} s & -\alpha_S & -\epsilon\omega_S \\ \alpha_S & s & \epsilon s \\ 0 & s & s+\gamma \end{bmatrix} \begin{bmatrix} \omega_x \\ \omega_y \\ \Omega \end{bmatrix} = \begin{bmatrix} 0 \\ 0 \\ \beta \end{bmatrix} e_M. \tag{6.6.11}$$

From Eq. 6.6.11 we can find the transfer functions $\omega_x(s)/e_M(s)$ and $\omega_y(s)/e_M(s)$. If one of the lateral angular velocities (e.g. ω_x) can be measured, then a proportional error signal can be applied to the electrical motor ($e_M = K\omega_x$) in order to provide

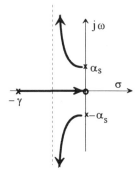

Figure 6.6.4 Root locus for Eq. 6.6.12: $L(s) = K's/[(s^2 + \alpha_S^2)(s+\gamma)]$.

nutational stability also for the case in which the spin is about the axis with minimum moment of inertia, $I_S < I_T$.

With the assumption that $I_w \ll I_T$ (which is normally the case in a practical system), it is easily found from Eq. 6.6.11 and the equality $e_M = K\omega_x$ that the open-loop transfer function of the nutation controller will be

$$L(s) = K\frac{\omega_x(s)}{e_M(s)} \approx K\frac{NI_S\omega_S}{RI_T^2}\frac{s}{(s^2+\alpha_S^2)(s+\gamma)} = K'\frac{s}{(s^2+\alpha_S^2)(s+\gamma)}, \quad (6.6.12)$$

where $K' = KNI_S\omega_S/(RI_T^2)$.

The roots of the characteristic equation of the closed loop are the roots of the polynomials in the numerator of $1 + L(s)$. From root-locus considerations (see Figure 6.6.4), we conclude that the roots of the closed loop will be stable for any α_S, which means for $I_S > I_T$ and also for $I_S < I_T$.

6.7 Dual-Spin Stabilization

The communication efficiency of a single-spin–stabilized satellite is very low. In fact, since the communications antenna spins together with the satellite body, the communications beam scans the earth globe for only a short time in one spin cycle. Most of the time the beam is directed toward outer space, where no use is made of the emitted communication energy.

A much better technique involves using a *dual-spin* stabilization concept. This is based on the principle that one part of the satellite is spinning fast in order to provide the necessary stabilizing angular momentum, while the second part revolves once per orbit so that the communication payload is constantly pointing toward the earth. The communication payloads are placed in the part of the satellite called the *platform*, and the rotating part is called the *rotor*. The attitude control of the Hughes *INTELSAT VI*, which is one of the largest communications satellites ever built, is based on the dual-spin principle.

6.7.1 *Passive Damping of a Dual-Spin–Stabilized Satellite*

The fuel needed to stabilize the s/c in its different life stages may be located in the platform or the rotor part of the satellite. Since both platform and rotor will

6.7 / Passive Damping of a Dual-Spin-Stabilized Satellite

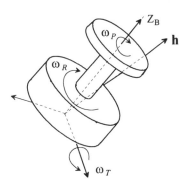

Figure 6.7.1 Dual-spin-stabilized satellite; reproduced from Agrawal (1986) by permission of Prentice-Hall.

generally exhibit some kind of energy dissipation, destabilization due to nutation should be expected and hence nutation damping control is imperative. In this section we will specify the conditions under which passive damping can be effective (see also Iorillo 1965, Likins 1967, Agrawal 1986).

To find the conditions for nutational stability, our analysis begins with the expressions of rotational kinetic energy T and the angular momentum H. For the special configuration of a dual-spinning s/c (see Figure 6.7.1), these expressions take the form

$$T = \tfrac{1}{2}[I_T \omega_T^2 + I_R \omega_R^2 + I_P \omega_P^2], \qquad (6.7.1)$$
$$H^2 = [I_P \omega_P + I_R \omega_R]^2 + [I_T \omega_T]^2,$$

where the subscript T denotes momentum in the *transverse* plane and P and R denote *platform* and *rotor*, respectively. With no external disturbances, $\dot{H} = 0$, so

$$2\dot{H}H = 0 = 2[I_P \omega_P + I_R \omega_R][I_P \dot\omega_P + I_R \dot\omega_R] + 2I_T^2 \omega_T \dot\omega_T;$$

hence,

$$I_T \omega_T \dot\omega_T = -\frac{I_P \omega_P + I_R \omega_R}{I_T}[I_P \dot\omega_P + I_R \dot\omega_R] = -\omega_n[I_P \dot\omega_P + I_R \dot\omega_R]. \qquad (6.7.2)$$

Differentiation of T with time leads to

$$\dot T = \dot T_R + \dot T_P = I_T \omega_T \dot\omega_T + I_R \omega_R \dot\omega_R + I_P \omega_P \dot\omega_P. \qquad (6.7.3)$$

Substituting Eq. 6.7.2 into Eq. 6.7.3 yields

$$\dot T = -I_P[\omega_n - \omega_P]\dot\omega_P - I_R[\omega_n - \omega_R]\dot\omega_R = -I_P \lambda_P \dot\omega_P - I_R \lambda_R \dot\omega_R. \qquad (6.7.4)$$

From Eq. 6.7.2 and Eq. 6.7.3 it follows that $I_P \dot\omega_P = -\dot T_P/\lambda_P$ and $I_R \dot\omega_R = -\dot T_R/\lambda_R$. Together with Eq. 6.7.3, we have

$$I_T \omega_T \dot\omega_T = \omega_n\left[\frac{\dot T_P}{\lambda_P} + \frac{\dot T_R}{\lambda_R}\right], \qquad (6.7.5)$$

where $\omega_n = (I_P \omega_P + I_R \omega_R)/I_T$, $\lambda_P = \omega_n - \omega_P$, and $\lambda_R = \omega_n - \omega_R$. On the other hand (see Section 4.6.2), for H_T defined as $H_T = I_T \omega_T$, we have $\sin(\theta) = I_T \omega_T/H$. Differentiating yields

$$\dot\theta \cos(\theta) = \frac{I_T \dot\omega_T}{H}. \qquad (6.7.6)$$

Using Eq. 6.7.5, we finally obtain

$$\dot{\theta} = \frac{I_T \dot{\omega}_T}{H\cos(\theta)} \frac{\omega_T}{\omega_T} = \frac{I_T \omega_T \dot{\omega}_T}{H\cos(\theta)[H\sin(\theta)(1/I_T)]} = \frac{2I_T}{H^2 \sin(2\theta)} \omega_n \left[\frac{\dot{T}_P}{\lambda_P} + \frac{\dot{T}_R}{\lambda_R}\right]. \quad (6.7.7)$$

Nutational stability requires $\dot{\theta} < 0$, which implies – using Eq. 6.7.7 – that

$$\frac{\dot{T}_P}{\lambda_P} + \frac{\dot{T}_R}{\lambda_R} < 0. \quad (6.7.8)$$

This means that, for stability, the sum of the energy dissipation in the rotor and in the platform must be negative. This is a necessary condition, but it is not sufficient. Obviously, $I_R \omega_R \gg I_P \omega_P$. This leads to

$$\lambda_P = \omega_n - \omega_P \approx \frac{I_R}{I_T} \omega_R - \omega_P \approx \frac{I_R}{I_T} \omega_R.$$

In the same way,

$$\lambda_R = \omega_n - \omega_R = \frac{I_P \omega_P + I_R \omega_R}{I_T} - \omega_R \approx \left[\frac{I_R}{I_T} - 1\right] \omega_R.$$

Given the two last approximations, the condition in Eq. 6.7.8 becomes

$$\frac{\dot{T}_P}{\lambda_P} + \frac{\dot{T}_R}{\lambda_R} = \frac{1}{\omega_R} \left[\frac{\dot{T}_P}{I_R/I_T} + \frac{\dot{T}_R}{I_R/I_T - 1}\right] < 0 \quad (6.7.9)$$

(obviously, for energy dissipation both $\dot{T}_P$ and $\dot{T}_R$ must be less than zero.) Two distinct cases must be checked for stability.

Case A: $I_R > I_T$. In this case, Eq. 6.7.9 is always satisfied. The satellite part in which the energy dissipation takes place is of no importance. This means that the passive nutation damper can be placed anywhere inside the satellite.

Case B: $I_R < I_T$. In this case, the second term on the right-hand side of Eq. 6.7.9 is positive. In order to effect nutational stability, the energy dissipation must take place in the platform. Moreover, we can find by what factor the energy dissipation rate in the platform must be larger than that in the rotor part of the satellite. From Eq. 6.7.9, the result is

$$|\dot{T}_P| > \left|\frac{\dot{T}_R (I_R/I_T)}{I_R/I_T - 1}\right|. \quad (6.7.10)$$

This means that, in order to stabilize the nutational motion, the dissipation capacity of a platform-mounted passive damper must satisfy Eq. 6.7.10.

6.7.2 Momentum Bias Stabilization

Dual-spin stabilization is closely connected to *momentum bias* stabilization. For nadir-pointing spacecraft (e.g., communications, meteorological, earth-scanning, etc.), this form of attitude stabilization is most effective. There are many variations on attitude control concepts associated with momentum-biased satellites. Chapter 8 will be dedicated to this important class of attitude control stabilization schemes.

6.8 Summary

Owing to its simplicity and efficiency, single-spin stabilization has been used extensively. For higher attitude accuracy, some passive or active nutation stabilization is necessary, especially if energy dissipation is part of the dynamics of the satellite. Attitude control and stabilization of a single-spin-stabilized spacecraft usually need some angular rate measurements. From the perspective of communications, however, dual-spin stabilization is notably superior.

References

Agrawal, B. N. (1986), *Design of Geosynchronous Spacecraft*. Englewood Cliffs, NJ: Prentice-Hall.

Bryson, A. (1983), "Stabilization and Control of Spacecraft," Microfiche supplement to the *Proceedings of the Annual AAS Rocky Mountain Guidance and Control Conference* (5-9 February, Keystone, CO). San Diego, CA: American Astronautical Society.

Devey, W., Field, C., and Flook, L. (1977), "An Active Nutation Control System for Spin Stabilized Satellites," *Automatica* 13: 161-72.

Fagg, A., and MacLauchlan, J. (1981), "Operational Experience on OTS-2," *Journal of Guidance and Control* 4(5): 551-7.

Fox, S. (1986), "Attitude Control Subsystem Performance of the RCA Series 3000 Satellite," Paper no. 86-0614-CP, AIAA 11th Communications Satellite System Conference (17-20 March, San Diego). New York: AIAA.

Garg, S. C., Farimoto, N., and Vanyo, Y. P. (1986), "Spacecraft Nutational Instability Prediction by Energy Dissipation Measurements," *Journal of Guidance, Control, and Dynamics* 9(3): 357-62.

Grasshoff, L. H. (1968), "An Onboard, Closed-Loop Nutation Control System for a Spin-Stabilized Spacecraft," *Journal of Spacecraft and Rockets* 5(5): 530-5.

Grumer, M., Komem, Y., Kronenfeld, J., Kubitski, O., Lorber, V., and Shyldkrot, H. (1992), "OFEQ-2: Orbit, Attitude and Flight Evaluation," 32nd Israel Annual Conference on Aviation and Astronautics (18-20 February, Tel Aviv). Tel Aviv: Kenes, pp. 266-78.

Iorillo, A. J. (1965), "Nutation Damping Dynamics of Axisymmetric Rotor Stabilized Satellites," ASME Winter Meeting (November, Chicago). New York: ASME.

Likins, P. (1967), "Attitude Stability Criteria for Dual-Spin Spacecraft," *Journal of Spacecraft and Rockets* 4(12): 1638-43.

Webster, E. A. (1985), "Active Nutation Control for Spinning Solid Motor Upper Stages," Paper no. 85-1382, 21st Joint Propulsion Conference, AIAA/SAE/ASME (8-10 July, Monterey, CA).

CHAPTER 7

Attitude Maneuvers in Space

7.1 Introduction

In the previous chapters dealing with satellite attitude control, the primary task of the control system was to stabilize the attitude of the satellite against external torque disturbances. Such disturbances are produced by aerodynamic drag effects, solar radiation and solar wind torques, parasitic torques created by the propulsion thrusters, and so on. We discussed gravity gradient stabilization, which succeeds because such torques tend to align the axis of least inertia with the nadir direction. Attitude controls that are based on spin have similar features: the spin principle tends to keep one axis of the satellite inertially stabilized in space. However, even for spacecraft that in their final mission stage are to be stabilized to some constant attitude relative to a reference frame, a number of prior tasks must be performed in which the satellite's attitude is maneuvered (see the introductory example in Chapter 1).

The primary mission tasks of some satellites require attitude maneuvers throughout their lifetime. Two well-known examples are the *Space Telescope* (Dougherty et al. 1982) and the *Rosat* satellite (Bollner 1991), scanning the sky for scientific observations. The capability to attitude-maneuver a satellite is based on using control torques. Control command laws using such torques are the subject of this chapter.

7.2 Equations for Basic Control Laws

In this section we shall write and analyze the control law equations, expressed in different attitude error terminologies. The most common include Euler angles for small attitude commands and, for large attitude maneuvers, direction cosine error and quaternion error terminologies.

7.2.1 Control Command Law Using Euler Angle Errors

The simplest *torque control law* is based on Euler angle errors. Suppose that the Euler angles, as defined in Section A.3, can be measured by the s/c instrumentation. As can be seen from Eqs. 4.8.14, for a satellite with a diagonal inertia matrix and small Euler angle rotations, the attitude dynamic equations can be approximated as

$$
\begin{aligned}
T_{dx} + T_{cx} &= I_x \ddot{\phi}, \\
T_{dy} + T_{cy} &= I_y \ddot{\theta}, \\
T_{dz} + T_{cz} &= I_z \ddot{\psi}.
\end{aligned}
\qquad (7.2.1)
$$

For such a simplified set of equations, the three-axis attitude dynamics can be separated into three one-axis second-order dynamics equations.

7.2 / Equations for Basic Control Laws

The simplest control law for stabilizing and attitude-maneuvering such a system may be stated as follows:

$$T_{cx} = K_x(\phi_{com} - \phi) + K_{xd}\dot{\phi} = K_x\phi_E + K_{xd}\dot{\phi},$$
$$T_{cy} = K_y(\theta_{com} - \theta) + K_{yd}\dot{\theta} = K_y\theta_E + K_{yd}\dot{\theta}, \qquad (7.2.2)$$
$$T_{cz} = K_z(\psi_{com} - \psi) + K_{zd}\dot{\psi} = K_z\psi_E + K_{zd}\dot{\psi},$$

where $\phi_{com}, \theta_{com}, \psi_{com}$ and ϕ_E, θ_E, ψ_E are the Euler command and error angles, respectively; $\dot{\phi}, \dot{\theta}, \dot{\psi}$ are the Euler angular rates.

Designing such a second-order control system is a trivial automatic control problem, treated in many basic texts on linear control theory (see e.g. D'Azzo and Houpis 1988 and Dorf 1989). All we need do is determine $K_x, K_{xd}, K_y, \ldots$ so that the three one-axis control systems about the $\mathbf{X}_B, \mathbf{Y}_B, \mathbf{Z}_B$ body axes have the desired dynamic characteristics, such as natural frequency ω_n and damping coefficients ξ, which will preferably be equal for all axes.

The problem is less trivial when large attitude maneuvers are considered, for three principal reasons. First, the simplified dynamics model of Eqs. 7.2.1 does not hold for large attitude maneuvers (see Eqs. 4.8.14). For attitude control systems requiring high accuracies and very short settling time, such terms as $I_{xy}\ddot{\theta}$ and $\omega_o(I_y - I_z - I_x)\dot{\psi}$ cannot be ignored; they must be taken into account in the design stage. Second, there is a control problem with regard to *saturation* – that is, the maximum achievable torques and angular velocities that the control driver can deliver to the satellite. These control difficulties necessitate the application of nonlinear automatic control design procedures (Junkins and Turner 1986); see Section 7.6.

The third reason is that, for large Euler attitude angles, the attitude kinematics equations can become singular. For example, in the Euler angle rotation $\theta \rightarrow \phi \rightarrow \psi$ (see Eqs. 4.7.8), the Euler kinematics equations become singular as ϕ approaches $90°$. As we shall see, this drawback can be alleviated by using more effective kinematics expressions for the attitude control laws.

7.2.2 Control Command Law Using the Direction Cosine Error Matrix

Suppose that the attitude of the satellite is expressed in terms of the direction cosine matrix $[\mathbf{A}_S]$ relative to the reference frame in which the attitude maneuver is to be commanded and achieved (see Appendix A). It is of no importance whether the reference frame is inertial or rotating with the orbit, so long as all the measurements and matrix transformations are performed in the same reference frame. Suppose then that a vector $\mathbf{a}$ has the components a_1, a_2, a_3 in the reference frame $\mathbf{a} = [a_1 \ a_2 \ a_3]^T$, and that the satellite is to be maneuvered so that its final direction cosine matrix will coincide with a known and defined matrix $[\mathbf{A}_T]$, called the *target matrix*.

According to the analysis in Appendix A, the vector $\mathbf{a}$ can be expressed in the satellite frame and in the target frame as $\mathbf{a}_S$ and $\mathbf{a}_T$ (respectively) in the following way:

$$\mathbf{a}_S = [\mathbf{A}_S]\mathbf{a}, \qquad (7.2.3)$$
$$\mathbf{a}_T = [\mathbf{A}_T]\mathbf{a}. \qquad (7.2.4)$$

Combining both equations, we have

$$\mathbf{a}_S = [\mathbf{A}_S][\mathbf{A}_T]^{-1}\mathbf{a}_T = [\mathbf{A}_S][\mathbf{A}_T]^T\mathbf{a}_T = [\mathbf{A}_E]\mathbf{a}_T. \quad (7.2.5)$$

The matrix $[\mathbf{A}_E]$, as defined in Eq. 7.2.5, has the following meaning: if the components of two noncollinear vectors **a** are identical in both the satellite frame S and the target frame T, then it is obvious that these frames coincide and that the satellite body axes have reached the desired target attitude in space. Hence, $[\mathbf{A}_E]$ is the *direction cosine error matrix*. When this matrix becomes the unit matrix, $[\mathbf{A}_S] = [\mathbf{A}_T]$ and the satellite has reached the desired attitude in space. To clarify the meaning of this statement, let us write explicitly the matrix multiplication of Eq. 7.2.5:

$$[\mathbf{A}_E] = \begin{bmatrix} a_{11S} & a_{12S} & a_{13S} \\ a_{21S} & a_{22S} & a_{23S} \\ a_{31S} & a_{32S} & a_{33S} \end{bmatrix} \begin{bmatrix} a_{11T} & a_{21T} & a_{31T} \\ a_{12T} & a_{22T} & a_{32T} \\ a_{13T} & a_{23T} & a_{33T} \end{bmatrix} = \begin{bmatrix} a_{11E} & a_{12E} & a_{13E} \\ a_{21E} & a_{22E} & a_{23E} \\ a_{31E} & a_{32E} & a_{33E} \end{bmatrix}. \quad (7.2.6)$$

For the last matrix to become diagonal, the off-diagonal elements must be zeroed and the diagonal elements must become unity.

To understand the meaning of zeroing the off-diagonal elements, let us examine Figure 7.2.1 and interpret correctly the meaning of the elements a_{ijE} in Eq. 7.2.6. For example, a_{12E} is the scalar dot product between the $\mathbf{X}_S$ and the $\mathbf{Y}_T$ axes; in Eq. A.2.2, $a_{12E} = \mathbf{X}_S \cdot \mathbf{Y}_T$. Hence, $a_{12E} = 0$ is equivalent to making the $\mathbf{X}_S$ axis perpendicular to the $\mathbf{Y}_T$ axis by increasing the angle α in Figure 7.2.1. This may be achieved by rotating the satellite about the $\mathbf{Z}_S$ axis until the following equality is satisfied:

$$a_{12E} = \mathbf{X}_S \cdot \mathbf{Y}_T = 0. \quad (7.2.7)$$

In the same way, it is easily seen that zeroing a_{13E} is equivalent to the scalar dot product

$$a_{13E} = \mathbf{X}_S \cdot \mathbf{Z}_T = 0, \quad (7.2.8)$$

which means geometrically that the satellite is to be rotated about its $\mathbf{Y}_S$ axis until the $\mathbf{X}_S$ satellite axis becomes perpendicular to the $\mathbf{Z}_T$ target axis. Finally, rotation of the satellite about the $\mathbf{X}_S$ axis will make the $\mathbf{Y}_S$ axis perpendicular to the $\mathbf{Z}_T$ axis, thus zeroing a_{23E}:

$$a_{23E} = \mathbf{Y}_S \cdot \mathbf{Z}_T = 0. \quad (7.2.9)$$

By similar reasoning, it can be shown that if both the target and the satellite axis frames coincide, then the elements of $[\mathbf{A}_E]$ that lie below the matrix diagonal are

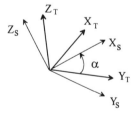

Figure 7.2.1 Geometrical interpretation of zeroing the off-diagonal elements of $[\mathbf{A}_E]$.

also zeroed. Thus, with the completion of the maneuver, the error matrix becomes the unit diagonal matrix.

Simultaneous satisfaction of Eq. 7.2.7, Eq. 7.2.8, and Eq. 7.2.9 tends to rotate the satellite axis frame so that it coincides with the desired target axis frame, thus achieving the desired attitude maneuver in space. Since the basic attitude dynamics of the satellite consists of two integrations per axis, rate terms must be used in order to stabilize the three axes, as in Eqs. 7.2.2. Finally, similarly to Eqs. 7.2.2, the following control laws can be written:

$$
\begin{aligned}
T_{cx} &= K_x a_{23E} + K_{xd} p, \\
T_{cy} &= K_y a_{13E} + K_{yd} q, \\
T_{cz} &= K_z a_{12E} + K_{zd} r.
\end{aligned}
\tag{7.2.10}
$$

In Eqs. 7.2.10, a_{ij} can be replaced with $-a_{ji}$ to obtain the same control results. The terms p, q, r are the angular velocities of the body axes in the reference axis frame, used for damping purposes.

In the beginning of the maneuver, the error elements may be quite large, depending on the initial attitude orientation of the satellite relative to the target reference frame. At the final stages of the attitude maneuver, when the s/c axes are closely aligned with the target reference frame axes, the error elements a_{12E}, a_{13E}, and a_{23E} approach the errors of the Euler angles. As mentioned in Appendix A, all the Euler transformations in Eqs. A.3.6–A.3.12 can be approximated to the transformation Eq. A.3.13 for small Euler angles. Therefore $a_{12E} \to \psi_E$, $a_{13E} \to -\theta_E$, and $a_{23E} \to \phi_E$ as expected. Hence, at the final stage of the large attitude maneuver in space, the small-angle approximation (Eq. 7.2.1) is justified. As in Section 7.2.1, the control gains should be devised so that, at the end of the large maneuver in space, the time responses will be well behaved. Also, sufficient stability margins in the frequency domain must be procured. These gains are dependent on the physical characteristics of the satellite (the inertia matrix).

7.2.3 Control Command Law about the Euler Axis of Rotation

The direction cosine matrix can be used in a similar way as in Section 7.2.2 to achieve a large attitude maneuver about the Euler axis of rotation (as defined in Section A.4). Using Eq. A.4.6, we can find the Euler angle about this axis with respect to the initial error matrix $[\mathbf{A}_E]$; namely,

$$\cos(\alpha) = \tfrac{1}{2}\{\text{trace}([\mathbf{A}_E]) - 1\}. \tag{7.2.11}$$

Using also the results in Eqs. A.4.10a–A.4.10c, we derive the following control command law:

$$
\begin{aligned}
T_{cx} &= -\tfrac{1}{2} K_x (a_{32E} - a_{23E}) + K_{xd} p, \\
T_{cy} &= -\tfrac{1}{2} K_y (a_{13E} - a_{31E}) + K_{yd} q, \\
T_{cz} &= -\tfrac{1}{2} K_z (a_{21E} - a_{12E}) + K_{zd} r.
\end{aligned}
\tag{7.2.12}
$$

Since the initial and final attitude orientations of the satellite are known in advance, using Eqs. 7.2.12 will enable rotation about the Euler axis of rotation, thus minimizing the angular path that will be traversed by the satellite in its angular motion.

This control law does have one shortcoming: the six off-diagonal elements of the error matrix must be computed continuously. This is done by integrating Eq. 4.7.11 in order to find $[A_S]$. At least six elements must be found by any integration algorithm in the onboard computer. The matrix $[A_T]$ is known by definition and so, finally, the error matrix (Eq. 7.2.6) must be computed. All these calculations result in a large number of operations for the onboard computer to perform in real time. However, the computational burden can be reduced by making use of quaternion terminology (see Section A.4).

7.2.4 Control Command Law Using the Quaternion Error Vector

In the previous section we defined the direction cosine matrix and gave control laws in terms of the elements of this matrix. There exists an equivalent *quaternion error vector* that expresses the attitude error between (i) the satellite attitude direction in space and (ii) the target direction toward which the satellite is oriented at the end of the attitude maneuver. To find this quaternion error vector we refer to Section A.4.4 as well as Glaese et al. (1976), Schletz (1982), and Wertz (1986).

According to Eq. A.4.20, we can write

$$[A(\mathbf{q}_E)] = [A(\mathbf{q}_T)][A(\mathbf{q}_S)]^{-1} = [A(\mathbf{q}_T)][A(\mathbf{q}_S^{-1})]. \tag{7.2.13}$$

In quaternion notation, Eq. 7.2.13 leads to

$$\mathbf{q}_S^{-1}\mathbf{q}_T = \mathbf{q}_E = \begin{bmatrix} q_{T4} & q_{T3} & -q_{T2} & q_{T1} \\ -q_{T3} & q_{T4} & q_{T1} & q_{T2} \\ q_{T2} & -q_{T1} & q_{T4} & q_{T3} \\ -q_{T1} & -q_{T2} & -q_{T3} & q_{T4} \end{bmatrix} \begin{bmatrix} -q_{S1} \\ -q_{S2} \\ -q_{S3} \\ q_{S4} \end{bmatrix}, \tag{7.2.14}$$

where $\mathbf{q}_E$, $\mathbf{q}_T$, and $\mathbf{q}_S$ are (respectively) the error, target, and spacecraft quaternions.

As explained in Section A.4.3, there is a one-to-one equivalence between the direction cosine matrix elements and the elements of the quaternion vector. The relations are given in Eqs. A.4.16. Inserting these relations into Eqs. 7.2.12, we obtain the following attitude control laws:

$$\begin{aligned} T_{cx} &= 2K_x q_{1E} q_{4E} + K_{xd} p, \\ T_{cy} &= 2K_y q_{2E} q_{4E} + K_{yd} q, \\ T_{cz} &= 2K_z q_{3E} q_{4E} + K_{zd} r. \end{aligned} \tag{7.2.15}$$

As in Section 7.2.3, the spacecraft quaternion vector is obtained by integrating Eqs. 4.7.13. Only three elements need to be integrated – namely, q_{1S}, q_{2S}, and q_{3S} – since q_{4S} is known from the relation $|\mathbf{q}| = 1$. Performing the computation of $\mathbf{q}_E$ in Eq. 7.2.14 requires fewer algebraic operations than computing the elements in $[A_E]$. This is one reason why the control law of Eqs. 7.2.15 is preferred to that of Eqs. 7.2.12, although they are equivalent from a physical point of view. See also Wie, Weiss, and Arapostathis (1989).

7.2.5 Control Laws Compared

In order to compare the qualities of the different torque control laws, a 6-DOF (six-degrees-of-freedom) time-domain simulation has been carried out for each

7.2 / Equations for Basic Control Laws

law. Since the control error definitions for the compared laws are different, some common variables need to be chosen. The Euler angles are the most natural physical variables to be compared where attitude control of the satellite is concerned. The evolution of the error angle α in Eq. 7.2.11 about the Euler axis of rotation is another indication of the quality of the control laws, because it shows the overall angle path that the satellite traverses during the maneuver. Its integral, named EULERINT ($=\int \alpha\, dt$), is shown in the figures to follow as the definitive criterion for comparing the different attitude control laws.

We present a simulated case for which the moments of inertia of the satellite are $I_x = 1{,}000$, $I_y = 500$, $I_z = 700$ kg-m^2. There are no control torque limits. The control gains are assumed to be identical for all the control laws compared. The open-loop transfer functions of the three axes are designed to obtain closed-loop natural frequencies of $\omega_n = 1$ rad/sec and closed-loop damping factors of $\xi = 1$. Given the listed moments of inertia, the proportional and derivative gains become: $K_x = 1{,}000$, $K_{xd} = 2{,}000$, $K_y = 500$, $K_{yd} = 1{,}000$, $K_z = 700$, and $K_{zd} = 1{,}400$. The satellite is submitted to step angular commands, starting at $t_0 = 1$ sec.

Figures 7.2.2 show time-domain responses for the Euler angles error control law as explained in Section 7.2.1 (Eqs. 7.2.2) - namely, responses of the Euler angles and also of the integral of the Euler axis rotation error (Figure 7.2.2.d). For comparison, the time-domain responses for the quaternion error control law explained in Section 7.2.4 (Eqs. 7.2.15) are shown in Figures 7.2.3. In these time-domain responses, the attitude commands are: $\psi_{\text{com}} = -6°$, $\theta_{\text{com}} = -4°$, and $\phi_{\text{com}} = 4°$. For these comparatively small attitude commands, the time responses for the two control laws are almost identical: EULERINT $= 16.2$ deg-sec (cf. Figures 7.2.2.d and 7.2.3.d). As explained in Section 7.2.2 and Appendix A, this results because, for small attitude changes, the direction cosine attitude errors approach the Euler angle errors.

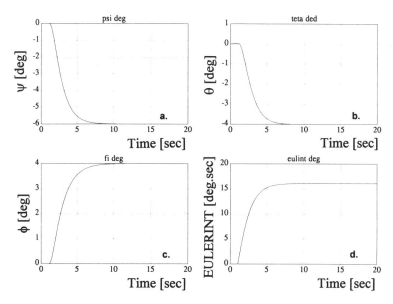

Figure 7.2.2 Euler angle step responses using the Euler angle error control law: $\psi_{\text{com}} = -6°$, $\theta_{\text{com}} = -4°$, and $\phi_{\text{com}} = 4°$.

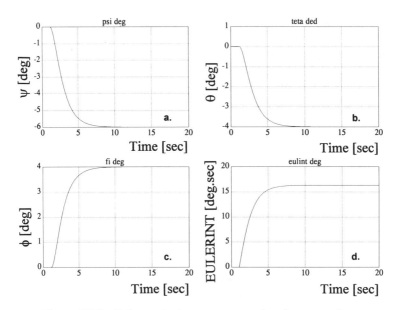

Figure 7.2.3 Euler angle time responses using the quaternion error control law: $\psi_{com} = -6°$, $\theta_{com} = -4°$, and $\phi_{com} = 4°$.

This approximate identity does not hold for large attitude maneuvers. The same simulations were repeated for the much larger attitude commands $\psi_{com} = -60°$, $\theta_{com} = -40°$, and $\phi_{com} = 40°$. Comparison between the time responses for the Euler angles control law (Figures 7.2.4) and for the quaternion control law (Figures 7.2.5) shows the clear superiority of the latter. In the first place, with the quaternion control law, the Euler attitude angles are well behaved and resemble their time responses for small attitude commands. Moreover, an oscillating behavior is observed for the Euler angles control law, where also the angular path of the Euler axis of rotation angle α is much larger.

The criterion factor EULERINT, which is the integral of the Euler axis-of-rotation error, is shown as part of Figure 7.2.4 and Figure 7.2.5 for large attitude maneuvers. For the simpler Euler angles control law, EULERINT = 262; for the quaternion control law, EULERINT = 157. This underscores the quaternion efficiency with respect to minimizing the length of the total angular path followed by the s/c in space.

7.2.6 Body-Rate Estimation without Rate Sensors

The three angular rates about the body axes of the spacecraft are included in the torque control laws of Eqs. 7.2.2, Eqs. 7.2.10, Eqs. 7.2.12, and Eqs. 7.2.15. In general, these rates can be obtained from rate gyro measurements or other equivalent instrumentation. Sometimes, in order to simplify the hardware of the attitude control system (ACS), such instrumentation is unavailable or, if it does exist, may – in order to prolong the life of the gyroscopic hardware – be reserved for those cases where (for example) highly accurate measurement of the attitude rates is mandatory in order to obtain the precise attitude of the s/c by integrating those rates.

7.2 / Equations for Basic Control Laws

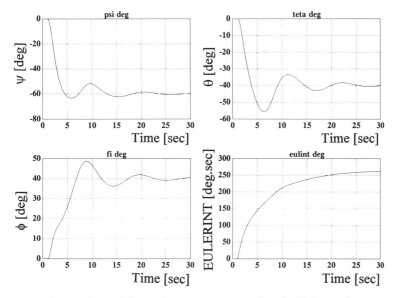

Figure 7.2.4 Euler angle step responses using the Euler angle error control law: $\psi_{com} = -60°$, $\theta_{com} = -40°$, and $\phi_{com} = 40°$.

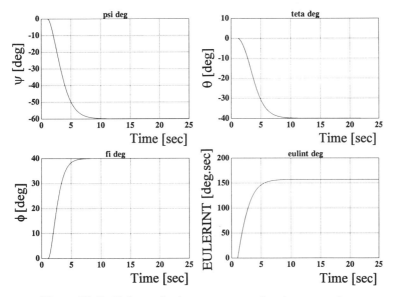

Figure 7.2.5 Euler angle time responses using the quaternion error control law: $\psi_{com} = -60°$, $\theta_{com} = -40°$, and $\phi_{com} = 40°$.

When the angular rates are not measured directly, they can be estimated from knowledge of the quaternion vector or the direction cosine matrix, which are obtained from attitude position measurements using horizon, sun, or star sensors. In this section we estimate the body rates p, q, and r from the quaternion vector $\mathbf{q}_S$. We will use Eqs. 4.7.13, which can be rewritten in the slightly different form

$$\dot{q}_i = \tfrac{1}{2}[\mathbf{Q}]\begin{bmatrix} p \\ q \\ r \end{bmatrix}, \quad i = 1, \ldots, 4 \tag{7.2.16}$$

(see Glaese et al. 1976).

With the definition

$$[\mathbf{Q}] = \begin{bmatrix} q_{S4} & -q_{S3} & q_{S2} \\ q_{S3} & q_{S4} & -q_{S1} \\ -q_{S2} & q_{S1} & q_{S4} \\ -q_{S1} & -q_{S2} & -q_{S3} \end{bmatrix}, \tag{7.2.17}$$

Eqs. 7.2.16 describe four equations with three unknowns. The matrix $[\mathbf{Q}]$ is not square and has no conventional inverse. But there does exist a *left pseudoinverse* matrix $[\mathbf{Q}_L]$ that can solve the problem. First define $\omega_{BR}^T = [p, q, r]$, where p, q, r are the body rates relative to the chosen reference frame. From Eq. 7.2.16, the solution for ω_{BR} becomes

$$\omega_{BR} = 2[\mathbf{Q}_L]\dot{\mathbf{q}}_i, \tag{7.2.18}$$

where $[\mathbf{Q}_L]$ is the left pseudoinverse matrix:

$$[\mathbf{Q}_L] = \{[\mathbf{Q}]^T[\mathbf{Q}]\}^{-1}[\mathbf{Q}]^T. \tag{7.2.19}$$

Since $\{[\mathbf{Q}]^T[\mathbf{Q}]\} = [\mathbf{1}]$, it follows that $[\mathbf{Q}_L] = [\mathbf{Q}]^T$, so

$$\begin{bmatrix} p \\ q \\ r \end{bmatrix} = 2 \begin{bmatrix} q_{S4} & q_{S3} & -q_{S2} & -q_{S1} \\ -q_{S3} & q_{S4} & q_{S1} & -q_{S2} \\ q_{S2} & -q_{S1} & q_{S4} & -q_{S3} \end{bmatrix} \begin{bmatrix} \dot{q}_{S1} \\ \dot{q}_{S2} \\ \dot{q}_{S3} \\ \dot{q}_{S4} \end{bmatrix}. \tag{7.2.20}$$

In Eq. 7.2.20, the spacecraft quaternion elements need to be differentiated. Since the calculated quaternion elements – obtained from position measurement sensors – are always noisy, their differentiation must be performed with adequate noise filters that are compatible with the bandwidth of the automatic control loops about the three body axes.

7.3 Control with Momentum Exchange Devices

The control torque laws of Section 7.2 assume the existence of control hardware that can generate the commanded torques. There are at least four distinct means of producing torque for the attitude control of spacecraft, based on:

(1) earth's magnetic field;
(2) reaction forces produced by expulsion of gas or ion particles;
(3) solar radiation pressure on spacecraft surfaces; and
(4) momentum exchange devices (rotating bodies inside the s/c).

The first three techniques listed are *inertial controllers,* in the sense that they change the overall inertial angular momentum of the satellite. Such controllers produce torques labeled $\mathbf{T}_c$ in Eq. 4.8.2.

7.3 / Control with Momentum Exchange Devices

Magnetic techniques provide continuous and smooth control. However, the level of torques that can be achieved with magnetic torqrods is normally low (in the range of 1–10 mN-m) and generally insufficient for fast attitude maneuvers (see also Section 7.4). Magnetic torques are also dependent on the chosen orbit inclination, on the altitude above earth of the satellite, and so on. Moreover, magnetic control is excluded for spacecraft not revolving about the earth, since it is based on the earth's magnetic field.

Reaction controllers are not linear, in the sense that they provide reaction torques of constant amplitude and modulated time duration. The level of control torques that can be achieved with reaction pulses is almost unbounded. However, no smooth control can be achieved owing to the inherent impulsive nature of reaction thrusters (see Chapter 9).

Torques obtained from solar pressure cannot be used for attitude maneuvering since the level of torques that can be produced (when the s/c sees the sun) are of the order of tens of μN-m only, which are clearly insufficient for attitude maneuvers. Moreover, they cannot produce torques about the three s/c axes. However, solar torques are sometime used in geostationary satellites to counteract the parasitic solar disturbances acting on the s/c and to provide active nutation damping to momentum bias–controlled satellites (see Section 8.6).

The remaining option is based on rotating masses inside the spacecraft body, so that angular momentum is transferred between different parts of the satellite without changing its overall inertial angular momentum. The resulting torque control devices are called *momentum exchange devices,* and include reaction wheels, momentum wheels, and control moment gyros (CMGs).

For very accurate attitude control systems and for moderately fast maneuvers, the reaction wheels are preferred because they allow continuous and smooth control with the lowest possible parasitic disturbing torques (see Appendix C). The level of torque that can be achieved with reaction wheels is of the order of 0.05–2 N-m. With control moment gyros (used in manned spacecraft), torques of 200 N-m are achievable. However, such CMGs are very heavy and are seldom used in the ACS of ordinary-sized satellites. (For more about the use of CMGs see Kaplan 1976 or Oh and Valadi 1991.) In this section we will focus on reaction wheels.

7.3.1 Model of the Momentum Exchange Device

Inside a spacecraft, a symmetrical rotating body produces angular torque when accelerated about its axis of rotation. The rotating body may have an initial constant momentum h_w. Since this momentum is internal to the spacecraft, its increase does not change the overall momentum of the system but instead merely transfers the momentum change (with negative sign) to the spacecraft (see Eq. 4.8.2). This is the principle of *conservation of angular momentum.*

The reaction wheel can be mounted in the satellite with its rotational axis in any direction relative to the satellite's axis frame. The momentum vector of all the momentum exchange devices inside the satellite body can be expressed with reference to the axes of the spacecraft body frame as $\mathbf{h}_w = [h_{wx}\ h_{wy}\ h_{wz}]^T$. For controlling the attitude in space, at least three reaction wheels are required. With this modification, Eqs. 4.8.14 were obtained for the general case of a satellite containing any number

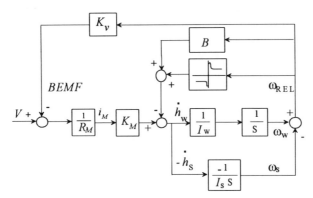

Figure 7.3.1 Basic model of a momentum exchange device.

of momentum exchange devices whose axes of rotation are not all coplanar and at least three of whose axes are noncollinear. In those equations, any device's momentum change (e.g. $\dot{h}_{wx}$) will produce an equal angular torque on the body about its X_B body axis, but opposed in direction.

In the simplified Eqs. 7.2.2 of Section 7.2.1, for instance, the following control torque commands will be applied:

$$\begin{aligned}\dot{h}_{wx} &= T_{cx} = K_x(\phi_{com}-\phi)+K_{xd}\dot{\phi},\\ \dot{h}_{wy} &= T_{cy} = K_y(\theta_{com}-\theta)+K_{yd}\dot{\theta},\\ \dot{h}_{wz} &= T_{cz} = K_z(\psi_{com}-\psi)+K_{zd}\dot{\psi}.\end{aligned} \quad (7.3.1)$$

The required angular torques can be achieved by accelerating the rotor of electrical motors whose axes of rotation are aligned with the body axes X_B, Y_B, Z_B; on their rotating axes are assembled flywheels having moments of inertia labeled I_{wx}, I_{wy}, I_{wz}. Of course, reality is not so simple, as will be seen in the analysis that follows. A complete dynamic model for the electrical motor to be used as the torque controller is shown in Figure 7.3.1. See also Section 6.6.3 for a similar, but incomplete, analysis.

In Figure 7.3.1, V is the input voltage to the electrical motor, R_M is the electrical resistance of the motor armature, and K_M is the torque coefficient of the motor. The term I_w denotes the overall moment of inertia of the rotor including the flywheel, the task of which is to produce the desired torque, and I_S is the moment of inertia of the satellite. Finally, ω_{REL} is the angular velocity of the rotating part of the motor relative to the stator and (equivalently) to the satellite body, since the stator is fixed to the body of the spacecraft.

The primary task of the electrical motor is to provide the necessary angular torque to the satellite. Assume there are no external disturbances T_d and no inertial control torques acting on the satellite. With this assumption, according to Euler's moment equation of angular motion we have

$$\dot{h}_S + \dot{h}_w = 0. \quad (7.3.2)$$

This means that, in order to apply a torque on the body about some axis, a torque in the opposite direction must be produced by the rotor of the electrical motor. Thus, $\dot{h}_S = -\dot{h}_w$.

7.3 / Control with Momentum Exchange Devices

The block B in Figure 7.3.1 is the viscosity damping coefficient sensed by the rotor, and the damping torque is proportional to the angular velocity of the rotor relative to the satellite body frame. The block describing the coulomb and dry friction will be omitted in the analysis so as to enable a linear transfer function of the complete dynamical model, including the satellite dynamics. With these assumptions it is easily found that

$$\frac{\dot{h}_w}{V} = \frac{I_w \dot{\omega}_w}{V} = \frac{s\left(\frac{K_M}{R_M}\right)}{s + \left(\frac{1}{I_w} + \frac{1}{I_S}\right)\left(\frac{K_V K_M}{R_M} + B\right)}. \tag{7.3.3}$$

Equation 7.3.3 indicates that a step in the input voltage to the electrical motor does not produce a pure angular torque, because there is a time constant in the denominator and a differentiator in the numerator. With the valid assumptions that $B \to 0$ and $I_w \ll I_S$, Eq. 7.3.3 reduces to

$$\frac{\dot{h}_w}{V} = \frac{s\frac{I_w}{K_V}}{1 + s\frac{I_w R_M}{K_V K_M}}. \tag{7.3.4}$$

Our true interest is in obtaining an immediate torque $\dot{h}_w$ as a response to a torque command T_c. The set-up in Figure 7.3.2 helps to achieve this characteristic: there is a feedback path from the motor current, which is proportional to the torque provided by the electrical motor. The transfer function between the torque command T_c and the achieved angular torque $\dot{h}_w$ becomes:

$$\frac{\dot{h}_w}{T_c} = \frac{\frac{K}{sR_M}}{1 + \frac{K}{sR_M}\left(1 + \frac{K_V K_M}{KI_w}\right)}. \tag{7.3.5}$$

If we choose $K \gg K_V K_M / I_w$ then

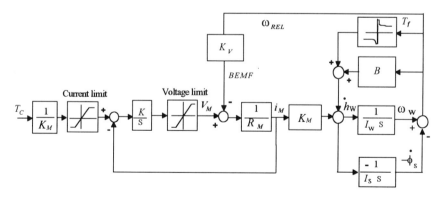

Figure 7.3.2 Use of a momentum exchange device in the torque command mode.

$$\frac{\dot{h}_w}{T_c} = \frac{1}{1+s(R_M/K)}. \tag{7.3.6}$$

This is the basic equation that converts an electric motor into what is known as a reaction wheel (RW) or a momentum wheel (MW). The added electronics complete the reaction wheel assembly (RWA). See Appendix C for technical information on reaction wheels.

In Eq. 7.3.6 there is still a small time constant, which for a good RWA may be of the order of milliseconds and thus can be neglected for practical purposes. Equation 7.3.6 gives a linear model for the momentum exchange device. Technically, such a device has torque and velocity limitations; when the attitude control of a satellite is designed, these limitations must be taken into consideration. The first stage in the design process is to size the wheel correctly, which means, first of all, deciding about the maximum torque and momentum that the wheel should be able to provide (see Section 7.7). In a later stage we will see that there is another characteristic deserving special attention: the inherent torque noises, *cogging* and *ripple torque,* that are due to the nonlinear physical components (e.g. stator and rotor poles) that compose the wheel. In future analysis, it will be explicitly stated if the linear model used for the wheel is sufficient or if additional disturbing elements must be added to it.

In the linearized dynamic model of the satellite, Eqs. 4.8.14, the linear model of the momentum exchange device is used with the additional assumption that the small time constant of the wheel assembly in Eq. 7.3.6 can be ignored, so that $\dot{h}_w \approx T_c$.

7.3.2 Basic Control Loop for Linear Attitude Maneuvers

The simplest attitude maneuver control loop is shown in Figure 7.3.3. For small attitude maneuvers around $\psi \approx 0$, $\theta \approx 0$, and $\phi \approx 0$, the quaternion, direction cosine, and Euler axis error control laws reduce to the simplest Euler angle error control law, Eqs. 7.2.2. With $K_1 = K_2 = 0$ in Figure 7.3.3, the transfer function between the error e and the attitude command θ_{com} is easily found to be

$$\frac{e}{\theta_{\text{com}}} = \frac{s\left(s+\dfrac{K_d}{I_S}\right)}{s^2+\left(\dfrac{K_d s+K}{I_S}\right)}. \tag{7.3.7}$$

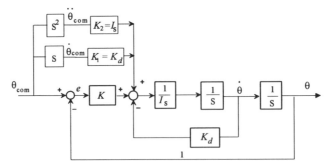

Figure 7.3.3 Basic one-axis attitude control loop with null steady-state error for position, velocity, and parabolic input commands.

7.3 / Control with Momentum Exchange Devices

For a step command $\theta_{\text{com}}(s) = |\theta_{\text{com s}}|/s$, we find that

$$e_{ss} = \lim_{t \to \infty} e(t) = \lim_{s \to 0} \frac{|\theta_{\text{com s}}|s^2(K_d/I_S)}{s(K/I_S)} = 0. \tag{7.3.8}$$

However, if a velocity input is expected then, with this control configuration, the steady-state error will not be null. For a velocity input $\theta(t) = |\theta_{\text{com v}}|t$, the steady-state error will become

$$\lim_{t \to \infty} e(t) = |\theta_{\text{com v}}| \frac{K_d}{K}. \tag{7.3.9}$$

There are two ways to decrease this steady-state error to zero: (1) we can incorporate an electronic integration in the loop; or (2) we can also add a feedforward command that is proportional to the differentiation of the input command θ_{com} and to K_1 in Figure 7.3.3. This latter approach is preferred because θ_{com} is a known function of time, the differentiation of which ($\dot{\theta}_{\text{com}}$) is also known analytically. Moreover, no noise is added to the input of the control loop. The error response for the velocity input $\theta_{\text{com}}(s) = |\theta_{\text{com v}}|/s^2$ becomes

$$\frac{e}{\theta_{\text{com}}} = \frac{I_S s^2 \left(1 + \frac{K_d - K_1}{s I_S}\right)}{I_S s^2 + K_d s + K}. \tag{7.3.10}$$

Using the final-value theorem of the Laplace transforms in Eq. 7.3.10 with $K_1 = K_d$, we have

$$e_{ss}(t) = \lim_{t \to \infty} e(t) = \lim_{s \to 0} \frac{|\theta_{\text{com v}}|}{s^2} \frac{s^3}{s^2 + (1/I_S)(K_d s + K)} = 0. \tag{7.3.11}$$

However, for a parabolic command input, the steady-state error will still remain finite. Another feedforward command can be added in order to eliminate this steady-state error. As in the previous discussion, we can feedforward an input proportional to $\ddot{\theta}_{\text{com}}(t)$ with a gain K_2. An analysis similar to the foregoing shows that if $K_2 = I_S$ then the steady-state error for a parabolic attitude command will be null.

At this point we should mention the error due to the nonlinear characteristics of the wheel. The wheel remains at rest so long as the commanded torque does not overcome the coulomb friction level; hence, an angular error will appear because the wheel does not react until the torque overcomes the friction. Therefore, the additional steady-state error will be: $e_{ss} = T_{\text{fric}}/K$, where T_{fric} is the friction torque of the wheel at zero angular velocity.

In this section it was assumed that the attitude input commands were small, so that the reaction wheel assembly works in its linear region. A time-optimal control solution will be analyzed in Section 7.6 for large attitude maneuvers in which torque and velocity saturation levels of the wheel assembly are reached. Saturation by momentum accumulation will be discussed next.

7.3.3 Momentum Accumulation and Its Dumping

One drawback of a momentum exchange device is that it cannot independently remove the angular momentum that accumulates, owing to external disturbances, in the satellite system. As we know, according to Euler's moment equations,

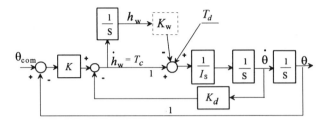

Figure 7.3.4 Momentum saturation and dumping of the reaction wheel.

any external torque disturbance acting on the body augments the angular momentum of the whole spacecraft system. With no active attitude control, the body accumulates an angular velocity as the angular momentum of the s/c increases, changing the attitude of the spacecraft. However, if reaction wheels are used to stabilize the s/c attitude, then the accumulated angular momentum will be transferred to the wheels. With harmonic external disturbances, the stored momentum will also be harmonic; as long as its level stays well inside the momentum saturation limits of the wheels, no control problem arises. However, with constant external disturbances, the momentum of the wheels will increase without limit, thus saturating them and precluding their ability to provide the necessary control torques. The unwanted accumulated momentum in the wheel must be removed from the momentum exchange device, a process called *dumping* of the momentum.

In the present analysis, we assume that the time constant in Eq. 7.3.6 is close to null, so that the controlling torque $\dot{h}_w$ equals the commanded control torque T_c. The set-up for the simplified dynamics of the spacecraft, the external disturbances, and the wheel controlling torque are shown in Figure 7.3.4. In this figure, with $K_w = 0$ (i.e., without the dumping control), an error control torque T_c will be required to counteract a constant disturbance T_d. But, as assumed and as shown in Figure 7.3.4, $\dot{h}_w = T_c$. If the open-loop transfer function is defined as

$$L_{in}(s) = \frac{[K_d s + K]}{I_S s^2} \frac{[s + K_w]}{s},$$

then

$$\frac{\dot{h}_w}{T_d} = \frac{L_{in}(s)}{1 + L_{in}(s)} = \frac{[K_d s + K]s}{I_S s^3 + [K_d s + K][s + K_w]}. \qquad (7.3.12)$$

Hence, for a constant disturbance $T_d = |T_d|/s$,

$$\dot{h}_w(s) = |T_d| \frac{K_d s + K}{I_S s^3 + [K_d s + K][s + K_w]}. \qquad (7.3.13)$$

Also,

$$h_w(s) = \frac{\dot{h}_w(s)}{s} = \frac{|T_d|}{s} \frac{K_d s + K}{I_S s^3 + [K_d s + K][s + K_w]}. \qquad (7.3.14)$$

To find the steady-state value of $h_w(t)$, we apply the final-value theorem:

$$\lim_{t \to \infty} h_w(t) = \lim_{s \to 0} \frac{s|T_d|}{s} \frac{[K_d s + K]}{I_S s^3 + [K_d s + K][s + K_w]}. \qquad (7.3.15)$$

7.3 / Control with Momentum Exchange Devices

If $K_w = 0$, which means that the momentum of the wheel is not actively controlled and consequently not limited, we find that $h_w(t)_{ss} = \lim_{t\to\infty} h_w(t) = \infty$; that is, the momentum accumulating in the wheel will increase infinitely. To be more specific, as $t \to \infty$:

$$h_w(t) \to |T_d|t. \tag{7.3.16}$$

However, if we choose $K_w \neq 0$ then

$$h_w(t)_{ss} = \lim_{t\to\infty} h_w(t) = \frac{|T_d|}{K_w}. \tag{7.3.17}$$

The gain coefficient K_w requires some explanation. The variable h_w is measured and an external torque to the spacecraft, proportional to h_w, is applied about the axis of rotation of the wheel. This external torque can be produced by reaction or magnetic means (see also Section 7.5). There must be applied an *external* inertial torque that cannot be supplied by the reaction wheel itself, as explained previously with regard to Euler's moment equations of motion. This is also observed in the block diagram of the control system in Figure 7.3.4: note that the dumping torque $h_w K_w$ is applied directly to the dynamics of the satellite. If the commanded torque $h_w K_w$ were erroneously applied via the wheel itself, then the transfer function of Eq. 7.3.14 would be different and the added $h_w K_w$ term would no longer be effective for dumping the excess momentum in the momentum exchange device.

7.3.4 A Complete Reaction Wheel-Based ACS

Three reaction wheels, with each one's rotational axis parallel to one of the satellite's body axes, make up the simplest control system. As the three body axes' dynamics are separated, the design can be carried out independently for each axis. However, if one of the assemblies becomes damaged then the satellite's attitude can no longer be adequately controlled. For this reason, a fourth RWA is installed in order to increase the reliability of the entire system (Fleming and Ramos 1979, Junkins and Turner 1986). The additional wheel is installed with its axis "off" the three principal s/c axes, enabling (reduced) torque control about any one of those axes. Thus, the incapacity of any one of the RWAs aligned with the satellite's principal axes can be compensated by the torque capabilities of the fourth wheel.

In this section we will analyze one possible geometrical configuration of a control system based on four reaction wheels; see Figure 7.3.5. The rotational axes of the

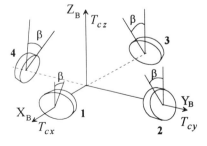

Figure 7.3.5 Attitude control system with four reaction wheels.

four wheels are inclined to the X_B-Y_B plane by an angle β. Because of this inclination, each wheel can apply torques and momentum in the Z_B direction also. The torques delivered by the wheels are called T_i ($i = 1, ..., 4$). The torques produced along the three body axes are $\hat{T}_{cx}, \hat{T}_{cy}, \hat{T}_{cz}$. Thus we have the following relations:

$$\begin{bmatrix} \hat{T}_{cx} \\ \hat{T}_{cy} \\ \hat{T}_{cz} \end{bmatrix} = \begin{bmatrix} T_{cx}/c\beta \\ T_{cy}/c\beta \\ T_{cz}/s\beta \end{bmatrix} = \begin{bmatrix} 1 & 0 & -1 & 0 \\ 0 & 1 & 0 & -1 \\ 1 & 1 & 1 & 1 \end{bmatrix} \begin{bmatrix} T_1 \\ T_2 \\ T_3 \\ T_4 \end{bmatrix} = [\mathbf{A}_w] \begin{bmatrix} T_1 \\ T_2 \\ T_3 \\ T_4 \end{bmatrix} = [\mathbf{A}_w]\mathbf{T}. \qquad (7.3.18)$$

Here, $c\beta = \cos(\beta)$ and $s\beta = \sin(\beta)$, with β the inclination angle of the wheel axes to the X_B-Y_B body plane.

The control vector $\hat{\mathbf{T}}_c$ is computed by any one of the control laws described in Section 7.2. We need to calculate the components T_i, which are the control torques to be applied by each one of the four wheels. Unfortunately, the matrix $[\mathbf{A}_w]$ in Eq. 7.3.18 is not square, and cannot be inverted. To find the vector components of T_i, we must assume some optimizing criterion. For instance, we might wish to minimize the norm of the vector $\mathbf{T} = [T_1 \ T_2 \ T_3 \ T_4]^T$, for which we define the Hamiltonian H as

$$H = \sum_{i=1}^{4} T_i^2. \qquad (7.3.19)$$

By definition,

$$\begin{aligned} \hat{T}_{cx} &= T_1 - T_2, \\ \hat{T}_{cy} &= T_2 - T_4, \\ \hat{T}_{cz} &= T_1 + T_2 + T_3 + T_4. \end{aligned} \qquad (7.3.20)$$

Let us define the functions

$$\begin{aligned} g_1 &= T_1 - T_3 - \hat{T}_{cx}, \\ g_2 &= T_2 - T_4 - \hat{T}_{cy}, \\ g_3 &= T_1 + T_2 + T_3 + T_4 - \hat{T}_{cz}. \end{aligned} \qquad (7.3.21)$$

The Lagrangian will be

$$L = H + \lambda_1 g_1 + \lambda_2 g_2 + \lambda_3 g_3 + \lambda_4 g_4, \qquad (7.3.22)$$

and the conditions for minimizing H will be:

$$\begin{aligned} \frac{\partial L}{\partial T_1} &= 2T_1 + \lambda_1 + \lambda_3 = 0, \\ \frac{\partial L}{\partial T_2} &= 2T_2 + \lambda_2 + \lambda_3 = 0, \\ \frac{\partial L}{\partial T_3} &= 2T_3 - \lambda_1 + \lambda_3 = 0, \\ \frac{\partial L}{\partial T_4} &= 2T_4 - \lambda_2 + \lambda_3 = 0. \end{aligned} \qquad (7.3.23)$$

7.3 / Control with Momentum Exchange Devices

From Eqs. 7.3.23 we derive the final condition:

$$\Delta T = T_1 - T_2 + T_3 - T_4 = 0, \qquad (7.3.24)$$

so that Eq. 7.3.18 may be rewritten in the following matrix form:

$$\begin{bmatrix} \hat{T}_{cx} \\ \hat{T}_{cy} \\ \hat{T}_{cz} \\ 0 \end{bmatrix} = \begin{bmatrix} 1 & 0 & -1 & 0 \\ 0 & 1 & 0 & -1 \\ 1 & 1 & 1 & 1 \\ 1 & -1 & 1 & -1 \end{bmatrix} \begin{bmatrix} T_1 \\ T_2 \\ T_3 \\ T_4 \end{bmatrix}. \qquad (7.3.25)$$

The square matrix in Eq. 7.3.25 has an inverse, which is easily found to be

$$\begin{bmatrix} T_1 \\ T_2 \\ T_3 \\ T_4 \end{bmatrix} = \frac{1}{2} \begin{bmatrix} 1 & 0 & \frac{1}{2} & \frac{1}{2} \\ 0 & 1 & \frac{1}{2} & -\frac{1}{2} \\ -1 & 0 & \frac{1}{2} & \frac{1}{2} \\ 0 & -1 & \frac{1}{2} & -\frac{1}{2} \end{bmatrix} \begin{bmatrix} \hat{T}_{cx} \\ \hat{T}_{cy} \\ \hat{T}_{cz} \\ 0 \end{bmatrix}. \qquad (7.3.26)$$

Equation 7.3.26 provides the needed transformation between the three body axes' command control torques and the four wheels' command control torques. The same result can be obtained by a right pseudoinverse transformation (see Wertz 1986). This transformation is defined by: $[\mathbf{A}_{wR}] = [\mathbf{A}_w]^T \{[\mathbf{A}_w][\mathbf{A}_w]^T\}^{-1}$ and leads to the same minimization of H in Eq. 7.3.19.

For the case where only three reaction wheels are used, with their axes of rotation not parallel to the three body principal axes, the body-to-wheel torque transformation is a 3×3 matrix. Assuming that the wheel axes are not collinear and not all in the same plane, the transformation matrix has the form

$$\begin{bmatrix} T_1 \\ T_2 \\ T_3 \end{bmatrix} = \begin{bmatrix} a_{w11} & a_{w21} & a_{w31} \\ a_{w12} & a_{w22} & a_{w32} \\ a_{w13} & a_{w23} & a_{w33} \end{bmatrix} \begin{bmatrix} T_{cx} \\ T_{cy} \\ T_{cz} \end{bmatrix}.$$

In this matrix, the a_{wij} elements are the direction cosines of the wheel axes with reference to the body frame.

Some clarification of Eq. 7.3.26 is needed. When an attitude maneuver is performed, the body axes command torques are divided between the four reaction wheels. Equation 7.3.19 assures that the norm of the wheel torque commands is minimized. The torque produced by a reaction wheel is proportional to the current in the stator of the motor. On the other hand, the power consumption in the RWA is proportional to the square of the current in the motor of the RW. Thus, our division of the control torque among the four reaction wheels also assures minimization of the electrical power consumption of the four RWAs. Unfortunately, this is not the entire story; we must also determine what happens to the momentum of the four wheels. As we shall see, it is important to minimize also the norm $|\mathbf{h}_w| = \sum h_{wi}^2$. Can we minimize both norms simultaneously? The answer will be given in the next section.

7.3.5 Momentum Management and Minimization of the $|\mathbf{h}_w|$ Norm

Viscosity torques produced on the constantly rotating rotor of the wheel increase power consumption of the RWA electronics. Of course, saving electrical power in all spacecraft subsystems is of utmost importance. Moreover, for satisfactory operation of a reaction wheel, under normal conditions its angular velocity should be as far as possible from saturation. These considerations dictate that, at the end of any attitude maneuver, the norm of the wheel angular velocities should be minimized.

If the four wheels are activated in their linear range – which means that no torque or momentum saturation levels are reached – and if no coulomb friction exists, then the angular velocities of each wheel at the end of any attitude maneuver will remain null (if they were so at the beginning of the maneuver). However, depending on the character of the attitude maneuver, if one of the foregoing conditions does not hold then the angular velocity norm of the four wheels will not be null at the end of the attitude maneuver.

Instead of controlling the attitude of the satellite with control torques as computed in Eqs. 7.2.2, we can control its attitude with *angular momentum commands* (see Fleming and Ramos 1979). In this case, instead of minimizing the norm of the control torque vector, we should minimize the norm of the *control momentum vector*. The results of the previous section will hold also for this control momentum case. In Eq. 7.3.25 and Eq. 7.3.26, $\hat{T}_{cx}, \hat{T}_{cy}, \hat{T}_{cz}, T_1, T_2, T_3, T_4$ can be exchanged with $\hat{h}_{cx}, \hat{h}_{cy}, \hat{h}_{cz}, h_1, h_2, h_3, h_4$, respectively, where $h_{w1} = I_{w1}\omega_{w1}$, $h_{w2} = I_{w2}\omega_{w2}$, and so on. As for the control torque law, here again minimization of the momentum norm requires the following condition:

$$\Delta h_w = h_{w1} - h_{w2} + h_{w3} - h_{w4} = 0. \tag{7.3.27}$$

Our task now is to measure Δh_w and to feed this error back to the commanded T_is, so that ΔT of Eq. 7.3.24 and Δh_w of Eq. 7.3.27 are simultaneously satisfied. See Figure 7.3.6 for the control set-up of the four-wheel configuration. The analysis to follow is due to E. Zemer (MBT, Israel Aircraft Industries).

It is important that the control of Δh_w not produce any change in the commanded body control torques $\hat{T}_{cx}, \hat{T}_{cy}, \hat{T}_{cz}$, which means that the following additional condition must be fulfilled:

$$\begin{bmatrix} \Delta\hat{T}_{cx} \\ \Delta\hat{T}_{cy} \\ \Delta\hat{T}_{cz} \end{bmatrix} = [\mathbf{A}_w] \begin{bmatrix} \Delta T_{c1} \\ \Delta T_{c2} \\ \Delta T_{c3} \\ \Delta T_{c4} \end{bmatrix} = [\mathbf{A}_w] \begin{bmatrix} \Delta h_w K_1 \\ \Delta h_w K_2 \\ \Delta h_w K_3 \\ \Delta h_w K_4 \end{bmatrix} = 0. \tag{7.3.28}$$

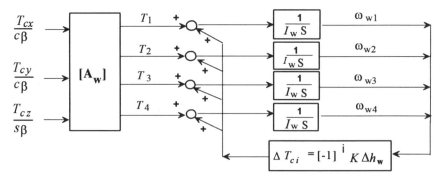

Figure 7.3.6 Momentum management control of the four reaction wheels.

7.3 / Control with Momentum Exchange Devices

Using the definition of $[\mathbf{A}_w]$ in Eq. 7.3.18, we obtain

$$\Delta h_w K_1 - \Delta h_w K_2 = 0,$$
$$\Delta h_w K_2 - \Delta h_w K_4 = 0, \tag{7.3.29}$$
$$\Delta h_w K_1 + \Delta h_w K_2 + \Delta h_w K_3 + \Delta h_w K_4 = 0.$$

These equations are satisfied for

$$K_1 = K, \quad K_2 = -K, \quad K_3 = K, \quad K_4 = -K. \tag{7.3.30}$$

This is our final result, which can also be put in the form

$$\Delta T_{ci} = [-1]^i K \Delta h_w, \quad i = 1, \ldots, 4, \tag{7.3.31}$$

with Δh_w as defined in Eq. 7.3.27.

In the present analysis, the wheel configuration of Figure 7.3.5 was taken as a design example. The same analysis could be carried out for any other "skewed" four-wheel configuration, in which no three wheel axes are coplanar and no two axes are collinear (see e.g. Azor 1993).

It remains to determine the value of K. Suppose that there is a nonzero initial condition in one of the wheel's angular velocities. The response of Δh_w to that initial condition will be:

$$\Delta h_w(s) = \frac{\omega_{w1}(0)}{s} \frac{1}{1 + 4K/(I_w s)} = \frac{\omega_{w1}(0)}{s + 4K/I_w}. \tag{7.3.32}$$

We must choose the time constant $I_w/4K$ so that it will be slower than that of the slowest attitude control loops.

A 6-DOF simulation was carried out to demonstrate the practicability of the momentum management control. In these simulations, the quaternion error control law was used (Eqs. 7.2.15). The moments of inertia [kg-m^2] of the satellite were chosen to be $I_x = 1,000$, $I_y = 500$, $I_z = 700$, and $I_w = 0.1$. The maximum torque that the reaction wheels can deliver is 0.5 N-m. The attitude control of the system was checked for the step angular inputs of: $\psi_{com} = 1°$, $\theta_{com} = 2°$, and $\phi_{com} = -5°$. The results are shown in Figures 7.3.7–7.3.9.

Figure 7.3.7 shows the Euler angle time responses for the system without torque saturation (Figure 7.3.7.a) and the response for the same command inputs with torque saturation of 0.5 N-m (Figure 7.3.7.b). Because of the limited torque capabilities of the reaction wheels, the time to reach the desired steady-state response increases drastically; see also Section 7.3.6.

Figure 7.3.8 shows the norms of the angular momentum of the four wheels, both *without* and *with* the momentum management feedback loop. Without momentum management (Figure 7.3.8.a), the norm of the angular momentum of the four wheels at steady state is 9.63 N-m-sec. With the momentum management feedback loop (Figure 7.3.8.b), the same norm is decreased to 0.182 N-m-sec after 55 sec. The minimizing factor Δh_w is also shown. With no momentum management ($K_w = 0$), it is seen that $\Delta h_w = 1.93$ N-m-sec at steady state (Figure 7.3.8.a). With momentum management ($K_w = 0.2$), Δh_w is reduced practically to zero (Figure 7.3.8.b).

The individual angular velocities of the four wheels, *without* and *with* momentum management, are shown in Figure 7.3.9.a and Figure 7.3.9.b, respectively. It is important to emphasize that the wheel momentum vector in Figure 7.3.9.a has not

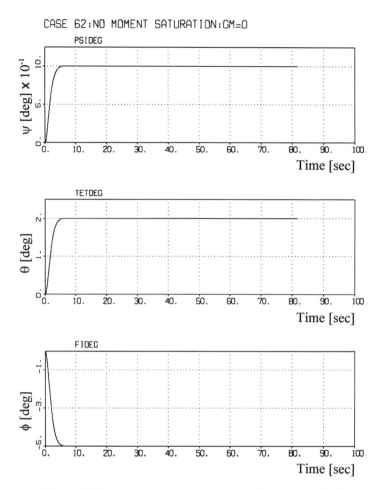

Figure 7.3.7.a Attitude time responses *without* torque limitations.

changed, since no external torques were applied on the satellite. Only the division of the angular momentum between the wheels on the same axis was altered.

7.3.6 *Effect of Noise and Disturbances on ACS Accuracy*

For accurate attitude control systems and moderately fast maneuvers, reaction wheels are well suited because they allow continuous and smooth control with comparatively low parasitic disturbing torques (see Appendix C). However, such disturbances have a strong influence on the quality of the attitude control, so we must take care to minimize their influence. The quality of the momentum exchange device is not the only factor that determines the capacity of the control design to achieve the desired attitude accuracy and stability. Attitude sensors are no less responsible for the quality and attitude accuracy of the ACS. Noise that is inherent in the various system sensors also influences the attitude accuracies that can be achieved.

7.3 / Control with Momentum Exchange Devices

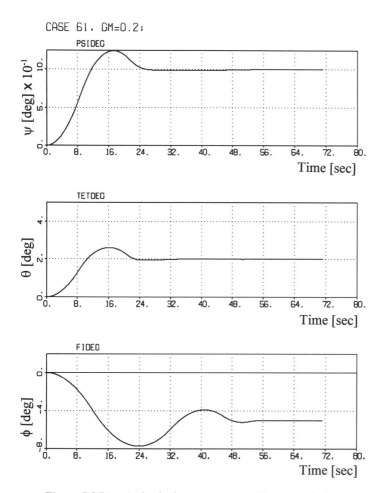

Figure 7.3.7.b Attitude time responses *with* torque limitations.

In this section we analyze the influence of disturbing torques and sensor noise on the overall behavior of the ACS, and examine some of their conflicting implications for the control loop bandwidth.

Principal Types of Attitude Control Sensors

Extremely accurate attitude control is required in spacecraft whose payloads are, for example, observation telescopes that operate in different light spectra (Boorg 1982, Dougherty et al. 1982). For such telescopes, attitude stability of several arc-sec/sec and positioning accuracy of 0.01° (or less) are commonly required. These standards must be achieved in spite of parasitic disturbances and sensor noise.

Figure 7.3.10 (p. 178) shows a block diagram of the basic scheme of the attitude control feedback system, including the inherent parasitic noise sources that render the control engineer's life difficult. In this figure, it is easily perceived that there are actually two dynamic states that must be controlled: (1) the attitude angle θ; and

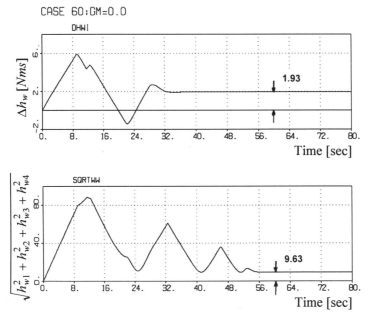

Figure 7.3.8.a Time histories of the norm of the angular momentums of the four wheels *without* momentum management.

(2) its attitude stability $\dot{\theta}$, which in some cases may be the more important variable. In order to control attitude and its stability, two principal types of attitude determination hardware are used: attitude sensors and angular velocity sensors (see also Appendix B).

Attitude Sensors The most common attitude sensor for earth-orbiting satellites is the earth sensor, which optically senses the globe contour and uses this information to calculate the attitude of the s/c with respect to the earth. Expected accuracies of these instruments are of the order of 0.02° for expensive versions and 0.5° for cheaper ones. Earth sensors have the drawback of being noisy, with RMS noise levels of about 0.03° or higher.

Sun sensors also are quite common on earth-orbiting satellites. For such satellites they are more accurate than earth sensors because they are based on measuring the angular distance from the sun disc, which is smaller than the earth disc. Sun sensors are also very efficient for spacecraft that are not orbiting the earth. Accuracies of about 0.01° can be achieved with very expensive instruments. The statistical noise level is quite low, of the order of 0.01° (RMS), depending also on the dynamic range of the instrument's output.

The star sensor is currently the most accurate attitude sensor, allowing accuracies of the order of 1 arc-sec. Unfortunately, star sensors are very expensive and less reliable than earth and sun sensors. The instrumentation is quite complex and difficult to handle, and their operation depends on complicated algorithms. For these reasons, star sensors are used as a last resort (see Pircher 1989).

7.3 / Control with Momentum Exchange Devices 175

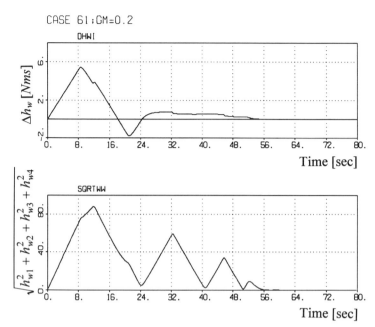

Figure 7.3.8.b Time histories of the norm of the angular momentums of the four wheels *with* momentum management.

Angular Velocity Sensors Rate integrating gyros (RIGs) are commonly used and very accurate; integrating the rate sensor output enables precise estimation of satellite attitude (Dougherty et al. 1982). Moreover, use of this instrument is not dependent on the s/c orientation in space. This is an advantage over horizon, star, and sun sensors, whose signal source must be within their optical fields of view (FOV), thereby reducing the spacecraft envelope of useful attitudes.

Conventional rate sensors are in principle less accurate than RIGs. They are used for rate sensing in various control tasks, including rate control (see Section 6.3) and damping in the ACS (Section 6.6.3). In general, the choice of particular sensors for attitude control will depend on the required accuracy and also on the specific tasks to be fulfilled by the spacecraft.

Conventional Attitude Control Configuration – Statistical Error Analysis

Figure 7.3.10 (p. 178) shows a control configuration using an attitude sensor of any kind together with a rate sensor. The parasitic noises pertaining to the sensors are indicated as WN_{RS} for the *rate sensor noise* and WN_{PS} for the attitude *position sensor noise*. In this case, the reaction wheel introduces its inherent *torque noise*, WN_{RW} (see Bosgra and Smilde 1982).

In the following analysis, it is theoretically possible to deal with ideal white noise, whose power spectral density (PSD) function is uniform and of amplitude WN_-. In nature, white noise does not exist; likewise, the noise existing in engineering problems is colored. We will assume a colored noise, which is the output of a first-order

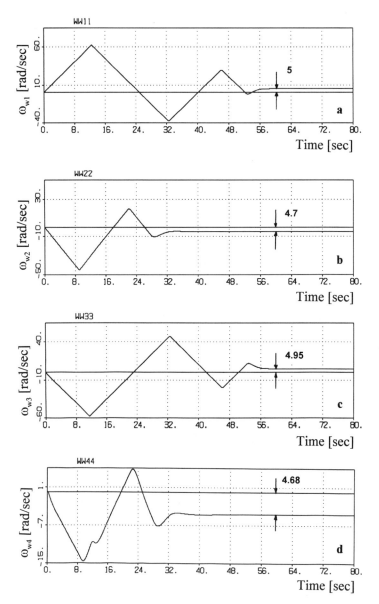

Figure 7.3.9.a Angular velocities of the four wheels *without* momentum management.

filter with corner frequency of ω_c at whose input was incorporated a white noise source. With this definition, the relation between the amplitude WN_ of the white noise source and the mean square (MS) value x^2 of the colored noise at the output of the filter is $x^2 = 0.5 \times \text{WN}_- \times \omega_c$. With this definition, x is the root mean square (RMS) value of the filter's output.

As discussed in Section 7.3.1, $\dot{h}_w$ is the idealized output of the reaction wheel assembly. In order to simplify the analysis, the torque disturbances, which generally

7.3 / Control with Momentum Exchange Devices

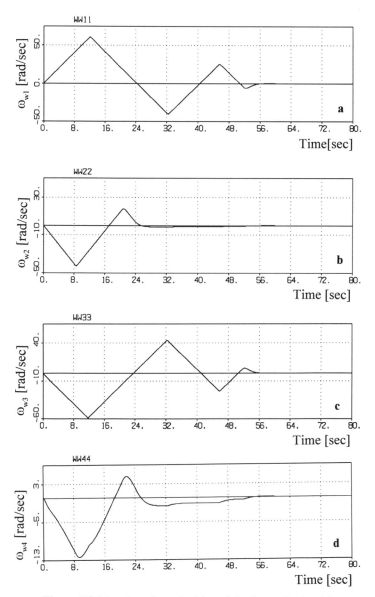

Figure 7.3.9.b Angular velocities of the four wheels *with* momentum management.

have harmonic characteristics (Bosga and Prins 1982, Bosga and Smilde 1982), have been modeled here as colored noise with a corner frequency ω_c; sensor noise is also modeled as colored noise. The corner frequencies for the three noise sources are not equal, but for convenience have been designated by the general term ω_c. The RMS *values* of the colored noises will be labeled RWT_N (reaction wheel torque noise), PS_N (position sensor noise), and RS_N (rate sensor noise).

Sensor noise and torque disturbances induce statistical errors in both the position attitude θ and its stability, the time derivative $\dot{\theta}$. These errors are functions of the

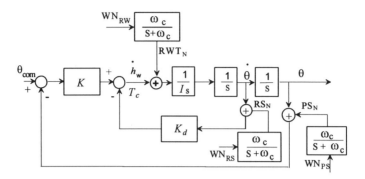

Figure 7.3.10 Conventional attitude control configuration with sensor noise and torque wheel disturbances.

bandwidth of the attitude control system, which can be represented in terms of the natural frequency ω_n and the damping coefficient ξ of a second-order pole. We make use of standard principles of statistical design for computing the RMS amplification from random sensor noises and parasitic control torques to the position and attitude outputs of the ACS (James, Nichols, and Phillips 1955, Solodovnikov 1960). The amplification of sensor noise to the input of the torque control command T_c in Figure 7.3.10 is also important. Exaggerated amplification of sensor noise may saturate a physical torque actuator (controller), thus precluding normal operation of the feedback control loop.

Let us deal first with noise amplification of the rate sensor (RS) noise. Given a white noise of amplitude WN_{RS}, the RMS value RS_N at the output of the coloring filter can be obtained from the relation $RS_N^2 = 0.5(WN_{RS}\omega_c)$. The procedure for calculating the mean square (MS) value of amplified noise at different outputs of the feedback control system of Figure 7.3.10 is as follows.

Amplification of RS Noise First, compute the transfer function:

$$\frac{\dot{\theta}}{WN_{RS}} = \frac{2\xi}{\omega_n} \frac{s}{\left[1 + \frac{2\xi}{\omega_n}s + \frac{s^2}{\omega_n^2}\right]\left[1 + \frac{s}{\omega_c}\right]}. \quad (7.3.33)$$

We find (see also James et al. 1955, Solodovnikov 1960) that

$$\dot{\theta}_N^2 = \frac{WN_{RS}\omega_n\xi}{1 + \frac{2\xi\omega_n}{\omega_c} + \frac{\omega_n^2}{\omega_c^2}} = \frac{2RS_N^2\omega_n\xi}{\omega_c} \frac{1}{1 + \frac{2\xi\omega_n}{\omega_c} + \frac{\omega_n^2}{\omega_c^2}}, \quad (7.3.34)$$

where WN_{RS} is the amplitude of the rate sensor white noise, $\dot{\theta}_N^2$ is the mean square of the amplified colored noise at $\dot{\theta}$, and RS_N is the RMS value of the RS colored noise.

For noncolored white noise, $\omega_c \to \infty$. Hence Eq. 7.3.34 reduces to

$$\dot{\theta}_N^2 \to WN_{RS}\omega_n\xi. \quad (7.3.35)$$

To find the amplification of the RS_N to the attitude position θ, one integration is added in Eq. 7.3.33, yielding

7.3 / Control with Momentum Exchange Devices

$$\theta_N^2 = \frac{WN_{RS}\xi}{\omega_n} \frac{1+\dfrac{2\xi\omega_n}{\omega_c}}{1+\dfrac{2\xi\omega_n}{\omega_c}+\dfrac{\omega_n^2}{\omega_c^2}} = \frac{2RS_N^2\xi}{\omega_c\omega_n} \frac{1+\dfrac{2\xi\omega_n}{\omega_c}}{1+\dfrac{2\xi\omega_n}{\omega_c}+\dfrac{\omega_n^2}{\omega_c^2}}. \tag{7.3.36}$$

Once more, for white noise we have

$$\theta_N^2 \to \frac{WN_{RS}\xi}{\omega_n}. \tag{7.3.37}$$

Amplification of PS Noise Next we ascertain the statistical amplification of the PS (position sensor) noise. In this case,

$$\frac{\dot\theta}{WN_{PS}} = \frac{s}{\left[1+\dfrac{2\xi}{\omega_n}s+\dfrac{s^2}{\omega_n^2}\right]\left[1+\dfrac{s}{\omega_c}\right]}. \tag{7.3.38}$$

The mean square of the amplified colored noise in $\dot\theta$ will be

$$\dot\theta_N^2 = \frac{WN_{PS}\omega_n^3}{4\xi} \frac{1}{1+\dfrac{2\xi\omega_n}{\omega_c}+\dfrac{\omega_n^2}{\omega_c^2}} = \frac{PS_N^2\omega_n^2}{2\xi}\frac{\omega_n}{\omega_c}\frac{1}{1+\dfrac{2\xi\omega_n}{\omega_c}+\dfrac{\omega_n^2}{\omega_c^2}}. \tag{7.3.39}$$

For noncolored white noise, the mean square of the amplified noise becomes

$$\dot\theta_N^2 = \frac{WN_{PS}\omega_n^3}{4\xi}. \tag{7.3.40}$$

Also, for the amplified noise in the angular position θ, we have

$$\theta_N^2 = \frac{WN_{PS}\omega_n}{4\xi} \frac{1+\dfrac{2\xi\omega_n}{\omega_c}}{1+\dfrac{2\xi\omega_n}{\omega_c}+\dfrac{\omega_n^2}{\omega_c^2}} = \frac{PS_N^2\omega_n}{2\xi\omega_c} \frac{1+\dfrac{2\xi\omega_n}{\omega_c}}{1+\dfrac{2\xi\omega_n}{\omega_c}+\dfrac{\omega_n^2}{\omega_c^2}}. \tag{7.3.41}$$

For white noise,

$$\theta_N^2 = \frac{WN_{PS}\omega_n}{4\xi}. \tag{7.3.42}$$

Amplification of the PS noise at the control torque input T_c will lead to

$$T_{cN}^2 = \frac{WN_{PS}I_S^2\omega_n^4}{4\xi} \frac{\omega_n+2\xi\omega_c}{1+\dfrac{2\xi\omega_n}{\omega_c}+\dfrac{\omega_n^2}{\omega_c^2}} = \frac{PS_N^2 I_S^2\omega_n^4}{2\xi\omega_c} \frac{\omega_n+2\xi\omega_c}{1+\dfrac{2\xi\omega_n}{\omega_c}+\dfrac{\omega_n^2}{\omega_c^2}}. \tag{7.3.43}$$

Amplification of RWT Noise Finally, we treat the reaction wheel parasitic statistical disturbance torques:

$$\frac{\dot\theta}{WN_{RW}} = \frac{1}{I_S\omega_n^2} \frac{s}{\left[1+\dfrac{2\xi}{\omega_n}s+\dfrac{s^2}{\omega_n^2}\right]\left[1+\dfrac{s}{\omega_c}\right]}. \tag{7.3.44}$$

As in the previous case, the amplified RW torque noise at $\dot{\theta}$ will be

$$\dot{\theta}_N^2 = \frac{WN_{RW}}{I_S^2 4\xi\omega_n} \frac{1}{1+\frac{2\xi\omega_n}{\omega_c}+\frac{\omega_n^2}{\omega_c^2}} = \frac{RWT_N^2}{I_S^2 2\omega_n\omega_c} \frac{1}{1+\frac{2\xi\omega_n}{\omega_c}+\frac{\omega_n^2}{\omega_c^2}}. \qquad (7.3.45)$$

With noncolored white noise,

$$\dot{\theta}_N^2 = \frac{WN_{RW}}{I_S^2 4\xi\omega_n}. \qquad (7.3.46)$$

To find the statistical amplification of the RWT_N disturbances to the position θ, we use the same procedure as in the previous stages of this analysis:

$$\theta_N^2 = \frac{WN_{RW}}{I_S^2 4\xi\omega_n^3} \frac{1+\frac{2\xi\omega_n}{\omega_c}}{1+\frac{2\xi\omega_n}{\omega_c}+\frac{\omega_n^2}{\omega_c^2}} = \frac{RWT_N^2}{I_S^2 2\xi\omega_n^3\omega_c} \frac{1+\frac{2\xi\omega_n}{\omega_c}}{1+\frac{2\xi\omega_n}{\omega_c}+\frac{\omega_n^2}{\omega_c^2}}. \qquad (7.3.47)$$

Finally, for white noise:

$$\theta_N^2 = \frac{WN_{RW}}{I_S^2 4\xi\omega_n^3}. \qquad (7.3.48)$$

Design Tradeoffs in Choosing the ACS Bandwidth

Various engineering realities and physical constraints influence the choice of the natural frequency ω_n of the attitude feedback control system. Four such factors are listed as follows.

(1) The bandwidth frequency of the attitude tracking control system is generally determined by the speed required of the satellite's payload in response to a defined attitude command. Consequently, a high corner frequency ω_c of the attitude control loop is desired.

(2) As seen in the preceding equations, sensor noise tends to spoil the quality of the attitude θ, and also of its stability $\dot{\theta}$, by factors of ω_n and ω_n^3, respectively (see e.g. Eq. 7.3.40 and Eq. 7.3.42). This means that, if we wish to be less sensitive to sensor noise, a lower bandwidth (ω_n) of the attitude control loop is imperative.

(3) According to Eqs. 7.3.46–7.3.48 and taking into account the statistical torque disturbances of the reaction wheel, the quality of the attitude and its stability are inversely proportional to the corner frequency ω_c of the attitude control loop. This means that, in order to decrease the influence of the disturbing torques, the bandwidth of the attitude control loop must be increased.

(4) Structural "bending" modes and liquid slosh dynamics, to be analyzed in Chapter 10, also tend to limit the bandwidth that can be achieved for the attitude control loops.

Such contradictory demands on the bandwidth ω_n of the attitude control loop will necessitate some tradeoffs in devising an optimal ACS. For example, suppose our goal is to maximize the quality of θ, despite the known (or expected) position sensor noise WN_{PS} and the torque noise disturbances WN_{RW} of the reaction wheel assembly.

7.3 / Control with Momentum Exchange Devices

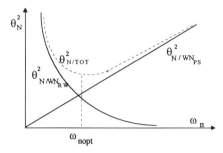

Figure 7.3.11 Tradeoff for achieving minimum noise level at the position output θ.

Suppose also that the noises are quasiwhite when compared to the corner frequency of the attitude control transfer function, $\omega_c \ggg \omega_n$. Figure 7.3.11 shows $\theta^2_{N/WN_{PS}}$ and $\theta^2_{N/WN_{RW}}$. The total mean square output amplitude of the disturbing noise will be $\theta^2_{N/TOT} = \theta^2_{N/WN_{PS}} + \theta^2_{N/WN_{RW}}$. The minimum value of $\theta^2_{N/TOT}$ will occur for some optimizing value of ω_n.

ACS without a Position Sensor

As we have seen, attitude accuracy depends very much on the accuracy of the attitude position sensor. For instance, the accuracy of a good earth sensor is of the order of 0.03°–0.2° (RMS). However, continuous use of an earth sensor is precluded if the satellite is to be aimed at objects located in geometrical configurations for which the earth is outside the sensor's field of view. Use of the star sensor is not practical when fast and large maneuvers are performed in space. For such situations, precise *inertial measuring units* (IMUs) are used.

Integrating the measured body rates by any one of the techniques described in Sections 4.7.3–4.7.5 allows us to obtain the position of the spacecraft in terms of Euler angles, a direction cosine matrix, quaternion vectors, or any other attitude presentation. Even with perfect RIGs, we still need some attitude position sensor to measure accurately the initial conditions required for use of the gyros' rate integration algorithm. Star sensors, if available, can be used for this purpose. For fast and large attitude maneuvers, it is common to use earth sensors together with some additional accurate sensor, such as a digital sun sensor (see Appendix B), to compute the initial attitude conditions required by the angular rate integration process. From here on, the ACS relies on the rate integrating gyros only, as shown in Figure 7.3.12.

In the following derivation, the only sensor noise is RIG noise. The accuracy of the satellite's attitude position is a function of this noise and of the RW torque disturbances.

Amplification of RIG Noise   According to Figure 7.3.12,

$$\frac{\theta}{WN_{RS}} = \frac{1 + \dfrac{2\xi s}{\omega_n}}{s\left[1 + \dfrac{s}{\omega_c}\right]\left[1 + \dfrac{2\xi s}{\omega_n} + \dfrac{s^2}{\omega_n^2}\right]}. \qquad (7.3.49)$$

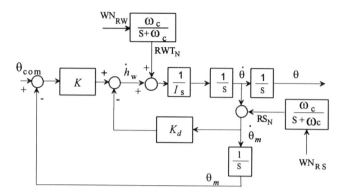

Figure 7.3.12 Attitude control system based on rate integrating gyros (the subscript m denotes "measured").

It follows that

$$\dot{\theta}_N^2 = \frac{\text{WN}_{\text{RS}}\omega_n}{4\xi} \frac{1 + 4\xi^2 + \frac{2\xi\omega_n}{\omega_c}}{1 + \frac{2\xi\omega_n}{\omega_c} + \frac{\omega_n^2}{\omega_c^2}}. \qquad (7.3.50)$$

For white noise, $\omega_c \to \infty$ and

$$\dot{\theta}_N^2 \to \frac{\text{WN}_{\text{RS}}\omega_n[1+4\xi^2]}{4\xi}. \qquad (7.3.51)$$

Computation of the mean square amplification of the RS noise with respect to the attitude position θ is not straightforward. The reason is as follows. Because

$$\frac{\theta}{\text{WN}_{\text{RS}}} = \frac{1 + \frac{2\xi s}{\omega_n}}{s\left[1 + \frac{s}{\omega_c}\right]\left[1 + \frac{2\xi s}{\omega_n} + \frac{s^2}{\omega_n^2}\right]} \qquad (7.3.52)$$

we have $\theta_N^2 \to \infty$, since the transfer function in Eq. 7.3.52 contains an integration. This phenomenon is due to (finite) sensor noise at very low frequencies. The physical interpretation is that, at such low frequencies, the sensor noise becomes the drift factor of the sensor. The lack of a true (physical) attitude position sensor causes the drift to be integrated by the kinematic integrator, and the attitude position error augments without bound. In order to obtain a reasonable useful model for the analysis at hand, the constant drift term of the sensor's noise must be eliminated by introducing a differentiator – accompanied by an additional first-order filter – into the model. The sensor noise model then becomes

$$\text{RS}_N = \frac{\text{WN}_{\text{RS}} s}{[1 + s/\omega_{c1}][1 + s/\omega_{c2}]}. \qquad (7.3.53)$$

From this model we obtain the required transfer function:

7.3 / Control with Momentum Exchange Devices

$$\frac{\theta}{WN_{RS}} = \frac{\left[1 + \dfrac{2\xi s}{\omega_n}\right]}{\left[1 + \dfrac{s}{\omega_{c1}}\right]\left[1 + \dfrac{s}{\omega_{c2}}\right]\left[1 + \dfrac{2\xi s}{\omega_n} + \dfrac{s^2}{\omega_n^2}\right]}. \quad (7.3.54)$$

EXAMPLE 7.3.1 So far, the control system of Figure 7.3.10 has been simulated to show the dependence of the variables θ, $\dot{\theta}$, and T_c on statistical noise of the sensors and of the torque controller. For the example at hand:

$I_S = 600$ kg-m^2, $\omega_n = 0.15$ rad/sec, $\omega_c = 100$ rad/sec,

$PS_N = 0.1°$ (RMS), $RWT_N = 0.2$ N-m (RMS).

The time-domain results are given in Figures 7.3.13–7.3.15. They agree well with the relevant Eq. 7.3.39 for Figure 7.3.13, Eq. 7.3.41 and Eq. 7.3.43 for Figure 7.3.14, and Eq. 7.3.45 and Eq. 7.3.47 for Figure 7.3.15. The results are summarized in Table 7.3.1. The difference between the analytical and the simulation results is due to the finite simulation time $T_f = 400$ sec and to the finite integration time of 0.001 sec.

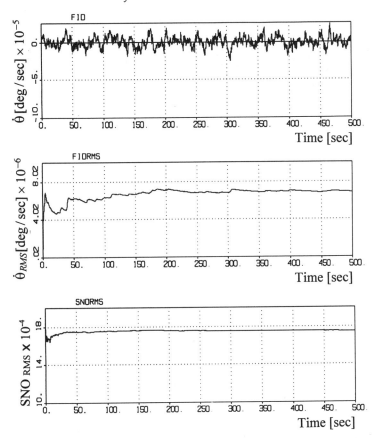

Figure 7.3.13 Response of the rate state $\dot{\theta}_N$ and its RMS value to position sensor noise.

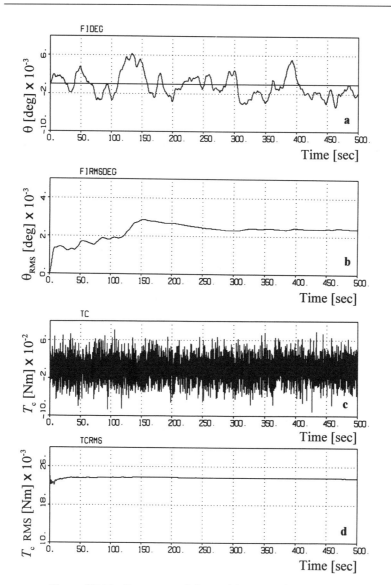

Figure 7.3.14 Responses of the position state θ_N, the control torque T_{cN}, and their RMS values.

Table 7.3.1 *Comparison between analytical and time-simulation results*

	$\dfrac{\dot{\theta}_N(\text{RMS})}{PS_N(\text{RMS})}$	$\dfrac{\theta_N(\text{RMS})}{PS_N(\text{RMS})}$	$\dfrac{T_{cN}(\text{RMS})}{PS_N(\text{RMS})}$	$\dfrac{\dot{\theta}_N(\text{RMS})}{RW_N(\text{RMS})}$	$\dfrac{\theta_N(\text{RMS})}{RW_N(\text{RMS})}$
Eq.(...)	$4.1\ 10^{-3}$	0.03	13.5	$3.043\ 10^{-4}$	$2.02\ 10^{-3}$
Simulation	$3.94\ 10^{-3}$	0.02	13.49	$2.726\ 10^{-4}$	$1.82\ 10^{-3}$

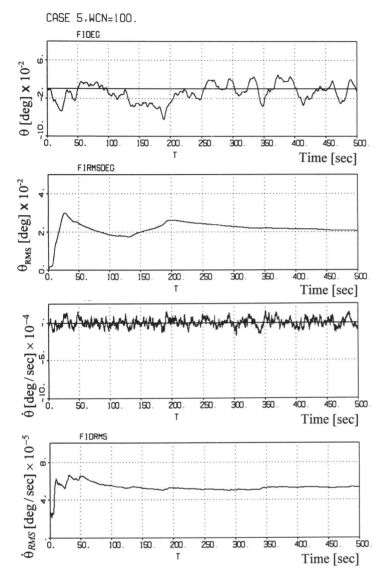

Figure 7.3.15 Time response of $\dot{\theta}_N$, θ_N, and their RMS values due to RW disturbing torque noise.

7.4 Magnetic Attitude Control

7.4.1 *Basic Magnetic Torque Control Equation*

Interaction between a magnetic moment generated within a spacecraft and the earth's magnetic field produces a mechanical torque acting on the spacecraft:

$$\mathbf{T}_B = \mathbf{M} \times \mathbf{B}, \tag{7.4.1}$$

where **M** is the generated magnetic moment inside the body and **B** is the earth's magnetic field intensity. Equation 7.4.1 can be written in matrix form as

$$\mathbf{T}_B = \begin{bmatrix} 1_x & 1_y & 1_z \\ M_x & M_y & M_z \\ B_x & B_y & B_z \end{bmatrix}. \tag{7.4.2}$$

Equation 7.4.2 can be rewritten as

$$\begin{bmatrix} T_{Bx} \\ T_{By} \\ T_{Bz} \end{bmatrix} = \begin{bmatrix} 0 & B_z & -B_y \\ -B_z & 0 & B_x \\ B_y & -B_x & 0 \end{bmatrix} \begin{bmatrix} M_x \\ M_y \\ M_z \end{bmatrix}. \tag{7.4.3}$$

For a desired torque $\mathbf{T}_B = \mathbf{T}_c$ to be applied on the spacecraft, we must generate the magnetic moment vector $\mathbf{M} = [M_x\ M_y\ M_z]^T$. However, we cannot invert the matrix in Eq. 7.4.3 because this matrix is singular (see also Section 5.3.4). The solution to our problem lies in replacing one of the magnetic torqrods with, for instance, a reaction wheel. By exchanging the $\mathbf{Y}_B$ magnetic torqrod with a RW, Eq. 7.4.3 is accordingly changed to

$$\begin{bmatrix} T_{Bx} \\ T_{By} \\ T_{Bz} \end{bmatrix} = \begin{bmatrix} 0 & 0 & -B_y \\ -B_z & 1 & B_x \\ B_y & 0 & 0 \end{bmatrix} \begin{bmatrix} M_x \\ \dot{h}_{wy} \\ M_z \end{bmatrix}. \tag{7.4.4}$$

The matrix in Eq. 7.4.4 does have an inverse, so

$$\begin{bmatrix} M_x \\ \dot{h}_{wy} \\ M_z \end{bmatrix} = \frac{1}{B_y^2} \begin{bmatrix} 0 & 0 & B_y \\ B_x B_y & B_y^2 & B_y B_z \\ -B_y & 0 & 0 \end{bmatrix} \begin{bmatrix} T_{cx} \\ T_{cy} \\ T_{cz} \end{bmatrix}. \tag{7.4.5}$$

Equation 7.4.5 is the basic equation for magnetic attitude control. In this equation B_x, B_y, B_z are the three components of the earth's magnetic field intensity, measured in the satellite body axes frame.

There is also the possibility of replacing the $\mathbf{X}_B$ or $\mathbf{Z}_B$ axis magnetic torqrod with a reaction (or momentum) wheel. For the second option, a reaction wheel with its axis aligned along the $\mathbf{Z}_B$ body axis, the control equations become

$$\begin{bmatrix} M_x \\ M_y \\ \dot{h}_{wz} \end{bmatrix} = \frac{1}{B_z^2} \begin{bmatrix} 0 & -B_z & 0 \\ B_z & 0 & 0 \\ B_x B_z & B_y B_z & B_z^2 \end{bmatrix} \begin{bmatrix} T_{cx} \\ T_{cy} \\ T_{cz} \end{bmatrix}. \tag{7.4.6}$$

A simple model of the earth's magnetic field approximates it as a dipole, passing N-S through the center of the earth's globe but deviating from the **Z** axis by 17° (see McElvain 1962). More sophisticated models are based on series expansions (Wertz 1986).

7.4.2 *Special Features of Magnetic Attitude Control*

The earth's magnetic field intensity is proportional to m/R^3, where R is the distance from the center of the earth and m is the magnetic dipole strength ($m = 8.1 \times 10^{15}$ Wb-m in 1962, but $m = 7.96 \times 10^{15}$ Wb-m in 1975). Thus, the strength of

7.4 / Magnetic Attitude Control

the magnetic field decreases strongly with the altitude of the satellite. In order to compensate for this loss in magnetic field intensity, the maximum obtainable magnetic dipole moment of the magnetic torqrods must be increased accordingly, with an inevitable increase in dimension and weight. To get an idea of the level of magnetic moment required to achieve a defined mechanical torque, Eq. 7.4.1 may also be written in the following form: $T_B = MB \sin(\alpha)$, where α is the angle between the earth's magnetic field and the artificial magnetic dipole moment produced inside the satellite. Assume an average angle of $\alpha = 30°$ during the attitude control stage. In this case, $T_B = 0.5 Mm/R^3$. At an altitude of 400 km, the mechanical torque on a satellite from a magnetic torqrod capable of a magnetic moment $M = 100$ A-m² will be only $T_B = 1.28 \times 10^{-3}$ N-m. The same magnetic torqrods will provide only $T_B = 5.23 \times 10^{-6}$ N-m at a geostationary altitude of $h = 35,786$ km. To obtain a torque of about 10^{-3} N-m at this high altitude would require a magnetic torqrod capable of 24,474 A-m², which is obviously not a practical possibility. Fortunately, geostationary satellites encounter disturbances of only about 15×10^{-6} N-m, so magnetic torqrods capable of 350 A-m² are sufficient for stabilization. (See also Chapter 8.) These numbers show clearly why an attitude control system based on magnetic control torques cannot achieve fast attitude maneuvers, especially at very high altitudes. With low-orbit satellites, it it still possible to achieve moderately fast attitude maneuvers using magnetic control.

Another drawback in magnetic attitude control is the dependence of the earth's useful magnetic field on orbit characteristics, and also on the location of the satellite within the orbit. A simplified model of the earth's magnetic field, as related to the orbit reference frame (see Section 4.7.2), is given in McElvain (1962). For the convenience of the reader, this model is reproduced here:

$$\begin{bmatrix} B_{xo} \\ B_{yo} \\ B_{zo} \end{bmatrix} = \frac{m}{R^3} \begin{bmatrix} \cos(\alpha - \eta_m) \sin(\xi_m) \\ \cos(\xi_m) \\ -2 \sin(\alpha - \eta_m) \sin(\xi_m) \end{bmatrix}. \tag{7.4.7}$$

In Eq. 7.4.7, $\alpha = \omega + \theta$ of Figure 2.6.2, η_m is a phase angle measured from the ascending node of the orbit relative to the earth's equator to the ascending node of the orbit relative to the geomagnetic equator, and ξ_m is the instantaneous inclination of the orbit plane to the geomagnetic equator. The functions η_m and ξ_m are both time-varying, but if $\omega_o \gg \omega_e$ then they can be assumed as constant over a small number of orbits (ω_o and ω_e denote the orbital and the earth's rotation frequencies). For equatorial orbits, ξ_m is constant (17° in 1962) and $\eta_m = \omega_o t$. In Eq. 7.4.7, it is clearly seen that the X and Z components of the earth's magnetic field intensity in the orbit reference frame are harmonic functions of the orbital period, whereas the Y component is a slowly varying function that can become negative for $\xi_m > 90°$ or for orbits with inclination greater than $90° - 17° = 73°$. This is important when we consider Eq. 7.4.5 and Eq. 7.4.6. For the control equation (Eq. 7.4.5), $M_x = T_{cz}/B_y$. If B_y does not change sign then there are no singularity problems in computing M_x; the same is true for M_z. On the other hand, using the control configuration defined by Eq. 7.4.6, with the reaction wheel axis aligned with the $\mathbf{Z}_B$ body axis, a singularity problem arises periodically when computing M_x and M_y since B_z changes sign, thus complicating the control algorithm. Moreover, according to Eq. 7.4.7, for an orbit inclined 40°

the maximum amplitudes of the harmonic components of M_x and M_z vary by a factor of at least 2, which renders the attitude magnetic control laws time-varying. In any case, a three-axis magnetometer is necessary when the magnetic control law is implemented.

As already mentioned, the magnetic field intensity is strongly dependent on the inclination of the orbit. For a dipole model of the earth's magnetic field, the magnitude of the field intensity is given by:

$$|\mathbf{B}| = \frac{m}{R^3}\sqrt{1 + 3\sin^2(\lambda)}, \qquad (7.4.8)$$

where λ is the elevation angle with respect to the plane that is perpendicular to the magnetic dipole axis. From Eq. 7.4.8 it is evident that the magnetic field for a polar orbit can be twice as strong as that of an equatorial orbit.

7.4.3 Implementation of Magnetic Attitude Control

The achievable levels of torques using magnetic torqrods are very limited. Saturation limits are easily attained, thus rendering the control law strongly nonlinear. The bandwidth of the control loops must be chosen accordingly, in order to prevent saturation of the controller. But since it is impossible to preclude saturation of the magnetic controllers entirely, it is advisable to scale down the inputs to the magnetic torqrods and the RW according to the level of the saturated one, which will at least keep the controlled Euler rotation axis unchanged. See Section 7.2.3.

EXAMPLE 7.4.1 A satellite in a circular orbit of 800-km altitude has the following moments of inertia: $I_x = 30$, $I_y = 40$, $I_z = 20$ kg-m². Saturation level of the magnetic torqrods is 150 A-m², and the reaction wheel has a torque capability of 0.2 N-m. The closed-loop bandwidths of the three axes were chosen to be $\omega_{nx} = 0.1$, $\omega_{ny} = 0.3$, and $\omega_{nz} = 0.1$; the damping coefficient $\xi = 1.0$.

The time-domain results following Euler angles command inputs of $\psi_{com} = 1°$, $\theta_{com} = -4°$, and $\phi_{com} = 3°$ are given in Figures 7.4.1–7.4.3. Figure 7.4.1 shows the Euler angle outputs. Figure 7.4.2 shows the control command inputs to the three satellite controllers – two magnetic torqrods aligned with the $\mathbf{X}_B$ and $\mathbf{Z}_B$ axes and one reaction wheel whose rotation axis is aligned with the $\mathbf{Y}_B$ body axis. As shown in the figure, the commanded torque on the $\mathbf{X}_B$ axis amounts to more than 10 mN-m, which the magnetic torqrods cannot provide. A saturation of the $\mathbf{Z}_B$ axis magnetic torqrod is present, as expected; see Figure 7.4.3. The $\mathbf{X}_B$ axis magnetic moment command M_x has been decreased proportionally to about 40 A-m², so that the direction in space of the commanded Euler axis of rotation remains constant (see Section 7.2.3). As expected, the torque produced by the magnetic torqrods is also saturated, to about 3.0 mN-m about the $\mathbf{X}_B$ body axis. The same proportional reduction is also performed on the reaction wheel torque command.

The time histories described in Example 7.4.1 show that satisfactory attitude control can be achieved by using magnetic torques, albeit with slower time responses.

7.5 / Magnetic Unloading of Momentum Exchange Devices

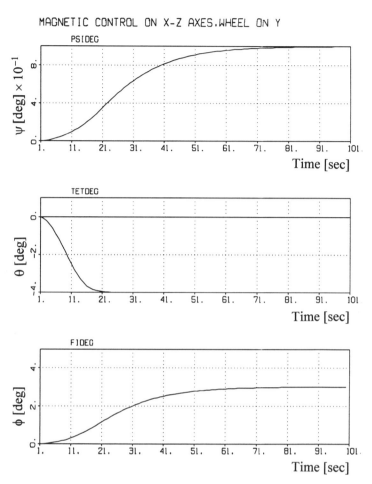

Figure 7.4.1 Euler angles attitude responses to angular step commands.

Assuming that this response is acceptable, such a configuration has the advantage of being much cheaper than an ACS incorporating three reaction wheels, since the cost of a magnetic torqrod is typically one tenth that of a reaction wheel.

It is important to realize that, as seen in Eq. 7.4.3, M_x and M_z produce also a parasitic torque about the Y_B axis. This disturbance torque on the Y_B axis is easily handled by the torque capabilities of the reaction wheel, which are greater than those of the magnetic torqrods by at least one order of magnitude.

7.5 Magnetic Unloading of Momentum Exchange Devices

7.5.1 *Introduction*

As explained in Section 7.3.3, external disturbances acting on the body of an attitude-controlled s/c induce accumulation of momentum in the momentum exchange devices. This excess momentum might bring the wheels to improper working

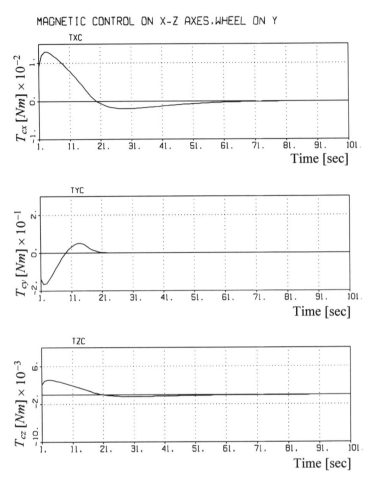

Figure 7.4.2 Control torque commands to the two magnetic torqrods and the reaction wheel.

conditions. Moreover, the existence of angular momentum in the satellite causes control difficulties when attitude maneuvers in space are executed, because this superfluous momentum provides the spacecraft with unwanted gyroscopic stability. For this reason, three-axis stabilized attitude-maneuvering spacecraft are basically *zero-bias-momentum* systems.

Excess momentum must be unloaded when it exceeds some predetermined limiting value. The two primary control hardware items used to dump the wheels are magnetic torqrods and reaction thrusters. We will focus our attention on the first of these options.

7.5.2 *Magnetic Unloading of the Wheels*

Magnetic torqrods generate magnetic dipole moments whose interactions with the earth's magnetic field produce the torques necessary to remove the excess

7.5 / Magnetic Unloading of Momentum Exchange Devices

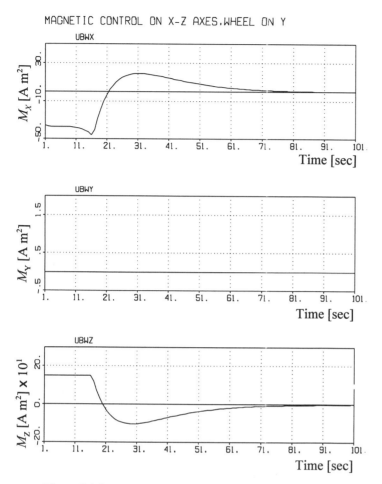

Figure 7.4.3 Magnetic dipoles produced by the magnetic torqrods.

momentum. This momentum unloading strategy is quite simple in principle, and was first proposed in the sixties (see McElvain 1962, Stickler and Alfriend 1976, Lebsock 1982).

The basic control equation for momentum unloading is

$$\mathbf{T} = -k(\mathbf{h} - \mathbf{h}_N) = -k\Delta\mathbf{h}. \tag{7.5.1}$$

In this vectorial equation, k is the unloading control gain, $\mathbf{h}$ is the wheel's momentum vector, $\mathbf{h}_N$ is the desired and nominal wheel momentum vector, and $(\mathbf{h} - \mathbf{h}_N) = \Delta\mathbf{h}$ is the excess momentum to be removed.

The magnetic torque equation was previously stated to be $\mathbf{T} = \mathbf{M} \times \mathbf{B}$. Together with Eq. 7.5.1 this yields

$$-k\Delta\mathbf{h} = \mathbf{M} \times \mathbf{B}. \tag{7.5.2}$$

However, the control magnetic dipole vector $\mathbf{M}$ cannot be computed from Eq. 7.5.2 (see Section 7.4.1). Using vector product by $\mathbf{B}$ on both sides, Eq. 7.5.2 becomes

$$\mathbf{B} \times (-k\Delta\mathbf{h}) = \mathbf{B} \times (\mathbf{M} \times \mathbf{B}) = B^2\mathbf{M} - \mathbf{B}(\mathbf{M}\cdot\mathbf{B}). \tag{7.5.3}$$

With some simplifying assumptions, we can find $\mathbf{M}$ from Eq. 7.5.3. Suppose that the applied $\mathbf{M}$ happens to be perpendicular to the earth's magnetic field $\mathbf{B}$. In this case $\mathbf{M}\cdot\mathbf{B}$ (which is a scalar product) is zeroed, and from Eq. 7.5.3 we have

$$\mathbf{M} = -\frac{k}{B^2}(\mathbf{B} \times \Delta\mathbf{h}). \tag{7.5.4}$$

This control magnetic moment produces a magnetic torque that is not exactly proportional to the excess momentum:

$$\mathbf{T} = -\frac{k}{B^2}[B^2\Delta\mathbf{h} - \mathbf{B}(\mathbf{B}\cdot\Delta\mathbf{h})]. \tag{7.5.5}$$

Physically, Eq. 7.5.2 states that no torque can be obtained about the earth's magnetic field $\mathbf{B}$ and, moreover, that if the excess momentum to be dumped is parallel to this vector then the wheels cannot be unloaded. Fortunately, for most inclined orbits this condition does not continue indefinitely: both the direction and the amplitude of $\mathbf{B}$ with respect to body axes will vary during an orbit, and an average removal of the excess momentum takes place. In any case, the effectiveness of momentum unloading depends very much on the specific orbit in which the satellite is moving. For equatorial orbits, the efficiency is quite low.

Equation 7.5.4 can be put in more explicit form:

$$\begin{bmatrix} M_x \\ M_y \\ M_z \end{bmatrix} = -\frac{k}{B^2} \begin{bmatrix} B_y\Delta h_z - B_z\Delta h_y \\ B_z\Delta h_x - B_x\Delta h_z \\ B_x\Delta h_y - B_y\Delta h_x \end{bmatrix}, \tag{7.5.6}$$

where B_x, B_y, B_z are the *measured* earth magnetic field strength components in the body axis frame. Hence, a three-axis magnetometer is imperative. The excess momentum $\Delta\mathbf{h}$ is known from measuring the wheel's angular velocities and transforming their components to body axes, so that the components of the magnetic dipole moment $\mathbf{M}$ can be computed also in body axes using Eq. 7.5.6. The saturation level of the magnetic torqrods must be such that they can provide torques larger than the external disturbing torques causing the momentum loading of the wheels; this was discussed in Section 7.4.2. Our remaining task is to evaluate k.

7.5.3 Determination of the Unloading Control Gain k

The control system is time-varying, because the components of the earth's magnetic field in the orbit reference frame are also time-varying and depend strongly on the orbit parameters. Moreover, the control law expressed by Eq. 7.5.6 was obtained under the dubious assumption that the commanded magnetic dipole moment $\mathbf{M}$ is always perpendicular to the earth's magnetic field $\mathbf{B}$. Hence, an analytic procedure to obtain the correct value of k does not seem to be feasible. In our analysis, k is obtained by the often useful "cut-and-try" method. Several simulations with different ks are performed until an acceptable steady-state excess momentum remains, with limited control magnetic dipole moments. We note that excessive magnetic torques are accompanied by larger power consumption and larger dimension and

7.5 / Magnetic Unloading of Momentum Exchange Devices

weight of the magnetic torqrods, and might also interfere with the primary task of achieving good attitude control.

EXAMPLE 7.5.1 A satellite has the following physical specifications: $I_x = 1,000$, $I_y = 500$, $I_z = 700$ kg-m^2. The satellite is in a circular orbit at an altitude of 400 km, and an inclination of 40°, with its Y_B axis perpendicular to the orbit plane. External disturbance torques of 10^{-3} N-m are foreseen about the Y_B body axis. The overall excess momentum in the satellite must not exceed 3 N-m-sec. The satellite's attitude is stabilized with the aid of three reaction wheels, whose rotational axes are aligned with the principal body axes.

Solution According to the discussion in Section 7.4.2, magnetic torqrods that produce 100 A-m^2 are sufficient at this low altitude. For a constant disturbance about the Y_B axis, the excess momentum of the wheel aligned with the Y_B axis will increase indefinitely ($k = 0$; see Figure 7.5.1). The excess angular momentum of the

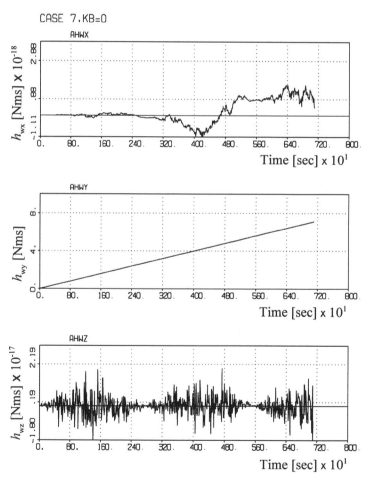

Figure 7.5.1 Increase of the excess momentum in the Y_B reaction wheel due to T_{dy}, $k = 0$.

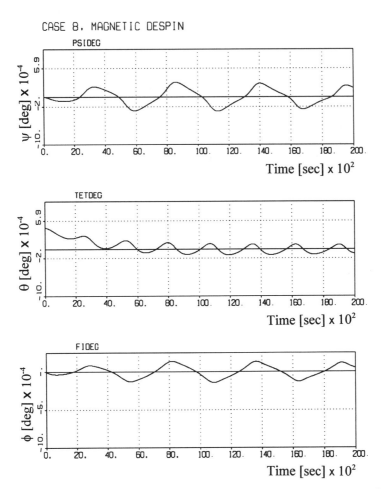

Figure 7.5.2 Attitude disturbances due to the momentum unloading control.

remaining wheels is obviously null. With a momentum dumping control gain $k = 0.001$ and the same disturbance of 10^{-3} N-m about the Y_B axis, Figure 7.5.2 shows that no noticeable attitude disturbances are developing. On the other hand, the excess momentum of the Y_B axis wheel is well controlled, and limited to less than 2 N-m-sec. The average momentum about the two remaining wheels is null, as expected, since there is no disturbance about the X_B and Z_B axes (see Figure 7.5.3). The magnetic dipole moments of the magnetic torqrods are shown in Figure 7.5.4.

Figure 7.5.5 shows the excess momentum in the three reaction wheels when an additional disturbance about the X_B axis is applied, $T_{dx} = 0.001$ N-m. In this figure, average excess momentums of about 0.5 N-m-sec in the X_B reaction wheel and about 0.8 N-m-sec in the Z_B reaction wheel are perceived. Because of the assumptions made in Section 7.5.2, an excess momentum accumulates also in the Z_B wheel, although no disturbance about this axis was applied.

7.6 / Time-Optimal Attitude Control

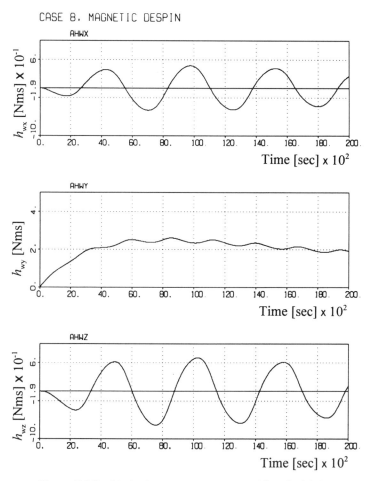

Figure 7.5.3 Limited excess momentum achieved with $k = 0.001$.

7.6 Time-Optimal Attitude Control

7.6.1 *Introduction*

Time-optimal control is used in spacecraft attitude control to minimize the time of attitude orientation of the satellite in space. For a sky observation satellite, it is important to be able to reorient the satellite from one to another part of the sky in minimum time, so that a maximum quantity of observations can be performed when the satellite is favorably located. The same criterion holds for earth-observing satellites; low-orbit s/c stay in the region to be observed for a short period of time and, in order to collect the maximum quantity of data, the satellite must be maneuvered in the shortest possible time. For a one-axis reorientation, the time-optimal control is a *bang-bang* control (Elgerd 1967, Bryson and Ho 1969). For a three-axis reorientation, the time-optimal control is not an *eigenaxis rotation* about a chosen control axis

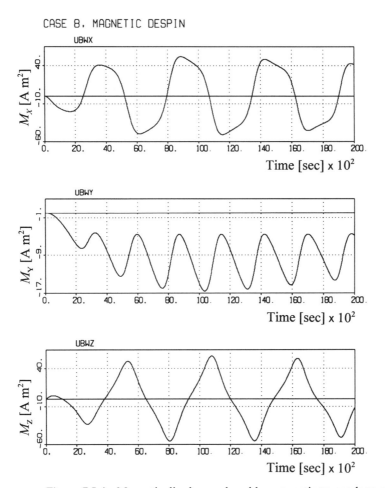

Figure 7.5.4 Magnetic dipoles produced by magnetic torqrods to counteract T_{dy}.

(Bilimora and Wie 1993). A complete treatment of time-optimal reorientation in space, based on optimal control theory, can be found in Junkins and Turner (1986).

The transfer function methods that are so indispensable in linear system analysis cannot be used for time-optimal control systems, where the controller is drawn into heavy saturation in order to deliver to the satellite the maximum physically obtainable angular accelerations. The satellite dynamics equations in time-optimal control are characterized by a linear second-order plant and a nonlinear *maximum-effort* or *on-off* controller. The resulting control problems are most efficiently analyzed in the phase plane.

Time-optimal control theory cannot in itself provide a practical time-optimal orientation without taking into consideration some practical physical constraints, such as control time delays, uncertainty in the levels of the maximum control acceleration and deceleration, additional time constants existing in the satellite hardware, structural dynamics, and so on. We will see that these physical constraints have a major impact on the achievable qualities of the time-optimal orientation control schemes.

7.6 / Time-Optimal Attitude Control

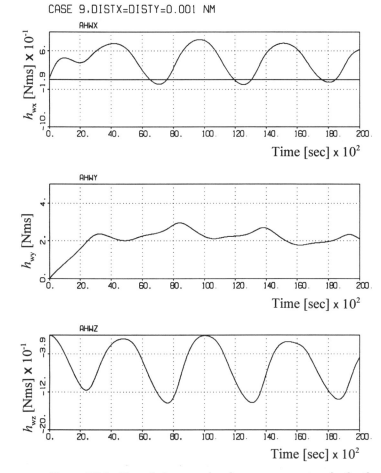

Figure 7.5.5 Bounded accumulated excess momentum in the three reaction wheels with disturbances about the Y_B and X_B axes.

7.6.2 Control about a Single Axis

The basic time-optimal control is a bang-bang control. The plant consists of two integrators, preceded by the moment of inertia I of the satellite about the orientation axis. We assume here that the torque controller is a reaction wheel, although the analysis can as well be carried out with other torque controllers (e.g., propulsion thrusters). The maximum torque about the axis depends on the reaction wheel capabilities $\dot{h}_{w\,max} = T_{max}$. If two wheels are aligned about the orientation axis, the maximum torque capabilities are augmented accordingly. The maximum angular acceleration and deceleration for one wheel are $U_{max} = \pm T_{max}/I$. (For simplicity, we shall use $u \equiv U_{max}$.) Let us also define

$$\epsilon\phi = e = \phi_{com} - \phi \quad \text{and} \quad \epsilon\dot{\phi} = \dot{e} = \dot{\phi}_{com} - \dot{\phi}, \qquad (7.6.1)$$

where ϕ_{com} and $\dot{\phi}_{com}$ are the Euler command angle and its derivative.

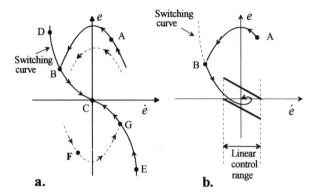

Figure 7.6.1 Ideal and practical time-optimal control solution.

Given these definitions, the time-optimal control behavior is shown in Figure 7.6.1. In this figure, we define the *switching curve;* any initial condition of $\dot{e}$ and e moves on this curve toward the origin, thus completing the optimal trajectory. The basic equations for acceleration and deceleration are

$$\ddot{\phi} I = \pm T_{max}, \tag{7.6.2}$$

$$\dot{\phi} I = \pm T_{max} t + \dot{\phi}_0 I, \tag{7.6.3}$$

$$\phi I = \pm T_{max} \frac{t^2}{2} + \phi_0 I + \dot{\phi}_0 It. \tag{7.6.4}$$

The switching curve is a parabolic curve passing through the origin. In this case, from Eq. 7.6.3 and Eq. 7.6.4 we obtain the equation for this curve:

$$\phi = -\frac{1}{2u} \dot{\phi} |\dot{\phi}|. \tag{7.6.5}$$

For any initial conditions in the phase plane, the motion from point A to point B is on an acceleration path. When point B on the switching curve is reached, the torque T (and accordingly the acceleration u also) changes sign. In Figure 7.6.1.a, in the deceleration motion on the arc BC, the angular motion reaches the origin C and the time-optimal cycle is completed. Because of time delays in the control system and since the moment of inertia I and the torque T are not exactly known, some "chattering" about the origin is inevitable. To eliminate this effect, a conventional linear control solution will replace the bang-bang solution near the origin.

In the linear range, the torque command T_c will have the following expression:

$$T_c = Ke + K_d \dot{e}, \tag{7.6.6}$$

where K and K_d are computed in the standard way for a second-order linear feedback system. If the closed-loop feedback system is to have a natural frequency ω_n and a damping coefficient ξ, then

$$K = \omega_n^2 I \quad \text{and} \quad K_d = 2\xi \omega_n I. \tag{7.6.7}$$

The linear range mode is entered when

$$|\dot{e}| < T_{max}/K_d. \tag{7.6.8}$$

7.6 / Time-Optimal Attitude Control

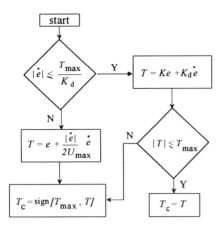

Figure 7.6.2 The quasi-time-optimal attitude control algorithm.

In Figure 7.6.1, the axes are e and $\dot{e}$. The phase plane shows paths for the initial values e_0 and $\dot{e}_0$. The complete control algorithm is summarized in the flow chart of Figure 7.6.2.

EXAMPLE 7.6.1 In this example, we suppose that $I = 600$ kg-m^2, $\xi = 1.0$, $\omega_n = 0.5$ rad/sec, $T_{max} = 0.4$ N-m, and $\phi_{com} = 2°$. With the nominal moment of inertia and no time delays in the feedback loop, the time-domain behavior is a quasi-time-optimal bang-bang solution (see Figures 7.6.3). With a virtually continuous control system ($T_{sam} = 0.01$ sec), the time behavior is the nominal bang-bang solution until the linear range in Figure 7.6.1.b is approached. The phase-plane behavior $\epsilon\dot{\phi}$ viz. $\epsilon\phi$ is shown in Figure 7.6.3.d.

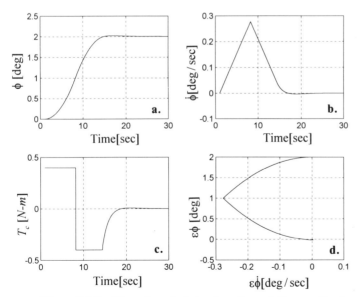

Figure 7.6.3 Time-domain behavior of quasi-time-optimal continuous control.

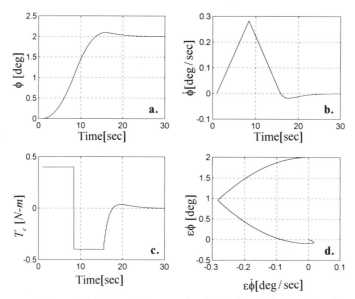

Figure 7.6.4 Quasi-time-optimal solution with sampling time $T_{sam} = 0.2$ sec.

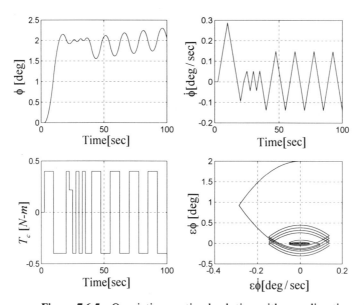

Figure 7.6.5 Quasi-time-optimal solution with sampling time of 2.5 sec.

The same control system has been simulated with $T_{sam} = 0.2$ sec, which is a comparatively short sampling time. The time behavior in Figures 7.6.4 shows clearly the difference from the case of time optimality, but still there is a single entrance to the linear range in Figure 7.6.4.d. An overshoot of 10% in ϕ may be discerned.

With the much larger sampling time of $T_{sam} = 2.5$ sec, the control solution is unacceptable; this is evident in the time responses shown in Figures 7.6.5. The large time delay, generated by the exaggerated sampling time, causes the decelerating path to miss the entrance to the linear range, resulting in a large oscillatory motion about the origin. This is unacceptable behavior, which can be overcome as described in the following section.

7.6.3 Control with Uncertainties

The time-optimal behavior in Figure 7.6.3 is achieved under ideal conditions: there are no time delays in the control loop, the moment of inertia of the satellite is known exactly, and the theoretical maximum torque T_{max} used in the control equations exactly equals the actual activating torque. In practice, the activating torque T might be larger (or smaller) than T_{max}. In general, owing to sampling features of the controller and to additional delays in the control loop, the actual time behavior of the control system will deviate from the nominal time-optimal solution as follows (see Pierre 1986):

(a) if $T < T_{max}$, then an overshoot in the time behavior is to be expected;
(b) if $T > T_{max}$, a time delay Δt in the control loop gives rise to a chatter along the switching line.

The path in the phase plane in Figure 7.6.1 is a parabola, until point B is reached. Because of the finite time delay, the path – after point B is reached – actually overshoots the nominal switching line DC by a small amount; then the torque reversal sends the path back across this line, reversal occurs once more, and so on, giving rise to the *chatter effect*. The smaller the time delay Δt, the smaller the chatter amplitude but also the larger its cycling frequency. With practical intermediate time delays, the amplitude of the chatter might be well pronounced, as shown in Figure 7.6.6 (path 2 in the phase plane).

Figure 7.6.6 simulates the behavior of the quasi–time-optimal control law of Example 7.6.1 in the phase plane for the following uncertainty conditions.

Path 1: The nominal time-optimal solution, with no time delays and exactly known activating acceleration $U_{max} = T_{max}/I$. (We have set $u \equiv U_{max}$.)

Path 2: $T = 1.5T_{max}$, $\Delta t = 0.2$ sec. The chatter effect is clearly perceived.

Path 3: $T = 0.5T_{max}$, $\Delta t = 0.0$ sec. The acceleration time is longer, and the time-optimal behavior is maintained in the accelerating period. However, in the deceleration period the activating torque is smaller than expected so the path does not follow the switching curve, and this leads to an overshoot, followed by an undershoot, et cetera, until the origin is reached. If the torque controller is a reaction wheel then this chatter will lead to a waste of electrical power, which is not too disturbing. But in the case of reaction torque control the chatter is accompanied by a large waste of fuel, which cannot be tolerated from the system engineering point of view; in these cases, chattering must be prevented.

7.6.4 Elimination of Chatter and of Time-Delay Effects

Time-optimal control is needed so that delays, as well as the uncertainty of the physical level of the applied torque in the feedback loop, are compensated for. Our analytical derivation follows closely that of D. Verbin (MBT, Israel Aircraft Industries).

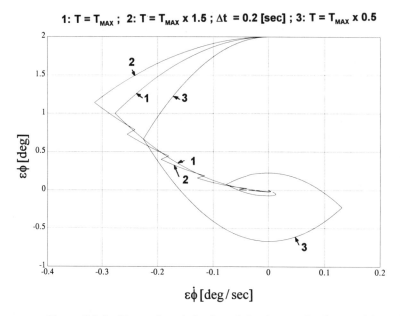

Figure 7.6.6 Phase-plane behavior of the time-optimal control law with different time delays and uncertainty in the level of the activating torques.

In Figure 7.6.7, when the switching curve is reached and the command for deceleration is issued, the path will not follow the switching curve exactly because of time delays in the control loop. Instead, a delayed path will be followed. To compensate for this delay, the switching command for deceleration must be issued earlier on the acceleration path: at point 2, before the switching curve is attained. Knowing the maximum time delay in the control loop, it is possible to *advance* the "switch to deceleration" command to point 2. Because of the delay, the new deceleration curve will reach point 3 – which is on the nominal switching curve – and the overshoot in the time response will be avoided.

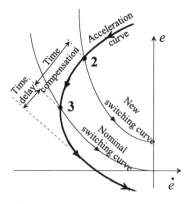

Figure 7.6.7 Definition of a new switching curve for compensating the time delays in the control loop.

7.6 / Time-Optimal Attitude Control

Points 2 and 3 are located on the path of maximum acceleration. Thus we have

$$\dot{\phi}_3 = \dot{\phi}_2 - u\Delta t, \tag{7.6.9}$$

$$\phi_3 = \phi_2 + \dot{\phi}_2 \Delta t - (u/2)\Delta t^2, \tag{7.6.10}$$

$$\dot{\phi}_3^2 = \dot{\phi}_2^2 - 2u(\phi_3 - \phi_2). \tag{7.6.11}$$

Point 3 also satisfies Eq. 7.6.5, since it must be located on the original switching curve. Hence

$$\phi_3 = \dot{\phi}_3^2/2u. \tag{7.6.12}$$

Together with Eq. 7.6.11, we have

$$\dot{\phi}_3^2 = (\dot{\phi}_2^2 + 2u\phi_2)/2. \tag{7.6.13}$$

From Eq. 7.6.13 and Eq. 7.6.9 we obtain, after squaring both sides,

$$(\dot{\phi}_2^2 + 2u\phi_2)/2 = \dot{\phi}_2^2 - 2u\Delta t \dot{\phi}_2 + u^2 \Delta t^2. \tag{7.6.14}$$

Equation 7.6.14 can be put in the following form:

$$\phi_2 = \dot{\phi}_2^2/2u - 2\Delta t \dot{\phi}_2 + u\Delta t^2; \tag{7.6.15}$$

this equation is true for $\dot{\phi} < 0$. In order to hold also for $\dot{\phi} > 0$, it should be rewritten as

$$\phi_2 = -\dot{\phi}_2|\dot{\phi}_2|/2u - 2\Delta t \dot{\phi}_2 - u\Delta t^2(\dot{\phi}_2/|\dot{\phi}_2|). \tag{7.6.16}$$

With no internal delays in the control loop, $\Delta t = 0$ and Eq. 7.6.16 is equivalent to Eq. 7.6.5. In order to eliminate chatter, we shall define a kind of linear solution between the switching curves expressed by these two equations. Using once more the definitions of Eq. 7.6.1, we can rewrite Eq. 7.6.5 and Eq. 7.6.16 as

$$e_1 = -\dot{e}|\dot{e}|/2u, \tag{7.6.17}$$

$$e_2 = -\dot{e}|\dot{e}|/2u - 2\Delta t \dot{e} - u\Delta t^2(\dot{e}/|\dot{e}|). \tag{7.6.18}$$

The control moment T_c is computed between the two paths by a linear interpolation. For $\dot{e} < 0$, we obtain

$$T_c = \frac{e - e_1}{e_2 - e_1}(+T_{\max}) + \frac{e_2 - e}{e_2 - e_1}(-T_{\max}) = \frac{2e - e_1 - e_2}{e_2 - e_1}T_{\max}. \tag{7.6.19}$$

For $\dot{e} > 0$,

$$T_c = \frac{e_1 + e_2 - 2e}{e_2 - e_1}T_{\max}. \tag{7.6.20}$$

For both $\dot{e} < 0$ and $\dot{e} > 0$, Eq. 7.6.19 and Eq. 7.6.20 become

$$T_c = \frac{e_1 + e_2 - 2e}{e_2 - e_1}\frac{\dot{e}}{|\dot{e}|}T_{\max}. \tag{7.6.21}$$

Insertion of e_1 and e_2 from Eq. 7.6.9 and Eq. 7.6.10 into Eq. 7.6.13 leads to:

$$T_c = \frac{-(|\dot{e}|/u)\dot{e} - 2\Delta t \dot{e} - u\Delta t^2(\dot{e}/|\dot{e}|) - 2e}{-2\Delta t \dot{e} - u\Delta t^2(\dot{e}/|\dot{e}|)}\frac{\dot{e}}{|\dot{e}|}T_{\max}. \tag{7.6.22}$$

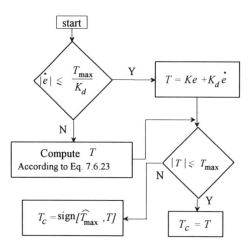

Figure 7.6.8 Time-optimal control with compensation for Δt and p, the uncertainty in $T_{\max}$.

In the foregoing analysis $u = T_{\max}/I$, where $T_{\max}$ is the nominal maximum torque that the reaction wheel can apply on the satellite. Since we assumed that the applied torque was no longer the nominal one, let us define $\hat{u} = (T_{\max}/I)(1-p)$, with p expressing the deviation of the applied acceleration from the nominal; p combines the uncertainties in both $T_{\max}$ and I, and so is a kind of margin factor. The value of p must be chosen such that, for the expected uncertainties in the existing torque level, chatter will not be present. However, after the margin factor p is selected, the solution will remain suboptimal only for the case in which $\hat{u}$ exactly suits $\hat{T}_{\max}$, the new physically applied torque.

With these definitions, the final equation for the computed torque becomes:

$$T = \left[\frac{|\dot{e}|}{\hat{u}\Delta t(2|\dot{e}| + \hat{u}\Delta t)}\dot{e} + \frac{2}{\Delta t(2|\dot{e}| + \hat{u}\Delta t)}e + \text{sign}(1,\dot{e})\right]T_{\max}. \qquad (7.6.23)$$

With the torque value to be applied as calculated via Eq. 7.6.23, we can modify the flowchart of the control algorithm shown in Figure 7.6.2; this modification is shown in Figure 7.6.8. Example 7.6.2 uses Eq. 7.6.23 to demonstrate the elimination of the chatter effect.

EXAMPLE 7.6.2 In this example, $T_{\text{sam}} = 0.5$ sec. We choose $p = -1$, which means that the applied torque can be twice the nominal maximum torque: $\hat{T}_{\max} = 2T_{\max}$. Simulation of the time-optimal control law with $\Delta t = 0.0$ sec (uncompensated) and an applied torque of $2T_{\max}$ shows a chatter in the time response of the system (see Figure 7.6.9). As expected, using $\Delta t = 0.5$ sec and $p = -1$ in Eq. 7.6.23 eliminates the chatter effect completely (see Figure 7.6.10). In Figure 7.6.11, the time responses of ϕ for the compensated and uncompensated cases are superimposed. The figure shows that elimination of the chatter effect does not appreciably compromise the optimality of the time-optimal control solution. On the other hand, with $0 < p < 1$, the time response will be slower and with an appreciable overshoot.

7.6 / Time-Optimal Attitude Control

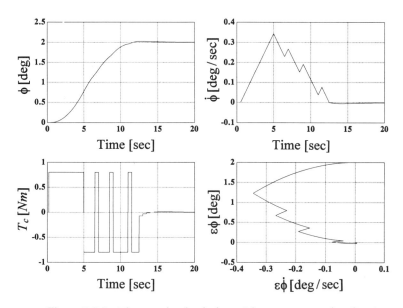

Figure 7.6.9 Time-optimal solution *without* compensation for the time delay and augmented applied maximum torque.

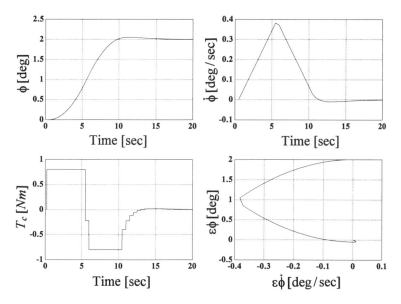

Figure 7.6.10 Time-optimal solution *with* compensation for the time delay and augmented applied maximum torque.

The preceding example concludes our analysis of single-axis time-optimal control. The three-axis time-optimal problem is outside the scope of this text; the interested reader is referred to Junkins and Turner (1986), Bikdash, Cliff, and Nayfeh (1993), and Bilimora and Wie (1993).

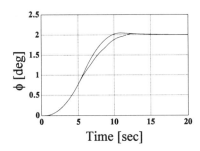

Figure 7.6.11 Time-optimal control with chatter compared to compensated time-optimal control without chatter.

7.7 Technical Features of the Reaction Wheel

We have used the RWA as the primary torque controller when accurate and time-optimal attitude control was mandatory. In practice, choice of the right wheel depends on the performance to be achieved by the satellite's ACS. There are some basic technical features that a reaction wheel must possess if the desired performance parameters of the satellite's ACS are to be achieved. These include: maximum achievable torque; maximum momentum capacity; low torque noise; and low coulomb friction torques.

The following simplified analysis will help define the first two necessary features, which pertain to the attitude change about a single axis of the satellite. The total momentum about one of the axes is $H_B = H_S + H_w$, where H_w is the momentum of the reaction wheel about the satellite rotation axis and H_S is the momentum of the satellite about the same axis. Since there are no external moments, $\dot{H}_I = \dot{H}_S + \dot{H}_w = 0$ and

$$\dot{H}_S = -\dot{H}_w. \tag{7.7.1}$$

As in the discussion of Section 7.6, time-optimal attitude control is obtained by delivering maximum angular accelerations and decelerations to the s/c by the reaction wheel. The angular motion of the satellite and of the reaction wheel caused by the application of torques by the reaction wheel is shown in Figure 7.7.1.

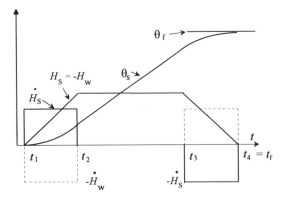

Figure 7.7.1 Relationship between the momentum delivered by the reaction wheel and the angular rotation of the satellite.

7.7 / Technical Features of the Reaction Wheel

To obtain a positive rotation of the satellite, a negative moment $-\dot{H}_w = \dot{H}_S$ must be delivered by the reaction wheel between t_1 and t_2. The momentum of the satellite will increase in that period by the following amount:

$$-\dot{H}_w \times (t_2 - t_1) = \dot{H}_S \times (t_2 - t_1) = H_S(t_2). \tag{7.7.2}$$

The satellite will rotate in the same period of time to $\theta(t_2)$:

$$-\dot{H}_w \frac{(t_2 - t_1)^2}{2} = \dot{H}_S \frac{(t_2 - t_1)^2}{2} = I_S \theta(t_2). \tag{7.7.3}$$

In a time-optimal trajectory, we must set $t_3 = t_2$. The final angular rotation is then $\theta_f = 2\theta(t_2)$, or

$$\theta_f = \frac{\dot{H}_S(t_1 - t_2)^2}{I_S} = -\frac{\dot{H}_w(t_2 - t_1)^2}{I_S}. \tag{7.7.4}$$

The performance of the attitude control demands that an angular rotation θ_f be achieved within a given time t_f. Hence, $t_2 - t_1 = t_f/2$. Finally,

$$\dot{H}_w = \frac{4 I_S \theta_f}{t_f^2}. \tag{7.7.5}$$

During the acceleration stage, $(t_2 - t_1) = 0.5 t_f$, the momentum of the satellite increases to

$$H_S(t_2) = -\dot{H}_w(t_2 - t_1) = -\dot{H}_w \frac{t_f}{2}; \tag{7.7.6}$$

the reaction wheel must be able to accumulate the same level of momentum, so

$$H_w = \frac{2 I_S \theta_f}{t_f}. \tag{7.7.7}$$

EXAMPLE 7.7.1 An attitude control system is designed to follow an attitude rotation of 0.2 rad in 10 sec, where the moment of inertia about the rotation axis is $I_S = 100$ kg-m². To specify the characteristics of the reaction wheel for this control specification, the following relations are used:

$$\text{maximum torque} = \dot{H}_w = \frac{4 \times 100 \times 0.2}{100} = 0.8 \text{ N-m};$$

$$\text{maximum momentum} = \frac{2 \times 100 \times 0.2}{10} = 4 \text{ N-m-sec}.$$

If the calculated parameters for the potential reaction wheel are too difficult to acquire with existing commercial and space-qualified wheels, the attitude control specifications must be lowered. Reconsidering Figure 7.7.1, this would involve an increase of the acceleration period $(t_2 - t_1)$ as well as a forceful prolongation of the final time t_f of the attitude maneuver. Another constraint in activating the reaction wheel is the heating of the stator of the electrical motor due to prolonged torque operation. This difficulty can be partially eliminated by torque control maneuvers in which the acceleration and deceleration stages are separated by a *no-control torque period*, as in Figure 7.7.1.

Until now, our calculations of maximal accumulation by the momentum wheel have been based solely on attitude-maneuvering specifications. However, there is another factor that must be considered. We have already treated the subject of momentum saturation and dumping of the reaction wheel (Section 7.3.3). Any external disturbance will add momentum to the wheel, which must be periodically dumped. Generally, because of mission constraints, there is some minimum period of time t_{dump} during which dumping is not allowed. The reaction wheel must be designed so that it can retain any momentum accumulated during t_{dump} without adverse effects.

Finally, specifying the maximum acceptable torque noise is based on our analysis in Section 7.3.6. An example of acceptable torque noise levels in different frequency spectrum ranges is shown in Table C.4.1 (p. 395).

7.8 Summary

The present chapter dealt with the critical subject of satellite attitude maneuvers. The two most important factors influencing maneuver quality are the characteristics of the torque controllers and of the attitude sensors. Torque controllers are characterized by the maximum torques they can produce, and by the level of parasitic disturbing torques. Attitude sensors are characterized by their accuracies, and also by their parasitic noises and biases.

Under the constraints of the physical characteristics of the control hardware, maximum performance was achieved by minimizing amplification of sensor noise and of the controllers' disturbing torque noises. Also, time-optimal techniques were used in order to take full advantage of the maximum torques that controllers can deliver. Finally, we examined momentum management of a multi-reaction wheel system, together with momentum dumping of the momentum accumulated in the wheels due to external torque disturbances.

References

Azor, R. (1993), "Momentum Management and Torque Distribution in a Satellite with Reaction Wheels," Israel Annual Conference on Aviation and Astronautics (24–25 February, Tel Aviv). Tel Aviv: Kenes, pp. 339–47.

Bikdash, M., Cliff, E., and Nayfeh, A. (1993), "Saturating and Time Optimal Feedback Controls," *Journal of Guidance, Control, and Dynamics* 16(3): 541–8.

Bilimora, K., and Wie, B. (1993), "Time-Optimal Three-Axis Reorientation of a Rigid Spacecraft," *Journal of Guidance, Control, and Dynamics* 16(3): 446–60.

Bollner, M. (1991), "On-board Control System Performance of the Rosat Spacecraft," *Acta Astronautica* 25(8/9): 487–95.

Boorg, P. (1982), "Validation for Spot Attitude Control System Development," *IFAC Automatic Control System Development*. Noordwijk, Netherlands, pp. 67–73.

Bosgra, J. A., and Prins, M. J. (1982), "Testing and Investigation of Reaction Wheels," *IFAC Automatic Control in Space*. Noordwijk, Netherlands, pp. 449–58.

Bosgra, J. A., and Smilde, H. (1982), "Experimental and Systems Study of Reaction Wheels. Part I: Measurement and Statistical Analysis of Forces and Torque Irregularities," National Aerospace Laboratory, NLR, Netherlands.

Bryson, A., and Ho, Y. (1969), *Applied Optimal Control*. Waltham, MA: Blaisdel.

D'Azzo, J. J., and Houpis, C. H. (1988), *Linear Control System Analysis and Design, Conventional and Modern*. New York: McGraw-Hill.

References

Dorf, C. R. (1989), *Modern Control Systems*. 5th ed. Reading, MA: Addison-Wesley.

Dougherty, H., Rodoni, C., Tompetrini, K., and Nakashima, A. (1982), "Space Telescope Control," *IFAC Automatic Control in Space*. Oxford, UK: Pergamon, pp. 15-24.

Elgerd, O. (1967), *Control System Theory*. New York: McGraw-Hill.

Fleming, A. W., and Ramos, A. (1979), "Precision Three-Axis Attitude Control via Skewed Reaction Wheel Momentum Management," Paper no. 79-1719, AIAA Guidance and Control (6-8 August, Boulder, CO). New York: AIAA, pp. 177-90.

Glaese, J. R., Kennel, H. F., Nurre, G. S., Seltzer, S. M., and Shelton, H. L. (1976), "Low-Cost Space Telescope Pointing Control System," *Journal of Spacecraft and Rockets* 13(7): 400-5.

James, H. M., Nichols, N. B., and Phillips, R. S. (1955), *Theory of Servomechanisms* (MIT Radiation Laboratory Series, vol. 25). New York: McGraw-Hill.

Junkins, J. L., and Turner, J. D. (1986), *Optimal Spacecraft Rotational Maneuvers*. Amsterdam: Elsevier.

Kaplan, M. (1976), *Modern Spacecraft Dynamics and Control*. New York: Wiley.

Lebsock, K. (1982), "Magnetic Desaturation of a Momentum Bias System," AIAA/AAS Astrodynamics Conference (9-11 August, San Diego). New York: AIAA.

McElvain, R. J. (1962), "Satellite Angular Momentum Removal Utilizing the Earth's Magnetic Field," Space Systems Division, Hughes Aircraft Co., El Segundo, CA.

Oh, H. S., and Valadi, S. R. (1991), "Feedback Control and Steering Laws for Spacecraft Using Single Gimbal Control Moment Gyros," *Journal of the Astronautical Sciences* 39(2): 183-203.

Pierre, A. D. (1986), *Optimization Theory with Applications*. New York: Dover.

Pircher, M. (1989), "Spot 1 Spacecraft in Orbit Performance," *IFAC Automatic Control in Space*. Noordwijk, Netherlands.

Schletz, B. (1982), "Use of Quaternion in Shuttle Guidance," Paper no. 82-1557, AIAA Conference on Navigation and Control (9-11 August, San Diego). New York: AIAA, pp. 753-60.

Solodovnikov, V. V. (1960), *Introduction to the Statistical Dynamics of Automatic Control Systems*. New York: Dover.

Stickler, A. C., and Alfriend, K. (1976), "Elementary Magnetic Attitude Control System," *Journal of Spacecraft and Rockets* 13(5): 282-7.

Wertz, J. R. (1986), *Spacecraft Attitude Determination and Control*. Dordrecht: Reidel.

Wie, B., Weiss, H., and Arapostathis, A. (1989), "Quaternion Feedback Regulator for Spacecraft Eigenaxis Rotations," *Journal of Guidance, Control, and Dynamics* 12(3): 375-80.

CHAPTER 8

Momentum-Biased Attitude Stabilization

8.1 Introduction

Momentum-biased satellites are dual-spin satellites that do *not* consist of two parts (platform and rotor) as described in Chapter 6. Here, constant angular momentum is provided by a momentum wheel – a momentum exchange device described in Chapter 7. Momentum-biased satellites are three-axis–stabilized as follows: (1) the momentum bias provides inertial stability to the wheel axis, which is perpendicular to the orbit plane; and (2) the torque capabilities of the wheel about the wheel axis are used to stabilize the attitude of the satellite in the orbit plane. Most communications satellites, especially those operating in geostationary altitudes, are momentum-biased.

As we shall see, the momentum bias is not sufficient to stabilize completely the momentum axis in space. Active control means are generally added to assure an accurate attitude stabilization, keeping the attitude errors within strict permitted limits. Common controllers are magnetic torqrods, reaction thrusters, or even an additional small reaction wheel.

An unusual and important feature of momentum-biased satellites is that their yaw attitude error need not be measured, rendering that difficult task unnecessary. Different control schemes based on the momentum-bias principle will be treated in this chapter. See also Dougherty, Lebsock, and Rodden (1971), Iwens, Fleming, and Spector (1974), Schmidt (1975), Lebsock (1980), and Fox (1986).

8.2 Stabilization without Active Control

The orbit reference frame was defined in previous chapters as follows: the Z_R axis points toward the center of mass of the earth; the Y_R axis is normal (perpendicular) to the orbit plane; the X_R axis completes a right-hand three-axis orthogonal

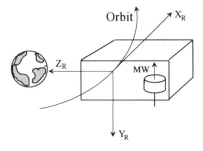

Figure 8.2.1 Definition of the reference frame and the direction of the MW axis.

8.2 / Stabilization without Active Control

frame, and points in the direction of the orbiting satellite. For a circular orbit, the $\mathbf{X}_R$ axis coincides with the velocity vector of the satellite. The axis of the momentum wheel (MW) is nominally in the direction of the normal to the orbit plane. (See Figure 8.2.1.) The angular dynamic equations of the system at hand emerge from Eqs. 4.8.14 with respect to a single momentum wheel, the one with its axis aligned on the $\mathbf{Y}_B$ body axis. This means that only the control variables h_{wy} and $\dot{h}_{wy}$ will remain in the equations.

For notational simplicity, we define

$$a = 4\omega_o^2(I_y - I_z), \quad b = -\omega_o(I_x + I_z - I_y), \quad c = \omega_o^2(I_y - I_x), \quad d = 3\omega_o^2(I_x - I_z).$$

Given our reference frame and the definitions just listed, Eqs. 4.8.14 become

$$\begin{aligned}
T_{dz} + T_{cz} &= I_z \ddot{\psi} + [-b + h_{wy}]\dot{\phi} + [c - \omega_o h_{wy}]\psi, \\
T_{dx} + T_{cx} &= I_x \ddot{\phi} + [a - \omega_o h_{wy}]\phi - [-b + h_{wy}]\dot{\psi}, \\
T_{dy} + T_{cy} &= I_y \ddot{\theta} + d\theta + \dot{h}_{wy},
\end{aligned} \qquad (8.2.1)$$

where h_{wy} stands for h_{wy0} of Eqs. 4.8.14.

The approximated Eqs. 8.2.1 show that it is possible to separate the first two equations, pertaining to the $\mathbf{X}_B$ and $\mathbf{Z}_B$ body axes, from the third equation, which is the pitch dynamics equation about the $\mathbf{Y}_B$ body axis. The pitch attitude is controlled with the torque capabilities of the momentum wheel, namely $\dot{h}_{wy}$, in exactly the same way as explained in Section 7.3. The momentum of the wheel, set initially at some nominal bias value, may change owing to external disturbances acting about the $\mathbf{Y}_B$ axis; the additional momentum must be dumped, as described in Section 7.3.3.

The more difficult attitude stabilization problem exists in the $\mathbf{X}_B$-$\mathbf{Z}_B$ plane of the s/c dynamics. Let us take Laplace transforms of the first two of Eqs. 8.2.1 for constant disturbances about the $\mathbf{X}_B$ and $\mathbf{Z}_B$ body axes, and also for initial conditions of the Euler angles and their derivatives. The constant h_{wy} in Eqs. 8.2.1 is, in any practical system, large enough to justify neglecting the terms a, b, and c. For instance, these terms become much smaller than the specified h_{wy} in a geostationary orbit, where $\omega_o = 7.39 \times 10^{-5}$ rad/sec. Even if the differences in the moments of inertia (in the terms a, b, c in Eqs. 8.2.1) are as high as 500 kg-m^2 and $h_{wy} = 20$ kg-m-sec, the errors introduced in the coefficients of ϕ and ψ are less than 1%. With these approximations, the equations can be put in the following matrix form:

$$\begin{bmatrix} s^2 - \dfrac{\omega_o h_{wy}}{I_x} & -\dfrac{s h_{wy}}{I_x} \\ \dfrac{s h_{wy}}{I_z} & s^2 - \dfrac{\omega_o h_{wy}}{I_z} \end{bmatrix} \begin{bmatrix} \phi \\ \psi \end{bmatrix} = \begin{bmatrix} \dfrac{T_{dx}}{sI_x} + s\phi(0) + \dot{\phi}(0) - \psi(0)\dfrac{h_{wy}}{I_x} \\ \dfrac{T_{dz}}{sI_z} + s\psi(0) + \dot{\psi}(0) + \phi(0)\dfrac{h_{wy}}{I_z} \end{bmatrix}. \qquad (8.2.2)$$

The solution is

$$\begin{bmatrix} \phi \\ \psi \end{bmatrix} = \dfrac{1}{\Delta(s)} \begin{bmatrix} s^2 - \dfrac{\omega_o h_{wy}}{I_z} & \dfrac{s h_{wy}}{I_x} \\ -\dfrac{s h_{wy}}{I_z} & s^2 - \dfrac{\omega_o h_{wy}}{I_x} \end{bmatrix} \begin{bmatrix} \dfrac{T_{dx}}{sI_x} + s\phi(0) + \dot{\phi}(0) - \psi(0)\dfrac{h_{wy}}{I_x} \\ \dfrac{T_{dz}}{sI_z} + s\psi(0) + \dot{\psi}(0) + \phi(0)\dfrac{h_{wy}}{I_z} \end{bmatrix}, \qquad (8.2.3)$$

where

$$\Delta(s) = s^4 + \left[-\omega_o h_{wy}\left(\frac{1}{I_x} + \frac{1}{I_z}\right) + \frac{h_{wy}^2}{I_x I_z}\right]s^2 + \frac{\omega_o^2 h_{wy}^2}{I_x I_z}. \tag{8.2.4}$$

With the practical assumption that $h_{wy}^2 \ggg \omega_o^2 I_x I_z$ and with $h_{wy} \gg \omega_o(I_x + I_z)$, the determinant in Eq. 8.2.4 can be put in the more compact form:

$$\Delta(s) \approx (s^2 + \omega_o^2)\left(s^2 + \frac{h_{wy}^2}{I_x I_z}\right). \tag{8.2.5}$$

In Eq. 8.2.5 we see that the determinant consists of two second-order poles: the first located at the orbital frequency ω_o and the second at the nutation frequency of the satellite, which is proportional to the momentum bias h_{wy} and the moments of inertia about the two transverse axes $\mathbf{X}_B$ and $\mathbf{Z}_B$ of the spacecraft.

Time Behavior

The momentum-biased satellite dynamics consist of two undamped second-order poles, so we cannot use the final-value theorem in Laplace transforms for evaluating the steady-state errors in ϕ and ψ owing to their harmonic behavior. However, we can assume that some momentary damping factors do exist in both second-order poles in Eq. 8.2.5. In this case, the steady-state errors will coincide with the average values of the harmonic motions of ϕ and ψ. According to Eq. 8.2.3 and Eq. 8.2.4, the average errors for the external disturbances T_{dx} and T_{dz} become

$$\phi_{av} = \frac{-T_{dx}}{\omega_o h_{wy}} + 0 T_{dz},$$
$$\psi_{av} = 0 T_{dx} + \frac{-T_{dz}}{\omega_o h_{wy}}. \tag{8.2.6}$$

We see that the average value of the Euler angles depends on the level of momentum bias supplied to the satellite and also on the orbital frequency ω_o. Note that the sign of the average values depends on the sign of h_{wy}. To find the exact amplitudes of the attitude harmonic motion, we must take the inverse Laplace transforms of Eq. 8.2.3. First, let us find the time behavior of $\phi(t)$ as a consequence of an external disturbance T_{dx}. From the preceding equations we find that

$$\frac{\phi(s)}{T_{dx}} = \frac{1}{I_x} \frac{\left(s^2 - \frac{\omega_o h_{wy}}{I_z}\right)}{s(s^2 + \omega_o^2)(s^2 + \omega_{nut}^2)} = A\frac{s}{s^2 + \omega_o^2} + B\frac{s}{s^2 + \omega_{nut}^2} + C\frac{1}{s}$$

$$= \left[\frac{1}{\omega_{nut}^2 - \omega_o^2}\right]\left[\frac{1}{I_x} + \frac{h_{wy}}{I_x I_z \omega_o}\right]\frac{s}{s^2 + \omega_o^2}$$

$$- \left[\frac{1}{\omega_{nut}^2 - \omega_o^2}\right]\left[\frac{1}{I_x} + \frac{\omega_o}{h_{wy}}\right]\frac{s}{s^2 + \omega_{nut}^2} - \frac{1}{\omega_o h_{wy}}\frac{1}{s}; \tag{8.2.7}$$

$$\phi(t) = T_{dx}[C + A\cos(\omega_o t) + B\cos(\omega_{nut} t)]. \tag{8.2.8}$$

The inverse Laplace transform consists of two harmonic cosine functions with frequencies ω_{nut} and ω_o of different amplitudes, and a constant that is the average value of the time response with amplitude $-1/\omega_o h_{wy}$, as also found in Eqs. 8.2.6. It is easily seen from Eq. 8.2.7 that the amplitude A of the harmonic motion with orbital frequency ω_o is much larger than the amplitude B of the harmonic motion with nutation frequency ω_{nut}.

8.2 / Stabilization without Active Control 213

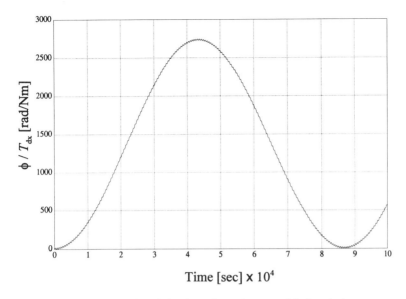

Figure 8.2.2 Time behavior of $\phi(t)$ in one orbital period.

EXAMPLE 8.2.1 In this example, we choose a geostationary satellite with $\omega_o = 7.3 \times 10^{-5}$ rad/sec. Also, $I_x = 800$ kg-m^2, $I_z = 1{,}000$ kg-m^2, and $h_{wy} = -10$ N-m-sec. With these satellite characteristics, we find that $\omega_{\text{nut}} = 0.0112$ rad/sec. It follows that $A = -1{,}357$, $B = -10$, and $C = 1{,}371$. With the values of A, B, and C:

$$\phi(t) = T_{dx}[1{,}371 - 1{,}357\cos(0.0000729t) - 10\cos(0.0112t)].$$

The nutation harmonic amplitude is smaller by a factor of 135 than that of the orbital harmonic amplitude. The time response of the roll error $\phi(t)$ is shown in Figure 8.2.2. In order to show the difference between the amplitudes of orbital and nutation harmonic motions, this time response is telescoped in Figure 8.2.3 (overleaf) and shown for the first 10,000 seconds only. A similar time response can be simulated for the yaw angle ψ.

Even if the disturbance amplitude is as small as 10^{-5} N-m, the maximum error in ϕ will be 1.5° – a tremendous error for a geostationary communications satellite, for which an acceptable roll error is only about 0.05°. This situation could be remedied by drastically increasing the value of the momentum bias, but such an approach would require large increases in the dimensions, weight, and power consumption of the momentum wheel assembly, which for practical reasons are usually not feasible.

Moreover, because the open-loop poles of the transfer functions in Eqs. 8.2.7 are not damped, harmonic disturbances having frequencies of ω_o or ω_{nut} will destabilize the system and hence the amplitude of the harmonic motion will increase linearly with time. We should keep in mind that various external disturbances acting on the satellite, such as solar pressure disturbing torques, may have harmonic components matching the basic orbital frequency ω_o. All these factors lead to the conclusion that active attitude stabilization is mandatory.

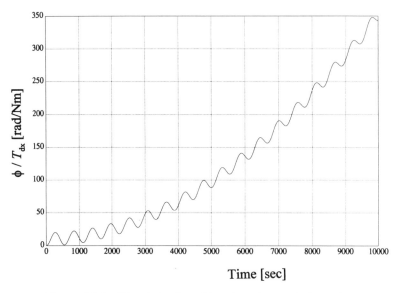

Figure 8.2.3 Time behavior of $\phi(t)$ in a time period that is 1/10 of the orbital period.

8.3 Stabilization with Active Control

The momentum bias is not itself sufficient to adequately attitude-stabilize the X_B–Z_B lateral plane of the satellite. For any three-axis–stabilized satellite orbiting a planet, the *horizon sensor* (also called an *earth sensor* for earth-orbiting s/c) enables measurement of the roll (ϕ) and the pitch (θ), but not the yaw (ψ), Euler angles. The pitch measurement allows control of the pitch attitude in the manner explained in Section 7.3. For roll–yaw attitude control, it is theoretically necessary to measure (or estimate) both the roll and the yaw angle errors, which cannot be done with a horizon sensor. One alternative is to measure the yaw angle with the aid of a sun sensor. Unfortunately, during eclipse the sun sensor is of no use. Moreover, even if the satellite is not within an eclipse of the sun, it might happen that the geometry between the sun vector and the nadir vector is such that the yaw angle cannot be measured with sufficient accuracy or even at all (see Wertz 1978). Another approach would use *magnetometers,* but fluctuations in the earth's magnetic field make it impossible to estimate the yaw angle with an accuracy better than 0.5°, which is generally not good enough for a geostationary communications satellite. A third possibility is to use *star sensors.* These sensors allow very accurate attitude determination, but they are quite complicated, sometimes unreliable, and data-intensive – compelling onboard control algorithms to store and access memory-intensive star catalogs for attitude reference. For low-inclination orbiting satellites, use of star sensors is easier because the Polaris star is a good bright reference, almost inertial with respect to the earth's north axis direction; hence simpler star catalogs of moderate complexity can be used (Maute et al. 1989). A star sensor with a field of view of 7–12° allows measuring the yaw angle of the satellite during all stages of its life, including the GTO-to-GEO transfer. Even so, star sensors are neither simple nor straightforward to use.

8.3 / Stabilization with Active Control

These considerations underlie our desire to control the s/c attitude *without* measuring the yaw angle. In a momentum-bias attitude-controlled satellite, the roll and yaw angles are related via the constant momentum, and this property is used to design a complete momentum-biased ACS without measuring the yaw angle. See Section 8.3.2 and Figure 8.3.1. In this section we shall analyze both possibilities – namely, control of the roll and yaw angles with and without yaw measurement.

Another technical problem is the need to use angular rate sensors. Such instruments are not very reliable when operated continuously for 7 to 10 years, the expected lifetime of modern geostationary satellites. Control engineers prefer not to use angular rate sensors except for very special tasks, and for short periods. In this chapter we shall do likewise and obtain the rate of the Euler angles by differentiating them, with adequate noise filters.

8.3.1 Active Control Using Yaw Measurements

Returning to Eq. 8.2.3 and Eq. 8.2.5, we see that the two second-order poles of the momentum-biased dynamics are undamped. The active control must damp these poles and also decrease the steady-state errors, as implied by Eqs. 8.2.6. The simplest torque control commands that can achieve these tasks are:

$$T_{cx} = -(k_x \phi + k_{xd} \dot{\phi}), \tag{8.3.1}$$

$$T_{cz} = -(k_z \psi + k_{zd} \dot{\psi}). \tag{8.3.2}$$

Returning now to the first two of Eqs. 8.2.1, after neglecting the terms a, b, c we have

$$T_{dx} = I_x \ddot{\phi} - \omega_o h_{wy} \phi - h_{wy} \dot{\psi} + k_x \phi + k_{xd} \dot{\phi}, \tag{8.3.3}$$

$$T_{dz} = I_z \ddot{\psi} - \omega_o h_{wy} \psi + h_{wy} \dot{\phi} + k_z \psi + k_{zd} \dot{\psi}. \tag{8.3.4}$$

The solution of these equations for external disturbances T_{dx} and T_{dz} is then

$$\begin{bmatrix} \phi(s) \\ \psi(s) \end{bmatrix} = \frac{1}{\Delta(s)} \begin{bmatrix} s^2 + \dfrac{1}{I_z}(k_z + k_{zd}s - \omega_o h_{wy}) & s\dfrac{h_{wy}}{I_x} \\ -s\dfrac{h_{wy}}{I_z} & s^2 + \dfrac{1}{I_x}(k_x + k_{xd}s - \omega_o h_{wy}) \end{bmatrix} \begin{bmatrix} \dfrac{T_{dx}}{I_x} \\ \dfrac{T_{dz}}{I_z} \end{bmatrix},$$

$$\tag{8.3.5}$$

where

$$\begin{aligned}\Delta(s)I_x I_z &= s^4 I_x I_z + s^3 (k_{xd} I_z + k_{zd} I_x) \\ &+ s^2 [k_{xd} k_{zd} + h_{wy}^2 + I_z(k_x - \omega_o h_{wy}) + I_x(k_z - \omega_o h_{wy})] \\ &+ s[k_{zd}(k_x - \omega_o h_{wy}) + k_{xd}(k_z - \omega_o h_{wy})] \\ &+ k_x k_z + (\omega_o h_{wy})^2 - \omega_o h_{wy}(k_x + k_z). \end{aligned} \tag{8.3.6}$$

For the attitude-controlled system to have stable roots, all coefficients of the polynomial in Eq. 8.3.6 must be positive. This means that we must choose $h_{wy} = -h$ with $h > 0$. With this assumption, the determinant of Eq. 8.3.6 takes the form

$$\begin{aligned}\Delta(s)I_x I_z &= s^4 I_x I_z + s^3 (k_{xd} I_z + k_{zd} I_x) \\ &+ s^2 [k_{xd} k_{zd} + h^2 + I_z(k_x + \omega_o h) + I_x(k_z + \omega_o h)] +\end{aligned}$$

$$+ s[k_{zd}(k_x + \omega_o h) + k_{xd}(k_z + \omega_o h)]$$
$$+ k_x k_z + (\omega_o h)^2 + \omega_o h(k_x + k_z). \tag{8.3.7}$$

For the roots of the closed-loop system to be stable, there is a necessary (but not sufficient) condition that the coefficients of the polynomial in Eq. 8.3.7 be positive. To stabilize the system unconditionally, additional conditions on the control parameters k_x, k_{xd}, k_z, and k_{zd} are necessary. The parameters k_x and k_z are primarily responsible for the steady-state errors in ϕ and ψ caused by disturbances T_{dx} and T_{dz}; we can calculate their values from the steady-state error requirements. From Eq. 8.3.5 and Eq. 8.3.7 we find that

$$\frac{\phi_{ss}}{T_{dx}} = \frac{k_z + \omega_o h}{k_x k_z + (\omega_o h)^2 + \omega_o h(k_x + k_z)} = \phi_{ssx}, \tag{8.3.8}$$

$$\frac{\psi_{ss}}{T_{dz}} = \frac{k_x + \omega_o h}{k_x k_z + (\omega_o h)^2 + \omega_o h(k_x + k_z)} = \psi_{ssz}. \tag{8.3.9}$$

From the same equations, using once more the final-value theorem in Laplace transforms, it is easily concluded that

$$\frac{\phi_{ss}}{T_{dz}} = 0; \quad \frac{\psi_{ss}}{T_{dx}} = 0. \tag{8.3.10}$$

Suppose that h has already been determined by considerations to be stated later; then, from the required values of ϕ_{ssx} and ψ_{ssz}, k_x and k_z can be calculated using Eq. 8.3.8 and Eq. 8.3.9. For k_x and k_z we finally obtain

$$k_x = \frac{1 - \omega_o h \phi_{ssx}}{\phi_{ssx}}; \quad k_z = \frac{1 - \omega_o h \psi_{ssz}}{\psi_{ssz}}. \tag{8.3.11}$$

We are left with the more difficult task of calculating k_{xd} and k_{zd}. This can be accomplished as follows. The determinant of Eq. 8.3.7 is of the fourth order. We can assume without loss of generality that this determinant consists of two second-order damped poles:

$$\Delta(s) = (s^2 + 2\xi_1 \omega_{n1} s + \omega_{n1}^2)(s^2 + 2\xi_2 \omega_{n2} s + \omega_{n2}^2)$$
$$= s^4 + s^3 2(\xi_1 \omega_{n1} + \xi_2 \omega_{n2}) + s^2(\omega_{n1}^2 + \omega_{n2}^2 + 4\xi_1 \xi_2 \omega_{n1} \omega_{n2})$$
$$+ s 2\omega_{n1} \omega_{n2}(\xi_1 \omega_{n2} + \xi_2 \omega_{n1}) + \omega_{n1}^2 \omega_{n2}^2. \tag{8.3.12}$$

The coefficients in the polynomial of Eq. 8.3.12 can be equated to those of Eq. 8.3.7 to solve for k_{xd} and k_{zd}. In fact, given ξ_1 and ξ_2, we have four equations with four unknowns: k_{xd}, k_{zd}, ω_{n1}, and ω_{n2}; however, we are interested in only the first two. The last two are byproducts with no special meaning, except that they state how close the closed-loop poles stay to the open-loop poles of the system. An example will clarify the complete procedure.

EXAMPLE 8.3.1 The satellite moments of inertia are $I_x = 800$ kg-m^2 and $I_z = 1,000$ kg-m^2. The orbit is geostationary, with $\omega_o = 7.236 \times 10^{-5}$ rad/sec. The maximum external disturbances to be expected are: $T_{dx} = 5 \times 10^{-6}$ N-m and $T_{dz} = 5 \times 10^{-6}$ N-m. The maximum permitted steady-state errors in roll and yaw are $\phi_{ss} = 0.05°$ and $\psi_{ss} = 0.2°$. The momentum bias was chosen as $h = 20$ N-m-sec.

8.3 / Stabilization with Active Control

Solution Using the procedure of this section, we find that:

$k_x = 0.00427$, $k_{xd} = 22.9$;

$k_z = -0.26 \times 10^{-4}$, $k_{zd} = 12.8$;

$\omega_{n1} = 0.0295$ rad/sec, $\omega_{n2} = 0.109 \times 10^{-3}$ rad/sec.

In Chapter 7, it was of utmost importance to analyze the attitude sensors' noise amplification. Such an analysis is not needed for the attitude control scheme based on measuring both the roll and the yaw errors, a control configuration that is seldom used. An analysis of sensor noise amplification will be carried out in the next section for a more practical (but also more difficult) attitude control scheme: only the roll and pitch angles are measured, so a yaw error cannot be used for control purposes.

8.3.2 Active Control without Yaw Measurements

The control configuration based on measuring the roll and pitch angles only is the most popular today. The earth sensor, which is based on sensing the horizon contour of the earth with respect to the satellite body frame, is currently the lone sensor used to measure directly the roll error for nadir-pointing satellites. Accuracies of the order of 0.02° are common with this technology. However, we must also consider the attendant statistical noise (about 0.03° RMS), an effect of utmost importance where noise amplification is concerned (Sidi 1992a).

The control torque command equations are as follows:

$$T_{cx} = -(k_x \phi + k_{xd} \dot{\phi}), \tag{8.3.13}$$

$$T_{cz} = aT_{cx}. \tag{8.3.14}$$

The last equation merits some explanation. Equation 8.3.14 is based on the fact that, for a momentum-biased satellite, the roll and yaw errors interchange every quarter of the orbit. This means that an accumulated yaw error will change to a roll error after a quarter of the orbit period; since the roll is measured, the accumulated yaw error will be controlled as a roll error, and attenuated accordingly. This effect is shown in Figure 8.3.1, where X_R, Y_R, Z_R are the orthogonal axes of the orbit reference frame. The momentum wheel (MW) axis is aligned with the Y_B axis. At position 1 in the orbit, the direction of the momentum vector **h**, which is aligned with the MW axis, is inclined to the Y_R axis, so that an angle ψ exists. During the motion

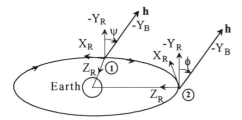

Figure 8.3.1 Change of the yaw error to roll error in a momentum-biased spacecraft.

of the satellite in the orbit, with no external disturbances, the momentum vector **h** remains stabilized in space, which means that the Y_B axis also remains constant in space. At position 2, after a quarter orbit, the ψ Euler angle has been transformed to a roll angle ϕ with the same amplitude as the previous yaw angle. Notice that the X_R and Z_R axes have interchanged in relation to the inertial frame. Consequently, with a time delay of one quarter of the orbit period, the whole yaw error will be sensed (and controlled) as a roll error, measured with the earth sensor. This is why the control command torque in Eq. 8.3.14 is effective in controlling the yaw error (Agrawal 1986). As in Section 8.3.1, using the first two of Eqs. 8.2.1 together with Eq. 8.3.13 and Eq. 8.3.14, we have:

$$T_{dx} = I_x \ddot{\phi} - \omega_o h_{wy} \phi - h_{wy} \dot{\psi} + k_x \phi + k_{xd} \dot{\phi}, \tag{8.3.15}$$

$$T_{dz} = I_z \ddot{\psi} - \omega_o h_{wy} \psi + h_{wy} \dot{\phi} + ak_x \phi + ak_{xd} \dot{\phi}. \tag{8.3.16}$$

These equations can be put in matrix form to yield the following solution:

$$\begin{bmatrix} \phi(s) \\ \psi(s) \end{bmatrix} = \frac{1}{\Delta(s)} \begin{bmatrix} s^2 - \frac{1}{I_z}\omega_o h_{wy} & s\frac{h_{wy}}{I_x} \\ -\frac{1}{I_z}(sh_{wy} + ak_x + sak_{xd}) & s^2 + \frac{1}{I_x}(-\omega_o h_{wy} + k_x + sk_{xd}) \end{bmatrix}$$

$$\times \begin{bmatrix} \frac{T_{dx}}{I_x} \\ \frac{T_{dz}}{I_z} \end{bmatrix}. \tag{8.3.17}$$

In this case,

$$\Delta(s)I_x I_z = s^4 I_x I_z + s^3 I_z k_{xd}$$
$$+ s^2[-\omega_o h_{wy}(I_x + I_z) + I_z k_x + h_{wy}^2 + ah_{wy}k_{xd}]$$
$$+ s(ah_{wy}k_x - \omega_o h_{wy}k_{xd}) + [(\omega_o h_{wy})^2 - \omega_o h_{wy}k_x]. \tag{8.3.18}$$

Once more, for stability reasons it is necessary (but not sufficient) that $h_{wy} = -h$ with $h > 0$. For the same reasons, we also choose $a < 0$. Set $a = -a_\psi$. With these definitions, the determinant in Eq. 9.3.18 will change to the following form:

$$\Delta(s)I_x I_z = s^4 I_x I_z + s^3 I_z k_{xd}$$
$$+ s^2[\omega_o h(I_x + I_z) + I_z k_x + h^2 + a_\psi h k_{xd}]$$
$$+ s(a_\psi h k_x + \omega_o h k_{xd}) + [(\omega_o h)^2 + \omega_o h k_x]. \tag{8.3.19}$$

In this determinant, k_{xd} must be positive in order not to violate the stability conditions of the coefficient of s^3, but k_x need not be positive. Equation 8.3.14 will be changed to

$$T_{cz} = -a_\psi T_{cx}. \tag{8.3.14'}$$

Next, we can proceed to find k_x, k_{xd}, and a_ψ. From Eq. 8.3.17 and Eq. 8.3.19, we have

$$\frac{\phi_{ss}}{T_{dx}} = \frac{1}{\omega_o h + k_x} = \phi_{ssx}, \tag{8.3.20}$$

8.3 / Stabilization with Active Control

$$\psi_{ss} = \frac{T_{dz}}{\omega_o h} + \frac{aT_{dx}k_x}{\omega_o h(k_x + \omega_o h)} \approx \frac{T_{dz} + aT_{dx}}{\omega_o h}. \quad (8.3.21)$$

since generally $k_x \gg \omega_o h$. We shall see that in practical cases, $a < 1$. With known and estimated external disturbances, and with limits on the maximum permitted steady-state error in yaw, we can find the needed minimum momentum bias h from the last term of Eq. 8.3.21. From Eq. 8.3.20, we can also find k_x:

$$k_x = \frac{T_{dx}}{\phi_{ss}} - \omega_o h. \quad (8.3.22)$$

We are left with the problem of finding k_{xd}. The same procedure as in Section 8.3.1 will be followed here. The desired determinant will have the form of Eq. 8.3.12. The coefficients of the polynomials of Eq. 8.3.12 and Eq. 8.3.19 will be equated. For assumed damping coefficients $\xi_1 = \xi_2 = \xi$ and for known moments of inertia and k_x, we can find $a = -a_\psi$ and k_{xd}. At this stage, an example would be instructive.

EXAMPLE 8.3.2 The satellite's moments of inertia are $I_x = 800$ kg-m² and $I_z = 1{,}000$ kg-m². The orbit is geostationary, with $\omega_o = 7.236 \times 10^{-5}$ rad/sec. The maximum external disturbances are expected to be $T_{dx} = T_{dz} = 5 \times 10^{-6}$ N-m. The maximum permitted steady-state errors in roll and yaw are $\phi_{ss} = 0.05°$ and $\psi_{ss} = 0.4°$.

Using the procedure of this section, we find according to Eq. 8.3.21 that the needed momentum bias is $h = 20$ N-m-sec. To find k_x, we solve Eq. 8.3.20 and find $k_x = 0.00427$. Equating the coefficients of the polynomials of Eq. 8.3.12 and Eq. 8.3.19, we find $k_{xd} = 42.8$ and $a = 0.888$. We also have the byproducts $\omega_{n1} = 0.0000848$ rad/sec and $\omega_{n2} = 0.0381$ rad/sec, which are the new closed-loop modal frequencies.

Figure 8.3.2 shows a block diagram of an ACS (attitude control system) composed of a MW and an earth sensor only. As already mentioned, the earth sensor (ES) can sense only the roll (ϕ) and the pitch (θ) Euler angles. Since no rate sensor is included, differentiation of the earth sensor outputs is necessary in order to implement the control laws in Eq. 8.3.13 and Eq. 8.3.14. The problem is that the earth sensor is noisy.

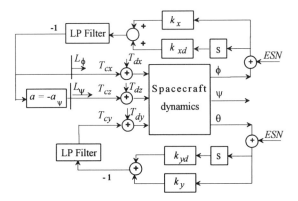

Figure 8.3.2 Mechanization of the attitude control system with an earth sensor only, and with no rate sensors for damping the control loops.

Without appropriate filtering, the RMS noise amplification from the sensor to the commanded torques T_{cx} and T_{cz}, with only $k_x + sk_{xd}$ as the control network, will become infinite for white noise. The LP filters in Figure 8.3.2 are low-pass filters included to prevent a high noise amplification.

Earth-Sensor Noise Amplification

It is important to analyze the amplification of ES noise to the torque commands at the input of the controllers, which could be reaction wheels, magnetic torqrods, or solar panels and flaps. A pure analytical procedure will not be followed here, because the complexity of the attitude dynamic equations of the satellite's X_B-Z_B plane precludes obtaining the RMS noise amplification as a function of a simple second-order closed-loop pole model as in Section 7.3.6. Instead, we will derive numerically the results for a realistic engineering example, such as Example 8.3.2. We will use a low-pass filter having two simple poles with corner frequencies of $\omega_c = 0.2$ rad/sec, about 10 times higher than that of the nutation frequency. The primary cause for the high amplification noise is the differentiation of ϕ, together with the high gain of k_{xd} needed to provide a high damping coefficient. The immediate consequence is that we will have to decrease this gain in order to decrease the ES RMS noise amplification. Naturally, the damping coefficients of the closed-loop poles will decrease also. The higher-frequency closed-loop pole, which is closer to the open-loop nutation pole, will be the most affected. The resulting new damping coefficients and amplification factors are shown in Table 8.3.1.

With the nominal case 1, the second damping factor ξ_2 is smaller than 0.7 owing to the LP filter, with the two corner frequencies at $\omega_c = 0.2$ rad/sec (see Figure 8.3.2). It is clear that decreasing the derivative gain does not appreciably change the damping coefficient of the smaller closed-loop pole, which is closer to the orbital frequency pole; however, the second damping coefficient, pertaining to the much higher nutation frequency pole, is drastically decreased with smaller k_{xd}. For case 4, we even see instability of the nutation closed-loop pole, since $\xi = -0.0004$. We shall return to these cases later. It is instructive to show the stability margins in the frequency domain on a Nichols chart. We can open the control loops at T_{cx} and at T_{cz} and compute the open-loop transfer functions L_ϕ and L_ψ defined in Figure 8.3.2; the results are shown in Figure 8.3.3.

We can now summarize the results. A good earth sensor is accompanied by a noise level of $0.03°$ (RMS). In the nominal case, since the noise amplification amounts to 2.09, the X_B-axis torque command will be accompanied by a noise level of $T_{cxN} = (2.09 \times 0.03)/57.321 = 0.001094$ N-m $= 1.094 \times 10^{-3}$ N-m (RMS). This level of amplified noise will affect the attitude control of a momentum-biased satellite.

Table 8.3.1 *Decrease of the noise amplification by decreasing the derivative gain k_{xd}*

Case No.	k_x	k_{xd}	ω_{n1}	ξ_1	ω_{n2}	ξ_2	T_{cx}/N_{ES}	T_{cz}/N_{ES}
1	4.27 10^{-3}	42.8	8.49 10^{-5}	0.7	0.07	0.52	2.09	1.86
2	4.27 10^{-3}	4.28	1.32 10^{-4}	0.65	0.03	0.08	0.19	0.17
3	4.27 10^{-3}	0.43	1.42 10^{-4}	0.65	0.02	0.00425	0.02	0.02
4	4.27 10^{-3}	0.21	1.43 10^{-4}	0.65	0.03	-4 10^{-4}	0.01	0.01

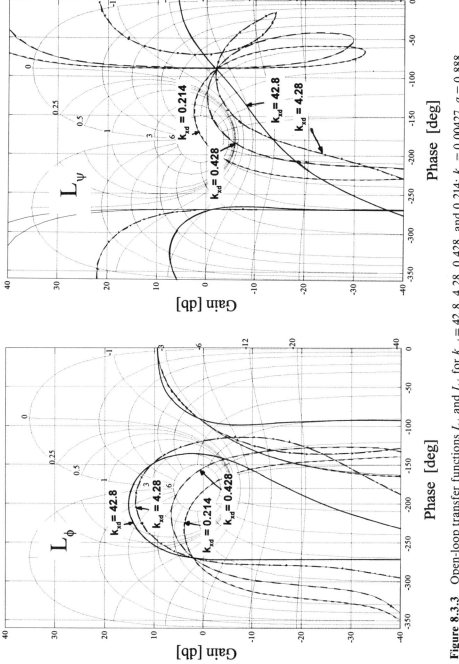

Figure 8.3.3 Open-loop transfer functions L_ϕ and L_ψ for $k_{xd} = 42.8, 4.28, 0.428,$ and 0.214; $k_x = 0.00427$, $a = 0.888$.

Noise Effects on Magnetic Attitude Control

At geostationary altitudes ($H_a = 35{,}786$ km), a large magnetic torqrod with a saturation level of 500 A-m^2 can achieve torque levels of only $T_B = 0.5 Mm/R^3 = 26.15 \times 10^{-6}$ N-m (see Section 7.4.2). Consequently, the magnetic torqrod will be constantly saturated by the noise, and the nominal solution (case 1 in Table 8.3.1) is not practically realizable. With case 3, the noise is amplified by a factor of 0.019. The noise amplified at the input to the magnetic torqrod will have an amplitude of $T_{cxN} = (0.019 \times 0.03)/57.321 = 9.9 \times 10^{-6}$ N-m. This noise level is acceptable with respect to saturation, but the nutation modal pole is almost undamped with a low damping coefficient of $\xi_2 = 0.00425$. In low orbits, on the other hand, control torques of the order of 6.4×10^{-3} N-m are achievable with the same magnetic torqrod. The amplified earth-sensor noise should be, as seen previously, of a much lower level. See also Schmidt and Muhlfelder (1981), Lebsock (1982), and Muhlfelder (1984).

Noise Effects on Solar-Torque Attitude Control

The situation is even worse if solar torques are used to control a momentum-biased satellite at geostationary altitudes, because the achievable control torque levels are even lower, of the order of 10–20×10^{-6} N-m (Lievre 1985, Sidi 1992b). With such low damping coefficients there is always the danger of nutational instability, which actually occurred with the *OTS* satellite (Benoit and Bailly 1987). Solar-torque attitude control thus requires active nutation damping via, for example, products of inertia (Phillips 1973; Devey, Field, and Flook 1977; Sidi 1992b). For low-orbit satellites, solar-torque attitude control is not practical because at low altitudes the disturbance torques are several orders of magnitude larger than the control torques that solar panels and flaps can provide. Mechanization of a solar-torque ACS is treated in Section 8.6.

Roll–Yaw Attitude Control with Momentum Exchange Devices

A straightforward way of stabilizing the roll–yaw attitude is to use additional momentum exchange devices – a reaction wheel, for instance, with its axis aligned parallel to the $\mathbf{X}_B$ body axis. This kind of control was treated in Chapter 7. The problem of noise amplification does not exist here, because the control torque levels of these devices are larger than 0.01 N-m.

Another possible solution, based on similar devices, is to use two momentum wheels that are slightly inclined to each other in a V geometry (Wie, Lehner, and Plescia 1985; Duhamel and Benoit 1991). In this configuration, the two inclined momentum wheels allow control of the $\mathbf{X}_B$ axis attitude, the roll angle ϕ, and the pitch angle θ about the $\mathbf{Y}_B$ axis while providing the system with the desired momentum stabilization. This control scheme will be treated in Section 8.7.

8.4 Roll–Yaw Attitude Control with Magnetic Torques

The realization of an ACS with magnetic torques is similar to that presented in Section 7.4. The only difference is that, in the present context, the wheel is in a momentum-bias condition. Equation 7.4.5 is used to derive the control inputs of the momentum wheel about the $\mathbf{Y}_B$ body axis, $\dot{h}_{wy}$, and also for the magnetic torqrod inputs creating the onboard magnetic control dipoles, M_x and M_z. In all the treated

8.4 / Roll-Yaw Attitude Control with Magnetic Torques

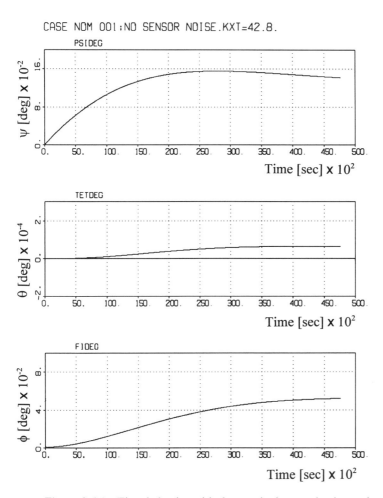

Figure 8.4.1 Time behavior with the nominal control gains and with no sensor noise.

examples, the pitch control gains are chosen so that the closed-loop bandwidth is $\omega_{ny} = 0.1$ rad/sec. Three examples will be simulated.

First, the nominal gains of case 1 (Table 8.3.1) were simulated with damping coefficients of 0.7 and 0.52 for the orbital and nutation modal frequencies, respectively. The steady-state amplitudes of the outputs of the magnetic torqrods are $M_{ssx} = -35$ A-m^2 and $M_{ssz} = -39.6$ A-m^2, as expected. The time behavior for a maximum disturbance $T_{dx} = 5 \times 10^{-6}$ N-m, with no sensor noise, is shown in Figure 8.4.1. The steady-state errors of the Euler angles are as theoretically expected, $\phi_{ss} = 0.05°$ and $\psi_{ss} = 0.15°$.

Next, the same nominal case was run with an RMS noise level of 0.03°. Figure 8.4.2 displays the time behavior of M_x and M_z, which are activated in almost a bang-bang mode owing to the much-amplified sensor noise. The RMS levels [A-m^2] of the control magnetic dipoles are in this case $M_x = 434$ and $M_z = 489$. The attitude control loop cannot behave correctly under these conditions of control torque saturation.

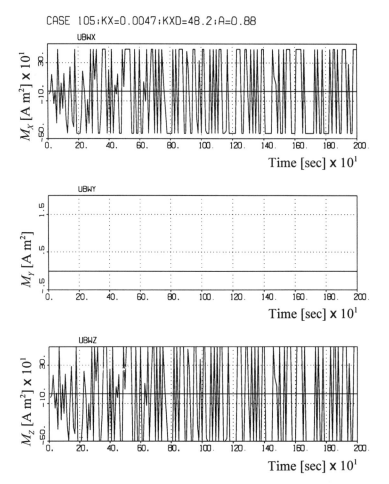

Figure 8.4.2 Time histories of the M_x and M_z control magnetic dipoles with nominal control gains and sensor noise.

The time-domain behavior of the Euler angles ϕ, ψ, and θ is shown in Figure 8.4.3. The steady-state errors are much higher than theoretically expected.

Finally, we simulate case 3 with $k_{xd} = 0.428$. In this case, the time histories of the ϕ and ψ errors show good agreement with the theoretical results: $\phi_{ss} \approx 0.05°$ and $\psi_{ss} \approx 0.15°$ (see Figure 8.4.4, page 226). But since the nutation modal pole is almost undamped, $\xi_{nut} = 0.00425$, time histories of the Euler angles clearly show the nutational motion. To exhibit this effect, the results in Figure 8.4.4 are shown telescoped for the first 2,000 seconds in Figure 8.4.5 (page 227): the nutation pole is clearly undamped. With the low gain $k_{xd} = 0.428$, the sensor noise is only slightly amplified, and the amplitudes of the magnetic control dipoles are well within their linear range; see Figure 8.4.6 (page 228). The RMS values of the magnetic dipoles are, despite the sensor noise, reasonably low: $M_x = 67.1$ A-m^2 and $M_z = 75.6$ A-m^2 (RMS). In many practical cases, virtually undamped nutation can be tolerated. However, such nutational motion can be destabilized by energy dissipation in the rotor of the momentum

8.5 / Active Nutation Damping via Products of Inertia 225

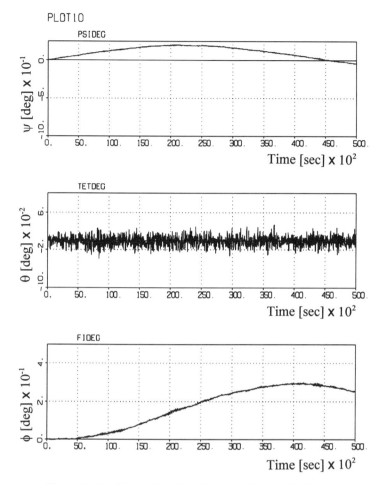

Figure 8.4.3 Time histories of ϕ, θ, ψ with nominal control gains and sensor noise.

wheel, which happened with the *OTS* satellite (Benoit and Bailly 1987; see also Section 6.7.1). This is unacceptable, so a way must be found to increase effectively the damping factor of the nutation modal pole without augmenting the noise amplification. Toward this end, a nutation control scheme using the products of inertia of the s/c will be introduced in the next section.

8.5 Active Nutation Damping via Products of Inertia

The use of products of inertia for active nutation damping was first suggested in the sixties (see Phillips 1973, Devey et al. 1977, Fox 1986). A control scheme using the torque capabilities of the momentum wheel, together with products of inertia, is shown in Figure 8.5.1 (page 229). The control network $G_{\text{nut}}(s)$ in this figure was added to that of Figure 8.3.2. The basic idea is to sense the roll angle ϕ and to extract from it the roll nutation error, which is then fed to the pitch control loop.

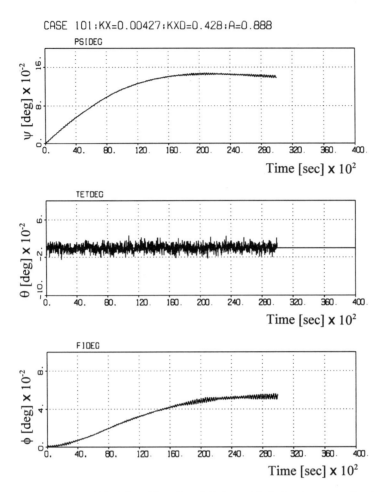

Figure 8.4.4 Time histories of ψ, θ, ϕ with derivative control gain $k_{xd} = 0.428$ and sensor noise.

Since there is a product of inertia (I_{yz} or I_{yx}; it is immaterial which one), the nutation control torque is transferred from the momentum wheel – whose torque capabilities are about the $\mathbf{Y}_B$ axis – to the $\mathbf{X}_B$-$\mathbf{Z}_B$ axes. Thus control of the nutation is achieved actively. The nutation control network must have the following basic form:

$$G_{\text{nut}}(s) = \frac{K_{\text{nut}} s}{(1+s\tau_1)(1+s\tau_2)}. \tag{8.5.1}$$

The control network $G_{\text{nut}}(s)$ must comprise at least one differentiation in order to prevent constant roll errors from being fed to the pitch control loop, which would introduce a disturbance error. On the other hand, the poles at $1/\tau_1$ and $1/\tau_2$ will pass only the roll error whose frequency spectrum components are in the vicinity of the nutation frequency, thus controlling (via the moment of inertia) only the nutation error component. Since there is now an interaction between the pitch control loop

8.5 / Active Nutation Damping via Products of Inertia

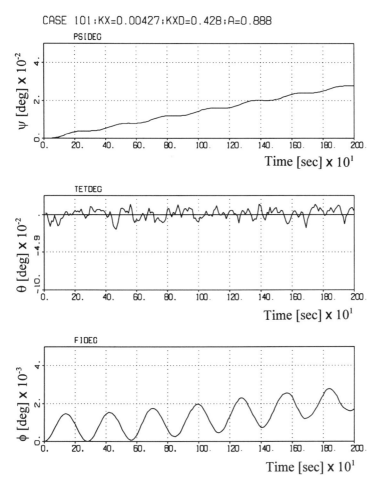

Figure 8.4.5 Time histories of ψ, θ, ϕ for $k_{xd} = 0.428$ (reduced time scale).

and the $\mathbf{X}_B$-$\mathbf{Z}_B$ plane control loop, care must be taken that the open-loop transfer function of the pitch loop does not deteriorate. Slight modification of the $G_{\text{nut}}(s)$ network might be necessary for a good shaping of the open-loop transfer function.

The basic frequency-domain shape of $G_{\text{nut}}(s)$ is given in Figure 8.5.2 (page 229). The nutation frequency ω_{nut} depends on the geometrical properties of the satellite, on I_x and I_z, and on the momentum bias h_{wy}. Hence the corner frequencies $1/\tau_1$ and $1/\tau_2$ also depend on these geometrical properties. Since the mass of the satellite decreases with time owing to fuel consumption, the moments of inertia change too. Furthermore, since external disturbances cause the momentum bias to increase (or decrease), h_{wy} may be expected to change by about 10–15% of its nominal value. Consequently, the nutation frequency is also expected to change around its nominal value by at least 10–15%. This means that due allowances should be made when designing the corner frequencies of $G_{\text{nut}}(s)$. Yet these frequencies must also be as close as possible to the nutation frequency, so that the bandpass filter will minimize noise amplification; hence a tradeoff occurs in determining the ideal corner frequencies.

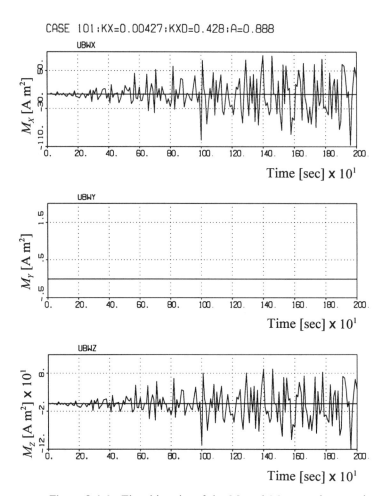

Figure 8.4.6 Time histories of the M_x and M_z control magnetic dipoles with $k_{xd} = 0.428$ and with sensor noise.

To show the effectiveness of the active nutation controller, we first simulate the control system of case 4 in Table 8.3.1. With $k_{xd} = 0.214$, there is instability at the nutation frequency, $\xi_2 = -0.0004$! With no active damping, the time histories of the roll and yaw Euler angles clearly show instability (see Figure 8.5.3, page 230). With a product of inertia of $I_{yz} = 20$ kg-m² and with

$$G_{\text{nut}}(s) = \frac{2{,}000s}{(1+s/0.03)(1+s/0.02)},$$

we obtain a well-damped nutational motion with practically no increase of sensor noise amplification. The resulting time histories with the well-damped nutation modal pole are shown in Figure 8.5.4 (page 231). By the way, since the derivative gain k_{xd} was decreased in the last control solution, the level of the amplified sensor noise at the input of the magnetic torqrods was diminished to RMS values of $M_x = 40.4$ and

8.6 / Roll-Yaw Attitude Control with Solar Torques

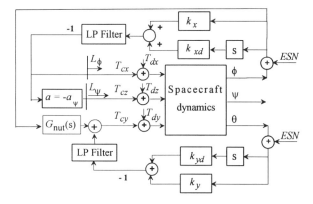

Figure 8.5.1 Control scheme for active nutation damping via products of inertia.

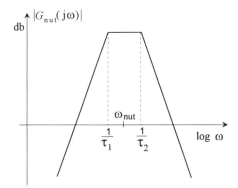

Figure 8.5.2 Basic frequency behavior of $G_{nut}(s)$.

$M_z = 45.5$ A-m^2. The active nutation damping scheme described here can be implemented with different control hardware, such as solar and reaction torque controllers, a subject to be treated in the next two sections.

8.6 Roll-Yaw Attitude Control with Solar Torques

Attitude control systems using solar control torques are known in the literature as "solar sailing" attitude control (Renner 1979, Lievre 1985). The technique is based on forces and torques produced by solar radiation pressure on a surface (Georgevic 1973, Forward 1990). Attitude control systems based on these control torques are feasible for satellites in which the solar panels can be rotated relative to the s/c body, as with all geostationary satellites. In Section 8.3 we remarked that achievable control torques using solar radiation pressure are quite low, in the range of 10–20 μN-m. Hence, the conclusions of Section 8.5 apply directly to this analysis also, and will not be repeated. Here we shall deal with the engineering realization of the control algorithm, and also with the physical constraints that preclude higher

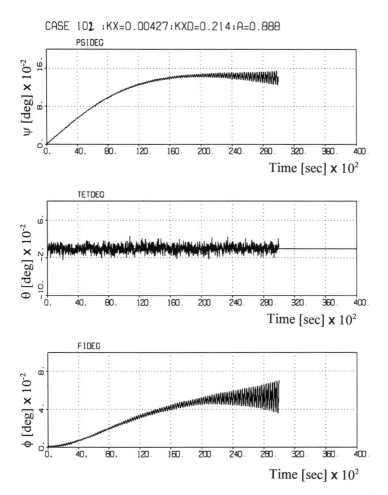

Figure 8.5.3 Time histories of the Euler angle errors *without* active nutation damping, $k_{xd} = 0.214$.

control torques. We must first write the dynamic equations arising from the radiation effect on the solar panels and flaps, which are necessary for achieving the desired control torques.

8.6.1 *Dynamic Equations for Solar Panels and Flaps*

In developing the equations relating control torques to the solar panel and flap positions we follow the presentation of Lievre (1985). The basic idea is to use the solar panels, whose primary function is to produce electrical power, to provide also the necessary torques about the X_B and Z_B body axes for attitude control purposes. In geostationary satellites there are always two symmetrical solar panels – the first directed toward the north, the second toward the south – in order to minimize the disturbance torque balance on the satellite produced by solar pressure on the panels.

8.6 / Roll-Yaw Attitude Control with Solar Torques

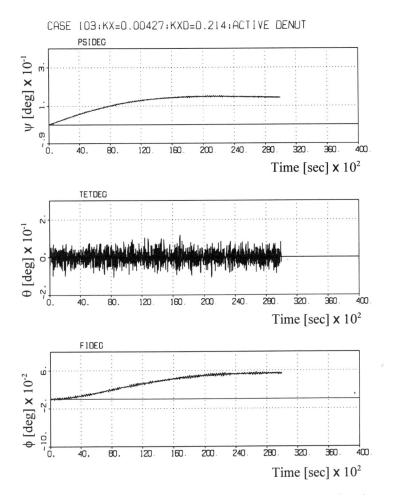

Figure 8.5.4 Time histories of the Euler angle errors *with* active nutation damping network $G_{\text{nut}}(s)$; $k_{xd} = 0.214$.

If the angular position of the two solar panels relative to the sun is not symmetrical, a *windmill torque* will be generated, which means that a torque about the line connecting the sun and the satellite will be created. In order to achieve also a torque about a line in the orbit plane and perpendicular to the sun direction, flaps must be appended to the solar panels; see Figure 8.6.1 (overleaf), which also shows the fixed flaps configuration. Figure 8.6.1.b displays the various geometrical notation pertaining to the panel and flap assembly. The North and South panels and flaps will be designated by N and S; d_p (resp. d_f) is the distance of the panel's (flap's) geometrical center from the cm of the satellite's body; and S_p and S_f are the (surface) areas of the panels and flaps, respectively. The angle δ shown in Figure 8.6.1.b is the deviation angle of the flaps from the normal to the panels.

Figure 8.6.2 (overleaf) depicts the three axis frames necessary for obtaining the directions of the torques on the satellite produced by solar pressure. As usual, $\mathbf{X}_R$,

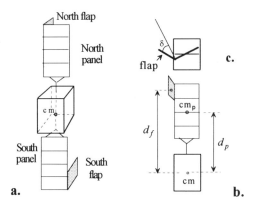

Figure 8.6.1 Geometry of the solar panels and flaps.

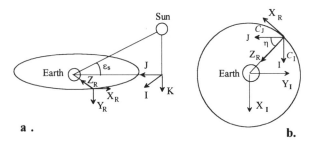

Figure 8.6.2 Inertial solar frame and the orbit reference frame of the satellite; reproduced from Lievre (1985) by permission of IFAC.

Y_R, and Z_R denote the orbit reference frame, centered in the cm of the satellite, and moving in the orbit with the satellite; X_I, Y_I, Z_I are the inertial frame axes, and I, J, K are the solar axes' frame. As shown in Figure 8.6.2, the two last frames coincide: the solar frame is defined as $I = X_I$, $J = -Y_I$, $K = Z_I$. Here ϵ_s is the sun elevation above the orbit plane. For the geostationary orbit, ϵ_s changes $\pm 23.5°$ during the year.

The solar panels are nominally directed toward the sun for maximum energy absorption. The angular motion of the panels for creating torques will therefore be defined in the solar frame. The *solar torque principle* is presented in Figure 8.6.3, where γ_N and γ_S are (small) angular displacements of the North and the South panels relative to the inertial solar frame (I, J, K). Figure 8.6.3 pertains to an orbit for which the η of Figure 8.6.2 is null, $\eta = 0°$. By definition, $\tan(\eta) = -X_I/Y_I$.

We begin by defining

$$\delta\gamma = \gamma_N - \gamma_S, \quad \gamma = \tfrac{1}{2}(\gamma_N + \gamma_S). \tag{8.6.1}$$

This notation enables a short physical explanation of the solar torque (sailing) principle as follows. Suppose for simplicity that the sun is in the orbit plane with its direction aligned with the J axis, perpendicular to the plane of Figure 8.6.3. If the deviations of the North and South panels in Figure 8.6.3 are antisymmetric with respect to the I axis, $\gamma_N = -\gamma_S$, a pure windmill torque C_J is generated about the J axis. If the deviations of the North and South panels are not antisymmetric, $\gamma_N \neq -\gamma_S$, then an *unbalance* torque C_I is generated about the I axis. It can be shown (see Section C.3) that an approximate set of equations for both torque components is

8.6 / Roll-Yaw Attitude Control with Solar Torques

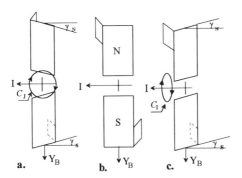

Figure 8.6.3 Solar torque control principle; reproduced from Lievre (1985) by permission of IFAC.

$$\frac{C_J}{\cos(\epsilon_s)} = B_3\delta\gamma + B_4 = B_3(\gamma_N - \gamma_S) + B_4,$$

$$\frac{C_I}{\cos(\epsilon_s)} = A_1\gamma\delta\gamma + A_2\gamma = \frac{A_1}{2}(\gamma_N + \gamma_S)(\gamma_N - \gamma_S) + \frac{A_2}{2}(\gamma_N + \gamma_S). \quad (8.6.2)$$

From Eqs. 8.6.2, if C_I and C_J are known then γ and $\delta\gamma$ can be calculated. According to Appendix C, the coefficients A_1, A_2, B_3, and B_4 depend on the physical properties of the panels and the flaps as follows:

$$A_1 = -PS_p d_p(1 + 5\eta_p) + PS_f d_f[4.5\eta_f \sin(3\delta) - \sin(\delta) + \eta_f \sin(\delta)/2], \quad (8.6.3)$$

$$A_2 = PS_f d_f[2\cos(\delta) - 3\eta_f \cos(3\delta) + \eta_f \cos(\delta)], \quad (8.6.4)$$

$$B_3 = 2PS_p d_p \eta_p + \tfrac{1}{2} PS_f d_f \eta_f [\sin(\delta) - 3\sin(3\delta)], \quad (8.6.5)$$

$$B_4 = PS_f d_f \eta_f [\cos(3\delta) - \cos(\delta)] \quad (8.6.6)$$

(see also Azor 1992, Sidi 1992a); η_p and η_f are the panel and flap reflexivities.

8.6.2 Mechanization of the Control Algorithm

The basic control laws to be used are Eq. 8.3.1 and Eq. 8.3.2; if no measurement of the yaw angle is feasible, Eq. 8.3.13 and Eq. 8.3.14 are used instead. A block diagram for the mechanization of the complete solar torque control principle is shown in Figure 8.6.4 (overleaf). The "control mechanization" block in this figure includes all the algorithms that must be implemented in the onboard computer. The roll angles (or the roll and the yaw angles, if adequate hardware for yaw angle measurement exists) are used to calculate the torque control command T_{cx} and T_{cz}, using Eq. 8.3.1 and Eq. 8.3.2 (or using Eq. 8.3.13 and Eq. 8.3.14 if ψ cannot be measured).

From Eqs. 8.6.2, if C_I and C_J are known then γ and $\delta\gamma$ can also be found:

$$\delta\gamma = \frac{C_J - B_4 \cos(\epsilon_s)}{B_3 \cos(\epsilon_s)},$$

$$\gamma = \frac{C_I}{(A_1\delta\gamma + A_2)\cos(\epsilon_s)}. \quad (8.6.7)$$

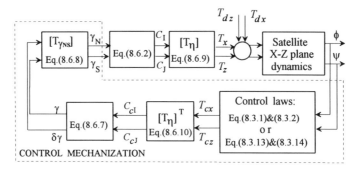

Figure 8.6.4 Mechanization of the solar torque control laws – the complete time-domain simulation.

The transformation between γ_N, γ_S and $\gamma, \delta\gamma$ is simply

$$\begin{bmatrix} \gamma_N \\ \gamma_S \end{bmatrix} = \begin{bmatrix} 1 & 0.5 \\ 1 & -0.5 \end{bmatrix} \begin{bmatrix} \gamma \\ \delta\gamma \end{bmatrix} = [\mathbf{T}_{\gamma NS}] \begin{bmatrix} \gamma \\ \delta\gamma \end{bmatrix}, \tag{8.6.8}$$

where γ_N and γ_S are the panels' control angular deviations necessary to achieve the desired torques. Here $[\mathbf{T}_{\gamma NS}]$ is the transformation matrix between the control variables $\gamma, \delta\gamma$ and the panels' deviations γ_N, γ_S.

In order to simulate the complete solar torque dynamics, we must also define a transformation between the solar torques C_I, C_J and the torques applied on the body axes $\mathbf{X}_B, \mathbf{Z}_B$. The transformation depends on the location of the satellite in the orbit, which is defined by the argument η (see Figure 8.6.2):

$$\begin{bmatrix} T_x \\ T_z \end{bmatrix} = \begin{bmatrix} -\cos(\eta) & \sin(\eta) \\ \sin(\eta) & \cos(\eta) \end{bmatrix} \begin{bmatrix} C_I \\ C_J \end{bmatrix} = [\mathbf{T}_\eta] \begin{bmatrix} C_I \\ C_J \end{bmatrix}. \tag{8.6.9}$$

If the desired control torques are known, then we can find the desired solar torques by using the transpose of the transformation defined in Eq. 8.6.9:

$$\begin{bmatrix} C_{cI} \\ C_{cJ} \end{bmatrix} = [\mathbf{T}_\eta]^T \begin{bmatrix} T_{cx} \\ T_{cz} \end{bmatrix}. \tag{8.6.10}$$

In Figure 8.6.4, the three blocks lying outside the control mechanization comprise the kinematics and dynamics of the satellite, panels, and flaps.

Control torque saturation due to sensor noise amplification is a problem here, too, and one that is worsened by the generally lower obtainable control torques as compared to magnetic control means. Solar sailing has the additional drawback that, to the extent the panels deviate from their nominal position relative to the sun direction, their *solar efficiency* decreases. This results in an electrical power loss, which cannot be permitted to exceed 1–2% on average per orbit. Typical solar torque capability for different solar efficiency losses is shown in Figure C.3.3 (p. 392).

EXAMPLE 8.6.1 Example 8.3.2, treated in the previous section with magnetic control torques, is now to be realized with solar control torques. The panel and flap

8.6 / Roll-Yaw Attitude Control with Solar Torques 235

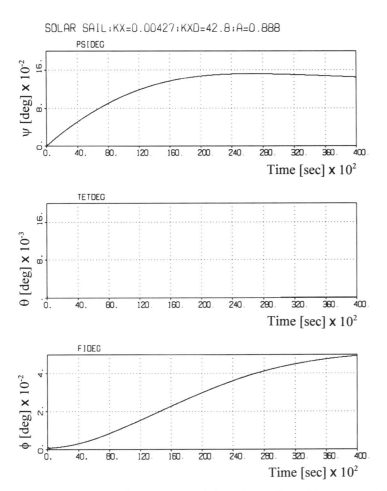

Figure 8.6.5 Time response of the Euler angles to T_{dx}, an external disturbance about the X_B axis.

characteristics were chosen as follows: $S_p = 6$ m², $\eta_p = 0.2$, $S_f = 2$ m², $\eta_f = 0.1$, $\delta = 15°$, $d_p = 4$ m, $d_f = 5.5$ m, $\epsilon_s = 0$.

With these parameters and using Eqs. 8.6.3–8.6.6, we obtain: $A_1 = -0.217 \times 10^{-3}$, $A_2 = 0.92 \times 10^{-4}$, $B_3 = 0.394 \times 10^{-4}$, $B_4 = -0.13 \times 10^{-5}$. There is no sensor noise. The results in the time domain are given in Figures 8.6.5–8.6.8. In this example, the maximum panel angular deviations are smaller than 7° (Figure 8.6.6), causing a maximum solar efficiency loss of 0.35%; see Figure 8.6.8.

Without sensor noise, there is no problem in achieving a good damping coefficient for the nutation frequency pole, with $\xi \approx 0.5$ for $k_{xd} = 42.8$. If sensor noise is present then it is unrealistic to have an adequate damping coefficient without some additional control means – for instance, an active nutation control scheme as explained in Section 8.5.

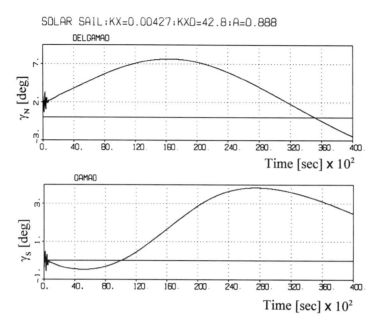

Figure 8.6.6 The panels' angular deviations from nominal.

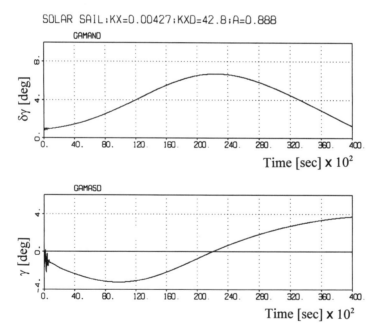

Figure 8.6.7 The computed control variables $\delta\gamma$ and γ.

8.7 / Roll-Yaw Attitude Control with Two Momentum Wheels

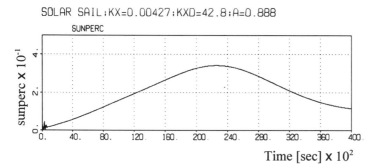

Figure 8.6.8 Efficiency loss of solar arrays.

8.7 Roll-Yaw Attitude Control with Two Momentum Wheels

8.7.1 Introduction

In the previous control schemes, a single momentum wheel was the primary control element responsible for ensuring gyroscopic stability of the satellite. Additional control equipment, such as magnetic torqrods or solar panels and flaps, was used to fine-tune the attitude accuracy by controlling nutational motion about the roll-yaw satellite axes.

The single-momentum wheel scheme has two principal drawbacks. First, with an earth sensor as the only sensor used to implement the control laws, amplification of sensor noise precludes adequate damping of the nutation pole. This is especially pronounced with high-orbit satellites, for which the maximum attainable magnetic or solar control torques are insufficient. The second inherent drawback is that the roll and yaw attitude angles can be stabilized close to null.

We found in Chapter 2 that the evolution of the inclination vector for geostationary satellites caused changes in the orbit's inclination angle of about 0.7°-0.9° per year. The task of the station keeping process, introduced in Chapter 3, is to restrict this inclination to within ±0.05°. However, some satellite missions require pointing the payload antennas at different ground targets, a task that can be accomplished by appropriate changes in roll and pitch attitudes.

Using two symmetrically inclined momentum wheels in a V configuration allows control of both the pitch and the roll angles, while exploiting the feature of inertial attitude stabilization that keeps the yaw error close to null without being forced to measure it (Wie et al. 1985). Moreover, sensor noise amplification is no longer a problem because (1) no magnetic or solar torques are needed for attitude control and (2) the torque control capabilities of momentum exchange devices are much higher than any amplified noise levels.

In the basic *two-momentum wheel* control configuration, we can establish adequate redundancy by using a system that consists of three wheels, as shown in Figure 8.7.1 (overleaf). A smaller momentum wheel MW_Z is located with its momentum axis aligned along the Z_B body axis. If one of the primary wheels (MW_1 or MW_2) fails then the third wheel (MW_Z), which is nominally held inactive, can now be used as an additional momentum bias to compensate for the lost wheel.

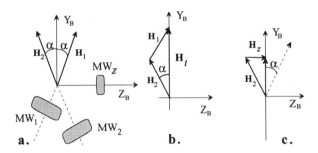

Figure 8.7.1 Two-wheel momentum bias arrangement, with a third wheel for backup.

8.7.2 Adapting the Equation of Rotational Motion

The generalized dynamic equation of motion, Eqs. 4.8.14, will be adapted in this section for the configuration in Figure 8.7.1. The two wheels MW_1 and MW_2 are the primary sources of angular momentum bias. Their momentum axes lie in the Y_B-Z_B body plane, and they deviate from the Y_B axis by an angle α, as shown in the figure. The net effect is that a momentum bias of $H_t = 2H_1 \cos(\alpha)$ is aligned along the Y_B axis. This is the nominal momentum bias treated in previous sections (see also Bingham, Craig, and Flook 1984; Wie et al. 1985).

If either MW_1 or MW_2 should fail, MW_Z can be activated to compensate for the lost wheel. The primary task of this compensation is to realign the wheels' total momentum with the Y_B body axis, as shown in Figure 8.7.1.c. The new system will have a reduced total momentum equal to half that of the nominal one.

In this section we shall adapt Eqs. 4.8.14 to the 2-MW control configuration. Using the torque capabilities of both MW_1 and MW_2, the torque command equations can be written in the following form:

$$T_{cy} = (\dot{H}_1 + \dot{H}_2) \cos(\alpha) = \dot{H}_{cy},$$
$$T_{cz} = (\dot{H}_1 - \dot{H}_2) \sin(\alpha) = \dot{H}_{cz}. \tag{8.7.1}$$

In matrix notation:

$$\begin{bmatrix} T_{cy} \\ T_{cz} \end{bmatrix} = \begin{bmatrix} \cos(\alpha) & \cos(\alpha) \\ \sin(\alpha) & -\sin(\alpha) \end{bmatrix} \begin{bmatrix} \dot{H}_1 \\ \dot{H}_2 \end{bmatrix}. \tag{8.7.2}$$

For the required torque commands (as determined by any adequate attitude control law), the wheels should provide the following torques:

$$\begin{bmatrix} \dot{H}_1 \\ \dot{H}_2 \end{bmatrix} = \frac{-1}{2\cos(\alpha)\sin(\alpha)} \begin{bmatrix} -\sin(\alpha) & -\cos(\alpha) \\ -\sin(\alpha) & \cos(\alpha) \end{bmatrix} \begin{bmatrix} T_{cy} \\ T_{cz} \end{bmatrix}$$
$$= \frac{1}{2} \begin{bmatrix} 1/\cos(\alpha) & 1/\sin(\alpha) \\ 1/\cos(\alpha) & -1/\sin(\alpha) \end{bmatrix} \begin{bmatrix} T_{cy} \\ T_{cz} \end{bmatrix} = [\mathbf{T}_{TH}] \begin{bmatrix} T_{cy} \\ T_{cz} \end{bmatrix}. \tag{8.7.3}$$

Transferring the momentum produced by the momentum wheels to the body axes and using the definitions of Chapter 4, we have

$$h_{wy} = (H_1 + H_2) \cos(\alpha), \qquad h_{wz} = (H_1 - H_2) \sin(\alpha), \tag{8.7.4}$$

8.7 / Roll-Yaw Attitude Control with Two Momentum Wheels

and also

$$\dot{h}_{wy} = (\dot{H}_1 + \dot{H}_2)\cos(\alpha), \qquad \dot{h}_{wz} = (\dot{H}_1 - \dot{H}_2)\sin(\alpha). \tag{8.7.5}$$

The total momentum of the body and momentum wheels, in body coordinates, is therefore

$$\mathbf{h} = \begin{bmatrix} h_x \\ h_y + h_{wy} \\ h_z + h_{wz} \end{bmatrix} = \begin{bmatrix} h_x \\ h_y + (H_1 + H_2)\cos(\alpha) \\ h_z + (H_1 - H_2)\sin(\alpha) \end{bmatrix}. \tag{8.7.6}$$

Equations 8.7.1 can be rearranged as follows:

$$\frac{T_{cy}}{\cos(\alpha)} = \dot{H}_1 + \dot{H}_2,$$

$$\frac{T_{cz}}{\sin(\alpha)} = \dot{H}_1 - \dot{H}_2. \tag{8.7.1'}$$

With this notation and with equal momentum biases in the wheels ($H_1(0) = H_2(0) = H(0)$), we have

$$H_1 + H_2 = H_1(0) + H_2(0) + \int \frac{T_{cy}}{\cos(\alpha)}\,dt = 2H(0) + \int \frac{T_{cy}}{\cos(\alpha)}\,dt, \tag{8.7.7}$$

$$H_1 - H_2 = H_1(0) - H_2(0) + \int \frac{T_{cz}}{\sin(\alpha)}\,dt = 0 + \int \frac{T_{cz}}{\sin(\alpha)}\,dt. \tag{8.7.8}$$

Next, we use Euler's moment equations (Eq. 4.8.2) with $h_{wx} \equiv 0$, since there is no momentum wheel projection on the $\mathbf{X}_B$ axis. This yields

$$T'_{cx} + T_{dx} = \dot{h}_x + \omega_y\omega_z(I_z - I_y) + \omega_y(H_1 - H_2)\sin(\alpha) - \omega_z(H_1 + H_2)\cos(\alpha), \tag{8.7.9}$$

$$T'_{cz} + T_{dz} = \dot{h}_z + \omega_x\omega_y(I_y - I_x) + (\dot{H}_1 - \dot{H}_2)\sin(\alpha) + \omega_x(H_1 + H_2)\cos(\alpha), \tag{8.7.10}$$

$$T'_{cy} + T_{dy} = \dot{h}_y + \omega_x\omega_z(I_x - I_z) + (\dot{H}_1 + \dot{H}_2)\cos(\alpha) - \omega_x(H_1 - H_2)\sin(\alpha). \tag{8.7.11}$$

Equations 8.7.9–8.7.11 are used to simulate the body attitude dynamics. They can also be linearized as in Chapter 4 to obtain formulas similar to Eqs. 4.8.14. To perform the linearization, the approximated Eq. 4.8.12 and Eqs. 4.8.13 must be used. A more direct way is to use Eqs. 4.8.14 together with Eqs. 8.7.4 and Eqs. 8.7.5. Remember also that by definition $h_{wx} = \dot{h}_{wx} \equiv 0$. In the present analysis we suppose that the products of inertia are null. With these assumptions, Eqs. 4.8.14 become

$$T'_{cx} + T_{dx} = I_x\ddot{\phi} + 4\omega_o^2(I_y - I_z)\phi + \omega_o(I_y - I_z - I_x)\dot{\psi} - \omega_o(H_1 - H_2)\sin(\alpha), \tag{8.7.12}$$

$$T'_{cz} + T_{dz} = I_z\ddot{\psi} + \omega_o(I_z + I_x - I_y)\dot{\phi} + \omega_o^2(I_y - I_x)\psi + (\dot{H}_1 - \dot{H}_2)\sin(\alpha), \tag{8.7.13}$$

$$T'_{cy} + T_{dy} = I_y\ddot{\theta} + 3\omega_o^2(I_x - I_z)\theta + (\dot{H}_1 + \dot{H}_2)\cos(\alpha). \tag{8.7.14}$$

Once again, the $\mathbf{Y}_B$ axis attitude dynamics is independent of the dynamics of the lateral plane attitude, $\mathbf{X}_B$ and $\mathbf{Z}_B$. The control laws have the general form

$$T_{cy} = (\theta_{\text{com}} - \theta)G_Y(s) \quad \text{and} \tag{8.7.15}$$

$$T_{cz} = (\phi_{\text{com}} - \phi)G_Z(s), \tag{8.7.16}$$

where θ_{com} and ϕ_{com} are the commanded pitch and roll angles to be tracked.

Having computed T_{cy} and T_{cz}, from Eq. 8.7.3 we can find the torque commands to both momentum wheels H_1 and H_2. This completes the design stage. The control networks $G_Y(s)$ and $G_Z(s)$ are designed using conventional frequency-domain linear control techniques. The transfer functions $\theta(s)/T_{cy}(s)$ and $\phi(s)/T_{cz}(s)$ are needed in this stage, and can be computed from Eqs. 8.7.12–8.7.14 and Eq. 8.7.3.

8.7.3 Designing the Control Networks $G_Y(s)$ and $G_Z(s)$

The first stage in the design process consists of determining the value of the momentum bias H of both momentum wheels. Since the yaw error is not measured, H depends on external disturbances (such as solar pressure for high-orbit satellites or aerodynamic drag for low-orbit satellites) and on the permitted error of the yaw angle ψ. For the nominal case, in which both principal (MW_1 and MW_2) wheels are active, the required total momentum can be calculated from the well-known equation $\psi = T_{dz}/[2H\cos(\alpha)\omega_o]$. If we want to assure the same maximum error in ψ in the event of a MW failure then, according to Figure 8.7.1, we must choose momentum wheels with double momentum capabilities.

The choice of the inclination angle α depends very much on the sensor noise amplification. According to Eq. 8.7.3, the noise level existing in T_{cy} and T_{cz} is amplified at $\dot{H}_1$ and $\dot{H}_2$ by the factor $1/\sin(\alpha)$. Accordingly, $\alpha = 25°$ is a reasonable choice. Moreover, α determines the control torque capabilities about the Z_B axis, which will be sufficient to deal with the anticipated level of external disturbances.

The design of the pitch loop ($G_Y(s)$) is straightforward and will not be repeated here; see Section 7.3. The design of $G_Z(s)$ can be carried out in the frequency domain using classical linear control techniques. The block diagram of the control system is shown in Figure 8.7.2.

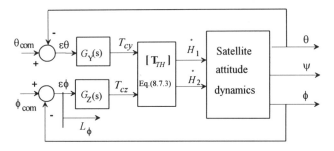

Figure 8.7.2 Block diagram of the 2-MW control system.

EXAMPLE 8.7.1 The satellite described in Example 8.3.1 will be used to demonstrate attitude control using two inclined momentum wheels in a V configuration. Here, $G_Y(s)$ has been designed so that the closed pitch control loop has a bandwidth of 0.05 rad/sec, and H_1 and H_2 have both been determined (according to the permitted yaw error) to be 10 N-m-sec.

The design of the L_ϕ open-loop transfer function shown in Figure 8.7.3 yields the control network

8.7 / Roll-Yaw Attitude Control with Two Momentum Wheels 241

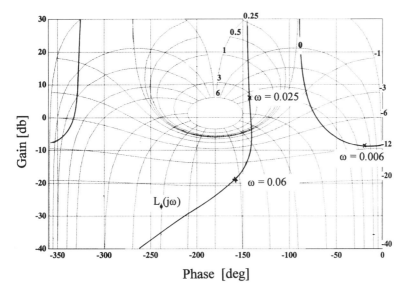

Figure 8.7.3 Open-loop transfer function of the roll control loop L_ϕ.

$$G_Z(s) = \frac{0.02(1+s/0.005)(1+s/0.01)}{(1+s/0.3)^4}.$$

Time responses of the closed-loop control system for the attitude command inputs $\theta_{\text{com}} = 2°$ and $\phi_{\text{com}} = 1°$ are shown in Figure 8.7.4 (overleaf).

8.7.4 *Momentum Dumping of the MW with Reaction Thrust Pulses*

In Section 7.3.3 we showed that external disturbances tend to change the momentum of momentum exchange devices. There are two principal reasons for dumping the momentum wheel. First, a minimum level of momentum bias must be retained in order to satisfy the attitude accuracy of ϕ and ψ. Second, the hardware of the momentum wheel is optimized to work at predetermined angular wheel velocity conditions because of power dissipation problems and other reliability constraints. Excess momentum must be dumped when the wheel momentum approaches the permitted limits. This desaturation can be accomplished with magnetic torqrods (as explained in Chapter 7) or with thrust pulses. To dump a wheel that has become saturated with momentum, a pulse from the correct thruster is fired so that a torque about the Y_B axis, with the correct sign, will effect the needed momentum dumping. In order to prevent large pitch attitude errors, the quantity of dumping per single thruster firing must be limited.

To exemplify the momentum dumping process we refer once more to Example 8.7.1, where the following external disturbances are applied to the satellite:

$$T_{dx} = 4 \times 10^{-6} + 2 \times 10^{-6} \sin(\omega_o t),$$
$$T_{dy} = 6 \times 10^{-6} + 3 \times 10^{-6} \sin(\omega_o t),$$

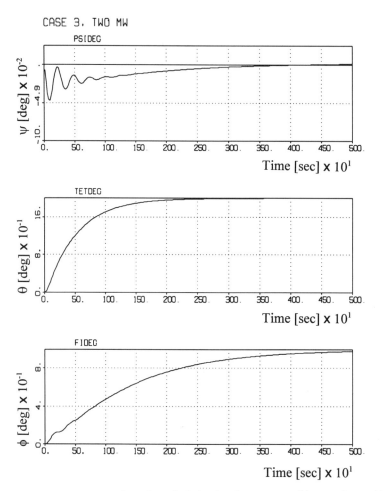

Figure 8.7.4 Time-domain behavior for command inputs $\theta_{com} = 2°$ and $\phi_{com} = 1°$.

$$T_{dz} = 3 \times 10^{-6} + 3 \times 10^{-6} \sin(\omega_o t).$$

The ES noise level is 0.03° (RMS).

Time histories for the Euler angles are shown in Figure 8.7.5. The normal steady-state errors in ϕ (for the roll) are smaller than 0.03°. Attitude errors larger than 0.13° are reached during momentum dumping, but their amplitude can be lessened by dumping smaller quantities of momentum per single thrust firing. Limits of about 3% of nominal were set for the permitted change in nominal MW momentum bias. The change in momentum bias of both wheels is shown in Figure 8.7.6.

8.8 Reaction Thruster Attitude Control

8.8.1 *Introduction*

This section deals with the attitude control of a momentum-biased three-axis–stabilized geostationary satellite using thrusters for active nutation control. In

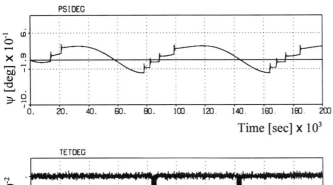

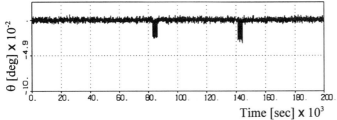

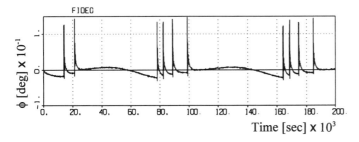

Figure 8.7.5 Time behavior of the Euler angle errors for external disturbances and sensor noise.

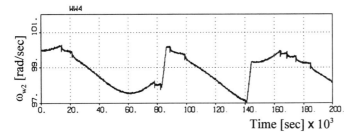

Figure 8.7.6 Time-domain behavior of MW momentum dumping.

continuous attitude control systems, which are based on solar or magnetic control torques, sensor noise precluded sufficient damping of nutation (see Sections 8.4–8.7). In this case, unforeseen or unexpected energy dissipation in the satellite's equipment (such as the momentum wheel rotor) may lead to nutation divergence (Benoit and Bailly 1987).

In thruster attitude control of a momentum-biased satellite, roll attitude is maintained within the specified roll "deadbeat" limits using two thruster impulses fired at the right time (Dougherty et al. 1968, Iwens et al. 1974, Bittner et al. 1977). This is the well-known "WHECON" (wheel control) attitude control concept. However, if for any reason the nutation is stimulated, it cannot be properly damped without the help of some continuous active damping scheme (Sidi 1992b). The situation is aggravated by sensor noise, which tends to activate unnecessary thruster firings.

In the ACS treated in this section, the roll attitude is kept within the allowed roll limits by firing (with the same thruster) two pulses that are separated by half the nutation period (Bryson 1983). With absolute knowledge of the nutation frequency and supposing that the two pulses are identical, no nutation will be excited. However, identical pulses and knowledge of the precise nutation frequency are impossible in practice, so the parasitic excited nutation must be damped. The most effective way to provide active damping is based on the torque capabilities of the existing single momentum wheel, via the products of inertia I_{yz} or I_{yx} (Phillips 1973).

8.8.2 Control of ϕ (Roll) and ψ (Yaw)

As usual, the pitch angle θ is controlled with the momentum wheel operated in torque mode, while the roll and yaw angles are controlled by torque thruster impulses. The minimum equipment needed for changing the momentum axis attitude of the inertially stabilizing momentum h_w by an amount $\delta\alpha$ is thruster 1 in Figure 8.8.1. However, since the roll (ϕ) and yaw (ψ) angles are controlled simultaneously, reaction thrusters – each one providing torques about both X_B and Z_B axes – are necessary; these are thrusters C and F in Figure 8.8.1.

In the latter case, two pulses from the same thruster (C or F, depending on the sign of the actual roll error), and separated by half the nutation period, will change the Euler angles and velocities as in Figure 8.8.2 (page 246; see also Section 4.5 and Bryson 1983). We use Euler's moment equations, with the following assumptions.

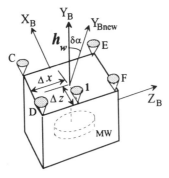

Figure 8.8.1 Reaction thrusters required for achieving roll attitude control.

(1) The satellite is inert about the Y_B axis, which is also the axis of the momentum wheel.
(2) The momentum wheel provides the momentum bias to the satellite, the direction of which is collinear with the Y_B axis.
(3) There are no momentum exchange devices in the satellite except for the momentum wheel mentioned in item (2).

The torques applied by the thrusters are external torques. Also, we will relabel the variables in Eq. 4.5.1: p now stands for ω_x and r for ω_z. The torques T_X and T_Z are applied about the X_B and Z_B axes, respectively, and h_w is the momentum bias of the wheel. With this notation and Eq. 4.5.1, the dynamic equations for the nutational motion become

$$T_x = \dot{p}I_x - rh_w,$$
$$T_z = \dot{r}I_z + ph_w. \tag{8.8.1}$$

According to Eqs. 8.8.1, two pulses fired from the same thruster (C or F) and separated by half the satellite's nutation period will change Euler angles ϕ and ψ, and the velocities p and q, as follows:

$$p(t) = \frac{F\Delta t \Delta x}{I_x} \cos(\omega_{nut} t) - \frac{F\Delta t \Delta z}{\sqrt{I_x I_z}} \sin(\omega_{nut} t), \tag{8.8.2}$$

$$r(t) = \frac{F\Delta t \Delta x}{\sqrt{I_x I_z}} \sin(\omega_{nut} t) + \frac{F\Delta t \Delta z}{I_z} \cos(\omega_{nut} t), \tag{8.8.3}$$

$$\phi(t) = \frac{F\Delta t \Delta x}{h_w} \sqrt{\frac{I_z}{I_x}} \sin(\omega_{nut} t) + \frac{F\Delta t \Delta z}{h_w} [\cos(\omega_{nut} t) - 1], \tag{8.8.4}$$

$$\psi(t) = -\frac{F\Delta t \Delta x}{h_w} [\cos(\omega_{nut} t) - 1] + \frac{F\Delta t \Delta z}{h_w} \sqrt{\frac{I_x}{I_z}} \sin(\omega_{nut} t), \tag{8.8.5}$$

where
F = thrust level,
$F\Delta t \Delta z$ = torque impulse bit for ϕ control,
$F\Delta t \Delta x$ = torque impulse bit for ψ control,
$\Delta x, \Delta z$ = torque arms,
h_w = constant momentum bias about the Y_B axis,
$\omega_{nut} = h_w/\sqrt{I_x I_z}$, the nutation frequency, and
$T_{nut} = 2\pi/\omega_{nut}$.

According to Eqs. 8.8.2–8.8.5, a desired change in the roll and yaw attitudes can be achieved whereby, at the end of this process, the roll and yaw rates (p and r) are both zeroed. See Figure 8.8.2 (overleaf), in which sketches of the time histories of these variables are shown. The change in $\Delta \phi$ will be:

$$\Delta \phi = 2 \frac{F\Delta t \Delta z}{h_w}. \tag{8.8.6}$$

Equation 8.8.6 shows that controlling smaller $\Delta \phi$ increments depends heavily on the minimum impulse bit $(F\Delta t)_{min}$ that the thruster can supply, as Δz and h_w are fixed by physical constraints. With an impulse bit of 40 mN-m-sec and with $h_w = 35$ N-m-sec, $\Delta \phi = 0.13°$; this means we can achieve a roll error deadbeat limit of $\sim 0.08°$.

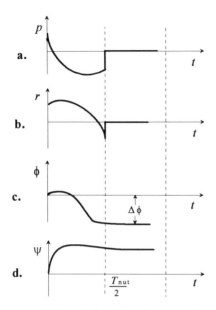

Figure 8.8.2 Development of $\Delta\phi$ with two impulses of the same thruster motor, the second one delayed by $T_{\text{nut}}/2$.

We have already seen that with momentum-biased satellites there is no need to measure or estimate the yaw attitude, since only the ϕ error is used for attitude control in the $\mathbf{X}_B$-$\mathbf{Z}_B$ plane. Moreover, a torque command proportional to the roll error ϕ, with a correct sign, is continuously applied about the $\mathbf{Z}_B$ axis: $T_{cz} = aT_{cx}$ (see Section 8.3.2). In principle, thrusters C and D (Figure 8.8.1) can each provide correct control torques about the $\mathbf{X}_B$ axis. However, only thruster C provides adequate control about the $\mathbf{Z}_B$ axis, satisfying the correct sign of a in Eq. 8.3.14. Hence, thrusters C and F are sufficient to provide the reaction pulses necessary for achieving both negative and positive changes in the roll attitude, simultaneously with the requisite attitude change about the $\mathbf{Z}_B$ axis.

Under ideal conditions, no parasitic nutational motion will be excited. However, given various imperfections in control hardware (Benoit and Bailly 1987), nutation divergence could arise and so necessitate appropriate nutation damping.

8.8.3 *Immunity to Sensor Noise*

Immunity to sensor noise, which would reduce unnecessary thruster firings, is achieved in a very simple way. A first-order low-pass filter is used to perform the necessary filtration of the earth sensor's output; this filter has a corner frequency that is slightly higher than the nutation frequency. A first-order analog filter with time constant τ, after discretization by the Tustin transform, will attenuate the RMS value σ_N of the sensor noise to the value

$$\sigma_0 = \sigma_N \sqrt{\left(1 + \frac{2\tau}{T_{\text{sam}}}\right)^{-1}} \tag{8.8.7}$$

(Sidi 1992b), where T_{sam} is the sampling time of the onboard control computer.

8.8 / Reaction Thruster Attitude Control

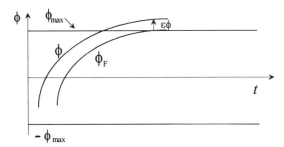

Figure 8.8.3 Delayed activation of reaction impulse caused by the sensor noise filter.

Suppose that $T_{sam} = 0.5$ sec, $\tau = 50$ sec, and $\sigma_N = 0.1°$ (RMS). With these inputs, σ_0 is attenuated to an acceptable value of only $\sigma_0 = 0.007°$. However, the measured roll error ϕ will be delayed, and the reaction impulse will be fired at an erroneous level of $\phi_{max} + \epsilon\phi$, as shown in Figure 8.8.3.

To find $\epsilon\phi$, suppose that the system is under the influence of a constant torque disturbance T_d. This disturbance will create an error: $\phi(t) = (T_d/h_w)t$. The measured and filtered roll $\phi(t)$ behaves as:

$$\phi_F(t) = \frac{T_d}{h_w}[t + \tau e^{-t/\tau} - \tau]. \tag{8.8.8}$$

The steady-state error in the measured roll then amounts to:

$$\epsilon\phi(\infty) = \lim_{t \to \infty}[\phi(t) - \phi_F(t)] = \frac{T_d}{h_w}\tau. \tag{8.8.9}$$

If we assume that $T_d = 10^{-5}$ N-m, $h_w = 35$ N-m-sec, and $\tau = 50$ sec, then $\epsilon\phi(\infty) = 0.0008°$ - a very small additive error in the measurement of ϕ.

When the roll error exceeds the roll deadbeat limits of $\pm\phi_{max}$, the first reaction impulse is fired. Automatically, half the nutation period later, the same thruster is fired again. Between the two firings, and also for at least 10 seconds following the second firing, the command control is inhibited in order to prevent unnecessary spurious firings. Meanwhile, the two thruster firings have brought the roll angle close to the opposite deadbeat limit. The simplified flowchart of the algorithm is shown in Figure 8.8.4 (overleaf).

8.8.4 Determining the Necessary Momentum Bias h_w

According to Eq. 8.8.6, if the minimum impulse bit is equal to 0.03 N-m-sec and $h_w = 20$ N-m-sec then $\Delta\phi_{min} = 0.171°$, which means that a roll overall band limit of $2 \times 0.1°$ is easily achievable with a maximum roll error of $\pm 0.1°$. To decrease this error in roll, we must either decrease the minimum impulse bit of the reaction control system or increase the constant momentum bias. The minimum impulse bit is an engineering constraint that in general cannot be improved. The torque arm Δz can be decreased, but this increases fuel consumption. Increasing the value of the wheel's momentum bias is the remaining factor - and the most easily and cheaply implemented - for increasing the roll accuracy.

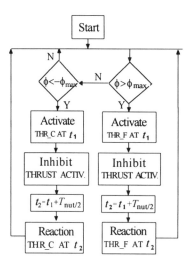

Figure 8.8.4 Flowchart of the roll attitude control.

Another important factor in choosing h_w is the acceptable yaw error magnitude for the expected secular disturbance torques. An analysis and resulting tables for choosing h_w according to desired requirements on ψ_{max} are presented in Iwens et al. (1974).

8.8.5 *Active Nutation Damping via Products of Inertia*

In principle, any spurious torque about the X_B or Z_B axis will excite the nutational motion. If the two consecutive thruster impulses are unequal or if the nutation time period is not accurately known, then the roll-controlled attitude of Section 8.8.3 will include a parasitic nutational motion. Also, any kind of energy dissipation, as in the rotor of the momentum wheel, could cause the already mentioned nutation divergence; see Benoit and Bailly (1987) and Section 6.7.1. To assure the damping of a possible nutation excitation, the technique of Section 8.5 can be used.

In a real control system, the reaction thrusters are activated about once every 30 minutes or so, depending on the level of external disturbances (see Figure 8.8.10). Between two thrusters' firing events, the reaction control system is not operative and hence the X_B–Z_B dynamics is virtually uncontrolled, as shown in Figure 8.8.5. On the other hand, pitch attitude is controlled continuously using the modified standard equation

$$T_{cy} = K_y\theta + k_{yd}\dot{\theta} + T_{cy\text{nut}}, \qquad (8.8.10)$$

where

$$T_{cy\text{nut}} = G_{\text{nut}}(s)\phi. \qquad (8.8.11)$$

As explained in Section 8.5 (see also Figure 8.5.2), only the nutation frequency band is processed, while adequate gain and phase margins are preserved. In fact, only the nutation part of the roll attitude error is processed and fed to the pitch control loop. In the frequency domain, we can see that the pitch control loop is only slightly disturbed at the nutation frequency spectrum band (see Figure 8.8.6).

8.8 / Reaction Thruster Attitude Control

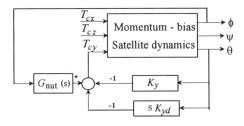

Figure 8.8.5 Block diagram of the pitch and the active nutation control loops.

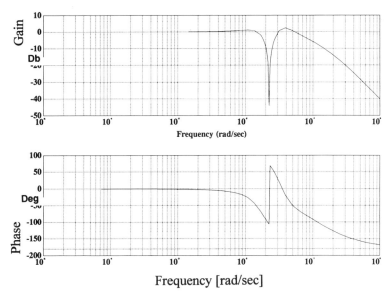

Figure 8.8.6 Closed-loop pitch control, with modulation alteration caused by the addition of the active nutation controller.

The level of the products of inertia must not be exaggerated, in order to limit the interaction between the pitch and the roll–yaw control loops. A level of 2%–5% (as compared to the principal moments of inertia) represents a good tradeoff.

EXAMPLE 8.8.1 The satellite moments of inertia [kg-m^2] are $I_{yz} = 20$, $I_x = 767$, and $I_z = 954$. With the control network

$$G_{\text{nut}}(s) = \frac{2{,}000s}{\left(\begin{array}{c} \xi = 1 \\ \omega_n = 0.03 \end{array}\right)},$$

a damping factor of 0.088 was obtained for the nutation modal pole, sufficiently high for the system at hand. Time-domain behavior of the active nutation controller is shown in Figure 8.8.7 (overleaf).

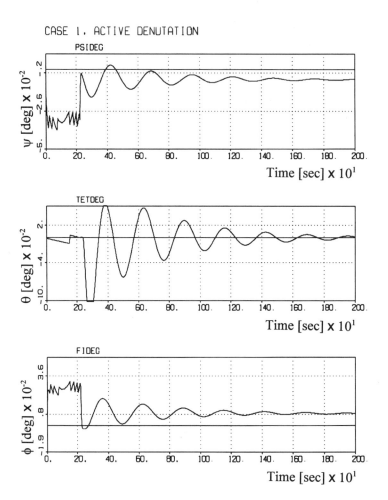

Figure 8.8.7 Time behavior of the active nutation control.

8.8.6 Wheel Momentum Dumping and the Complete Attitude Controller

As explained in Section 7.3.3, external disturbances continuously increase the momentum of momentum exchange devices, such as reaction or momentum wheels. This momentum must be dumped in order to avoid saturation of the MW. It is common to dump it by use of magnetic torques, as explained in Section 7.5.2, or reaction impulses, as explained in Section 8.7.4. Reaction momentum dumping was added to the control system treated in Example 8.8.1. Time results of the complete attitude control system, including the MW dumping, are shown in Figures 8.8.8–8.8.10. The system was submitted to the disturbances shown in Figure 8.8.11. The accuracies of the Euler angles shown in Figures 8.8.8 and 8.8.9 are undisturbed by the dumping process, and the velocity of the MW is kept within 3% of its nominal speed, as seen in Figure 8.8.10.

8.8 / Reaction Thruster Attitude Control 251

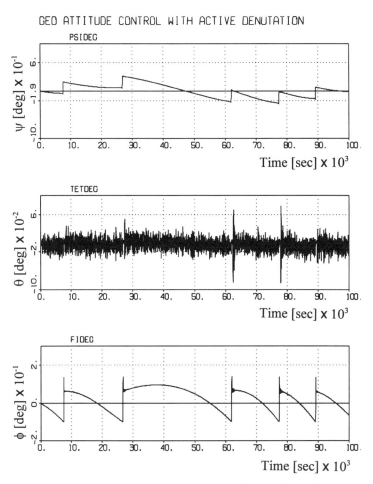

Figure 8.8.8 Time-domain behavior of the Euler angle errors.

8.8.7 *Active Nutation Damping without Products of Inertia*

As explained in Section 8.5, active nutation damping via products of inertia was developed, simulated, and evaluated in order to be used as the primary mode in the GEO mission stage of the Israeli *Amos* geostationary satellite. Although the test results were quite satisfactory, difficulties in obtaining the necessary product of inertia I_{yz} or I_{yx} precluded this elegant control option. Instead, a new version of active denutation control was suggested by D. Verbin (MBT, Israel Aircraft Industries) to assure at least a partial solution to the problem of attenuating any parasitic nutational motion to an acceptable level. The basic idea follows.

We showed in Section 8.8.2 that, by activating two reaction impulses separated by half the nutation period, a desired change in the roll attitude could be achieved without exciting the nutation mode of the momentum-bias control configuration. We will now see that, if the two pulses are applied at a precomputed time separation, then any remaining nutation can be completely removed, at least in theory.

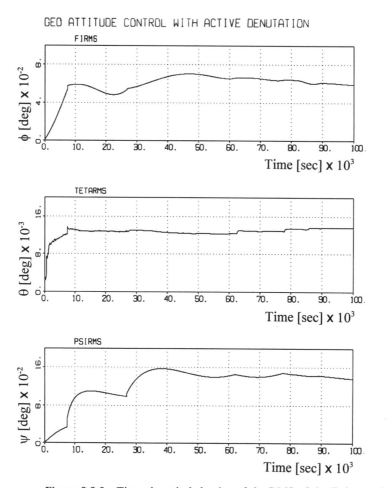

Figure 8.8.9 Time-domain behavior of the RMS of the Euler angle errors.

The first stage in this control scheme is to identify the parameters of the nutational motion, which is included in the measured roll angle. In practice, the nutation component of the roll angle is linearly superimposed on the slowly varying component of the roll error caused by the disturbance torques acting about the $\mathbf{X}_B$ and $\mathbf{Z}_B$ axes of the satellite. Since the nutation period T_{nut} is known, one simple way is to find the Fourier coefficients of the measured roll error. The resulting linearized form of the roll angle error will be

$$\phi(t) = \phi_{\text{CT}} + \phi_{\text{nut}} \sin(\omega_{\text{nut}} t + \alpha), \qquad (8.8.12)$$

where "CT" denotes "almost constant with respect to the nutational motion."

Having obtained the coefficients ϕ_{CT}, ϕ_{nut}, and α, with ω_{nut} known we can simultaneously control both the CT error and the nutation error by applying two consecutive reaction pulses at the correct timings t_1 and t_2 referenced to α. The analytical development follows.

Suppose that two pulses are activated with an angular interval of δ. According to Eqs. 8.8.1, the steady-state time history of the produced roll control will be:

8.8 / Reaction Thruster Attitude Control

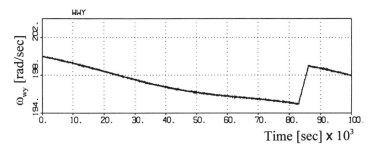
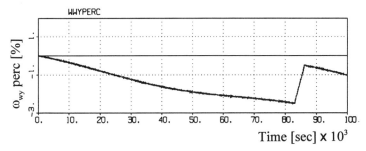

Figure 8.8.10 Time behavior of the momentum dumping process.

$$\phi_c = \frac{\Delta h_x}{h_w} \sqrt{\frac{I_z}{I_x}} \sin(\omega_{\text{nut}}t) + \frac{\Delta h_z}{h_w}[\cos(\omega_{\text{nut}}t) - 1]$$

$$+ \frac{\Delta h_x}{h_w} \sqrt{\frac{I_z}{I_x}} \sin(\omega_{\text{nut}}t - \delta) + \frac{\Delta h_z}{h_w}[\cos(\omega_{\text{nut}}t - \delta) - 1]$$

$$= \sin(\omega_{\text{nut}}t)\left\{\frac{\Delta h_x}{h_w} \sqrt{\frac{I_z}{I_x}}[1 + \cos(\delta)] + \frac{\Delta h_z}{h_w}\sin(\delta)\right\}$$

$$+ \cos(\omega_{\text{nut}}t)\left\{-\frac{\Delta h_x}{h_w} \sqrt{\frac{I_z}{I_x}} \sin(\delta) + \frac{\Delta h_z}{h_w}[1 + \cos(\delta)]\right\} - 2\frac{\Delta h_z}{h_w}. \quad (8.8.13)$$

This equation can be put in the more compact form

$$\phi_c = A \sin(\omega_{\text{nut}}t) + B \cos(\omega_{\text{nut}}t) + C$$
$$= \sqrt{A^2 + B^2} \sin(\omega_{\text{nut}}t + \alpha_c) + C, \quad (8.8.14)$$

with A, B, and C defined as follows:

$$A = \frac{\Delta h_x}{h_w} \sqrt{\frac{I_z}{I_x}}[1 + \cos(\delta)] + \frac{\Delta h_z}{h_w} \sin(\delta),$$

$$B = -\frac{\Delta h_x}{h_w} \sqrt{\frac{I_z}{I_x}} \sin(\delta) + \frac{\Delta h_z}{h_w}[1 + \cos(\delta)], \quad (8.8.15)$$

$$C = -2\frac{\Delta h_z}{h_w}.$$

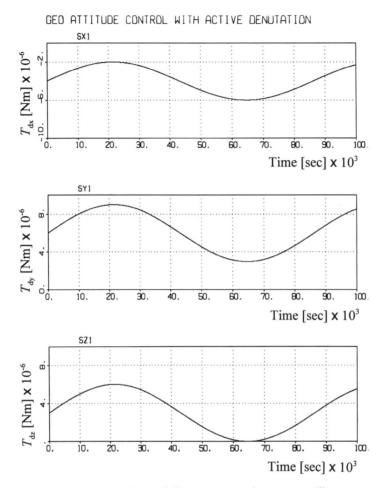

Figure 8.8.11 External disturbances acting on a satellite.

Let us also define

$$D = \left(\frac{\Delta h_x}{h_w}\right)^2 \frac{I_z}{I_x} + \left(\frac{\Delta h_z}{h_w}\right)^2. \tag{8.8.16}$$

Then

$$A^2 + B^2 = \phi_{\text{nut}c}^2 = 2D[1 + \cos(\delta)]. \tag{8.8.17}$$

In order to cancel the existing nutation, we must produce a controlled nutation with an amplitude $\phi_{\text{nut}c} = \phi_{\text{nut}}$. Using Eq. 8.8.17, we can compute the angular distance between the two thruster pulses:

$$\delta = \cos^{-1}\left(\frac{\phi_{\text{nut}c}^2}{2D} - 1\right). \tag{8.8.18}$$

We can also find the time separation between them:

$$\Delta t = \delta/\omega_{\text{nut}}. \tag{8.8.19}$$

8.8 / Reaction Thruster Attitude Control

It is important to notice that the maximum nutation amplitude that can be canceled is limited to $\phi_{\text{nut}c} \leq 2\sqrt{D}$.

Finally, we must calculate at what time the first pulse is to be applied. We know that

$$\alpha_c = \tan^{-1}(B/A). \tag{8.8.20}$$

If we define t_0 as the time at which $\phi(t) - \phi_{\text{CT}} = 0$ in Eq. 8.8.12, then the first reaction pulse should be applied at

$$t_1 = \alpha_c/\omega_{\text{nut}}. \tag{8.8.21}$$

The second pulse should then be applied at

$$t_2 = t_1 + \Delta t. \tag{8.8.22}$$

With these results, the control algorithm is straightforward.

Unlike the algorithm explained in Sections 8.8.2 and 8.8.5, the algorithm of this section can never completely cancel the nutation because of uncertainties in our knowledge of the physical parameters of the satellite's equipment (e.g., the minimum impulse bit, the momentum bias of the momentum wheel, the moments of inertia of the satellite, and so on). With the former algorithm, which was based on active denutation via products of inertia and on the torque capabilities of the momentum wheel, any initial or spurious nutation would be completely damped.

The term C in Eq. 8.8.15 is exactly the CT value of the roll error that will be corrected, as in Section 8.8.2. An example will clarify the obtainable results.

EXAMPLE 8.8.2 In this example, $I_x = 645$ kg-m^2, $I_z = 745$ kg-m^2, $h_w = 35$ N-m-sec, $\omega_{\text{nut}} = 0.0505$ rad/sec, $\Delta h_x = 0.0276$ N-m-sec, $\Delta h_z = 0.0436$ N-m-sec, and $C = 0.143°$. Suppose that a nutation amplitude of 0.06° is to be eliminated, with the aid of a constant negative $\Delta \phi$ that must cancel part of an existing ϕ_{CT}.

Solution Since ϕ_{nut} is known in Eq. 8.8.12, by use of Eq. 8.8.18 we find that $\delta = 139°$. Next, we must find the time at which to apply the two reaction pulses. From Eq. 8.8.20, we have $\alpha_c = -13.9°$. Translated to time, $t_1 = \alpha_c/\omega_{\text{nut}} = -4.82$ sec, referenced to α in Eq. 8.8.12. The second pulse must be applied at $t_2 = t_1 + \delta/\omega_{\text{nut}} = -4.82 + 48.2 = 43.38$ sec.

The results are given in Figure 8.8.12 (overleaf). Figure 8.8.12.a shows the existing nutation roll angle that is to be canceled. In Figures 8.8.12.b and 8.8.12.c we see the resulting nutation excited by the application of the two reaction pulses, 48.2 sec apart. The final roll angle ϕ is shown in Figure 8.8.12.d. With the application of the second reaction pulse, the nutation ϕ_{nut} in the original roll angle disappears completely, and a CT component of $C = -0.143°$ is added, as expected.

A 6-DOF simulation of the complete ACS is shown in Figure 8.8.13 (page 257). There are uncertainties in the physical parameters of the satellite. The initial nutation is canceled, but some angular nutational motion always remains because of the uncertainties in the control parameters. The boundary limits on the roll angle are $\phi_{\text{max}} = \pm 0.11°$; see also Figure 8.8.8.

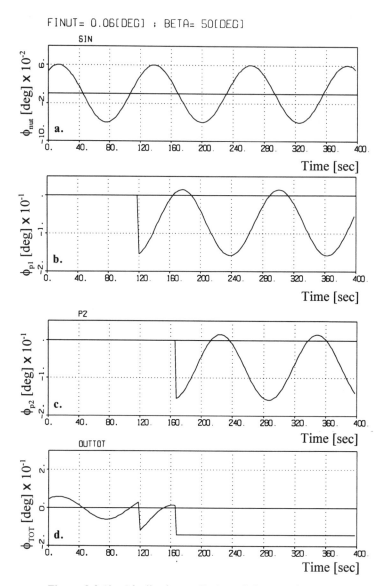

Figure 8.8.12 Idealized cancellation of the nutation components ϕ_{nut}.

For a slightly different realization of active nutation damping without products of inertia, see Azor (1995).

8.9 Summary

This chapter dealt with momentum-bias control systems. They have some nice features: (1) only one momentum exchange device is needed to attitude-stabilize the satellite, and (2) the yaw attitude error need not be measured, thus simplifying greatly the onboard attitude determination hardware and algorithms.

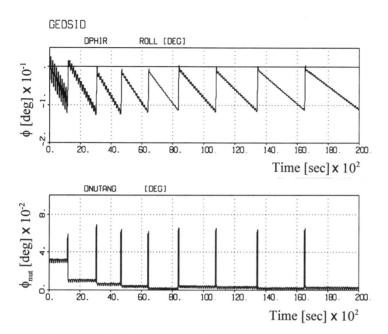

Figure 8.8.13 Complete 6-DOF simulation of the reaction control denutation scheme.

The basic drawback of such attitude control systems is that they do not allow execution of attitude maneuvers, except for small controlled attitude changes if two momentum wheels in a skewed V configuration are used. External disturbances tend to increase the roll and yaw Euler angles error. Active control of these errors can be achieved by use of reaction, magnetic, or solar control techniques.

References

Agrawal, B. N. (1986), *Design of Geosynchronous Spacecraft*. Englewood Cliffs, NJ: Prentice-Hall.

Azor, R. (1992), "Solar Attitude Control Including Active Denutation Damping in a Fixed Momentum Wheel Satellite," AIAA Guidance Navigation and Control Conference (10-12 August, Hilton Head Island, SC). Washington, DC: AIAA, pp. 226-35.

Azor, R. (1995), "Roll/Yaw Attitude Control and Denutation by Thrusters," Israel Annual Conference on Aviation and Astronautics (February, Tel Aviv). Tel Aviv: Kenes, pp. 86-94.

Benoit, A., and Bailly, M. (1987), "In-Orbit Experience Gained with the European OTS/ECS/TELECOM 1 Series of Spacecraft" (AAS 87-054), *Proceedings of the Annual Rocky Mountain Guidance and Control Conference* (31 January - 4 February 1987, Keystone, CO). San Diego, CA: American Astronautical Society, pp. 525-42.

Bingham, N., Craig, A., and Flook, L. (1984), "Evolution of European Telecommunication Satellite Pointing Performance," Paper no. 84-0725, AIAA, New York.

Bittner, H., Bruderle, E., Roche, Ch., and Schmidts, W. (1977), "The Attitude Determination and Control Subsystem of the Intelsat V Spacecraft" (ESA SP-12), *Proceedings of AOCS Conference* (3-6 October, Noordwijk, Netherlands). Paris: European Space Agency.

Bryson, A. (1983), "Stabilization and Control of Spacecraft," Microfiche supplement to the *Proceedings of the Annual AAS Rocky Mountain Guidance and Control Conference* (5-9 February, Keystone, CO). San Diego, CA: American Astronautical Society.

Devey, W. J., Field, C. F., and Flook, L. (1977), "An Active Nutation Control System for Spin Stabilized Satellites," *Automatica* 13: 161-72.

Dougherty, H. J., Lebsock, K. L., and Rodden, J. J. (1971), "Attitude Stabilization of Synchronous Communications Satellite Employing Narrow-Beam Antennas," *Journal of Spacecraft and Rockets* 8: 834-41.

Dougherty, H. J., Scott, E. D., and Rodden, J. J. (1968), "Analysis and Design of WHECON - An Attitude Control Concept," Paper no. 68-461, AIAA 2nd Communications Satellite Systems Conference (8-10 April, San Francisco).

Duhamel, T., and Benoit, A. (1991), "New AOCS Concepts for ARTEMIS and DRS," *Space Guidance, Navigation and Control Systems* (Proceedings of the First EAS International Conference, ESTEC, 4-7 June, Noordwijk, Netherlands). Paris: European Space Agency, pp. 33-9.

Forward, R. L. (1990), "Grey Solar Sails," *Journal of the Astronautical Sciences* 38(2): 161-85.

Fox, S. (1986), "Attitude Control Subsystem Performance of the RCA Series 3000 Satellite," Paper no. 86-0614-CP, AIAA 11th Communications Satellite System Conference (17-20 March, San Diego, CA).

Georgevic, R. M. (1973), "The Solar Radiation Pressure Force and Torques Model," *Journal of the Astronautical Sciences* 20(5): 257-74.

Iwens, R. P., Fleming, A. W., and Spector, V. A. (1974), "Precision Attitude Control with a Single Body-Fixed Momentum Wheel," Paper no. 74-894, AIAA Mechanics and Control of Flight Conference (5-9 August, Anaheim, CA).

Lebsock, K. L. (1980), "High Pointing Accuracy with a Momentum Bias Attitude Control System," *Journal of Guidance and Control* 3(3): 195-202.

Lebsock, K. L. (1982), "Magnetic Desaturation of a Momentum Bias System," Paper no. 82-1468, AIAA/AAS Astrodynamics Conference (9-11 August, San Diego, CA).

Lievre, J. (1985), "Solar Sailing Attitude Control of Large Geostationary Satellite," *IFAC Automatic Control in Space*. Oxford, UK: Pergamon, pp. 29-33.

Maute, P., Blancke, B., Jahier, J., and Alby, F. (1989), "Autonomous Geostationary Stationkeeping System Optimization and Validation," *Acta Automatica* 20: 93-101.

Muhlfelder, L. (1984), "Attitude Control System Evolution of Body-Stabilized Communication Spacecraft," Paper no. 84-1839, AIAA Guidance and Control Conference (20-22 August, Seattle). Washington, DC: AIAA, pp. 55-62.

Phillips, K. (1973), "Active Nutation Damping Utilizing Spacecraft Mass Properties," *IEEE Transactions on Aerospace and Electronic Systems* 9(5): 688-93.

Renner, U. (1979), "Attitude Control by Solar Sailing - A Promising Experiment with OTS-2," *ESA Journal* 3(1): 35-40.

Schmidt, G. E. (1975), "The Application of Magnetic Attitude Control to a Momentum-Biased Synchronous Communication Satellite," Paper no. 75-1055, AIAA Guidance and Control Conference (20-22 August, Boston).

Schmidt, G. E., and Mulhfelder, L. (1981), "The Application of Magnetic Torquing to Spacecraft Attitude Control" (AAS 81-002), *Proceedings of the Annual Rocky Mountain Guidance and Control Conference* (31 January - 4 February, Keystone, CO). San Diego, CA: American Astronautical Society.

Sidi, M. (1992a), "Attitude Stabilization of Bias-Momentum Satellites," Israel Annual Conference on Aviation and Astronautics (18-20 February, Tel Aviv). Tel Aviv: Kenes, pp. 333-8.

Sidi, M. (1992b), "Reactive Thrust Cruise for a Geosynchronous Bias-Momentum Satellite with Active Denutation," 12th FAC Symposium in Aerospace Control (7–11 September, Ottobrunn, Germany). Oxford, UK: Pergamon, pp. 201–5.

Wertz, J. R. (1978), *Spacecraft Attitude Determination and Control*. Dordrecht: Reidel.

Wie, B., Lehner, J., and Plescia, C. (1985), "Roll/Yaw Control of Flexible Spacecraft Using Skewed Bias Momentum Wheels," *Journal of Guidance, Control, and Dynamics* 8(4): 447–51.

CHAPTER 9

Reaction Thruster Attitude Control

9.1 Introduction

In Chapters 5–8, various control laws were presented for attitude stabilization and maneuvering. The hardware used to implement the control laws were principally momentum exchange devices as well as magnetic and solar torque controllers. Such controllers work in a linear continuous mode. The torques that they can provide are in the range of 0.02–1 N-m for momentum exchange devices, 10^{-2}–10^{-3} N-m for magnetic torque controllers, and 10^{-5}–10^{-6} N-m for solar torque controllers.

This form of attitude control has two major disadvantages. First, the speed of attitude maneuvering is limited by the low-level maximal torques that can be delivered to the ACS. The second but no less important difficulty was encountered in orbit-maneuvering tasks. The high-level liquid thrusters (or solid propulsion motors) used for orbit changes induce parasitic torques due to physical irregularities of the propulsion system. The level of induced parasitic torques is of the order of several newton-meters. The only way to control the attitude of the spacecraft under such disturbance conditions is to use reaction thruster controllers (see also Section 8.8).

Reaction thrusters used in attitude control are activated in a *pulsing mode* only. There are no linear, continuous reaction thrust controllers. This fact somehow complicates the analytical treatment of attitude control systems using them as torque controllers. However, they can provide almost any torque level, as surveyed in Appendix C. Reaction torque levels ranging between 0.01 N-m and 30 N-m are very common in most spacecraft. For practical considerations, it is convenient to use thrusters of the same thrust level for all control tasks in the satellite, but if this is not feasible then thrusters with different thrust levels can be incorporated as part of a unified propulsion system.

This chapter deals with the analysis and design of reaction thruster attitude control. It also covers two principal difficulties caused by the pulsing mode of thruster firing: the limits on attitude accuracy that can be achieved with a given thruster, and the fuel penalty associated with sensor noise. The quality of an ACS using propulsion torque controllers is strongly influenced by the specifications of the reaction thrusters; an introduction to propulsion hardware can be found in Appendix C.

9.2 Set-Up of Reaction Thruster Control

Reaction thrusters must be viewed in the context of a unified control system. In general, six thrusters are needed to allow attitude maneuvers in space, although some highly sophisticated systems claim to achieve the same space maneuvers with only four thrusters, strategically located on the satellite body (see Section 9.5). But for various practical reasons, six or more thrusters are necessary to complete a reaction control system.

9.2 / Set-Up of Reaction Thruster Control

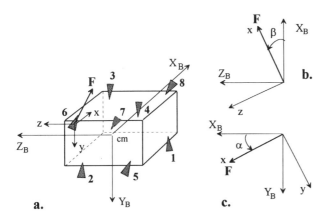

Figure 9.2.1 Eight-thruster arrangement, with direction details for thruster 6.

The level of torque that a reaction thruster can apply about a satellite axis depends not only on its thrust level but also on the torque-arm length about the axis. This statement suggests that correct thruster use depends primarily on its location on the satellite, and also on its inclination to the satellite body axes. Needless to say, different torque levels might be needed about the three principal body axes, so the location of the thrusters and their direction must be carefully studied before a final physical set-up is adopted for the propulsion system. The location and direction of the thrusters is also influenced by the location of the optical sensors and solar panels, which must not be damaged by the thrust flow. In the following analysis we will outline the different tradeoffs made in choosing the location of a thruster and the direction of its thrust axis relative to the body frame.

Figure 9.2.1 shows a potential arrangement of low-thrust satellite thrusters. First, it is clear that they must provide both positive and negative control torques about each of the satellite's body axes. (This arrangement is different from that in Figure 6.5.1, where e.g. thrusters TH2 and TH5 apply pure positive and negative torques about the X_B body axis.) In the next section we calculate the torque components applied by a thruster about each body axis as a function of the thruster's location and direction, denoted in the figure by the elevation and azimuth angles α and β, respectively.

9.2.1 Calculating the Torque Components of a Single Thruster

If the thrust vector is **F**, then the torque about the center of mass (cm) of the spacecraft will be $\mathbf{M} = \mathbf{r} \times \mathbf{F}$, where **r** is the vector distance of the thruster from the cm. The components of **r** are r_x, r_y, and r_z in the body axis frame; the thrust level is F. The direction of the thrust is defined by the elevation and azimuth angles α and β.

Suppose that initially **F** is in the direction of X_B. After two rotations – first about the **y** axis of the thruster by an angle β, and then about the **z** axis of the thruster by an angle α – we find that the components of **F** along the body axes are

$$F_x = F\cos(\alpha)\cos(\beta), \quad F_y = F\sin(\alpha), \quad F_z = F\cos(\alpha)\sin(\beta). \qquad (9.2.1)$$

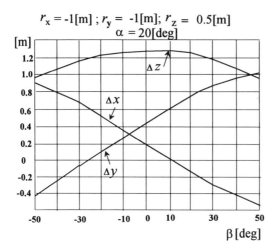

Figure 9.2.2 Equivalent torque arms for thruster 6.

The position **r** of the thruster can be expressed as:

$$\mathbf{r} = \mathbf{i} r_x + \mathbf{j} r_y + \mathbf{k} r_z. \tag{9.2.2}$$

With these relations, the torque components of **M** are

$$\mathbf{M} = \begin{bmatrix} M_x \\ M_y \\ M_z \end{bmatrix} = \mathbf{r} \times \mathbf{F} = \begin{bmatrix} r_y \sin(\beta)\cos(\alpha) - r_z \sin(\alpha) \\ r_z \cos(\alpha)\cos(\beta) - r_x \cos(\alpha)\sin(\beta) \\ r_x \sin(\alpha) - r_y \cos(\alpha)\cos(\beta) \end{bmatrix} F = \begin{bmatrix} \Delta x \\ \Delta y \\ \Delta z \end{bmatrix} F. \tag{9.2.3}$$

Equation 9.2.3 establishes the equivalent torque arms $\Delta x, \Delta y, \Delta z$ of the thrust **F** about the three body axes. Figure 9.2.2 exhibits the dependence of $\Delta x, \Delta y, \Delta z$ on the elevation and azimuth angles α and β for the location of thruster 6 at $r_x = -1$ m, $r_y = -1$ m, and $r_z = 0.5$ m; the value of the elevation angle is $\alpha = 20°$. Since $\alpha > 0$ (see Figure 9.2.1), Δz is always positive. But Δx and Δy can change sign, depending on the value of the azimuth angle β.

The importance of the results displayed in Figure 9.2.2 lies in the fact that the torque arms about any body axis can be decreased at will, thus enabling low torque levels even at relatively high-thrust levels. This in turn enables fine attitude control, but at the expense of low fuel efficiency.

Chapter 8 dealt with a number of control configurations in which the minimum impulse bit of a single reaction shot plays an important factor in achievable roll and yaw accuracies (see Eq. 8.8.6). The minimum impulse bit of the thruster is generally determined by the reaction thruster characteristics, and cannot be decreased at will. On the other hand, correct location and inclination of the thruster relative to the spacecraft body can decrease the minimum Δz as desired by affecting the equivalent torque arms of the thruster.

Most attitude control laws calculate the torques to be applied about the body axes. In Section 7.3.4 we learned how to distribute these torque commands to the four reaction wheels. The algorithm used to realize the torque transformation was very simple, as Eq. 7.3.26 showed.

9.2 / Set-Up of Reaction Thruster Control

The algorithm that transforms the command control torques about the body axes to reaction thruster activations is much more complicated, for two reasons. (1) Reaction thrusters are not linear controllers, since the level of the thrust output is constant. Consequently, the equivalent torque that the thruster will produce depends on the time period in which the thruster is activated. (2) A thruster is capable of producing one-signed torques only. In order to achieve a torque about the same axis with the opposite sign, a different thruster must be activated about the same axis in the opposite direction (see also Section 9.5). These two factors drastically complicate the algorithm that transforms body torque commands into thruster activation commands.

9.2.2 Transforming Torque Commands into Thruster Activation Time

This section describes a basic algorithm that uses the pulse width modulation principle to transform the torque commands into correctly timed activation of the relevant thrusters. The algorithm as described here is not optimized in terms of simplifying to a minimum its onboard version. We prefer to make the algorithm more comprehensible by showing clearly all its sequential operations.

The algorithm naturally depends on the physical set-up of the thrusters. We will base the demonstration algorithm on the set-up shown in Figure 9.2.3, which is a potential set-up for a geostationary satellite. The thrusters are arranged so that they provide the necessary torques for attitude control about the three body axes as well as the necessary thrust for station keeping. Thrusters 3–6 are used for N–S (inclination) SK (Section 3.5.1); thrusters 1 and 2 are used for E–W (longitude) SK (Section 3.4.4) and also for eccentricity corrections (Section 3.5.2).

In addition to station-keeping tasks, the set-up in Figure 9.2.3 allows for 3-DOF attitude control. Thrusters 1 and 2 provide the negative and the positive pitch control torques (respectively) about the Y_B axis; thruster 3 provides positive torque about the X_B axis but simultaneously a negative torque about the Z_B axis; and so on. Formally,

$$Ty+ = Th2, \qquad Ty- = Th1;$$
$$Tx+ = Th3 + Th5, \qquad Tx- = Th4 + Th6; \qquad (9.2.4)$$
$$Tz+ = Th5 + Th6, \qquad Tz- = Th3 + Th4.$$

Here the + and − signs indicate the sign of the produced torques about the body axes, and *Thi* denotes thruster *i*.

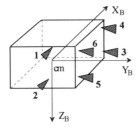

Figure 9.2.3 Possible thruster set-up for a geostationary satellite.

In order to simplify our analysis, the thrusters are located symmetrically about the body axes, with equal torque arms about the same axis; the directions of the thruster axes are parallel to the body axes. The torque efficiency of each thruster depends on the thrust level F, and also on the torque arms $\Delta x, \Delta y, \Delta z$ (Section 9.2.1). We define the torque constants as $G_X = F\Delta x$, $G_Y = F\Delta y$, and $G_Z = F\Delta z$. With this notation, we can express the torques about the body axes as

$$Tx = [Th5 + Th3 - Th4 - Th6]G_X,$$
$$Tz = [Th5 + Th6 - Th3 - Th4]G_Z, \qquad (9.2.5)$$
$$Ty = [Th2 - Th1]G_Y.$$

Since the produced torques are momentarily determined by G_X, G_Y, and G_Z, the *average torque* provided during a sampling time depends upon the time that the thrusters are *on* relative to the sampling time T_{sam}. This is the principle of pulse width modulation (PWM), which is treated in subsequent sections. First, we normalize the body control torques to

$$\hat{T}x = \frac{Tx}{G_X}, \quad \hat{T}y = \frac{Ty}{G_Y}, \quad \hat{T}z = \frac{Tz}{G_Z},$$

and define Ti as the ratio between the thruster "on-time" and the sampling time for thruster Thi. Thus the first two of Eqs. 9.2.5 can be rewritten in the following form:

$$\begin{bmatrix} \hat{T}x \\ \hat{T}z \end{bmatrix} = \begin{bmatrix} 1 & -1 & 1 & -1 \\ -1 & -1 & 1 & 1 \end{bmatrix} \begin{bmatrix} T3 \\ T4 \\ T5 \\ T6 \end{bmatrix}. \qquad (9.2.6)$$

(This is similar to Section 7.3.4, where four reaction wheels were used to provide torques about three body axes.) Although the matrix of Eq. 9.2.6 is not square, it does have a right pseudoinverse, yielding

$$\begin{bmatrix} T3 \\ T4 \\ T5 \\ T6 \end{bmatrix} = \frac{1}{4} \begin{bmatrix} 1 & -1 \\ -1 & -1 \\ 1 & 1 \\ -1 & 1 \end{bmatrix} \begin{bmatrix} \hat{T}x \\ \hat{T}z \end{bmatrix}. \qquad (9.2.7)$$

According to the values of $\hat{T}x$ and $\hat{T}z$, Ti can have negative values, which cannot be physically realized with the thruster i. A negative on-time means that a torque of opposite sign is to be produced with thruster Thi, which is not physically possible. This situation can be remedied by activating, for an identical duration, another thruster providing the same torque but with a *positive* on-time. For instance, in Figure 9.2.3, suppose that $T3$ comes out to be negative. In this case, $Th6$ should be activated instead for the same on-time $T3$, assuming that the Δx and Δz torque arms of $Th3$ and $Th6$ are equal.

If the torque arms are not equal then G_X, G_Y, G_Z must be defined separately for each thruster, and Eqs. 9.2.5 must be rewritten accordingly. The algorithm is listed as follows.

$T3 = [+\hat{T}_x - \hat{T}_z]/4$
$T4 = [-\hat{T}_x - \hat{T}_z]/4$
$T5 = [+\hat{T}_x + \hat{T}_z]/4$
$T6 = [-\hat{T}_x + \hat{T}_z]/4$

$TT6 = T6 - T3$; $TT3 = 0$

IF($TT6$.LT.0) THEN $TT3 = T3 - T6$; $TT6 = 0$

$TT4 = T4 - T5$; $TT5 = 0$

IF($TT4$.LT.0) THEN $TT5 = T5 - T4$; $TT4 = 0$
IF($\hat{T}_y$.GT.0) THEN $TT2 = \hat{T}_y$; $TT1 = 0$
IF($\hat{T}_y$.LT.0) THEN $TT1 = \text{Abs}(\hat{T}_y)$; $TT2 = 0$

Here *TTi* stands for the time duration of the *on* condition for each thruster *i*. In essence, the algorithm realizes a kind of pulse width modulation, upon which many pulsed attitude control schemes are based.

Algorithms for other reaction thruster set-ups, such as those shown in Figure 9.2.1 and Figure 6.5.1, can be written in a similar way. With asymmetrical location of the thrusters and different thrust levels or directions of their thrust axes, the algorithm might become quite complicated. Still, the technique of formulating it remains the same. The following sections deal with realization of attitude control loops based on the reaction thruster set-up of Figure 9.2.3.

9.3 Reaction Torques and Attitude Control Loops

9.3.1 *Introduction*

The control laws to be used for reaction attitude control loops are the same laws treated in Section 7.2. Unfortunately, reaction controllers do not possess the same linear relationship between the input to the controller and its output torque. In fact, they are activated in an on–off mode. Nonetheless, they can be used in a quasi-linear mode by modulating the width of the activated reaction pulse proportionally to the level of the torque command input to the controller. This is the often used *pulse width modulation* (PWM) principle. A related design technique is based on the well-known Schmidt trigger, which implements a *pulse width-pulse frequency modulation* (PWPFM) in which the distance between the pulses is also modulated. Both modulation techniques will be analyzed in this chapter as part of attitude feedback control loops.

By definition, attitude control loops based on reaction controllers are *sampled,* with all the implications attendant upon such systems. There are numerous automatic control textbooks dealing with sampled control systems; see for instance Saucedo and Schiring (1968), Kuo (1970), Franklin and Powell (1980), Houpis and Lamont (1985), or D'Azzo and Houpis (1988).

With certain assumptions, the "area" of the reaction pulse can be approximated as an impulse (Figure 9.3.1, overleaf). This allows the feedback control loops to be analyzed as a conventional digital control system, for which the classical tools of automatic control theory hold. Design techniques based on the Nyquist, Bode, Nichols,

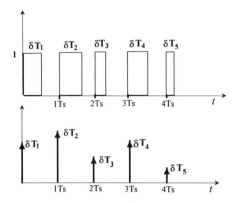

Figure 9.3.1 Approximation of PW-modulated pulses to impulses.

and root-locus theories apply nicely to these systems, as we shall see in the present chapter. There is, however, a significant deficiency in sampled systems. The bandwidth of the closed loop that can be achieved is limited – in a very determined way – by the sampling time (Sidi 1980).

9.3.2 Control Systems Based on PWPF Modulators

Apparently owing to technical heritage from the "analog age" of onboard computers, pulse width–pulse frequency modulators are still predominant in attitude and orbit control systems (AOCS). They emerged from the widely used analog monostable Schmidt trigger, a basic element in analog pulse techniques (Millman and Taub 1956, Joice and Clarke 1961).

The standard practical realization of a PWPF modulator control loop is shown in Figure 9.3.2, where K_x and K_{xd} are the usual proportional and derivative gains. The dynamics of the sensor and of the sensor noise filter have been omitted in order to simplify the analysis. The attitude and its derivative errors (Er in the figure) are transformed into a burst of idealized rectangular pulses.

Figure 9.3.2.b shows a variation of the PWPF modulator, the so-called pseudo rate (PR) modulator. Its special characteristic lies in the fact that the time-constant network is located in the feedback path of the modulator, thus giving the modulator "lead–lag" compensating abilities. Because of the dead-zone region of the hysteresis block, the lead phase is achieved with a lesser amplification of the sensor noises at the input to the modulator. For stabilization of spacecraft with large flexible appendages, control and modulator parameters must be carefully matched to ensure stability of structural modes of vibration (Bittner, Fisher, and Surauer 1982).

A detailed analysis of the PWPF modulator follows. Our aim is to obtain a relationship between the input $Er = In$ and the frequency- and width-modulated values of the output pulse sequence, characterized by the *on* and *off* time periods.

Calculating t_{on} and t_{off} for the PWPF Modulator

The calculation of these variables is based on the time behavior shown in Figure 9.3.3. Our presentation will be broken down into five steps.

9.3 / Reaction Torques and Attitude Control Loops

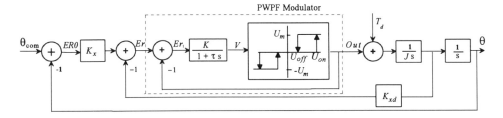

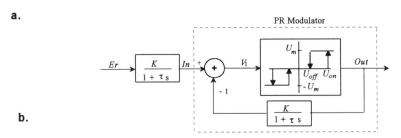

Figure 9.3.2 Basic attitude control loops using PWPF and PR modulators; adapted from Bittner, Fisher, and Surauer (1982) by permission of IFAC.

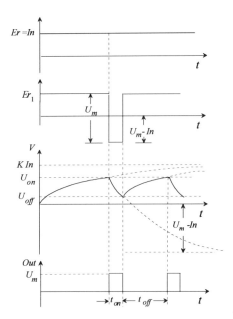

Figure 9.3.3 Time behavior of the PWPF modulator used for calculating t_{on} and t_{off}.

(1) As long as $V < U_{on}$, the system is quiet. The effective dead zone is U_{on}/K, where K is the DC gain of the time constant network of the modulator (Figure 9.3.2.a). The hysteresis block in the figure is the Schmidt trigger. The smallest input that can activate the hysteresis block is

$$In = U_{on}/K. \tag{9.3.1}$$

(2) Compute the time behavior of V as a result of Er_1. The dynamic equation is simply

$$\dot{V}\tau + V = KEr_1. \tag{9.3.2}$$

Taking the Laplace transform of this equation, we obtain

$$V(s) = \frac{\tau V(0)}{1+s\tau} + \frac{KEr_1}{s(1+s\tau)}. \tag{9.3.3}$$

The time-domain solution is simply

$$V(t) = V(0)e^{-t/\tau} + KEr_1[1-e^{-t/\tau}]. \tag{9.3.4}$$

(3) We now compute t_{on}. According to Figure 9.3.2, $V(t)$ starts at U_{on} and decreases asymptotically to $In - U_m$, which means that $V(0) = U_{\text{on}}$ and $KEr_1 = K(In - U_m)$. This decrease will stop at U_{off}, so that

$$U_{\text{off}} = V = U_{\text{on}} e^{-t_{\text{off}}/\tau} + K(In - U_m)(1 - e^{-t_{\text{on}}/\tau})$$
$$= K(In - U_m) + (U_{\text{on}} - KIn + KU_m)e^{-t_{\text{on}}/\tau}. \tag{9.3.5}$$

It follows that

$$e^{-t_{\text{on}}/\tau} = \frac{U_{\text{off}} - KIn + KU_m}{U_{\text{on}} - KIn + KU_m} = 1 - \frac{U_{\text{on}} - U_{\text{off}}}{U_{\text{on}} - KIn + KU_m}. \tag{9.3.6}$$

For small t_{on}, $e^{-t_{\text{on}}/\tau} \approx 1 - t_{\text{on}}/\tau$. This yields the first-order approximation

$$t_{\text{on}} \approx \tau \frac{U_{\text{on}} - U_{\text{off}}}{KU_m - KIn + U_{\text{on}}}. \tag{9.3.7}$$

(4) To calculate t_{off} we first observe that, according to Figure 9.3.2, $V(t)$ tends toward KIn; it starts at $V(0) = U_{\text{off}}$ but its increase is halted at U_{on}. Hence we have

$$V(t) = e^{-t_{\text{off}}/\tau} U_{\text{off}} + KIn(1 - e^{-t_{\text{off}}/\tau}) = U_{\text{on}}, \tag{9.3.8}$$

from which follows the final result:

$$e^{-t_{\text{off}}/\tau} = \frac{U_{\text{on}} - KIn}{U_{\text{off}} - KIn} = \frac{KIn - U_{\text{on}} + U_{\text{off}} - U_{\text{off}}}{KIn - U_{\text{off}}} = 1 - \frac{U_{\text{on}} - U_{\text{off}}}{KIn - U_{\text{off}}}. \tag{9.3.9}$$

For small t_{off}, a first-order approximation may likewise be used:

$$t_{\text{off}} \approx \tau \frac{U_{\text{on}} - U_{\text{off}}}{KIn - U_{\text{off}}}. \tag{9.3.10}$$

Some results concerning the time-domain behavior of the PWPF modulator with respect to our analytical equations are shown in Figure 9.3.4 and Figure 9.3.5. Figure 9.3.5.c (page 270) shows the resulting average output torque due to the input to the PWPF modulator in Figure 9.3.4.a. Except for the scaling factor of both input and output, the output adequately follows the input, as predicted. If we wish to use the reaction system with a lower fuel penalty, then wider pulses have some advantage because they have a higher average specific impulse I_{sp}. According to Eq. 9.3.7 and 9.3.10, the ratio $t_{\text{on}}/t_{\text{off}}$ becomes larger if U_{off} is zeroed.

(5) Equation 9.3.6 and Eq. 9.3.9 (or the simplified Eq. 9.3.7 and Eq. 9.3.10) can be used to determine the five modulator parameters K, τ, U_{on}, U_{off}, and U_m. It is imper-

9.3 / Reaction Torques and Attitude Control Loops 269

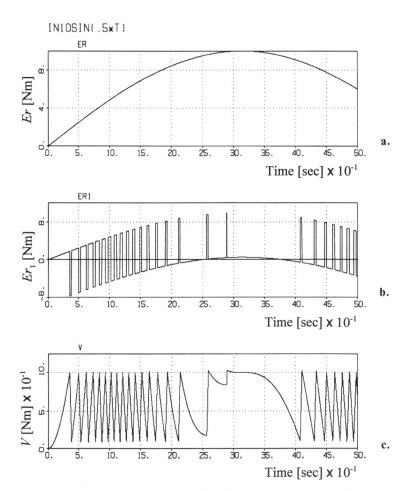

Figure 9.3.4 Time histories of the input and of the internal variables of the PWPF modulator.

ative to determine t_{on} and t_{off} based on physical considerations. The hysteresis coefficients are also important in determining the minimum dead zone that will give sufficient immunity to the sensor noise reaching the modulator input.

EXAMPLE 9.3.1 A PWPF modulator is implemented in a single-axis attitude control loop. The control gains K_x and K_{xd} in Figure 9.3.2 are chosen so that the control system has a closed-loop natural frequency of $\omega_n = 1$ rad/sec and a damping coefficient of $\xi = 1$. To satisfy some practical engineering requirements, the modulator parameters were chosen as follows: $K = 2$, $\tau = 0.5$, $U_{on} = 1$, $U_{off} = 0.1$, $U_m = 9.5$.

Figure 9.3.6 (page 271) shows the time responses for an angular input of 0.2°. The input Er and the output (Out) of the modulator are also shown on Figure 9.3.6. Figure 9.3.7 (page 272) shows the time response for an input of 1°.

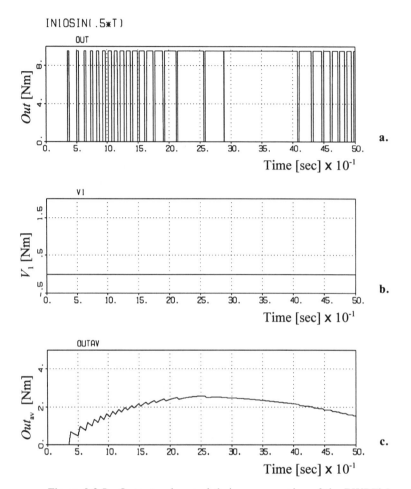

Figure 9.3.5 Output pulses and their average value of the PWPFM.

Calculating t_{on} and t_{off} for the PR Modulator

An analysis similar to that carried out for the PWPF modulator leads to the following results:

$$t_{on} = \tau \ln \frac{KU_m - In + U_{on}}{KU_m - In + U_{off}}, \tag{9.3.11}$$

$$t_{off} = \tau \ln \frac{In - U_{off}}{In - U_{on}}. \tag{9.3.12}$$

Equation 9.3.11 and Eq. 9.3.12 can be used to determine the parameters of the pseudo rate pulse modulator for efficient use in reaction attitude control systems.

9.3.3 Control Loop Incorporating a PWPF Modulator

Even though PWPF modulators are by definition nonlinear, they are easily incorporated in attitude control loops, whereupon the entire feedback control loop

9.3 / Reaction Torques and Attitude Control Loops

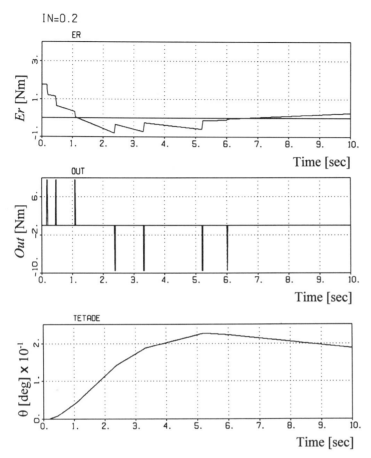

Figure 9.3.6 Time behavior of a PWPF modulator realization of an attitude control feedback loop, with an input of 0.2°.

can be analyzed by linear control theory. As can be seen from Figure 9.3.4.a and Figure 9.3.5.c, the average output of the modulator tracks the input quite accurately. In fact, the input–output characteristics of pulse modulators can be represented as in Figure 9.3.8 (see also Wie and Plescia 1984). Figure 9.3.8 (overleaf) shows the linear behavior of the modulator input–output characteristics except at very low and high inputs, where the nonlinear characteristics are purposely introduced to solve practical problems of sensor noise and to limit structural oscillation of panels and fuel slosh in a limit cycling mode (Vaeth 1965, Bittner et al. 1982).

Analog implementation of PWPF modulators is straightforward because analog (continuous) electronics technology is the natural medium for realizing these modulators, implemented by the Schmidt trigger scheme. However, PWPF modulation causes some practical problems with today's onboard microprocessors. Digital microprocessors work in a synchronous timing created by an electronic clock: the onboard computer sends control commands at equal time intervals, and pulse frequency modulation cannot be easily implemented. The equations of the previous section become invalid for a *discrete* "pulse frequency" modulator. This kind of modulator,

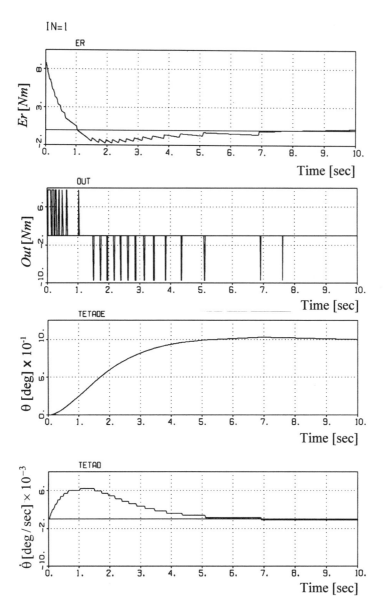

Figure 9.3.7 Time behavior of a PWPF modulator realization of an attitude control feedback loop, with an input of 1°.

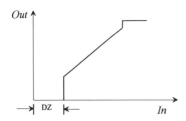

Figure 9.3.8 Input-output characteristics of a PWPF modulator.

when used in an onboard microprocessor, must be synchronized with the sampling time of that processor. Digital implementation of a PWPF modulator was the subject of a Ford Aerospace patent application (Chan 1982).

The nice characteristics that could be achieved with PWPF modulators are complicated by the practical difficulties just mentioned. It seems that, from the basic perspective of input–output characteristics, modulators based solely on pulse width modulation are simpler to apply in a microprocessor-based onboard computer. Attitude feedback control loops based on PWM will be analyzed in the next section.

9.4 Reaction Attitude Control via Pulse Width Modulation

9.4.1 *Introduction*

There are no fundamental differences between reaction pulse control loops using pulse width–pulse frequency modulation (PWPFM or PRM) and pulse width modulation (PWM). Both are based on the Schmidt trigger, but there do exist small differentiating nuances between these two kinds of control schemes.

In attitude feedback control loops based on PWM, the sampling frequency is constant and the reaction pulses are applied at equal time intervals. If the dynamics of the plant contains additional structural dynamics with low damping coefficients and eigenfrequencies equal to the sampling frequency of the control loop, then the structural dynamics might be excited, thus degrading the quality of the feedback control loop. Special precautions, such as adequate structural filters, must be incorporated to account for this phenomenon.

9.4.2 *Feedback Control Loop of a Pulsed Reaction System*

In previous chapters, in order to simplify the theoretical analysis, controllers were presumed to be analog. With today's technology, onboard control computers are no longer analog, and all feedback control loops are of the sampled type. For such continuous devices as reaction wheels or magnetic torque controllers, the onboard microprocessor approximates continuousness via a sampler followed by a *zero-order hold device*. With a sampling frequency that is reasonably high with respect to the bandwidth of the open-loop transfer function, the zero-order hold output is quasicontinuous. The sampler and the zero-order hold circuit have the net effect of adding some delay to the open-loop transfer function of the feedback system, which should be taken into consideration when designing with continuous control analysis techniques; design with discrete (digital) control techniques automatically takes care of this inherent delay. With *pulsed controllers,* in which a zero-hold device does not exist, the analysis and design must be carried out using discrete control techniques.

The basic control scheme of a pulsed attitude feedback control loop about one body axis is shown in Figure 9.4.1 (overleaf). With ideal torque impulses, the block of the PWM can be omitted. In this way, linear analysis of the sampled control loop is performed in the usual manner. With small *on* periods (as compared to the sampling period) of the reaction thrusters, omission of the PWM block is justified. This means that the amplitude of the impulses at the output of $G_1(z)$ and of $G(z)$ in Fig-

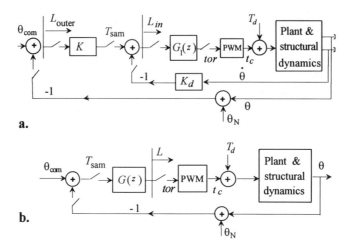

Figure 9.4.1 Pulsed controller attitude feedback control loop.

ure 9.4.1.a and Figure 9.4.1.b (respectively) takes the value of the rectangular pulses $F\Delta(\Delta t)$ of Section 9.2.2, where Δ is the torque arm and Δt is the pulse width.

Figure 9.4.1.a shows the most general block diagram for a single-axis attitude control loop in which the attitude and their derivatives can both be measured. Here K and K_d are the usual position and derivative controller gains, and $G_1(z)$ contains the required structural mode control networks, sensor noise filters, and an integrator (if one is needed to nullify the steady-state error to disturbances or to attitude command inputs). As mentioned previously, unless precise attitude control is imperative there is a tendency to avoid continuous use of rate sensors, such as the common rate gyro. In this case, the scheme of Figure 9.4.1.b is relevant, with the self-evident disadvantage of increasing the sensitivity of the system to position sensor noise.

The term T_d denotes the equivalent disturbance torque, comprising internal and external parasitic torques whose sources are irregularities in or bad modeling of the reaction thrusters, uncertainties in knowledge of the vehicle mass properties, and so forth. For instance, in the ABM stage – in which the transfer orbit is to be circularized at the apogee – a large disturbance torque may be present if the high-thrust vector does not pass exactly through the satellite's center of mass. The attitude reaction control loop must counteract this disturbance torque in order to prevent a large attitude error of the satellite, which would cause an incorrect $\Delta \mathbf{V}$ increment (see Section 3.4.3). Another example of the importance in decreasing ACS sensitivity to T_d is in the N-S station keeping of geostationary satellites. Keeping the inclination of the orbit inside permitted limits is achieved by adding linear velocity increments to the s/c perpendicularly to its orbit (as explained in Section 3.5.1). During this stage, which may continue for several minutes, the satellite attitude error cannot exceed permitted tolerances without degrading the communication mission. In today's communications satellites, attitude error tolerances are of the order of 0.03°-0.1°, a difficult standard for the control engineer to achieve.

Naturally, an increase in the gain bandwidth of the control loop will decrease the attitude errors due to the external disturbances, but the existence of structural dynamics will preclude very high loop gains. Hence, a careful design must procure the highest possible gains while still keeping the necessary gain and phase margins for the structural dynamics, including stability of the sloshing modes (see also Section 7.3.6). Another important control aspect is the behavior of the attitude output θ to control input θ_{com}. For all attitude control tasks, sensor noise is an important factor to be taken into account. For the set-up of Figure 9.4.1.b, in which it is assumed that no rate sensor is included in the control hardware, the noise of the position sensor might be strongly amplified; this precludes high open-loop gains (see also Section 7.3.6).

Sensitivity to External Disturbances

At the beginning of the analysis we shall assume that there are no structural filters, $G_1(z) = 1$ in Figure 9.4.1.a. We also know that, with no structural dynamics in the plant block, the simplified plant reduces to $P(s) = 1/Js^2$, where J is the satellite's moment of inertia. Also, without the pulse width modulator in the loop, PWM = 1. If we assume that the disturbance T_d is a continuous and constant step function with an amplitude of D, then

$$\theta(z) = \frac{DT_{sam}^2(z/2)(z+1)}{J(z-1)^3 + T_{sam}z(z-1)\left(K + K_d\frac{z-1}{z}\right)}, \quad (9.4.1)$$

where T_{sam} is the sampling time and z is the Z transform variable. Using the final-value theorem for sampled systems, the steady-state value of the output amounts to

$$\lim_{n\to\infty} \theta(nT_{sam}) = \lim_{z\to 1} \frac{z-1}{z}\theta(z) = \frac{DT_{sam}}{K} \quad (9.4.2)$$

and finally

$$TF_d = \frac{\theta_{ss}}{D} = \frac{T_{sam}}{K}. \quad (9.4.3)$$

Equation 9.4.3 shows that the steady-state error of the output angle θ depends on both the sampling time T_{sam} and the DC gain K. This gain is itself determined by the closed-loop natural frequency ($K = \omega_n^2 J$), and K_d is proportional to the damping coefficient ($K_d = 2\xi\omega_n J$). For the magnitude of the sensitivity ratio TF_d see Table 9.4.1, in which $J = 500$ kg-m^2 and $\xi = 0.7$.

Table 9.4.1 *Steady-state error as function of the sampling time and the closed-loop bandwidth*

ω_n [r/s]	K	K_d	T_{sam}[sec]	TF_d [rad/N-m]	D [N-m]	θ_{ss} [deg]
0.5	125	350	1.0	8·10^{-3}	1	0.46
0.5	125	350	0.5	4·10^{-3}	1	0.23
0.5	125	350	0.25	2·10^{-3}	1	0.114
0.25	31.25	175	1.0	32·10^{-3}	1	1.83
0.25	31.25	175	0.5	16·10^{-3}	1	0.92
0.25	31.25	175	0.25	8·10^{-3}	1	0.46

In this analysis of the discrete feedback control loop, the outputs of $G_1(z)$ or $G(z)$ in Figure 9.4.1 are impulses of amplitude *tor*, proportional to the processed attitude error. In practice, thrusters supply torques of constant amplitude and varying time length:

$$tp = tor/F\Delta, \qquad (9.4.4)$$

where F is the thrust level and Δ the torque arm. This time transformation is performed with the PWM included in Figure 9.4.1.

In order to prevent exaggerated detrimental influence of sensor noise on control system performance, at least one first-order filter must be incorporated into the feedback control loop. The time constant of the filter must be low enough to leave the control loop with satisfactory gain and phase margins. Using the Tustin transformation to render the analog control network $G(s)$ discrete, we obtain

$$G(z) = \frac{KT_{\text{sam}} + 2k_d + (kT_{\text{sam}} - 2k_d)z^{-1}}{T_{\text{sam}} + 2\tau + (T_{\text{sam}} - 2\tau)z^{-1}}. \qquad (9.4.5)$$

EXAMPLE 9.4.1 In this example, $\omega_n = 0.5$, $\xi = 0.7$, $J = 500$ kg-m^2, and either $T_{\text{sam}} = 0.25$ sec or $T_{\text{sam}} = 1$ sec; the disturbance $D = 1$ N-m. Using conventional frequency design techniques, the time constant of the filter was fixed to $\tau = 0.3$, with the resulting open-loop transfer function shown in Figure 9.4.2. The time-domain simulation results are shown in Figures 9.4.3 and 9.4.4 for a T_{sam} of 0.25 sec and of 1 sec, respectively. The steady-state error for the case with $T_{\text{sam}} = 0.25$ sec is 0.11°, as per Eq. 9.4.2. In Figure 9.4.4, with $T_{\text{sam}} = 1$ sec, the steady-state angular error is 0.456°; this also satisfies Eq. 9.4.2.

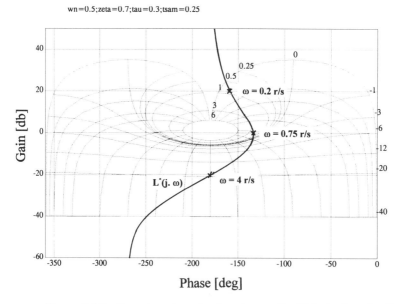

Figure 9.4.2 Open-loop transfer function on the Nichols chart for Example 9.4.1.

9.4 / Reaction Attitude Control via Pulse Width Modulation

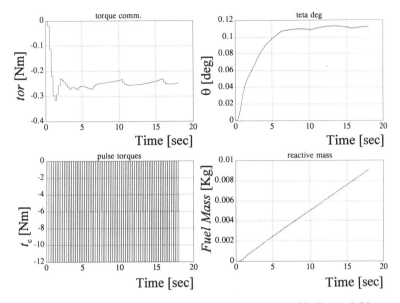

Figure 9.4.3 Time-domain results for the case with $T_{sam} = 0.25$ sec.

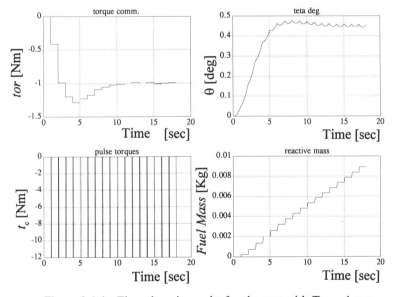

Figure 9.4.4 Time-domain results for the case with $T_{sam} = 1$ sec.

Equation 9.4.2 shows that the steady-state error depends on the sampling time T_{sam}. However, it would be more convenient to have a control law in which the steady-state error is *not* dependent on the sampling time. This is easily accomplished by arguing as follows. With a zero-order hold device, the processed output of the control network $G(z)$ is constant during the complete sampling period; thus, an angular momentum ($T_{sam} tor$) is delivered to the satellite. It is suggested that the reaction thruster deliver the same angular momentum, which means: $tp \times F\Delta = T_{sam} tor$, or

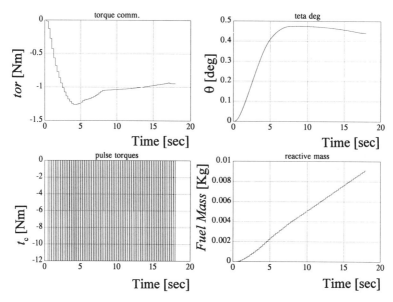

Figure 9.4.5 Time-domain results for the control system based on a PW modulator that renders the steady-state error independent of the sampling time T_{sam}.

$$tp = T_{sam} tor/F\Delta, \tag{9.4.6}$$

where *tor* is the commanded torque. The effect is equivalent to multiplying $G(z)$ by T_{sam}. With this assumption, Eq. 9.4.1 and Eq. 9.4.2 take the new form

$$\theta(z) = \frac{DT_{sam}^2(z/2)(z+1)}{J(z-1)^3 + T_{sam}z(z-1)\left(K + K_d\frac{z-1}{z}\right)T_{sam}} \tag{9.4.7}$$

and

$$\lim_{n \to \infty} \theta(nT_{sam}) = \lim_{z \to 1} \frac{z-1}{z}\theta(z) = \frac{D}{K}. \tag{9.4.8}$$

Equation 9.4.8 shows that the steady-state angular error is no longer dependent on the sampling period T_{sam}.

Figure 9.4.5 presents a simulation of the control system with a PWM based on Eq. 9.4.5 (see Example 9.4.1). In this case $T_{sam} = 0.25$ sec, but the steady-state error is now equal to $0.456°$, as predicted by Eq. 9.4.8.

Adding an Integrator to Nullify the Steady-State Error In orbit-maneuvering control tasks in which high thrust is activated for long periods (i.e., tens of minutes), it is of utmost importance to minimize the steady-state angular errors. This enables efficient orbit control with a minimal waste of fuel, as explained in Section 3.4.3.

The steady-state error due to the disturbance can be decreased to null by adding integration control error to the already existing position and derivative control errors. It is possible to add an integrator to the open-loop transfer function of the control loop without changing appreciably the high-range frequency characteristics.

9.4 / Reaction Attitude Control via Pulse Width Modulation

The integrator is added with an appropriate low-frequency zero, as is usual with frequency design techniques.

EXAMPLE 9.4.1 *(Continued)* Since the crossover is $\omega_{co} = 0.75$ rad/sec (Figure 9.4.2), it is appropriate to add an integrator with a zero located at $\omega_z = 0.05$ rad/sec so that the higher-frequency range of the open-loop gain is not compromised. After the Tustin transformation, the revised discrete control network becomes

$$G_1 = \frac{z(1/\omega_z + T_{sam}) - 1/\omega_z + T_{sam}}{(z-1)/\omega_z} = \frac{z20.25 - 19.75}{20(z-1)}.$$

The results are given on the Nichols chart of Figure 9.4.6, where it can be clearly seen that the gain and phase margins have not deteriorated as a result of adding the integrator.

The time responses for the 1-N-m disturbance are shown in Figure 9.4.7 (overleaf). The time constant of the integrator control network is of the order of 20 sec, as expected ($\omega_z = 0.05$ rad/sec). The error is comparatively large only for the first 20 sec, which is a negligible time in view of the total duration of the ABM stage. For the rest of the time, the attitude error is null.

Sensor Noise Amplification One of the major problems in the design of feedback control systems is the amplification of sensor noise. This can result in saturation of torque controllers and thus prevent the feedback control system from operating satisfactorily. The effect is common to all controllers, but is seriously aggravated when the controller is based on liquid reaction thrusters. Unpredictable parasitic

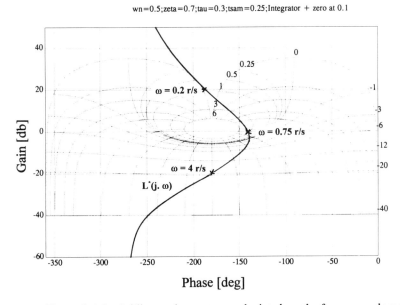

Figure 9.4.6 Adding an integrator as depicted on the frequency-domain chart.

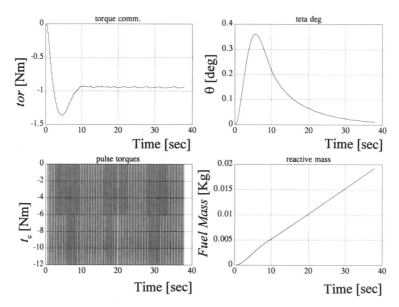

Figure 9.4.7 Time-domain results with the integrator added in order to nullify steady-state angular error.

waste of fuel shortens the mission lifetime of the satellite, which is intolerable from the system engineering point of view. Hence, even in the preliminary stages of designing a reaction attitude control loop, the amplification of sensor noise must be carefully considered and attenuated.

The first natural way to decrease noise amplification is to reduce the bandwidth of the feedback control system (at the expense of increasing the attitude errors due to external disturbances). This solution has the advantage of increasing the gain margins of the structural modes, a subject to be treated in Chapter 10. The second way to deal with the problem is to precede the PWM with an adequate dead zone, or to limit the minimum pulse width of the modulator output. There is also the possibility of using both techniques in conjunction. A compromise between attitude accuracy and fuel flow per unit time for a given level of sensor noise will help to fix the final bandwidth of the open-loop transfer function as well as the level of dead zone to be used (if any). These tradeoffs are clarified in the following example.

EXAMPLE 9.4.2 As in the previous example, $J = 500$ kg-m^2. We shall use a PW modulator to compute the on-time of the thrusters according to Eq. 9.4.6, so that T_{sam} has no influence on the steady-state angular error of the feedback control loop. No integrator is included in the control network. The sensor noise level of a good earth sensor (see Appendix B) – used to sense the attitude of the satellite in the orbit reference frame – ranges from 0.03° to 0.1° (RMS). In order to emphasize the effect of amplifying this noise, we will assume an exaggerated noise level of 0.3° (RMS). In the analysis to follow, $T_{\text{sam}} = 0.5$ sec and the disturbance level is $D = 1$ N-m.

In the first try, choose $\omega_n = 0.5$ rad/sec, $\xi = 0.7$, $\tau = 0.3$ sec, and $T_{\text{sam}} = 0.5$ sec. The open-loop gain of the solution is shown in Figure 9.4.8. With no sensor noise, the time history is shown in Figure 9.4.9. The steady-state error is 0.456°, as calcu-

9.4 / Reaction Attitude Control via Pulse Width Modulation

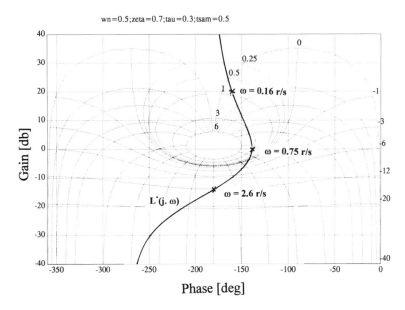

Figure 9.4.8 Open-loop transfer function for the feedback control with $\omega_n = 0.5$ rad/sec, $\xi = 0.7$, $\tau = 0.3$ sec, and $T_{sam} = 0.5$ sec.

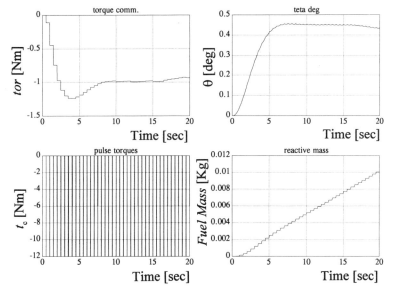

Figure 9.4.9 Time-domain results for a disturbance of 1 N-m without sensor noise; $\omega_n = 0.5$ rad/sec, $\xi = 0.7$, $\tau = 0.3$ sec, and $T_{sam} = 0.5$ sec.

lated from Eq. 9.4.8. The theoretical noise amplification was calculated to be $tor/$ noise $= 11.848$ (RMS). The fuel consumption rate is 0.5 g/sec. Amplification of sensor noise augments the attitude error as shown in Figure 9.4.10 (overleaf), where peak errors as high as 0.9° may be perceived. Even if so much angular error is acceptable, quadrupling the rate of fuel mass consumption (to 2 g/sec) is not.

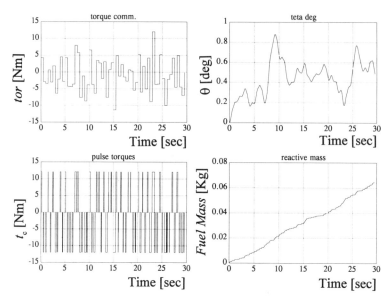

Figure 9.4.10 Time-domain results for a disturbance of 1 N-m with sensor noise of 0.3° (RMS); $\omega_n = 0.5$ rad/sec, $\xi = 0.7$, $\tau = 0.3$ sec, and $T_{sam} = 0.5$ sec.

Let us diminish the sensor noise amplification, and also the fuel consumption, by reducing the bandwidth of the attitude control loop to $\omega_n = 0.3$ rad/sec. Of course, we can increase the time constant of the noise filter to $\tau = 0.7$ sec. The theoretical noise amplification is now only tor/noise = 2.6 (RMS), a reduction by a factor of 4.5. With no noise, the consumption of fuel remains exactly as for the case with $\omega_n = 0.5$ rad/sec, since the fuel consumption rate depends only on the level of disturbance and not on the bandwidth of the control loop. The steady-state error due to the disturbance of 1 N-m is now higher: 1.21°, as predicted by Eq. 9.4.8 (see also Figure 9.4.11). With application of the same sensor noise, the attitude error now increases to about 1.7° with fuel consumption of 0.75 g/sec, which is higher by only 50% than the fuel consumption without sensor noise (see Figure 9.4.12).

The second proposal for decreasing fuel consumption was to insert a dead zone before the PWM. Suppose we use a dead zone of 3 N-m. The results in the time domain with $\omega_n = 0.5$ rad/sec show that the fuel consumption rate is now only 1.1 g/sec, compared to 2 g/sec without the dead zone, and with no significant increase in peak attitude errors.

We conclude that tradeoffs similar to those described here are necessary to achieve the best possible design results for the attitude control problems given under various mission constraints. The results of our analysis are summarized in Table 9.4.2.

Input–Output Behavior of a Reaction ACS

In classical continuous feedback theory, there is a high correlation between (a) the time response $\theta(t)$ of the output to a step input and (b) the closed-loop frequency response $T(s) = \theta(s)/\theta_{com}(s)$, where $\theta_{com}(s)$ is a step command input (see Horowitz 1963, Sidi 1973). This relationship is seriously distorted for sampled systems

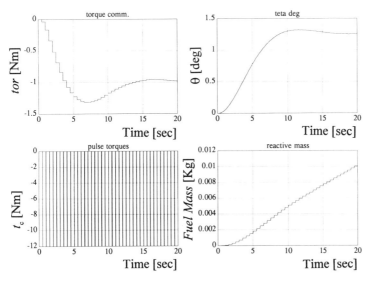

Figure 9.4.11 Time-domain results for a disturbance of 1 N-m without sensor noise; $\omega_n = 0.3$ rad/sec, $\xi = 0.7$, $\tau = 0.7$ sec, and $T_{sam} = 0.5$ sec.

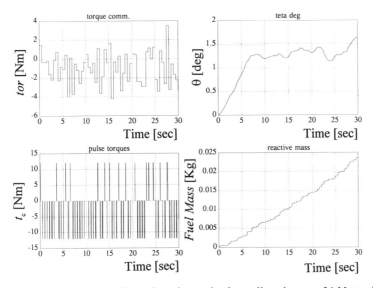

Figure 9.4.12 Time-domain results for a disturbance of 1 N-m with sensor noise of 0.3° (RMS); $\omega_n = 0.3$ rad/sec, $\xi = 0.7$, $\tau = 0.7$ sec, and $T_{sam} = 0.5$ sec.

Table 9.4.2 *Results of tradeoff to reduce sensor noise amplification* ($\xi = 0.7$, disturbance $= D = 1$ N-m)

Figure	ω_n [r/sec]	τ [sec]	K	K_d	Dead Zone [Nm]	Sensor Noise RMS [deg]	Noise Amplific.	Peak Error [deg]	Fuel Rate Consumption [gr/sec]
9.4.9	0.5	0.3	125	350	0	0	---	0.46	0.5
9.4.10	0.5	0.3	125	350	0	0.3	11.85	0.9	2.18
9.4.11	0.3	0.7	45	210	0	0	---	1.3	0.5
9.4.12	0.3	0.7	45	210	0	0.3	2.6	1.7	0.7
	0.5	0.3	125	350	3	0.3	---	1.05	1.17

if no special care is taken when defining the input–output transfer function. The transfer function for a digital, sampled feedback control loop can be designed in various equivalent ways (see Saucedo and Schiring 1968, Sidi 1977). We can put:

$$T(z)_{z=e^{sT_{sam}}} \equiv T^*(s)_{s=j\omega} \equiv T(w)_{w=(z-1)/(z+1)}, \qquad (9.4.9)$$

where * denotes "sampling process" in the continuous frequency domain, "s" is the Laplace transform variable, z is the Z transform variable, and w is the W (bilinear transformation) variable. We also have

$$\frac{\theta(z)}{\theta_{com}(z)} \equiv \frac{\theta^*(s)}{\theta^*_{com}(s)} \equiv \frac{\theta(w)}{\theta_{com}(w)}. \qquad (9.4.10)$$

The frequency response of the sampled transfer function in Eq. 9.4.10 does not correlate well with the step time response, as in continuous systems (see Horowitz 1963). We can alleviate this difficulty by defining a mixed transfer function:

$$T = \frac{\theta(s)}{\theta^*_{com}(s)}. \qquad (9.4.11)$$

The output of interest in Figure 9.4.1 is the continuous output $\theta(t)$, not $\theta^*(t)$. With this practical definition, it has been shown (Sidi 1977) that the approximate relationship between the time response of the output and the frequency response of the transfer function remains as good as for continuous feedback systems. With this assumption, it is much easier to define the desired input–output transfer function in the w frequency domain for a desired time response of a sampled feedback control loop (see Saucedo and Schiring 1968). This approach will be followed here.

Dead Zone and Minimum Impulse Bit of the Reaction Pulses Because of engineering constraints, there is a minimum pulse duration Δt_{min} that reaction thrusters can deliver. The minimum impulse bit $F(\Delta t_{min})$, multiplied by the torque arm Δ of the thrusters, is the minimum torque impulse bit (MTIB) delivered to the satellite at the sampling instance: $\Delta F(\Delta t_{min})$. Between two sampling events, the attitude control system is in open loop. During the period that the torque exists (designated by tp), angular velocity will increase by

$$\Delta\dot{\theta} = \frac{F\Delta tp}{J}, \qquad (9.4.12)$$

where F is the thrust of the reaction thruster, Δ the torque arm, and J the moment of inertia about the axis of rotation. Until the next sampling (in T_{sam} sec), the attitude will increase by

$$\Delta\theta = \frac{F\Delta tp}{J}(T_{sam} - tp). \qquad (9.4.13)$$

This equation is self-explanatory: $\Delta\theta$ is the minimum possible magnitude change of θ. However, it is important to emphasize that precise attitude control will demand very low impulse bits.

Unfortunately, things are not quite so simple. The basic relationship for attainable accuracy stems from Eq. 9.4.4 or Eq. 9.4.6. We shall concentrate on Eq. 9.4.6, the equation of the PW modulator. This modulator will deliver a pulse torque proportional to the torque command calculated in $G(z)$ of Figure 9.4.1. After the transient

9.4 / Reaction Attitude Control via Pulse Width Modulation

period in response to the step command (or whatever command it may be), the steady-state error is $e_{ss} = \theta_{com} - \theta_{ss}$. This error will be translated by $G(z)$ to a steady-state torque command, tor_{ss}. In these special circumstances – when the error is not continuous because of the sampling process – we shall rather speak of a minimum error e_{min} and a minimum torque tor_{min}. For a minimum pulse width tp_{min}, the minimum torque that will activate the PW modulator is

$$tor_{min} = \frac{tp_{min} F \Delta}{T_{sam}} = e_{min} K. \qquad (9.4.14)$$

From this equation it readily follows that

$$e_{min} = \frac{tp_{min} F \Delta}{T_{sam} K} = \frac{\text{MTIB}}{T_{sam} K}. \qquad (9.4.15)$$

In Eq. 9.4.15, MTIB is the *minimum torque impulse bit* that the reaction thruster can deliver to the satellite about the controlled axis. This term deserves some special attention: MIB = $tp_{min} F$ is a technical characteristic of the thruster – namely, its minimum impulse bit. The torque period tp is highly dependent on the chemical composition of the propellant (see Appendix C). The thrust level F also cannot be chosen at will; for instance, F is of the order of 10 N for bipropellant propulsion systems. For monopropellant propulsion systems such as hydrazine thrusters, much lower thrusts are common. Still, the term $tp_{min} F$ is of utmost importance, as can be seen in Eq. 9.4.15. The minimum tp_{min} is equivalent to a torque dead zone, with all the relevant implications such as limit cycling.

The torque impulse bit is related to the torque arm Δ. Practically, this term can be adapted as desired by correctly locating the thruster on the satellite, as explained in Section 9.2.1. The error term e_{min} of Eq. 9.4.15 can also be expressed in terms of ω_n, the natural frequency of the second-order model feedback control system of Figure 9.4.1, in which $K = \omega_n^2 J$.

EXAMPLE 9.4.3 In this example, $J = 500$ k-m^2. We will show the relationship between the steady-state attitude error and the MTIB factor. In order to perceive clearly this error, the closed-loop control system will be subjected to a small attitude command, let us say $\theta_{com} = 0.1°$. We choose $\omega_n = 1$ rad/sec, $\xi = 0.7$, and $\tau = 0.3$ sec, so that $K = 500$, $K_d = 700$, and $T_{sam} = 0.25$ sec.

Figure 9.4.13 (overleaf) shows the time response results for $F = 6$ N, $\Delta = 1$ m, and $tp_{min} = 0.01$ sec: MTIB = $6 \times 0.01 = 0.06$ N-m-sec. With this data, according to Eq. 9.4.15, $e_{min} = 0.06/(0.25 \times 500) = 4.8 \times 10^{-4}$ rad = $0.0275°$. This analytical result is clearly confirmed by the time-domain simulation, the results of which are shown in the figure.

To increase the accuracy, we could decrease F or Δ or both. Suppose that $F = 2$ N. Hence MTIB = 0.02 N-m-sec and, by Eq. 9.4.15, e_{min} is reduced to $e_{min} = 0.0092°$. The time-response results agree with Eq. 9.4.15 (see Figure 9.4.14, overleaf). Suppose once again that $F = 2$ N but that now $\theta_{com} = 1°$. This makes the transient in the error torque command relatively large (as compared to the maximum torque that the reaction thruster can provide), so the torque controller is saturated and hence the time response is more sluggish; see Figure 9.4.15 (page 287). (Of course, the situation could be partially improved by increasing Δ and hence also the torque reaction level

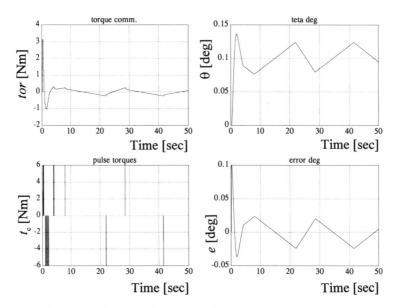

Figure 9.4.13 Time-domain results for the case in which $F\Delta = 6$ N-m, $tp_{min} = 0.01$ sec, $\omega_n = 1$ rad/sec, $K = 500$, $T_{sam} = 0.25$ sec, and $\theta_{com} = 0.1°$.

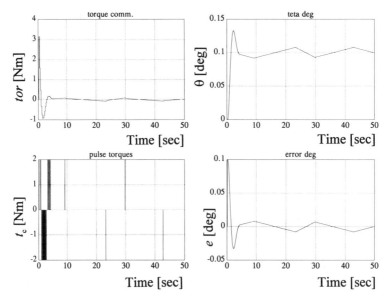

Figure 9.4.14 Time-domain results for the case in which $F\Delta = 2$ N-m, $tp_{min} = 0.01$ sec, $\omega_n = 1$ rad/sec, $K = 500$, $T_{sam} = 0.25$ sec, and $\theta_{com} = 0.1°$.

$F\Delta$. But doing so would increase the steady-state error, which is contrary to our primary desire for attitude accuracy.) To overcome this phenomenon and improve the time response, we must use nonlinear design techniques such as those presented in Section 7.6.

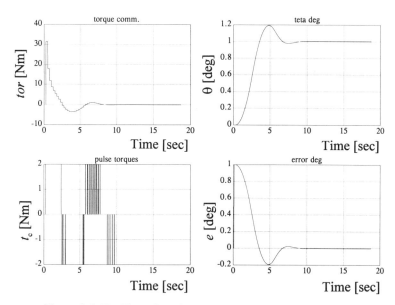

Figure 9.4.15 Time-domain results for the case in which $F\Delta = 2$ N-m, $tp_{min} = 0.01$ sec, $\omega_n = 1$ rad/sec, $K = 500$, $T_{sam} = 0.25$ sec, and $\theta_{com} = 1°$.

It is important to notice that the attitude error and output responses need not follow exactly the time histories shown in Figures 9.4.13, 9.4.14, and 9.4.15. The minimum values of the errors are guaranteed (as per Eq. 9.4.15), but these errors can reach their limiting values at different times according to the initial values $\theta(0)$ and $\dot{\theta}(0)$ of the system and according also to the level of the attitude input command.

9.5 Reaction Control System Using Only Four Thrusters

In Section 9.2 (see esp. Figure 9.2.3), six thrusters were used to implement a reaction control system for attitude and orbit control. For these two purposes, the six thruster jets provided the necessary positive and negative torques about the three body axes (for attitude control) as well as linear velocity changes along the X_B and the $-Y_B$ axes (for orbit control). It is possible to achieve three-axis attitude control by using only four thrusters. However, with such a reaction control system, the possibility of achieving linear velocity augmentation in desired body directions is no longer available. This is the principal drawback in using a limited number of thrusters (fewer than six).

Figure 9.5.1 (overleaf) shows the torque vector arrangement obtained by using four thrusters. As shown in the figure, each thruster provides a positive torque in a given direction with respect to the satellite body; these torques are denoted T_1, T_2, T_3, and T_4. In Figure 9.5.1, with directions of the torques produced by each one of the thrusters defined in the body frame, every vector control command torque $\mathbf{T}_c$ can be achieved with three of the four available torques. Since the torques produced by each thruster have constant torque levels, the on-time must be computed for each of the thrusters Ti_i, $i = 1, \ldots, 4$ (cf. Section 9.2). The transformation between the control torque components in the body axis frame and the on-time of each thruster can be written in the following matrix form:

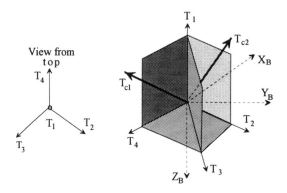

Figure 9.5.1 Tetrahedral torque configuration of four reaction thrusters.

$$\begin{bmatrix} T_{cx} \\ T_{cy} \\ T_{cz} \end{bmatrix} = \begin{bmatrix} a_{x1} & a_{x2} & a_{x3} & a_{x4} \\ a_{y1} & a_{y2} & a_{y3} & a_{y4} \\ a_{z1} & a_{z2} & a_{z3} & a_{z4} \end{bmatrix} \begin{bmatrix} Ti_1 \\ Ti_2 \\ Ti_3 \\ Ti_4 \end{bmatrix} = [\mathbf{A}] \begin{bmatrix} Ti_1 \\ Ti_2 \\ Ti_3 \\ Ti_4 \end{bmatrix}, \quad (9.5.1)$$

where the elements a_{xi}, a_{yi}, a_{zi} of matrix $[\mathbf{A}]$ are the torque component coefficients of each thruster along the body axes.

For a given torque control command vector $\mathbf{T}_c$, we can find the on-time of each thruster by inverting the matrix equality of Eq. 9.5.1. Matrix $[\mathbf{A}]$ is not square, but has the pseudoinverse (see Section 7.3.4)

$$\begin{bmatrix} Ti_1 \\ Ti_2 \\ Ti_3 \\ Ti_4 \end{bmatrix} = \begin{bmatrix} b_{x1} & b_{y1} & b_{z1} \\ b_{x2} & b_{y2} & b_{z2} \\ b_{x3} & b_{y3} & b_{z3} \\ b_{x4} & b_{y4} & b_{z4} \end{bmatrix} \begin{bmatrix} T_{cx} \\ T_{cy} \\ T_{cz} \end{bmatrix} = [\mathbf{B}] \begin{bmatrix} T_{cx} \\ T_{cy} \\ T_{cz} \end{bmatrix}, \quad (9.5.2)$$

where the matrix $[\mathbf{B}]$ defines the transformation between the torque command components in the body frame and the on-time of each thruster. In the general case, for any desired $\mathbf{T}_c$, only one of the Ti_i can be negative. Negative values cannot be physically realized, since the thruster cannot be activated for a negative time and there is no thruster able to produce a torque of opposite direction as in the six-thruster system analyzed in Section 9.2.

In Figure 9.5.1, the four thruster torques divide the space around the satellite's cm into four regions, each one bounded by three surfaces defined by three thruster torque vectors. Any control vector $\mathbf{T}_c$ can be located in only one of the four space regions, and can be expressed by only the positive components of the torques defining that region. This means that the $\mathbf{T}_c$ can be achieved by using just three thrusters, demanding only positive on-times. Consequently, by eliminating one of the thrusters and its related row in matrix $[\mathbf{A}]$ of Eq. 9.5.1, we can obtain four matrix relations of reduced order 3×3, each pertaining to one of the four regions of the torque space. The pertinent matrices will be called $[\mathbf{A}_j]$ and their inverses will be called $[\mathbf{C}_j]$, where the index j stands for the thruster j that was eliminated, $j = 1, ..., 4$. For example, $[\mathbf{C}_3]$ will have the form

$$\begin{bmatrix} Ti_1 \\ Ti_2 \\ Ti_4 \end{bmatrix} = \begin{bmatrix} c_{x1} & c_{y1} & c_{z1} \\ c_{x2} & c_{y2} & c_{z2} \\ c_{x3} & c_{y3} & c_{z3} \end{bmatrix} \begin{bmatrix} T_{cx} \\ T_{cy} \\ T_{cz} \end{bmatrix} = [\mathbf{C}_3] \begin{bmatrix} T_{cx} \\ T_{cy} \\ T_{cz} \end{bmatrix}. \qquad (9.5.3)$$

With these definitions, there are two ways to prepare an operational algorithm. In the first, we must determine which one of the space regions contains the desired control vector $\mathbf{T}_c$ to be produced; then use is made of the matrix transformation $[\mathbf{C}_j]$ pertinent to that region, as in Eq. 9.5.3 for the case $j = 3$. A more straightforward method is to use the four previously prepared transformations $[\mathbf{C}_j]$ and compute the on-times for each. For a given $\mathbf{T}_c$, only one of the transformations will come out with three positive Tis; this will be the desired solution. For example: in Figure 9.5.1, $\mathbf{T}_{c1}$ is achieved by use of thrusters 1, 3, and 4, whereas $\mathbf{T}_{c2}$ is achieved by use of thrusters 1, 2, and 3.

9.6 Reaction Control and Structural Dynamics

Structural dynamics is especially important with reaction control because the commanded torques are in many missions strong impulses, that is, pulses that inherently excite the structural vibration modes. (In general, the same problem exists also for other kinds of torque controllers, with the relieving condition that the commanded torques are not impulsive.) The structural dynamics effect in attitude control systems will be treated in Chapter 10.

9.7 Summary

This chapter treated the most basic control problems in reaction attitude control feedback systems. An important advantage of reaction controllers is the high level of torques that can be obtained, which are dearly needed in certain control tasks. With careful design approaches, accurate attitude tracking is also achievable. Design techniques include decreasing the dead zones and the minimum impulse bit of the reaction controller as well as increasing the bandwidth of the control loop.

The principal handicap of reaction controllers is their excessive consumption of fuel mass, which cannot be replenished once the s/c is placed into orbit.

References

Bittner, H., Fisher, H., and Surauer, M. (1982), "Design of Reaction Jet Attitude Control Systems for Flexible Spacecraft," *IFAC Automatic Control in Space*. Noordwijk, Netherlands: IFAC, pp. 373-98.
Chan, N. (1982), Ford Aerospace and Communications Corporation Patent Application for "Digital PWPF Three-Axis Spacecraft Attitude Control," US(SN) 407-196, filed 11 August.
D'Azzo, J., and Houpis, C. (1988), *Linear Control System Analysis and Design, Conventional and Modern*. New York: McGraw-Hill.
Franklin, F., and Powell, J. (1980), *Digital Control of Dynamic Systems*. Reading, MA: Addison-Wesley.
Horowitz, I. (1963), *Synthesis of Feedback Systems*. New York: Academic Press.
Houpis, H., and Lamont, G. (1985), *Digital Control Systems: Theory, Hardware, and Software*. New York: McGraw-Hill.

Joice, M., and Clarke, K. (1961), *Transistor Circuit Analysis.* Reading, MA: Addison-Wesley.

Kuo, C. (1970), *Discrete-Data Control Systems.* Englewood Cliffs, NJ: Prentice-Hall.

Millman, J., and Taub, H. (1956), *Pulse and Digital Circuits.* New York: McGraw-Hill.

Saucedo, R., and Schiring, E. (1968), *Introduction to Continuous and Digital Control Systems.* New York: Macmillan.

Sidi, M. (1973), "Synthesis of Feedback Systems with Large Plant Ignorances for Prescribed Time-Domain Tolerances," Ph.D. dissertation, Weizman Institute of Science, Rehovot, Israel.

Sidi, M. (1977), "Synthesis of Sampled Feedback Systems for Prescribed Time Domain Tolerances," *International Journal of Control* 26(3): 445–61.

Sidi, M. (1980), "On Maximization of Gain-Bandwidth in Sampled Systems," *International Journal of Control* 32(6): 1099–1109.

Vaeth, J. (1965), "Compatibility of Impulse Modulation Techniques with Attitude Sensor Noise and Spacecraft Maneuvering," *IEEE Transactions on Automatic Control* 10: 67–76.

Wie, B., and Plescia, C. (1984), "Attitude Stabilization of Flexible Stationkeeping Maneuvers," *Journal of Guidance, Control, and Dynamics* 7(4): 430–6.

CHAPTER 10

Structural Dynamics and Liquid Sloshing

10.1 Introduction

A typical spacecraft structure consists of two principal parts. The first one is the *body* of the spacecraft, which contains all the payload instrumentation and control hardware pertaining to the Attitude and Orbit Control System (AOCS). Its structure must be very rigid in order to withstand mechanical loads during the launch stage, and also to assure correct positioning of the control torquers and attitude sensors for achieving the necessary pointing accuracy in the s/c mission stage. The necessity to save on weight leads to mechanical design tradeoffs between weight and rigidity, which results in a body that is only *quasirigid* and in which structural vibration modes should be anticipated.

The second part of the spacecraft structure consists of large flexible appendages: parabolic antennae, large synthetic-aperture radar, and very large flexible solar arrays built from light materials in order to reduce their weight. Spacecraft structures are now becoming extremely complicated because of these appendages, which also induce structural oscillation under the excitation of external torques and forces. Finally, we must likewise consider vibrations due to the liquid contained in fuel tanks. The vibrational dynamics evolving from flexible appendages and liquid sloshing interferes strongly with the attitude control dynamics, and puts severe limitations on the achievable qualities of the attitude control system (ACS).

This chapter is primarily concerned with this second class of structural vibrations. First, we must write simplified structural dynamics models for solar panels and liquid sloshing before coupling them to the rigid-body attitude dynamics of spacecraft. An analytical model for the rigid body and the flexible structural modes will allow us to evaluate the limitations on the bandwidth of the ACS.

10.2 Modeling Solar Panels

Although the techniques of structural modeling described in this chapter are applicable to any nonrigid appendage to the solid body, we will concentrate on the structural dynamics of solar panels, which are common to almost all types of satellites. Our basic task is to create a simplified model that adequately represents the real structure and is easy to use for design purposes.

10.2.1 *Classification of Techniques*

There exist four basic modeling techniques: (1) distributed parameter modeling; (2) discrete parameter modeling; (3) *N*-body modeling; and (4) finite element modeling. A short conceptual explanation of these techniques follows (see also Williams and Wood 1989).

Distributed Parameter Modeling In many cases, the structure of the satellite can be seen as a rigid central body to which one or more solar panels, resembling cantilever plates, are attached. Deformations of these plates are described in terms of distributed coordinates. Excellent texts on vibration of cantilever plates include Thomson (1988) and Weaver, Timoshenko, and Young (1990). With this technique of modeling the entire structure, the equations of motion are expressed in terms of partial differential equations for the cantilever plates and by ordinary differential equations for the motion of the overall spacecraft.

Discrete Parameter Modeling In this method, the flexible appendages are modeled as a number of point masses, interconnected by spring elements. The stiffness of the elements is expressed in terms of influence coefficients, which are easily evaluated for simple structures such as cantilever rods and plates. Inversion of the matrix of influence coefficients provides the structural stiffness matrix, which we will use in this chapter and demonstrate with some simple examples.

Structural deformations are regarded as small, so that the equations of motion are linear ordinary differential equations. In this context, the equations of motion for liquid sloshing are easily incorporated within the remaining equations of motion of the rigid body and the solar panel dynamics.

N-Body Modeling With this technique, the structure is seen as a series of connected rigid bodies. Each body is modeled as a mass with a rotational inertia.

Finite Element Modeling In this approach, the previous two techniques for modeling are used conjointly. The finite elements may be modeled separately as lumped spring–mass elements, or as "distributed parameters."

In the next section, modeling of the solar panels is treated with several examples. It will be instructive to solve them using various analytic approaches, in order to familiarize ourselves with the techniques involved. A generalized approach for modeling structural and sloshing dynamics together will be presented at the end of this chapter.

10.2.2 *The Lagrange Equations and One-Mass Modeling*

As a first introduction to structural dynamics, let us solve the dynamic equations of the discrete model shown in Figure 10.2.1. In this simple example, the panels

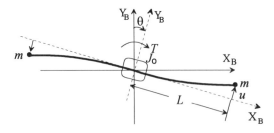

Figure 10.2.1 Discrete model consisting of a rigid body and two symmetrical panels, each modeled by a single mass m; reproduced from Williams (1976) by permission of C. G. Williams.

10.2 / Modeling Solar Panels

attached to the rigid body are modeled by a single discrete mass m, positioned at a distance L from the center of mass (cm) of the rigid body. The moment of inertia of the rigid body is J_0. A torque T is applied about the axis of rotation, thus exciting only the antisymmetric elastic mode. (With an additional force applied along the Y_B body axis, the symmetric elastic mode will also be excited; this model is not analyzed in the present example.) Since the panel is not rigid, a deformation u with respect to the rigid-body axis is to be anticipated. The mass m will experience two motions:

(1) a motion θ with the rigid body, with linear velocity $\dot{\theta}L$; and
(2) a deformation u from the rigid-body axis X_B, with velocity $\dot{u}$.

For small deformations, both velocities will be collinear; hence the mass m will have a velocity of

$$v = \dot{u} + L\dot{\theta}. \tag{10.2.1}$$

We shall solve this example by using Lagrange's method. Thus, the kinetic and potential energies E_k and E_p are to be written for the entire system, including the two panels (represented by two masses m) and the rigid body (represented by its moment of inertia J_0):

$$E_k = \tfrac{1}{2}J_0\dot{\theta}^2 + 2(\tfrac{1}{2}mv^2). \tag{10.2.2}$$

Together with Eq. 10.2.1, we have

$$E_k = \tfrac{1}{2}J_0\dot{\theta}^2 + m(\dot{u} + L\dot{\theta})^2 \quad \text{and} \tag{10.2.3}$$

$$E_p = 2\left[\frac{u^2}{2}K\right]. \tag{10.2.4}$$

We shall also define a dissipation function D, proportional to half the rate at which energy is dissipated. Taking into account both panels, we have

$$D = 2\left(\frac{\dot{u}^2}{2}K_d\right), \tag{10.2.5}$$

where K_d is the mechanical dissipative constant.

The most general form of Lagrange's equations is

$$\frac{d}{dt}\left(\frac{\partial E_k}{\partial \dot{q}_i}\right) - \frac{\partial E_k}{\partial q_i} + \frac{\partial E_p}{\partial q_i} + \frac{\partial D}{\partial \dot{q}_i} = Q_i \quad (i = 1, \ldots, n), \tag{10.2.6}$$

where the q_i are generalized coordinates and Q_i denotes the generalized forces or torques acting on the i station. Lagrange's equations for this example are as follows:

$$\frac{\partial E_k}{\partial \dot{\theta}} = J_0\dot{\theta} + 2mL(\dot{u} + L\dot{\theta}), \tag{10.2.7}$$

$$\frac{d}{dt}\left(\frac{\partial E_k}{\partial \dot{\theta}}\right) = J_0\ddot{\theta} + 2mL(\ddot{u} + L\ddot{\theta}) = (J_0 + 2mL^2)\ddot{\theta} + 2mL\ddot{u}, \tag{10.2.8}$$

$$\frac{\partial E_k}{\partial \dot{u}} = 2m(\dot{u} + L\dot{\theta}), \tag{10.2.9}$$

$$\frac{d}{dt}\frac{\partial E_k}{\partial \dot{u}} = 2m(\ddot{u} + L\ddot{\theta}), \tag{10.2.10}$$

$$\frac{\partial E_p}{\partial u} = 2uK, \tag{10.2.11}$$

$$\frac{\partial D}{\partial \dot{u}} = 2\dot{u}K_d, \tag{10.2.12}$$

$$\frac{\partial E_k}{\partial \theta} = \frac{\partial E_k}{\partial u} = \frac{\partial E_p}{\partial \theta} = \frac{\partial D}{\partial \dot{\theta}} = \frac{\partial D}{\partial u} = 0. \tag{10.2.13}$$

From Eqs. 10.2.7–10.2.13, the following equations follow:

$$(J_0 + 2mL^2)\ddot{\theta} + 2mL\ddot{u} = T, \tag{10.2.14}$$

$$m\ddot{u} + K_d\dot{u} + Ku = -mL\ddot{\theta}. \tag{10.2.15}$$

In Eq. 10.2.14, J_0 is the moment of inertia of the rigid body. We shall denote by $J_p = mL^2$ the moment of inertia of one of the two (panel) masses about the cm. Let us define $J = J_0 + 2mL^2$. This is the moment of inertia of the entire system, including the two panels. With this definition, Eq. 10.2.14 can be rewritten as:

$$J\ddot{\theta} + 2mL\ddot{u} = T, \tag{10.2.14'}$$

where $J = J_0 + 2mL^2 = J_0 + 2J_p$.

Equation 10.2.14 and Eq. 10.2.15 are both linear ordinary differential equations, so they can be solved by Laplace transformations. Simple algebraic manipulation leads to the final solution for $\theta(s)$ as a transfer function in the Laplace domain ("s" is the Laplace variable):

$$\frac{\theta(s)}{T(s)} = \frac{1}{J_0 s^2} \frac{s^2 + \frac{K_d}{m}s + \frac{K}{m}}{s^2 + \frac{J}{J_0}\frac{K_d}{m}s + \frac{J}{J_0}\frac{K}{m}} = \frac{1}{J_0 s^2} \frac{s^2 + 2\xi\sigma s + \sigma^2}{s^2 + 2\xi_S\sigma_S s + \sigma_S^2}. \tag{10.2.16}$$

Let us define as $\sigma = \sqrt{K/m}$ the *cantilever natural frequency* of the panel, and let $\xi = (K_d/2)\sqrt{1/Km}$ be its damping coefficient. In the transfer function of Eq. 10.2.16, the cantilever natural modal frequency appears in the numerator. This modal frequency is independent of the rigid-body characteristics. The pole in the denominator is the modal frequency of the entire system, which depends on the ratio between the moments of inertia of the rigid body plus panels to the moment of inertia of the rigid body only. We label this modal system frequency σ_S, and set $\sigma_S = \sqrt{J/J_0}\sigma$. Also, $\xi_S = \sqrt{J/J_0}\xi$ is the damping coefficient of the denominator pole. The higher the ratio, the higher the modal frequency σ_S and the damping coefficient ξ_S.

According to these definitions, $\sigma_S > \sigma$ and also $\xi_S > \xi$. From the automatic control perspective, the second-order dipole in Eq. 10.2.16 has a destabilizing effect, which increases with the panel's moment of inertia. With very small moments of inertia of the panels, the dipole disappears. Moreover, the low-frequency (DC) gain of the system's transfer function decreases owing to the existence of the added panels, which means that the equivalent moment of inertia increases from J_0 to $J = J_0(1 + 2J_p/J_0)$.

We have assumed that the damping parameter K_d is known from analytical data. This is generally not so; in practice, the damping coefficient ξ is derived from physical tests performed on the cantilever panel. Practical values for ξ are generally very low, of the order of $1-5 \times 10^{-3}$, depending on the amplitude of oscillations.

10.2 / Modeling Solar Panels

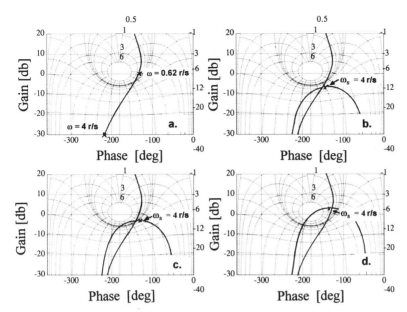

Figure 10.2.2 Frequency response of the open-loop transfer function, including one elastic mode at $\omega_{EM} = 4$ rad/sec and $\xi = 0.003, 0.002$, and 0.001 in parts (b), (c), and (d).

EXAMPLE 10.2.1 Let us choose a satellite with $J_0 = 720$ kg-m² to which are appended two solar panels, weighing 20 kg each, whose centers of mass are located 2 m from the satellite's rigid-body center of mass. We will find the transfer function of the satellite's angular motion about its cm to an applied external torque.

First we compute the panels' moments of inertia $J_p = mL^2 = 40$ kg-m²; adding the rigid body's moment of inertia yields $J = J_0 + (2 \times 40) = 800$ kg-m². If the cantilever's modal eigenvalue is $\sigma = \sqrt{K/m} = 4$ rad/sec, then $k = 320$ kg-rad²/sec². Suppose also that the damping coefficient of the panel was measured and found to be $\xi = 0.003$. With this information, for the structural model we obtain $K_d = 2\xi\sigma m = 0.48$ kg-rad/sec. According to Eq. 10.2.16, the denominator will consist of a second-order pole with an eigenfrequency of $\sigma_S = \sigma\sqrt{J/J_0} = 1.541$ with $\sigma = 4.2164$ rad/sec and a damping coefficient of $\xi_S = \xi\sqrt{J/J_0} = 0.00316$.

To acquire a feel for the complications emerging from the structural dipole now appearing in the attitude control loop, we set the open-loop crossover frequency at $\omega_{co} = 0.5$ rad/sec. With a conventional frequency domain design and without the existing structural dipole, it is easy to obtain an adequate loop gain as shown in Figure 10.2.2.a.

A simple control network $G(s) = 2,000(s+0.1)/(s^2+4s+4)$ provides the necessary gain and phase margins. The existence of the structural mode produces a peak at the modal frequency, thus decreasing the gain margin but still maintaining an acceptable control design (Figure 10.2.2.b). However, the situation is drastically aggravated by a decrease in the damping coefficient of the panels' cantilever model. In Figure 10.2.2.c, $\xi = 0.002$ and the system is scarcely stable. If the damping coefficient is decreased to $\xi = 0.001$ then the system verges on instability (Figure 10.2.2.d).

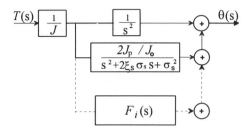

Figure 10.2.3 Dynamic block diagram, with the option of adding additional transfer functions pertaining to higher-order structural modes.

In practical situations the damping coefficient is not known exactly, so an uncertainty of the panel model must be taken into consideration when designing an ACS. It is feasible to add the necessary phase lag at the frequency of the peak in Figure 10.2.2, and to assure stability. A single elastic mode can be adequately controlled, especially if its characteristics are known exactly. Unfortunately, they are not; a satellite's moment of inertia J is altered during its life in space because of fuel consumption. This changes the dipole's peak location in the frequency domain.

Moreover, the situation is very much aggravated by the existence of additional structural modes and by sloshing dynamics (to be treated in subsequent sections). In this case, adjusting the location of the structural mode's peak in the frequency domain, as shown in the figure, might be problematic owing to the existence of additional oscillating modes. A tradeoff, satisfying stability margins for all the structural modes, generally compels the designer to decrease the overall bandwidth of the control system, with an inevitable increase in sensitivity to external disturbances and a decrease in attitude accuracy (see also the examples to follow).

In Example 10.2.1 we presented the panel as a single point mass located in the geometrical center of the panel. A single mass represents one single structural mode. It is common to express the transfer function of Eq. 10.2.16 in partial fraction form. The result is

$$\frac{\theta(s)}{T(s)} = \frac{1}{Js^2} + \frac{2J_p}{JJ_0}\frac{1}{s^2 + 2\xi_S\sigma_S s + \sigma_S^2} = \frac{1}{J}\left[\frac{1}{s^2} + \frac{2J_p/J_0}{s^2 + 2\xi_S\sigma_S s + \sigma_S^2}\right]. \quad (10.2.17)$$

A block-diagram representation of Eq. 10.2.17 is shown in Figure 10.2.3. Equation 10.2.17 can be rewritten in a more compact form as

$$\frac{\theta(s)}{T(s)} = \frac{1}{J}\left[\frac{1}{s^2} + F_1(s)\right]. \quad (10.2.18)$$

The definition of $F_i(s)$ is self-evident from Eq. 10.2.17 and Figure 10.2.3. The figure clearly shows how additional higher-order structural modes (explicated in Section 10.2.3) can be added to the attitude dynamics equations of motion.

10.2.3 The Mass–Spring Concept and Multi-Mass Modeling

In this technique, each flexible panel will be represented by two (or more) discrete masses, m_1 and m_2, as shown in Figure 10.2.4. Modeling the panel with two

10.2 / Modeling Solar Panels

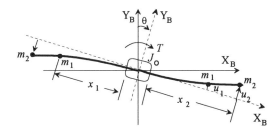

Figure 10.2.4 Discrete model consisting of a rigid body and two symmetrical panels, each modeled by two masses; reproduced from Williams (1976) by permission of C. G. Williams and British Aircraft Corp.

discrete masses is an appropriate way to show how the "influence coefficients" in vibration theory are defined and calculated (see e.g. Weaver et al. 1990). This will also allow us to introduce the formal matrix solution for structural dynamics.

In Figure 10.2.4, each of the panels is represented by a cantilever beam, further detailed in Figure 10.2.5.a. It is assumed that the beam is massless, with *flexural rigidity* of EI, and with two concentrated masses m_1 and m_2 located at x_1 and x_2. The masses can be located anywhere along the beam, but for simplicity we shall assume that $x_1 = L/2$ and $x_2 = L$.

The force-deflection equation of a massless cantilever beam is:

$$y = \frac{F x_f^3}{6EI}\left[3\left(\frac{x}{x_f}\right)^2 - \left(\frac{x}{x_f}\right)^3\right], \tag{10.2.19}$$

where F is the force applied at station x_f and y is the deflection at a distance x. This equation holds for $x_f > x$. (The variables y and u will be used interchangeably; y is sometimes used when compliance with terminology from existing literature on vibrational dynamics adds some clarity to the explanations.)

The flexibility coefficients are defined as the deflection y of a point located at x that is caused by a unit force F applied at a point x_f, with all other forces equal to

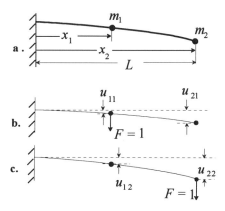

Figure 10.2.5 Idealized panel modeled as a cantilever beam with two discrete masses; adapted from Weaver, Timoshenko, and Young (1990) by permission of John Wiley & Sons.

zero (see Thomson 1988). To derive the required flexibility coefficients, we first apply a force $F_1 = 1$ at m_1, where $x_f = L/2$ for point 1 and $x_2 = L$ for point 2 (see Figure 10.2.5.b). Use of Eq. 10.2.19 yields the deflections

$$y_{11} = \frac{L^3}{24EI} \quad \text{and} \quad y_{21} = \frac{5L^3}{48EI}. \tag{10.2.20}$$

Next, the unit force $F = 1$ is applied at the mass m_2, where $x_f = L$; we obtain

$$y_{12} = \frac{5L^3}{48EI} \quad \text{and} \quad y_{22} = \frac{L^3}{3EI} \tag{10.2.21}$$

(see Figure 10.2.5.c). According to the reciprocity theorem (Thomson 1988), $y_{ij} = y_{ji}$. Hence $y_{21} = y_{12}$ from Eq. 10.2.21. With the calculated flexibility coefficients, we can write the *flexibility matrix*

$$[\mathbf{Y}] = \frac{L^3}{48EI} \begin{bmatrix} 2 & 5 \\ 5 & 16 \end{bmatrix}. \tag{10.2.22}$$

The stiffness coefficients are the reciprocal of the flexibility coefficients, and the *stiffness matrix* $[\mathbf{K}]$ is the inverse of the flexibility matrix:

$$[\mathbf{K}] = [\mathbf{Y}]^{-1} = \frac{48EI}{7L^3} \begin{bmatrix} 16 & -5 \\ -5 & 2 \end{bmatrix}. \tag{10.2.23}$$

Having found the stiffness matrix, we can proceed directly to calculate the dynamic equations of the body plus solar panels. For the satellite–panel system we have

$$J\ddot{\theta} = T + 2(m_1 x_1 \ddot{u}_1 + m_2 x_2 \ddot{u}_2). \tag{10.2.24}$$

For the appended panels only, in matrix form we have

$$\begin{bmatrix} m_1 & 0 \\ 0 & m_2 \end{bmatrix} \begin{bmatrix} \ddot{u}_1 \\ \ddot{u}_2 \end{bmatrix} + \frac{48EI}{7L^3} \begin{bmatrix} 16 & -5 \\ -5 & 2 \end{bmatrix} \begin{bmatrix} u_1 \\ u_2 \end{bmatrix} = \ddot{\theta} \begin{bmatrix} m_1 & 0 \\ 0 & m_2 \end{bmatrix} \begin{bmatrix} x_1 \\ x_2 \end{bmatrix}. \tag{10.2.25}$$

Equation 10.2.25 can be put in a more compact form as follows:

$$[\mathbf{M}][\ddot{\mathbf{u}}] + [\mathbf{D}][\dot{\mathbf{u}}] + [\mathbf{K}][\mathbf{u}] = \ddot{\theta}[\mathbf{M}][\mathbf{x}], \tag{10.2.26}$$

where $[\mathbf{M}]$ is the mass matrix and a dissipative matrix $[\mathbf{D}]$ has been added to account for the damping factors of the cantilever beam (i.e., the solar panels in our case). Equation 10.2.24 and Eq. 10.2.26 can be solved simultaneously to calculate the dynamics of the rigid body and structural elements together. Since two masses have been stipulated in the model, only two structural vibration modes will emanate from the solution of the two equations.

The matrices $[\mathbf{M}]$, $[\mathbf{D}]$, and $[\mathbf{K}]$ can be augmented to include the dynamics of Eq. 10.2.24. Equation 10.2.26 will become

$$[\bar{\mathbf{M}}] \begin{bmatrix} \ddot{\theta} \\ \ddot{u}_1 \\ \ddot{u}_2 \end{bmatrix} + [\bar{\mathbf{D}}] \begin{bmatrix} \dot{\theta} \\ \dot{u}_1 \\ \dot{u}_2 \end{bmatrix} + [\bar{\mathbf{K}}] \begin{bmatrix} \theta \\ u_1 \\ u_2 \end{bmatrix} = \begin{bmatrix} T \\ 0 \\ 0 \end{bmatrix}, \tag{10.2.27}$$

where $[\bar{\mathbf{M}}]$, $[\bar{\mathbf{D}}]$, and $[\bar{\mathbf{K}}]$ are the augmented matrices. This differential matrix equation is linear, and can be solved by use of Laplace transforms. Taking the Laplace transforms of the variables θ, u_1, and u_2, we have:

$$\{s^2[\bar{\mathbf{M}}] + s[\bar{\mathbf{D}}] + [\bar{\mathbf{K}}]\}[\theta \ u_1 \ u_2]^T = [T \ 0 \ 0]^T. \tag{10.2.28}$$

The result is a 3 × 3 matrix whose terms are polynomials in s, of up to second order, from which the vector $[\theta \ u_1 \ u_2]^T$ can be solved. The determinant of the solution will be a polynomial in s of the sixth order, pointing to the existence of two integrators of the rigid body, together with two second-order poles of the structural vibration modes.

10.3 Eigenvalues and Eigenvectors

The structural analysis in Section 10.2.2 is based on modeling the panels as concentrated masses located along the panels' axes. The stiffness coefficients have a meaning similar to the elasticity coefficient k of a spring, also called the *stiffness constant*. In fact, many additional structural problems connected to satellite dynamics can be modeled in terms of spring–mass elements, for instance, the fuel slosh dynamics treated in Section 10.4. In this section we analyze a simple two-body mass–spring problem in order to demonstrate and reinforce the important notions of eigenvalues and eigenvectors.

Consider the two mass–spring elements in Figure 10.3.1. The equations of motion of the two bodies are easily written for m_1 and m_2 as follows:

$$m_1 \ddot{u}_1 + k_1 u_1 - k_2(u_2 - u_1) = F_1,$$
$$m_2 \ddot{u}_2 + k_2(u_2 - u_1) = F_2. \tag{10.3.1}$$

Equations 10.3.1 can be put in the following matrix form:

$$\begin{bmatrix} m_1 & 0 \\ 0 & m_2 \end{bmatrix} \begin{bmatrix} \ddot{u}_1 \\ \ddot{u}_2 \end{bmatrix} + \begin{bmatrix} k_1 + k_2 & -k_2 \\ -k_2 & k_2 \end{bmatrix} \begin{bmatrix} u_1 \\ u_2 \end{bmatrix} = \begin{bmatrix} F_1 \\ F_2 \end{bmatrix}; \tag{10.3.2}$$

alternatively,

$$[\mathbf{M}][\ddot{\mathbf{u}}] + [\mathbf{K}][\mathbf{u}] = [\mathbf{F}]. \tag{10.3.2'}$$

As before, [**K**] is the stiffness matrix consisting of the *stiffness influence coefficients* k_{ij}; [**M**] is the *mass matrix*, which contains in its diagonal the masses m_1 and m_2. The k_{ij} coefficients can be evaluated in a systematic way. They can be computed by inducing unit displacements in each of the displacement coordinates u_i one at a time, and calculating the required holding forces F_i as in Section 10.2.2. To illustrate the process, suppose that a unit displacement u_1 is induced, while $u_2 = 0$. The static holding forces required for this situation will be k_{11} and k_{21}. It is easily seen from the figure that $k_{11} = k_1 + k_2$ and $k_{21} = -k_2$. They constitute the first column of the stiffness matrix. To find k_{22} and k_{12}, suppose that $u_1 = 0$ and $u_2 = 1$. In this case, the holding forces are $k_{12} = -k_2$ and $k_{22} = k_2$. Note that $k_{12} = k_{21}$. Equation 10.3.2' is sometimes referred to as a set of *action equations of motion* (see Weaver et al. 1990).

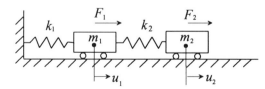

Figure 10.3.1 A simple two-body vibration problem.

The dynamic equations can also be written in terms of the flexibility coefficients and the flexibility matrix [**Y**]; see Section 10.2.2 and Eq. 10.2.22. To find the elements of the flexibility matrix, it is convenient to work with *displacement equations of motion* instead of the action equations. To illustrate this approach, suppose first that, in Figure 10.3.1, a unit force $F_1 = 1$ is applied to m_1 while $F_2 = 0$. The static displacements u_1 and u_2 are easily computed to be $u_1 = 1y_{11} = 1/k_1$ and $u_2 = 1y_{21} = 1/k_1$. Next, in Figure 10.3.1, induce $F_2 = 1$ while $F_1 = 0$. In this case, the flexibilities are $y_{12} = 1u_1 = 1/k_1$ and $y_{22} = u_1 + u_2 = 1/k_1 + 1/k_2$. Knowledge of the elements y_{ij} allows us to find the inverse of the flexibility matrix $[\mathbf{Y}]^{-1} = [\mathbf{K}]$; see Eq. 10.3.2.

Let us return to Eq. 10.3.2′. It is easy to solve this equation using Laplace transformations. The resulting matrix in the Laplace domain becomes:

$$\begin{bmatrix} s^2 m_1 + k_1 + k_2 & -k_2 \\ -k_2 & s^2 m_2 + k_2 \end{bmatrix} \begin{bmatrix} u_1(s) \\ u_2(s) \end{bmatrix} = \begin{bmatrix} F_1(s) \\ F_2(s) \end{bmatrix}. \quad (10.3.3)$$

The solution for the vector [**u**] is obtained simply by inversion of the matrix:

$$\begin{bmatrix} u_1(s) \\ u_2(s) \end{bmatrix} = \frac{1}{\Delta(s)} \begin{bmatrix} s^2 m_2 + k_2 & k_2 \\ k_2 & s^2 m_1 + k_1 + k_2 \end{bmatrix} \begin{bmatrix} F_1(s) \\ F_2(s) \end{bmatrix}, \quad (10.3.4)$$

where the determinant is of the fourth order,

$$\Delta(s) = s^4 m_1 m_2 + s^2 (m_1 k_2 + m_2 k_1 + m_2 k_2) + k_1 k_2. \quad (10.3.5)$$

Since the coefficients of the uneven power terms in s are zero, the roots can be easily found. The roots are the *eigenvalues* of the system. For $s^2 = \lambda$ with $s = j\omega$, we find

$$\lambda_{1,2} = s_{1,2}^2 = -\omega_{1,2}^2 = \frac{-b \pm \sqrt{b^2 - 4ac}}{2a}, \quad (10.3.6)$$

where $a = m_1 m_2$, $b = m_1 k_2 + m_2(k_1 + k_2)$, and $c = k_1 k_2$. The expression under the square root sign is always positive ($b^2 - 4ac > 0$), since by simple algebraic manipulation it can be expressed as $[m_1 k_2 - m_2(k_1 + k_2)]^2 + 4m_1 m_2 k_2^2 > 0$. It follows that $\lambda_{1,2}$ is always real. Also, $4ac > 0$, so both roots are negative. As a result, the determinant of Eq. 10.3.5 can be put into the form

$$\Delta(s) = m_1 m_2 (s^2 + s_1^2)(s^2 + s_2^2). \quad (10.3.7)$$

Thus, the characteristic equation comprises two natural frequencies of angular oscillations, ω_1 and ω_2, that depend on the physical properties of the system (m_1, m_2, k_1, and k_2). These are the *modal* frequencies.

We now investigate the free vibration of this system. Set $F_1 = F_2 = 0$, and for each of the eigenvalues assume a solution of the form $u_1 = A \sin(\omega_i t + \delta_i)$ and $u_2 = B \sin(\omega_i t + \delta_i)$. For instance, for the first modal frequency ω_1, substituting u_1 and u_2 into Eq. 10.3.1 yields the ratio of the A and B amplitudes:

$$r_1 = \frac{A_1}{B_1} = \frac{k_2}{k_1 + k_2 - m_1 \omega_1^2} = \frac{k_2 - m_2 \omega_1^2}{k_2}. \quad (10.3.8)$$

Similarly, for the second eigenvalue we obtain the ratio

$$r_2 = \frac{A_2}{B_2} = \frac{k_2}{k_1 + k_2 - m_1 \omega_2^2} = \frac{k_2 - m_2 \omega_2^2}{k_2}. \quad (10.3.9)$$

The amplitude ratios r_1 and r_2 represent the mechanical shapes of the two natural modes of vibration of the system. Using Eq. 10.3.1 for the two different natural frequencies of oscillation, we cannot calculate the amplitudes A and B but only their ratio. In fact, the vector $[A\ B]^T$ is the eigenvector for the eigenvalue. There are two such vectors – namely, $[A_1\ B_1]^T$ pertaining to ω_1 and $[A_2\ B_2]^T$ pertaining to ω_2. The *eigenvectors* are the modal shapes of the relevant eigenvalues. The following numerical example will demonstrate the notion of an eigenvector.

EXAMPLE 10.3.1 Suppose that $m_1 = 2m_2 = 2$ kg and $k_1 = k_2 = 1$ kg/sec^2. We find that $a = 2 \times 1 = 2$, $b = -[2 \times 1 + 1 \times (1+1)] = -4$, and $c = 1 \times 1 = 1$; hence

$$\omega_{1,2}^2 = \frac{4 \pm \sqrt{16-8}}{4} = 1 \pm \tfrac{1}{2}\sqrt{2},$$

$\omega_1^2 = 0.293$, and $\omega_2^2 = 1.707$. Finally, we use Eq. 10.3.8 and Eq. 10.3.9 to calculate $r_1 = 0.707$ and $r_2 = -0.707$. Interpretation of the displacements r_1 and r_2 is as follows. We normalize the amplitudes of u_1 and u_2 so that $B = 1$, which means that if $B = 1$ at ω_1 then the amplitude $A = 0.293$. For the second eigenvalue and eigenvector, if $B = 1$ then $A = -0.707$.

The purpose of Sections 10.2 and 10.3 was to present the reader with some basic principles of the structural dynamics modeling of solar panels. In the next section, a simplified mathematical model for liquid sloshing will be developed.

10.4 Modeling of Liquid Slosh

10.4.1 *Introduction*

With today's very large satellite structures, a substantial mass of fuel is necessary to place them into orbit and perform orbit corrections. As we have seen in Chapter 3, the mass of fuel contained in the tanks of a geosynchronous satellite amounts to approximately 40% of its total initial mass in the GTO. When the fuel containers are only partially filled and under translational acceleration, large quantities of fuel move uncontrollably inside the tanks and generate the *sloshing* effect.

The dynamics of motion of the fuel interacts with the solid-body and the appendage dynamics of the spacecraft. The interaction of sloshing with the ACS tends to produce attitude instability. Several methods have been employed to reduce the effect of sloshing, such as introducing baffles inside the tanks or dividing a large container into a number of smaller ones. These techniques, although helpful in some cases, do not succeed in canceling the sloshing effects. Hence the ACS must provide an adequate solution: first, to assure stability during the thrusting stage; and second, to achieve good attitude control despite the existence of sloshing dynamics.

10.4.2 *Basic Assumptions*

Modeling of the sloshing phenomenon was initiated in the early sixties. The models of sloshing motion presented in Abramson (1961) have not changed much over the years.

An exact analytical model of fluid oscillatory motion inside a moving container is extremely difficult – in fact, impossible. After some simplifying assumptions, reasonably adequate models have been obtained that are in good agreement with experimental results. These assumptions include:

(1) small displacements, velocities, and slopes of the liquid-free surfaces;
(2) a rigid tank;
(3) nonviscous liquid; and
(4) incompressible and homogeneous fluid.

With these assumptions, the sloshing dynamics model can be written using an infinite number of small masses. The results obtained with such models must be checked with experimental measurements, after which even greater accuracy is possible if the model parameters incorporate these test values (see Pocha 1986).

10.4.3 *One-Vibrating Mass Model*

A basic explanation of the sloshing effect follows. The fuel tank is under the action of a force **F**. The nonmoving parts of the container are concentrated in the mass M_0, which is located at the center of mass of the entire system. The moment of inertia about the center of mass is J_0. As a consequence of the applied force, an acceleration $g = F/(M_0+m)$ will act on the spacecraft. The moving mass m in the container experiences also an acceleration component in a direction opposite to the force **F**.

Because of a lateral force **f**, the mass m will tend to move away from the neutral position prior to the application of this force, and a pendulum analogy is evident. The simplest model in this context represents the sloshing mass by a pendulum, or a mass–spring element, as shown in Figure 10.4.1.a and Figure 10.4.1.b, respectively. As is well known, the frequency of oscillation of a pendulum is proportional to the acceleration constant g and inversely proportional to the length of the arm L:

$$\omega_{\text{osc}} = \sqrt{g/L}. \tag{10.4.1}$$

The mass–spring analogy is related to the pendulum analogy, in which the oscillation frequency of the mass–spring element in Figure 10.4.1.b is

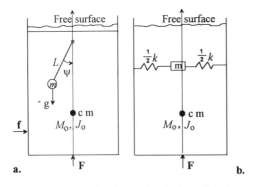

Figure 10.4.1 Simple mechanical models for pendulum and spring–mass analogies to the sloshing problem; reproduced from Abramson (1966).

10.4 / Modeling of Liquid Slosh

$$\omega_{\rm osc} = \sqrt{k/m}. \tag{10.4.2}$$

The difficult part in modeling the sloshing effect according to both analogies is to find (analytically and/or experimentally) the equivalent arm length L, the equivalent mass m, and the equivalent spring constant k. These parameters depend on the shape and other geometrical parameters of the container, the characteristics of the fluid, the fill ratio of the container (i.e., the ratio between fluid quantity and container volume), and so on (see Abramson 1961, Unruh et al. 1986). As we shall see, the oscillation frequency $\omega_{\rm osc}$ for both models is of utmost importance. Equating Eq. 10.4.1 and Eq. 10.4.2, we obtain that $k \approx g = F/(M_0 + m)$ and that the spring coefficient k is proportional to the applied force $\mathbf{F}$. From these two equations it also follows that

$$k = mg/L = m\omega_{\rm osc}^2; \tag{10.4.3}$$

L, the equivalent arm for the pendulum model, depends on the container geometry, the fill ratio of liquid, and so on. We can now write the equations of motion for the complete system, including the rigid-body part of the satellite and also the fixed and the moving (sloshing) part of the fuel. To be more precise, the model system illustrated in Figure 10.4.2 will be used.

In Figure 10.4.2, $\mathbf{Z}_S$ is the geometrical axis of the satellite, which is assumed to be a principal axis; $\mathbf{Z}_I$ is the inertial axis of the system before the lateral disturbance has been applied. The force $\mathbf{F}$ produces a linear acceleration that may be calculated as

$$g = \frac{F}{M_0 + m}. \tag{10.4.4}$$

This acceleration approximates the spring coefficient k.

If only one thruster is fired (e.g. Th1), the applied force $\mathbf{f}$ can be divided into a linear side force f – parallel to $\mathbf{f}$ and acting on the cm of the satellite – and a torque $T = fd$ about the cm. On the other hand, the force $\mathbf{f}$, together with the spring coefficient k as in Figure 10.4.2, produces a lateral acceleration $\mathbf{a}$ with amplitude

$$a = \frac{f + kx}{M_0}. \tag{10.4.5}$$

Next, we can write the moment equation about the cm:

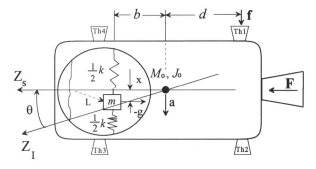

Figure 10.4.2 Model with one spring–mass system for writing the dynamics equation of motion; adapted from Bryson (1983) by permission of the American Astronautical Society.

$$J_0 \ddot{\theta} = df - bkx - mgx. \tag{10.4.6}$$

The resulting linear acceleration of the mass m satisfies the equation

$$m(a + \ddot{x} - b\ddot{\theta}) = -kx. \tag{10.4.7}$$

The simultaneous solution of Eqs. 10.4.5–10.4.7 is straightforward. If we define

$$k_1 = bk + mg \quad \text{and} \quad k_2 = k + mk/M_0 = k(1 + m/M_0),$$

then

$$J_0 \ddot{\theta} + k_1 x = df \quad \text{and} \tag{10.4.8}$$

$$m\ddot{x} + k_2 x - mb\ddot{\theta} = -mf/M_0. \tag{10.4.9}$$

Equation 10.4.8 and Eq. 10.4.9 are linear ordinary equations that can be solved with the Laplace transform:

$$\begin{bmatrix} J_0 s^2 & k_1 \\ mbs^2 & -m(s^2 + k_2/m) \end{bmatrix} \begin{bmatrix} \theta(s) \\ x(s) \end{bmatrix} = \begin{bmatrix} df \\ mf/M_0 \end{bmatrix}. \tag{10.4.10}$$

Exactly as in our analysis of solar panel structural dynamics, a damping factor ξ must be incorporated into the equations of angular motion. This is easily done by exchanging the term $m(s^2 + k_2/m)$ with the term $m(s^2 + 2\xi\sqrt{k_2/m}\,s + k_2/m)$. Equation 10.4.10 is thereby altered to read

$$\begin{bmatrix} J_0 s^2 & k_1 \\ mbs^2 & -m(s^2 + 2\xi\sqrt{k_2/m}\,s + k_2/m) \end{bmatrix} \begin{bmatrix} \theta(s) \\ x(s) \end{bmatrix} = \begin{bmatrix} df \\ mf/M_0 \end{bmatrix}. \tag{10.4.10'}$$

The solution is

$$\begin{bmatrix} \theta(s) \\ x(s) \end{bmatrix} = \frac{1}{\Delta(s)} \begin{bmatrix} -m(s^2 + 2\xi\sqrt{k_2/m}\,s + k_2/m) & -k_1 \\ -mbs^2 & J_0 s^2 \end{bmatrix} \begin{bmatrix} df \\ mf/M_0 \end{bmatrix}, \tag{10.4.11}$$

where the determinant is

$$\Delta(s) = -mJ_0 s^2 [s^2 + 2\xi\sqrt{k_2/m}\,s + k_2/m + bk_1/J_0]. \tag{10.4.12}$$

We are interested in the angular motion θ of the satellite. However, we must differentiate between the applied torque $\mathbf{T} = df$ and the applied side force $\mathbf{f}$. Two basic cases arise: applying pure torque versus applying torque with a side force.

Application of a Pure Torque

A pure torque T can be applied about the center of mass by firing two appropriate thrusters – for example, Th1 and Th3 to achieve a positive pure moment about the cm, or Th2 and Th4 for a negative pure torque. In this case, there is no application of a side force $\mathbf{f}$, and the transfer function becomes:

$$\frac{\theta(s)}{T(s)} = \frac{\theta(s)}{df} = \frac{s^2 + 2\xi\sqrt{k_2/m}\,s + k_2/m}{J_0 s^2 [s^2 + 2\xi\sqrt{k_2/m}\,s + k_2/m + k_1 b/J_0]}. \tag{10.4.13}$$

Remember that $g = F/(M_0 + m)$. Define also the equivalent sloshing frequency

$$\omega_{SL}^2 = \omega_z^2 = \frac{k_2}{m} = k\left(\frac{1}{m} + \frac{1}{M_0}\right) = \omega_{osc}^2\left(1 + \frac{m}{M_0}\right), \tag{10.4.14}$$

where ω_z is the zero of the transfer function. With these definitions, Eq. 10.4.13 can be rewritten as

10.4 / Modeling of Liquid Slosh

$$\frac{\theta(s)}{T(s)} = \frac{1}{J_0 s^2} \frac{s^2 + 2\xi\omega_{SL}s + \omega_{SL}^2}{s^2 + 2\xi\sqrt{k_2/m}\,s + \omega_{SL}^2 + b^2 k/J_0 + mbg/J_0}$$

$$= \frac{1}{J_0 s^2} \frac{s^2 + 2\xi\omega_z s + \omega_z^2}{s^2 + 2\xi_{psys}\omega_{psys}s + \omega_{psys}^2}. \tag{10.4.15}$$

In Eq. 10.4.15,

$$\omega_{psys}^2 = \omega_z^2 + \omega_{osc}^2 \frac{b^2 m}{J_0} + \frac{mbg}{J_0} = \omega_{osc}^2 \left[1 + \frac{m}{M_0} + \frac{b^2 m}{J_0}\right] + \frac{mbg}{J_0} \tag{10.4.16}$$

and

$$\xi_{psys} = \frac{\xi\omega_z}{\omega_{psys}}. \tag{10.4.17}$$

It is important to comment on the physical meaning of Eq. 10.4.14. The term ω_{osc} is the oscillation frequency of the mass–spring sloshing element; ω_z is the zero of the transfer function; and ω_{psys} in Eq. 10.4.16 is the natural frequency of the entire dynamic system, including the satellite.

Inspecting the determinant of the preceding transfer function, there are two important questions to be answered from the ACS perspective:

(1) Can the system pole be unstable? (b can be negative)
(2) Can the system pole value be smaller than the value of the zero?

In order to answer these questions, we write ω_{psys}^2 of Eq. 10.4.15 in terms of the basic parameters b, m, J_0, M_0, k (from Eq. 10.4.3), and g (from Eq. 10.4.4):

$$\omega_{psys}^2 = k\left[\frac{1}{m} + \frac{1}{M_0} + \frac{b^2}{J_0}\right] + \frac{mb}{J_0}g$$

$$= \omega_{osc}^2 \left[1 + \frac{m}{M_0} + \frac{mb^2}{J_0}\right] + \frac{bmg}{J_0}. \tag{10.4.16'}$$

According to Figure 10.4.2, b can have a positive or a negative sign, depending on the location of the fuel tank inside the satellite relative to the dry center of mass. Even if $b < 0$ but with $\omega_{psys} > \omega_z$, stabilization of the satellite – despite existence of the slosh dipole – is straightforward. A simple lead–lag compensator is sufficient to stabilize the two rigid-body integrators. The slosh dipole needs no special compensation, since the closed-loop root of the sloshing mode is inherently stabilized (Figure 10.4.3.a). However, should a negative b lead to the condition $\omega_{psys} < \omega_z$, an additional

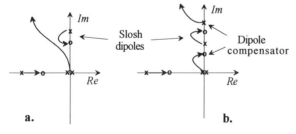

Figure 10.4.3 Root locus of the satellite's closed loop, including the slosh dipole.

compensating dipole will be necessary in order to secure a stable angular motion despite the existence of sloshing (see Figure 10.4.3.b).

Equation 10.4.15 is very similar to Eq. 10.2.16. As with the panel's structural model, the damping coefficient for the sloshing mass is usually determined by experimental measurements with a partially filled tank. For more sophisticated models that incorporate additional sloshing masses, the modeling should be similar to the panel represented by several structural modes as in Figure 10.2.3 (see also Section 10.4.3). This reasoning is best appreciated if the analysis is carried out on a Nichols chart or on Bode plots, as the following numerical example illustrates.

EXAMPLE 10.4.1 As in Example 10.2.1, we choose $J_0 = 720$ kg-m^2. Suppose also that $M_0 = 600$ kg, $m = 100$ kg, and $\omega_{osc} = 1.5$ rad/sec; $F = 500$ N. We shall find the transfer function for $b = +0.25$ m and then for $b = -0.25$ m, with $\xi = 0.002$ and also $\xi = 0.001$. For both cases, $g = 500/700 = 0.714$ m/sec^2 (Eq. 10.4.4).

Case 1: $b = +0.25$ According to Eq. 10.4.14, $\omega_z = 1.5\sqrt{1 + 100/600} = 1.6202$ rad/sec. Next, we find ω_{psys}:

$$\omega_{psys} = \sqrt{\left[1.5^2\left(1 + \frac{100}{600} + \frac{100 \times 0.25^2}{720}\right) + \frac{0.25 \times 100 \times 0.714}{720}\right]} = 1.634 \text{ rad/sec.}$$

In this case $\omega_{psys} > \omega_z$. Also, $\xi_{psys} = (0.002 \times 1.6202)/1.634 = 0.001983$.

Case 2: $b = -0.25$ In this case,

$$\omega_{psys} = \sqrt{\left[1.5^2\left(1 + \frac{100}{600} + \frac{100 \times 0.25^2}{720}\right) - \frac{0.25 \times 100 \times 0.714}{720}\right]} = 1.6186 \text{ rad/sec}$$

and $\omega_{psys} < \omega_z$; $\xi_{psys} = 0.002002$.

We will use the same feedback control solution as in Example 10.2.1. On the Nichols chart in Figure 10.4.4 are shown the open-loop gains for both positive and negative b and also for two different damping coefficients, $\xi = 0.002$ and $\xi = 0.001$.

At this point, some additional comments are in order.

(1) In Figure 10.2.2 it is clear that, in order to improve stability margins for the low-damping coefficient cases, some lag must be added at the panel's elastic modal frequencies, so that the peaks of the open-loop gains will be lowered away from the $-180°$ stability phase line in the Nichols chart. This will increase the phase at the sloshing modal frequencies too, so an additional decrease of the phase margins at the slosh modal frequencies will follow. Moreover, in multitank fuel systems, sloshing dipoles for both positive and negative b may be present simultaneously, thus gravely complicating the design of the feedback control loop (see Figure 10.4.4.c and Figure 10.4.4.d). The usual design-stage tradeoffs in the frequency domain will lead to a decrease in the crossover of the open-loop transfer function, with an inevitable degradation in the quality of the ACS.

(2) During the apogee insertion stage, the fill ratio of the fuel tanks decreases continuously, thus changing the parameters of the sloshing model. There is in any case

10.4 / Modeling of Liquid Slosh

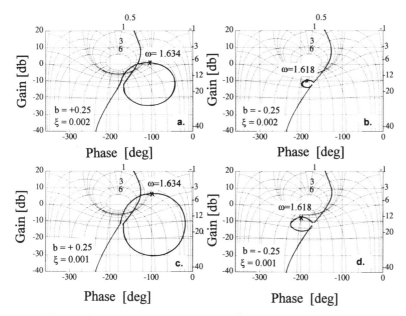

Figure 10.4.4 Frequency response of the loop gains, including one sloshing mode at $\omega_{SL} = 1.5$ rad/sec, $b = 0.25$ m, $\xi = 0.002$ and 0.001.

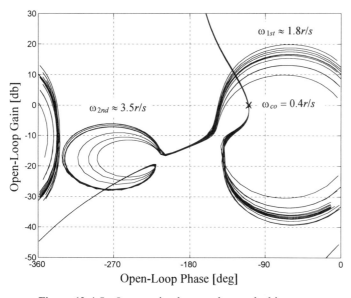

Figure 10.4.5 Loop gain changes due to sloshing parameter variations.

a large uncertainty about these parameters, such as the moment arm b and ω_{osc}. A family of uncertain plants will emerge, and the feedback control system must provide a satisfactory solution for all of them. An example of open-loop transfer functions for a family of such plants is shown in Figure 10.4.5. A systematic design technique for similar control problems was developed many years ago and can be used to obtain a satisfactory feedback control solution; see Sidi and Horowitz (1973) and Sidi (1976).

(3) In Example 10.2.1, only the panel structural model was introduced; in Example 10.4.1, only the slosh dynamics model was implemented. In Section 10.5, a systematic procedure will be derived to implement simultaneously all structural and slosh dynamics in the model.

Simultaneous Application of Pure Torque and a Side Force
In this case, the transfer function in terms of the applied force **f** will become

$$\frac{\theta(s)}{f} = \frac{-1}{\Delta(s)}\left[(ms^2+k_2)d + \frac{k_1 m}{M_0}\right] = \frac{-m}{\Delta(s)}\left[s^2 + \left(\frac{k_2 d}{m} + \frac{k_1}{M_0}\right)\right] \quad (10.4.18)$$

(cf. Eq. 10.4.11). It is clear from Eq. 10.4.13 and Eq. 10.4.18 that use of a pure torque **T** and a side force **f** enables changing the location of the sloshing model's complex zero in the s domain, thus helping to obtain a more favorable (from a root-locus viewpoint) location of the complex poles and zeros of the sloshing mode. Our aim is to decrease the value of the sloshing zero *below* the value of the sloshing pole. This simplifies the feedback control solution, as in Figure 10.4.3.a. In practical situations this effect is obtained by using a side thruster providing the side force **f**, a thruster different from those providing the pure torque **T**.

10.4.4 Multi-Mass Model

As with all structural models, an infinity of structural modes exists but only the most pronounced are taken into consideration. For the sloshing model, two to three masses at most are sufficient to characterize the sloshing dynamics. Generally, it is more convenient to define the modes by their oscillating frequency ω_{osci} (see Eq. 10.4.2) and their assumed damping coefficient ξ_i. For the multi-mass sloshing model, Figure 10.4.2 is readapted to yield Figure 10.4.6.

In fact, the multiple-mass sloshing model can be presented in block-diagram form as in Figure 10.2.3 for the solar panel structural modes. Formulating the dynamics model for the panels and the sloshing oscillatory dynamics in a generalized model will be explained in the next section.

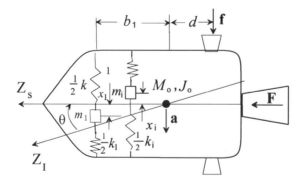

Figure 10.4.6 Model with multiple sloshing masses; adapted from Bryson (1983) by permission of the American Astronautical Society.

10.5 Generalized Modeling of Structural and Sloshing Dynamics

The task of previous sections was to introduce basic notions of structural and sloshing dynamics. The emphasis was on the physical interpretation of these models. In practice, the number of structural modal frequencies is quite large, as is the number of sloshing masses that influence the satellite's dynamics. Consequently, the procedures explained in previous sections seem impractical for modeling the overall dynamics of a s/c with a large number of oscillating modes. In this section we systematize writing the complete set of dynamics equations for angular motion of the spacecraft.

Before we can present a complete analysis of the dynamics of the satellite, including solar panels and fluid in the fuel containers, some discussion of the hardware being incorporated will be necessary.

10.5.1 *A System of Solar Panels*

In the most general case, a solar panel might be appended to the satellite by a two-axis gimbaled system, so that the panel can be rotated toward the sun from any orbit location. To simplify the analysis, we assume that the panel has only one degree of rotational freedom. This is true of the *Spot* satellite, the *Hubble* space telescope, and all geostationary satellites.

Suppose we are dealing with a geostationary satellite, so that the solar panels must rotate once per day about one of the satellite body axes in order to be continuously directed toward the sun. A schematic of the satellite–panels hardware is shown in Figure 10.5.1. As shown in the figure, the panel can rotate about the Y_B axis by an angle α, so that the sun axis will be located in the X_p-Y_p plane for maximum sun absorption.

Since the sun panel is not rigid, it can have three modes of deflection about its axes X_p, Y_p, Z_p. The deflection about the Z_p axis is called the *out-of-plane deflection mode*. The deflection about the X_p axis is called the *in-plane* deflection mode. The last mode, deflection around the Y_p axis, is the *torsional* deflection mode. It is easily perceived that the out-of-plane mode is in general the most pronounced and has the lowest structural eigenfrequency. The influence of all three panel deflection modes

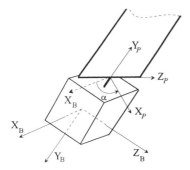

Figure 10.5.1 Geometry of a single-axis rotating solar panel.

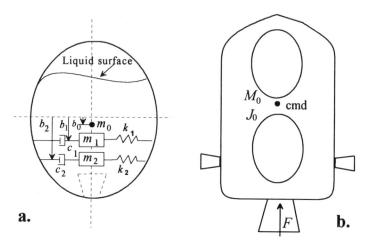

Figure 10.5.2 Definition of (a) the geometrical parameters of a single fuel tank and (b) the geometry of the rigid body plus two fuel tanks; part (a) reproduced from Abramson (1966).

on the rigid part of the satellite depends on the rotation angle α of the panel relative to the rigid body's axis frame.

10.5.2 *A System of Fuel Tanks*

A propulsion system may consist of a number of fuel tanks. With bipropellant propulsion systems (see Appendix C), at least two tanks are mandatory. A number of smaller tanks are generally preferred over a single bigger one in order to lessen sloshing effects, although such systems are more expensive. In any case, the geometrical parameters of each tank must be defined independently of the overall satellite conglomeration.

In Figure 10.5.2, m_0 is the mass of the nonsloshing part of the fuel inside the tank and m_1, m_2 are the two assumed sloshing masses. These three masses are distanced b_0, b_1, and b_2 from a reference geometrical location, in this case the geometrical center of the ellipsoid-shaped tank. The terms k_1, k_2 and c_1, c_2 are the spring coefficients and damping parameters responsible for the damping coefficients defined in Section 10.4. In Figure 10.5.2.b, "cmd" is the *dry* center of mass of the rigid-body part of the s/c, including the dry weight of the fuel tanks; M_0 is the nonsloshing mass of the satellite, and J_0 is the moment of inertia of the satellite excluding the sloshing masses.

With these definitions and the known location of the tanks inside the satellite, the parameters of the sloshing masses inside each tank can be defined according to the fill ratio of the tanks and the translatory acceleration imparted to the s/c in the direction of **F** (Abramson 1961). A nonmoving cm can then be calculated, and the torque arm of each mass referenced to it can be evaluated, thus enabling us to write the dynamics equations of the overall satellite system.

10.5.3 *Coupling Coefficients and Matrices*

Examining Eq. 10.2.14 and Eq. 10.2.15 reveals immediately the existence of a coupling factor between the rigid-body dynamics and the solar panels' structural dynamics; this is the term mL in these equations. The order of the structural dynamics equations depends on the number of appended structural bodies and on the number of assumed structural modes per body.

All structural mode equations for the panels and the sloshing modes can be expressed in a canonical form of second-order dynamics, with known eigenfrequencies and fairly well-anticipated damping coefficients. The coupling coefficients allow us to augment the dynamics equations of rigid-body angular motion with structural and sloshing model dynamics. A generalized approach follows in the next section.

10.5.4 Complete Dynamical Modeling of Spacecraft

In this section we assume that the translational and rotational dynamics are uncoupled. See also Prins (1983) and Williams and Wood (1989). With this assumption, it suffices to augment the dynamics of Euler's moment equations (Eq. 4.8.1) with the structural modes and sloshing dynamics. In Euler's equations, $\mathbf{h}$ is the momentum matrix, including all products of inertia and momentum biases, if existing in the satellite.

The augmented Euler equations of motion become:

$$\mathbf{T} = \mathbf{T}_c + \mathbf{T}_d = \dot{\mathbf{h}}_I + \boldsymbol{\omega} \times \{\mathbf{h} + [\mathbf{B}]\dot{\boldsymbol{\eta}} + [\mathbf{D}]\dot{\boldsymbol{\sigma}}\} + [\mathbf{B}]\ddot{\boldsymbol{\eta}} + [\mathbf{D}]\ddot{\boldsymbol{\sigma}}, \qquad (10.5.1)$$

$$[\mathbf{U}]\ddot{\boldsymbol{\eta}} + 2[\boldsymbol{\xi}_\eta][\boldsymbol{\Omega}_\eta]\dot{\boldsymbol{\eta}} + [\boldsymbol{\Omega}_\eta]^2 \boldsymbol{\eta} = [\mathbf{B}]^T \dot{\boldsymbol{\omega}}, \qquad (10.5.2)$$

$$[\mathbf{E}]\ddot{\boldsymbol{\sigma}} + 2[\boldsymbol{\xi}_\sigma][\boldsymbol{\Omega}_\sigma]\dot{\boldsymbol{\sigma}} + [\boldsymbol{\Omega}_\sigma]^2 \boldsymbol{\sigma} = [\mathbf{D}]^T \dot{\boldsymbol{\omega}}, \qquad (10.5.3)$$

with $[\mathbf{B}] = \{[\mathbf{K}][\mathbf{C}]\}^T$. In Eqs. 10.5.1–10.5.3, $\boldsymbol{\omega}$, $\mathbf{h}$, $\boldsymbol{\eta}$, and $\boldsymbol{\sigma}$ are vectors. The rotational matrix $[\mathbf{C}]$ is the usual rotational transformation about the $\mathbf{Y}_B$ axis in Figure 10.5.1 by an angle α:

$$[\mathbf{C}] = \begin{bmatrix} \cos(\alpha) & 0 & -\sin(\alpha) \\ 0 & 1 & 0 \\ \sin(\alpha) & 0 & \cos(\alpha) \end{bmatrix}. \qquad (10.5.4)$$

Formal definitions of the different matrices are as follows:

- $[\mathbf{U}]$ – unit matrix $(m \times m)$;
- $[\boldsymbol{\xi}_\eta]$ – flexible modes damping matrix, diagonal $(m \times m)$;
- $[\boldsymbol{\Omega}_\eta]$ – flexible modes frequency matrix, diagonal $(m \times m)$;
- $[\mathbf{B}]$ – flexible modes coupling matrix $(3 \times m)$;
- $[\mathbf{K}]$ – matrix of coupling coefficients $(m \times 3)$;
- $[\mathbf{E}]$ – unit matrix $(n \times n)$;
- $[\boldsymbol{\xi}_\sigma]$ – sloshing modes damping matrix, diagonal $(n \times n)$;
- $[\boldsymbol{\Omega}_\sigma]$ – sloshing modes frequency matrix, diagonal $(n \times n)$; and
- $[\mathbf{D}]$ – sloshing modes coupling matrix $(3 \times n)$.

With regard to matrix dimensions, m denotes the number of flexible modes of the solar panels for the three different panel axes; n is the number of sloshing masses. Any element of $[\boldsymbol{\Omega}]$ represents the natural frequency of a mode, structural or sloshing; any element of $[\boldsymbol{\xi}]$ represents the damping coefficient of a single mode. The elements of $[\mathbf{B}]$ define the angular momentum coefficients of the ith flexure mode at the cm of the satellite; the elements of $[\mathbf{D}]$ define the coupling coefficients of a sloshing mass to the cm of the satellite. The element η_i of η is the amplitude of the ith flexure mode, and the element σ_j of σ is the amplitude of the jth liquid sloshing mode.

As far as the structural dynamics of the solar panels is concerned, the modes should be defined according to the panel axes $\mathbf{X}_p, \mathbf{Y}_p, \mathbf{Z}_p$. For instance,

$$[\Omega_\eta] = \mathrm{diag}[\Omega_{xp1}, \Omega_{xp2}, \ldots, \Omega_{yp1}, \Omega_{yp2}, \ldots, \Omega_{zp1}, \Omega_{zp2}, \ldots],$$

$$[\xi_\eta] = \mathrm{diag}[\xi_{xp1}, \xi_{xp2}, \ldots, \xi_{yp1}, \xi_{yp2}, \ldots, \xi_{zp1}, \xi_{zp2}, \ldots].$$

Also, $\eta^\mathrm{T} = [\eta_{xp1}, \eta_{xp2}, \ldots, \eta_{yp1}, \eta_{yp2}, \ldots, \eta_{zp1}, \eta_{zp2}, \ldots]$. With these definitions,

$$[\mathbf{K}]^\mathrm{T} = \begin{bmatrix} k_{xp1}, k_{xp2}, \ldots & 0 & 0 \\ 0 & k_{yp1}, k_{yp2}, \ldots & 0 \\ 0 & 0 & k_{zp1}, k_{zp2}, \ldots \end{bmatrix}. \tag{10.5.5}$$

To find $[\mathbf{B}]$ and $[\mathbf{B}]^\mathrm{T}$ explicitly, suppose that there exist two structural modes for each of the panel's axes. In this case,

$$[\mathbf{B}]^\mathrm{T} = [\mathbf{K}][\mathbf{C}] = \begin{bmatrix} k_{xp1}\cos(\alpha) & 0 & -k_{xp1}\sin(\alpha) \\ k_{xp2}\cos(\alpha) & 0 & -k_{xp2}\sin(\alpha) \\ & k_{yp1} & \\ & k_{yp2} & \\ k_{zp1}\sin(\alpha) & 0 & k_{zp1}\cos(\alpha) \\ k_{zp2}\sin(\alpha) & 0 & k_{zp2}\cos(\alpha) \end{bmatrix}. \tag{10.5.6}$$

Equations 10.5.1–10.5.3, together with Eq. 10.5.6, can be integrated to yield body rates, which can be integrated once more to obtain the attitude angles of the rigid spacecraft, on which is superimposed the angular motion due to the sloshing and structural dynamics.

Once again, some explanatory comments are in order.

(1) The equations of the structural and the sloshing dynamics are linearized, so they are correct for small amplitudes only. For more sophisticated, nonlinear dynamics models, the reader is referred to Peterson, Crawley, and Hansman (1988).

(2) The augmented Euler moment equations of Eq. 10.5.1 are nonlinear. For linear control design of the ACS, a linearization procedure (similar to that used in Section 4.8.3) must be carried out, but this time including the structural and sloshing dynamics equations.

(3) Equation 10.5.6 makes it clear that, when $\mathbf{X}_p$ and $\mathbf{X}_B$ in Figure 10.5.1 are parallel, the out-of-plane mode of the panel will exist only about the $\mathbf{Z}_B$ axis. A 90° rotation of the panel will transfer the out-of-plane mode to the $\mathbf{X}_B$ axis dynamics. At intermediate rotation angles of the panel, the out-of-plane and the in-plane modes will exist partially in both $\mathbf{X}_B$ and $\mathbf{Z}_B$ dynamics.

10.5.5 *Linearized Equations of Motion*

To develop the linearized equations that include structural and sloshing modes, we can proceed from Eqs. 4.8.14. The influence of structural and sloshing dynamics is most pronounced when the ACS is controlled with the aid of reaction torques (see Chapter 9), which generally apply torque impulses of large amplitudes and thus excite the motion of the structural and sloshing modes.

With this assumption, terms including the low orbital frequency ω_o can be omitted, and the torques produced by momentum exchange devices can be assimilated by the external control torques (as we assumed in Section 7.3, Eqs. 7.3). This assumption is very reasonable from a practical point of view, and greatly simplifies the analysis in different applications. Equation 10.5.2 and Eq. 10.5.3 remain unchanged because they are already linear. Only Eq. 10.5.1 needs to be simplified:

$$\mathbf{T}_c = [\mathbf{I}]\dot{\omega} + [\mathbf{B}]\ddot{\eta} + [\mathbf{D}]\ddot{\sigma}, \tag{10.5.7}$$

where $[\mathbf{I}]$ is the inertia matrix.

Given the assumptions and simplifications just listed, it is now easier to calculate the transfer functions of the satellite dynamics, including the structural and sloshing modes, for the purpose of linear feedback control analysis and design. Equation 10.5.2, Eq. 10.5.3, and Eq. 10.5.7 can be rearranged so that standard analytical software programs can be used to solve them. It is easy to write these matrix equations in the form $[\mathbf{F}]\dot{\mathbf{x}} = [\bar{\mathbf{A}}]\mathbf{x} + [\bar{\mathbf{B}}]\mathbf{u}$, which can then be transformed to the more usual form $\dot{\mathbf{x}} = [\mathbf{A}]\mathbf{x} + [\mathbf{B}]\mathbf{u}$. This is the standard form used in many dynamics simulation programs.

10.6 Constraints on the Open-Loop Gain

10.6.1 *Introduction*

We have seen in previous chapters that it is generally better to have a high open-loop gain, for three important reasons. First, the higher the open-loop gain of the attitude feedback control system, the smaller the tracking attitude error. Second, a high open-loop gain renders the attitude control system less sensitive to external disturbances. Third, an ACS with high open-loop gain responds faster to attitude commands.

We have also seen that there are a number of practical physical problems that make the achievement of a high-gain ACS difficult. These difficulties are due mostly to irregularities in the attitude sensors and control hardware. The finite control torques that the control hardware can produce limit the speed with which the ACS can react to attitude commands. Even if the control hardware is capable of producing the needed control torques, the open-loop gain still cannot be chosen high enough owing to the effects of sensor noise amplification. Moreover, these attitude and control hardware deficiencies are exacerbated by the effects of structural and sloshing dynamics, which exist to some extent in all satellites – and certainly in any satellite using extendable solar panels or liquid fuel. In the next section we explore limitations on the gain bandwidth of an attitude feedback control loop that are due to structural dynamics.

10.6.2 *Limitations on the Crossover Frequency*

Structural dynamics is best treated in the frequency domain. Sloshing is sensed in the lower-frequency range of the satellite dynamics, while structural effects are more pronounced in the higher-frequency range. It is therefore logical that the structural dynamics will interfere with the gain margin (GM) of the open-loop transfer function, thus limiting the bandwidth of the attitude control system.

First, let us obtain a rough idea of the maximal achievable loop gain and of its *crossover frequency* ω_{co}, defined as the frequency at which $|L(j\omega)| = 0$ db on a Bode graph or Nichols chart. In Figure 10.2.2.b, for instance, the GM is about 6 db, whereas in Figure 10.2.2.d there is no GM at all and the system verges on instability. In practice, the peak of the structural mode can be moved to the left by introducing into the open-loop transfer function $L(j\omega)$ a control network that will add the necessary lag phase at the peak frequency, thus significantly increasing the GM. However, even if this process is adopted, the peak of the loop gain $|G(j\omega)|$ should not be larger than some predetermined level on the Nichols chart. A practical reason for this is as follows. Since the frequencies of the structural modes are comparatively high relative to the other dynamics of the attitude feedback system, any small unpredictable delay in the loop will be translated to a large additional phase lag at the frequency of the structural mode peak, and the open-loop gain at this frequency might then move toward the instability point ($-180°$, 0 db). So let us accept in our analysis that the peaks in the open-loop gain near the structural mode frequency will satisfy a gain margin $GM_{\sigma S}$, in accordance with various practical engineering arguments. With these assumptions, the maximum feasible bandwidth of the open-loop transfer function $L(j\omega)$ will depend on the damping factors and the modal natural frequency characteristics that define the structural mode. Naturally, there exists more than one structural mode dipole, but to simplify the analysis we will assume the existence of only the dipole with the smallest modal frequency.

For this situation, the peak in the frequency response due to the structural mode can be calculated from parameters of the structural mode dipole expressed in Eq. 10.2.16 at $\omega = \sigma_S$. Using the equalities $\sigma = \sigma_S \sqrt{J_0/J}$ and $\xi_S = \xi \sqrt{J/J_0}$ in Eq. 10.2.16, the gain of the peak at the pole frequency of the dipole at $\omega = \sigma_S$ becomes

$$|G_{\text{dipole}}(j\sigma_S)| = \sqrt{\left(\frac{J_0}{J} - 1\right)^2 + 4\xi^2 \frac{J_0}{J}} \bigg/ 2\xi \sqrt{\frac{J}{J_0}}. \tag{10.6.1}$$

It is easy to relate this peak to the obtainable bandwidth ω_{co} of the open-loop gain of the attitude feedback control loop (see Figure 10.6.1).

The phase of the open-loop transfer function $L(j\omega)$ must be less than $-180°$ in the vicinity of the ω_{co} frequency region, in order to provide the necessary phase margin for stability. In the frequency region above ω_{co} and up to σ_S, the phase can be higher than $180°$ but not by much. To simplify the analysis, we assume that in the

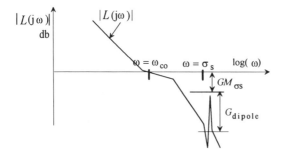

Figure 10.6.1 Approximate calculation of the maximum ω_{co} for a given modal dipole and a minimum defined $GM_{\sigma S}$.

10.6 / Constraints on the Open-Loop Gain

region ω_{co} to σ_S there exists an average phase lag of $\alpha\pi$ rad, $\alpha < 1$. Such a phase characteristic is accompanied by an average constant slope of the open-loop gain $|L(j\omega)|$ of -12α db/oct (see Bode 1945). For given $\text{GM}_{\sigma S}$, G_{dipole}, and α, the following relation holds: $\text{GM}_{\sigma S}$ [db] $+ G_{\text{dipole}}$ [db] $= 12\alpha y$, from which it follows that

$$y = \frac{\text{GM}_{\sigma S} + G_{\text{dipole}}}{12\alpha}. \tag{10.6.2}$$

In Eq. 10.6.2, y is the number of octaves that separate ω_{co} from σ_S (see also Sidi 1980). Finally,

$$\omega_{co} = \frac{\sigma_S}{2^y} = \frac{\sqrt{J/J_0}\,\sigma}{2^y}. \tag{10.6.3}$$

The meaning of Eq. 10.6.3 is illustrated by the following example.

EXAMPLE 10.6.1 Given the elastic structural mode of Example 10.2.1, find the approximate maximum achievable crossover frequency ω_{co} of the open-loop gain $|L(j\omega)|$. Assume that $G_{\text{dipole}} = 6$ db and $\alpha = 0.8$.

Solution From Eq. 10.6.1 we find that $|G_{\text{dipole}}| = 15.8 = 24$ db. From Eq. 10.6.2, $(24+6)/(12\times0.8) = 3.1$. Finally, from Eq. 10.6.3 we have $\omega_{co} = 4.2164/2^{2.7} = 0.48$ rad/sec.

The resulting $L(j\omega)$ is shown in Figure 10.6.2, where $\omega_{co} = 0.54$ rad/sec. Taking into account the almost arbitrary choice of $\alpha = 0.8$, this is fairly close to the approximated value of 0.48 rad/sec.

Equation 10.6.3 estimates the maximum bandwidth that can be achieved with the existence of a single structural dynamics pole. It gives a good "rule of thumb"

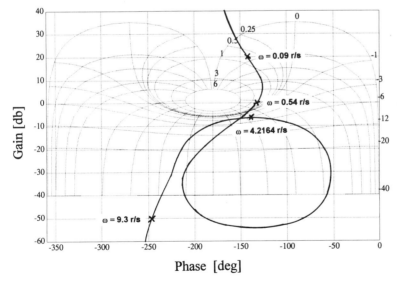

Figure 10.6.2 Open-loop gain of the feedback control system of Example 10.6.1.

concerning the achievable gain bandwidth. The existence of additional structural or sloshing modal dynamics may only make things worse.

Sloshing modal dynamics are generally located at lower frequencies. Let us assume a sloshing mode at a frequency close to ω_{co} of Example 10.6.1. In order to accommodate this mode and prevent decreasing the existing adequate phase margin, some lead-lag network will be required to provide the necessary lead phase. However, such networks tend to decrease the gain margin of the higher-frequency structural elastic mode. Maintaining the original gain margin will require a lower ω_{co}.

When several structural and sloshing modal poles are present, the usual "cut-and-try" techniques in the frequency domain will – it is hoped – yield the maximum achievable gain bandwidth.

10.7 Summary

Structural and sloshing dynamics are a serious problem in the design of attitude feedback control systems. It can be quite difficult to obtain a reliable model, especially where sloshing dynamics is concerned. It is advisable to perform some experimentation and measurements with the actual hardware, when this is feasible, in order to increase confidence about the model obtained by theoretical analysis. Considerable experimentation is also advisable when designing a feedback control system with sufficient gain and phase margins, especially in the presence of structural and sloshing modal poles whose parameters are not known exactly.

References

Abramson, H. (1961), "The Dynamic Behavior of Liquids in Moving Containers," NASA SP-106 (N67-15884), Scientific and Technical Information Division, National Aeronautics and Space Administration, Washington, DC.

Bode, H. W. (1945), *Network Analysis and Feedback Amplifier Design*. New York: Van Nostrand.

Bryson, A. (1983), "Stabilization and Control of Spacecraft," Microfiche supplement to the *Proceedings of the Annual AAS Rocky Mountain Guidance and Control Conference* (5-9 February, Keystone, CO). San Diego, CA: American Astronautical Society.

Peterson, D., Crawley, E., and Hansman, R. (1988), "Nonlinear Fluid Slosh Coupled to the Dynamics of a Spacecraft," *AIAA Journal* 27(9): 1230-40.

Pocha, J. (1986), "An Experimental Investigation of Spacecraft Sloshing," Paper no. 255, *Proceedings of the Second International Symposium on Spacecraft Flight Dynamics* (20-23 October, Darmstadt, Germany). Paris: European Space Agency.

Prins, J. (1983), "Trends in AOCS Testing," IFAC Workshop (August, Stanford, CA).

Sidi, M. (1976), "Feedback Synthesis with Plant Ignorance, Non-minimum-Phase, and Time-Domain Tolerances," *Automatica* 12: 265-71.

Sidi, M. (1980), "On Maximization of Gain-Bandwidth in Sampled Systems," *International Journal of Control* 32(6): 1099-1109.

Sidi, M., and Horowitz, I. (1973), "Synthesis of Feedback Systems with Large Plant Ignorance for Prescribed Time Domain Tolerances," *Sensitivity, Adaptivity and Optimality* (Proceedings of the Third IFAC Symposium, 18-23 June, Ischia, Italy). Oxford, UK: Pergamon, pp. 202-5.

Thomson, W. (1988), *Theory of Vibration with Applications*. Englewood Cliffs, NJ: Prentice-Hall.

Unruh, J., Kana, F., Dodge, F., and Fey, T. (1986), "Digital Data Analysis Techniques for Extraction of Slosh Model Parameters," *Journal of Spacecraft and Rockets* 23(2): 171–7.

Weaver, W., Timoshenko, S., and Young, D. (1990), *Vibration Problems in Engineering*. New York: Wiley.

Williams, C. G. (1976), "Dynamics Modelling and Formulation Techniques for Non-Rigid Spacecraft," Electronic and Space Systems, British Aircraft Corporation, Bristol, UK.

Williams, C., and Wood, T. (1989), "Dynamics and Formulation Techniques for Non-Rigid Spacecraft," ESTEC contract no. 8085/88/NL/MAC,TP9257, British Aircraft Corporation, Bristol, UK.

APPENDIX A

Attitude Transformations in Space

A.1 Introduction

The purpose of this appendix is to provide a short description of various aspects of attitude transformation in space. The attitude of a three-dimensional body is most conveniently defined with a set of axes fixed to the body. This set of axes is generally a triad of orthogonal coordinates, and is normally called a *body coordinate frame*. The attitude of a body is thought of as a coordinate transformation that transforms a defined set of reference coordinates into the body coordinates of the spacecraft.

This appendix presents some basic attitude transformation techniques; the treatment is not extensive, but rather a summary of the material required in the main chapters of the text. For further reading on the subject, see Sabroff et al. (1965) or Wertz (1978).

A.2 Direction Cosine Matrix

A.2.1 Definitions

The basic three-axis attitude transformation is based on the direction cosine matrix. Any attitude transformation in space is actually converted to this essential form. In Figure A.2.1, the axes **1, 2**, and **3** are unit vectors defining an orthogonal, right-handed triad. This triad is chosen as the reference inertial frame. Next, a similar orthogonal triad is attached to the center of mass of a moving body, defined by the unit vectors **u, v**, and **w**.

In the context of Figure A.2.1, we define the matrix [**A**] as follows:

$$[\mathbf{A}] = \begin{bmatrix} u_1 & u_2 & u_3 \\ v_1 & v_2 & v_3 \\ w_1 & w_2 & w_3 \end{bmatrix}. \quad (A.2.1)$$

In this matrix, u_1, u_2, u_3 are the components of the unit vector **u** along the three axes **1, 2, 3** of the reference orthogonal system: $\mathbf{u} = [u_1 \ u_2 \ u_3]^T$. In a similar way, **v** and **w** have components v_1, v_2, v_3 and w_1, w_2, w_3 along the same reference axes: $\mathbf{v} = [v_1 \ v_2 \ v_3]^T$ and $\mathbf{w} = [w_1 \ w_2 \ w_3]^T$. The direction cosine matrix [**A**], also called the *attitude matrix*, has the important property of mapping vectors from the reference frame to the body frame. Suppose that a vector **a** has components a_1, a_2, a_3 in the reference frame: $\mathbf{a} = [a_1 \ a_2 \ a_3]^T$. The following matrix vector multiplication expresses the components of the vector **a** in the body frame:

A.2 / Direction Cosine Matrix

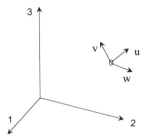

Figure A.2.1 Definition of the orientation of the spacecraft axes $\mathbf{u}, \mathbf{v}, \mathbf{w}$ in the reference frame $\mathbf{1}, \mathbf{2}, \mathbf{3}$.

$$[\mathbf{A}]\mathbf{a} = \begin{bmatrix} u_1 & u_2 & u_3 \\ v_1 & v_2 & v_3 \\ w_1 & w_2 & w_3 \end{bmatrix} \begin{bmatrix} a_1 \\ a_2 \\ a_3 \end{bmatrix} = \begin{bmatrix} \mathbf{u} \cdot \mathbf{a} \\ \mathbf{v} \cdot \mathbf{a} \\ \mathbf{w} \cdot \mathbf{a} \end{bmatrix} = \begin{bmatrix} a_u \\ a_v \\ a_w \end{bmatrix} = \mathbf{a}_B, \qquad (A.2.2)$$

where $\mathbf{a}_B$ is the vector $\mathbf{a}$ mapped into the body frame.

Since $\mathbf{u}$ is a unit vector, it follows that the scalar product $\mathbf{u} \cdot \mathbf{a}$ is the component a_u of the vector $\mathbf{a}$ along the unit vector $\mathbf{u}$. By the same reasoning, the components of the vector $\mathbf{a}$ on the remaining unit vectors of the body triad are a_v and a_w.

A.2.2 Basic Properties

Some basic properties of the matrix $[\mathbf{A}]$ may be stated as follows.

(1) Each of its elements is the cosine of the angle between a body unit vector and a reference axis; its name is derived from this property.
(2) Each of the vectors $\mathbf{u}, \mathbf{v}, \mathbf{w}$ are vectors with unit length; hence:

$$\sum_{i=1}^{3} u_i^2 = 1, \quad \sum_{i=1}^{3} v_i^2 = 1, \quad \sum_{i=1}^{3} w_i^2 = 1.$$

(3) The unit vectors $\mathbf{u}, \mathbf{v}, \mathbf{w}$ are orthogonal to each other; hence:

$$\sum_{i=1}^{3} u_i v_i = 0, \quad \sum_{i=1}^{3} u_i w_i = 0, \quad \sum_{i=1}^{3} v_i w_i = 0.$$

(4) The relationships in (2) and (3) lead to the useful identity $[\mathbf{A}][\mathbf{A}]^T = \mathbf{1}$, or $[\mathbf{A}]^T = [\mathbf{A}]^{-1}$. Of course, transposition of a matrix is a much simpler process than inversion of the same matrix.
(5) It is well known that $\det[\mathbf{A}] = \mathbf{u} \cdot (\mathbf{v} \times \mathbf{w})$. Since $\mathbf{u}, \mathbf{v}, \mathbf{w}$ form a cubic orthogonal triad, it follows that $\det[\mathbf{A}] = 1$. Thus,

$$\mathbf{a} = [\mathbf{A}]^T \mathbf{a}_B. \qquad (A.2.3)$$

We conclude that $[\mathbf{A}]$ is a proper real orthogonal matrix. It is shown in Wertz (1978) that such a matrix transformation preserves the lengths of vectors and also the angles between them, and thus represents a rotation. The product of two proper real orthogonal matrices $[\mathbf{A}] = [\mathbf{A}_2][\mathbf{A}_1]$ is the result of two successive rotations, first by $[\mathbf{A}_1]$ and then by $[\mathbf{A}_2]$. A chain of successive rotations is common in attitude transformations.

A.3 Euler Angle Rotation

The Euler angle rotation is defined as successive angular rotations about the three orthogonal frame axes. Suppose we define the three orthogonal axes of the body frame by **i**, **j**, and **k**, and those of the reference frame by **I**, **J**, and **K**. There is a multitude of order combinations by which the rotation can be performed. For instance, we might first perform a rotation about the **i**, then about the **j**, and finally about the **k** axis. The order of rotation could also be about **j**, **i**, **k**, and so on.

There are two distinct types of rotations.

(1) Successive rotations about each of the three axes **i**, **j**, **k**. There are six possible orders of such a rotation: 1-2-3, 1-3-2, 2-1-3, 2-3-1, 3-1-2, and 3-2-1.

(2) First and third rotations about the same axis with the second rotation about one of the two remaining axes. Again we have six possibilities: 1-2-1, 1-3-1, 2-1-2, 2-3-2, 3-1-3, and 3-2-3.

The second type of rotation sequences can be useful in special situations (see Section 2.6.1), and sometimes helps to solve problems in which successive rotations about three distinct axes may give rise to singularities.

Which specific rotation order is chosen depends on the situation at hand. It is common to define the Euler roll angle (ϕ) as a rotation about the **x** body axis, the pitch angle (θ) about the **y** body axis, and the yaw angle (ψ) about the **z** body axis. However, any other definition is acceptable as long as it remains consistent with the analytical development. For reference, an example of a complete rotation will be given next.

Suppose we want to perform the transformation $\psi \to \theta \to \phi$ successively about the z, y, and x body axes. First, the body triad undergoes an angular rotation ψ about the z body axis. In Figure A.3.1, R is the distance of a point from the origin of both Cartesian systems [**I**, **J**] and [**i**, **j**]. System [**i**, **j**] is rotated by an angle ψ with respect to system [**I**, **J**]. The components of **R** are X, Y and x, y (respectively) in the two coordinate systems; **I**, **J** are unit vectors in system $[X, Y]$ and **i**, **j** are unit vectors in $[x, y]$. For a transformation in the plane,

$$\mathbf{R} = X\mathbf{I} + Y\mathbf{J} = x\mathbf{i} + y\mathbf{j}. \tag{A.3.1}$$

Taking the scalar product by the vector **i** yields $X\mathbf{I}\cdot\mathbf{i} + Y\mathbf{J}\cdot\mathbf{i} = x\mathbf{i}\cdot\mathbf{i} + y\mathbf{j}\cdot\mathbf{i}$. We know that $\mathbf{I}\cdot\mathbf{i} = \cos(\psi)$, $\mathbf{J}\cdot\mathbf{i} = \sin(\psi)$, $\mathbf{i}\cdot\mathbf{i} = 1$, and $\mathbf{i}\cdot\mathbf{j} = 0$; hence $x = X\cos(\psi) + Y\sin(\psi)$. Next, taking the scalar product of Eq. A.3.1 by **j** yields $X\mathbf{I}\cdot\mathbf{j} + Y\mathbf{J}\cdot\mathbf{j} = x\mathbf{i}\cdot\mathbf{j} + y\mathbf{j}\cdot\mathbf{j}$. We also know that $\mathbf{I}\cdot\mathbf{j} = -\sin(\psi)$, $\mathbf{J}\cdot\mathbf{j} = \cos(\psi)$, $\mathbf{i}\cdot\mathbf{j} = 0$, and $\mathbf{j}\cdot\mathbf{j} = 1$; therefore $y =$

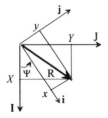

Figure A.3.1 The first rotation is about the z axis.

A.3 / Euler Angle Rotation

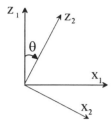

Figure A.3.2 The second rotation is about the **y** axis.

$X[-\sin(\psi)] + Y\cos(\psi)$. The rotation is about axis **K**, perpendicular to both the **I** and **J** axes. Looking on that transformation as a three-dimensional transformation in space with Z along the **K** axis and z along the **k** axis, we have $Z = z$.

Finally, let us label the new values of the body axes as x_1, y_1, z_1. In matrix form we can write

$$\begin{bmatrix} x_1 \\ y_1 \\ z_1 \end{bmatrix} = \begin{bmatrix} \cos(\psi) & \sin(\psi) & 0 \\ -\sin(\psi) & \cos(\psi) & 0 \\ 0 & 0 & 1 \end{bmatrix} \begin{bmatrix} X \\ Y \\ Z \end{bmatrix} = [\mathbf{A}_\psi] \begin{bmatrix} X \\ Y \\ Z \end{bmatrix}. \quad (A.3.2)$$

Initially the body axes were aligned with the reference axes $\mathbf{X}, \mathbf{Y}, \mathbf{Z}$. In Eq. A.3.2, $[\mathbf{A}_\psi]$ is the first angular rotation about the **z** body axis. The next rotation will be about the $\mathbf{y}_1$ axis by an angle θ. Care must be taken as to the direction of rotation so that the right-hand properties of rotation are preserved (see Figure A.3.2).

To simplify the notation, we shall abbreviate $\cos(\_)$ to $c_\_$ and $\sin(\_)$ to $s_\_$. With this convention, the second transformation will take the form

$$\begin{bmatrix} x_2 \\ y_2 \\ z_2 \end{bmatrix} = \begin{bmatrix} c\theta & 0 & -s\theta \\ 0 & 1 & 0 \\ s\theta & 0 & c\theta \end{bmatrix} \begin{bmatrix} x_1 \\ y_1 \\ z_1 \end{bmatrix} = [\mathbf{A}_\theta] \begin{bmatrix} x_1 \\ y_1 \\ z_1 \end{bmatrix}. \quad (A.3.3)$$

The last rotation will about the $\mathbf{x}_2$ axis; the result is

$$\begin{bmatrix} x_3 \\ y_3 \\ z_3 \end{bmatrix} = \begin{bmatrix} 1 & 0 & 0 \\ 0 & c\phi & s\phi \\ 0 & -s\phi & c\phi \end{bmatrix} \begin{bmatrix} x_2 \\ y_2 \\ z_2 \end{bmatrix} = [\mathbf{A}_\phi] \begin{bmatrix} x_2 \\ y_2 \\ z_2 \end{bmatrix}. \quad (A.3.4)$$

Finally,

$$\begin{bmatrix} x \\ y \\ z \end{bmatrix} = \begin{bmatrix} x_3 \\ y_3 \\ z_3 \end{bmatrix} = [\mathbf{A}_\phi][\mathbf{A}_\theta][\mathbf{A}_\psi] \begin{bmatrix} X \\ Y \\ Z \end{bmatrix} = [\mathbf{A}_{\psi\theta\phi}] \begin{bmatrix} X \\ Y \\ Z \end{bmatrix}. \quad (A.3.5)$$

After we multiply the matrices in Eq. A.3.5, we find

$$[\mathbf{A}_{321}] = [\mathbf{A}_{\psi\theta\phi}] = \begin{bmatrix} c\theta\, c\psi & c\theta\, s\psi & -s\theta \\ -c\phi\, s\psi + s\phi\, s\theta\, c\psi & c\phi\, c\psi + s\phi\, s\theta\, s\psi & s\phi\, c\theta \\ s\phi\, s\psi + c\phi\, s\theta\, c\psi & -s\phi\, c\psi + c\phi\, s\theta\, s\psi & c\phi\, c\theta \end{bmatrix}, \quad (A.3.6)$$

$$[\mathbf{A}_{\psi\theta\phi}] = [\mathbf{A}_\phi][\mathbf{A}_\theta][\mathbf{A}_\psi]. \quad (A.3.7)$$

The meaning of $[\mathbf{A}_{\alpha\beta\gamma}]$ is that the first transformation is about the axis with the appropriate α Euler angle, and so on. The remaining five attitude transformations of the first type are as follows:

$$[\mathbf{A}_{231}] = [\mathbf{A}_{\theta\psi\phi}] = \begin{bmatrix} c\psi\, c\theta & s\psi & -c\psi\, s\theta \\ -c\phi\, s\psi\, c\theta + s\phi\, s\theta & c\phi\, c\psi & c\phi\, s\psi\, s\theta + s\phi\, c\theta \\ s\phi\, s\psi\, c\theta + c\phi\, s\theta & -s\phi\, c\psi & -s\phi\, s\psi\, s\theta + c\phi\, c\theta \end{bmatrix}; \quad (A.3.8)$$

$$[\mathbf{A}_{213}] = [\mathbf{A}_{\theta\phi\psi}] = \begin{bmatrix} c\psi\, c\theta + s\psi\, s\phi\, s\theta & s\psi\, c\phi & -c\psi\, s\theta + s\psi\, s\phi\, c\theta \\ -s\psi\, c\theta + c\psi\, s\phi\, s\theta & c\psi\, c\phi & s\psi\, s\theta + c\psi\, s\phi\, c\theta \\ c\phi\, s\theta & -s\phi & c\phi\, c\theta \end{bmatrix}; \quad (A.3.9)$$

$$[\mathbf{A}_{132}] = [\mathbf{A}_{\phi\psi\theta}] = \begin{bmatrix} c\psi\, c\theta & c\theta\, s\psi\, c\phi + s\theta\, s\phi & c\theta\, s\psi\, s\phi - s\theta\, c\phi \\ -s\psi & c\psi\, c\phi & c\psi\, s\phi \\ s\theta\, c\psi & s\theta\, s\psi\, c\phi - c\theta\, s\phi & s\theta\, s\psi\, s\phi + c\theta\, c\phi \end{bmatrix}; \quad (A.3.10)$$

$$[\mathbf{A}_{123}] = [\mathbf{A}_{\phi\theta\psi}] = \begin{bmatrix} c\psi\, c\theta & c\psi\, s\theta\, s\phi + s\psi\, c\phi & -c\psi\, s\theta\, c\phi + s\psi\, s\phi \\ -s\psi\, c\theta & -s\psi\, s\theta\, s\phi + c\psi\, c\phi & s\psi\, s\theta\, c\phi + c\psi\, s\phi \\ s\theta & -c\theta\, s\phi & c\theta\, c\phi \end{bmatrix}; \quad (A.3.11)$$

$$[\mathbf{A}_{312}] = [\mathbf{A}_{\psi\phi\theta}] = \begin{bmatrix} c\theta\, c\psi - s\theta\, s\phi\, s\psi & c\theta\, s\psi + s\theta\, s\phi\, c\psi & -s\theta\, c\phi \\ -c\phi\, s\psi & c\phi\, c\psi & s\phi \\ s\theta\, c\psi + c\theta\, s\phi\, s\psi & s\theta\, s\psi - c\theta\, s\phi\, c\psi & c\theta\, c\phi \end{bmatrix}. \quad (A.3.12)$$

The remaining six transformations (of the second type) can be found in several textbooks (see e.g. Wertz 1978, Hughes 1988).

All the $[\mathbf{A}_-]$ matrices are direction cosine matrices, with the appropriate characterisics of such matrices - in particular, the identity $[\mathbf{A}_-]^{-1} = [\mathbf{A}_-]^T$. It is also important to note that, for small Euler angles, all six Euler transformations (Eq. A.3.6 and Eqs. A.3.8–A.3.12) have the same approximated form. Taking $\sin(\psi) \approx \psi$ and $\cos(\psi) \approx 1$ for small ψ, and with similar approximations for the remaining Euler angles θ and ϕ, we have

$$[\mathbf{A}_{\alpha\beta\gamma}] \approx \begin{bmatrix} 1 & \psi & -\theta \\ -\psi & 1 & \phi \\ \theta & -\phi & 1 \end{bmatrix}. \quad (A.3.13)$$

Attitude rotations derived on the basis of Euler angles necessitate dealing with nine elements of the direction cosine matrix, and each element may include several trigonometric functions. Equivalent, but simpler, transformations can be obtained based on the *quaternions,* to be explained in the next section.

A.4 The Quaternion Method

A.4.1 *Definition of Parameters*

The quaternion's basic definition is a consequence of the properties of the direction cosine matrix $[\mathbf{A}]$. It is shown by linear algebra that a proper real orthogonal 3×3 matrix has at least one eigenvector with eigenvalue of unity. This means that, since one of the eigenvalues λ_i ($i = 1, 2, 3$) is unity, the eigenvector is unchanged by the matrix $[\mathbf{A}]$ (Hildebrand 1968):

$$[\mathbf{A}]\mathbf{e}_1 = 1\mathbf{e}_1. \tag{A.4.1}$$

The eigenvector $\mathbf{e}_1$ has the same components along the body axes and along the reference frame axes. The existence of such an eigenvector is the analytical demonstration of Euler's famous theorem about rotational displacement: *The most general displacement of a rigid body with one point fixed is a rotation about some axis.* In this case, the rotation is about the eigenvector $\mathbf{e}_1$. It will be demonstrated that any attitude transformation in space by consecutive rotations about the three orthogonal unit vectors of the coordinate system can be achieved by a single rotation about the eigenvector with unity eigenvalue.

The quaternion is defined as a vector in the following way (Hamilton 1866, Goldstein 1950, Dalquist 1990):

$$\mathbf{q} = q_4 + \mathbf{i}q_1 + \mathbf{j}q_2 + \mathbf{k}q_3 \equiv q_4 + \boldsymbol{q}, \tag{A.4.2}$$

where the unit vectors $\mathbf{i}, \mathbf{j}, \mathbf{k}$ satisfy the following equalities:

$$\begin{aligned}
\mathbf{i}^2 &= \mathbf{j}^2 = \mathbf{k}^2 = -1, \\
\mathbf{ij} &= -\mathbf{ji} = \mathbf{k}, \\
\mathbf{jk} &= -\mathbf{kj} = \mathbf{i}, \\
\mathbf{ki} &= -\mathbf{ik} = \mathbf{j}.
\end{aligned} \tag{A.4.3}$$

Equation A.4.3 shows that the order of multiplication is important. We also define the *conjugate* of $\mathbf{q}$ as

$$\mathbf{q}^* = q_4 - \mathbf{i}q_1 - \mathbf{j}q_2 - \mathbf{k}q_3. \tag{A.4.4}$$

In the definition of the quaternion $\mathbf{q}$, q_4 is a scalar; $\boldsymbol{q}$ will be defined as the vector part of the quaternion:

$$\mathbf{q} = (q_4, \boldsymbol{q}), \tag{A.4.5}$$

where $\boldsymbol{q} = \mathbf{i}q_1 + \mathbf{j}q_2 + \mathbf{k}q_3$.

A.4.2 Euler's Theorem of Rotation and the Direction Cosine Matrix

An examination of Eq. A.3.3 and Eq. A.3.4 reveals the following equality:

$$\text{tr}[\mathbf{A}_\alpha] = 1 + 2\cos(\alpha). \tag{A.4.6}$$

In the previous section, successive rotations were performed about axes of the body coordinate frame. In general, the rotation can be executed about any direction in the body. Suppose that the body rotates about the Euler axis of rotation, defined by the principal eigenvector, having the eigenvalue of unity; Eq. A.4.6 will hold in this case, too.

Let us call $\mathbf{e}$ the *eigenvector of rotation*, with components $\mathbf{e} = [e_1 \ e_2 \ e_3]^T$. In this case the direction cosine matrix, in terms of the vector $\mathbf{e}$ and the angle of rotation α, will have the following form:

$$[\mathbf{A}_\alpha] = \cos(\alpha)\mathbf{1} + [1 - \cos(\alpha)]\mathbf{e}\mathbf{e}^T - \sin(\alpha)[\mathbf{E}], \tag{A.4.7}$$

where **1** is the unit matrix. We define [**E**] as

$$[\mathbf{E}] = \begin{bmatrix} 0 & -e_3 & e_2 \\ e_3 & 0 & -e_1 \\ -e_2 & e_1 & 0 \end{bmatrix}. \tag{A.4.8}$$

The detailed Eq. A.4.7 then becomes

$$[\mathbf{A}_\alpha] = \begin{bmatrix} c\alpha + e_1^2(1-c\alpha) & e_1 e_2(1-c\alpha) + e_3 s\alpha & e_1 e_3(1-c\alpha) - e_2 s\alpha \\ e_1 e_2(1-c\alpha) - e_3 s\alpha & c\alpha + e_2^2(1-c\alpha) & e_2 e_3(1-c\alpha) + e_1 s\alpha \\ e_1 e_3(1-c\alpha) + e_2 s\alpha & e_2 e_3(1-c\alpha) - e_1 s\alpha & c\alpha + e_3^2(1-c\alpha) \end{bmatrix}. \tag{A.4.9}$$

where again we abbreviate $\cos(\_)$ to $c\_$ and $\sin(\_)$ to $s\_$.

Equation A.4.6 is clearly satisfied by Eq. A.4.9. The matrix $[\mathbf{A}_\alpha]$ is a direction cosine matrix with elements a_{ij}. For a nonvanishing α, we can find the elements of the eigenvector of rotation in terms of a_{ij}:

$$e_1 = [a_{23} - a_{32}]/[2\sin(\alpha)], \tag{A.4.10a}$$

$$e_2 = [a_{31} - a_{13}]/[2\sin(\alpha)], \tag{A.4.10b}$$

$$e_3 = [a_{12} - a_{21}]/[2\sin(\alpha)]. \tag{A.4.10c}$$

A.4.3 Quaternions and the Direction Cosine Matrix

The elements of the quaternions, sometimes called the *Euler symmetric parameters,* can be expressed in terms of the principal eigenvector **e** (see Sabroff et al. 1965). They are defined as follows:

$$q_1 = e_1 \sin(\alpha/2), \tag{A.4.11a}$$

$$q_2 = e_2 \sin(\alpha/2), \tag{A.4.11b}$$

$$q_3 = e_3 \sin(\alpha/2), \tag{A.4.11c}$$

$$q_4 = \cos(\alpha/2). \tag{A.4.11d}$$

Clearly,

$$q_1^2 + q_2^2 + q_3^2 + q_4^2 = 1, \quad |\mathbf{q}| = 1. \tag{A.4.12}$$

Using Eqs. A.4.11 and Eqs. A.4.10, the direction cosine matrix can be expressed in terms of the quaternions:

$$[\mathbf{A}(\mathbf{q})] = (q_4^2 - q^2)\mathbf{1} + 2qq^\mathrm{T} - 2q_4[\mathbf{Q}], \tag{A.4.13}$$

where

$$[\mathbf{Q}] = \begin{bmatrix} 0 & -q_3 & q_2 \\ q_3 & 0 & -q_1 \\ -q_2 & q_1 & 0 \end{bmatrix} \quad \text{and} \tag{A.4.14}$$

$$[\mathbf{A}(\mathbf{q})] = \begin{bmatrix} q_1^2 - q_2^2 - q_3^2 + q_4^2 & 2(q_1 q_2 + q_3 q_4) & 2(q_1 q_3 - q_2 q_4) \\ 2(q_1 q_2 - q_3 q_4) & -q_1^2 + q_2^2 - q_3^2 + q_4^2 & 2(q_2 q_3 + q_1 q_4) \\ 2(q_1 q_3 + q_2 q_4) & 2(q_2 q_3 - q_1 q_4) & -q_1^2 - q_2^2 + q_3^2 + q_4^2 \end{bmatrix}. \tag{A.4.15}$$

A.4 / The Quaternion Method

Equation A.4.15 allows us to express the quaternions in terms of the direction cosine matrix. It is shown in Chapter 4 that both the quaternions and the elements of the direction cosine matrix can be found independently by integrating the angular rates measured about the three principal body axes. Having calculated one set of parameters (e.g., the set of elements of the direction cosine matrix), the set of quaternions can also be found, and vice versa (Klumpp 1976).

A total of four such sets can be defined for the quaternions. For the first set, q_4 is found first by summing the diagonal elements a_{11}, a_{22}, and a_{33}. To find q_1, q_2, and q_3, take the sums $a_{23}-a_{32}$, $a_{31}-a_{13}$, and $a_{12}-a_{21}$, respectively. The solution is

$$\begin{aligned} q_4 &= \pm 0.5\sqrt{1+a_{11}+a_{22}+a_{33}}, \\ q_1 &= 0.25(a_{23}-a_{32})/q_4, \\ q_2 &= 0.25(a_{31}-a_{13})/q_4, \\ q_3 &= 0.25(a_{12}-a_{21})/q_4. \end{aligned} \quad (A.4.16)$$

In those cases where q_4 is a very small number, annoying numerical inaccuracies in calculating the remaining components of the quaternion vector may preclude using Eqs. A.4.16. This problem may be overcome by using a different solution for the quaternions in terms of a_{ij}. This yields three more possible solution sets, as follows:

$$\begin{aligned} q_1 &= \pm 0.5\sqrt{1+a_{11}-a_{22}-a_{33}}, \\ q_2 &= 0.25(a_{12}+a_{21})/q_1, \\ q_3 &= 0.25(a_{13}+a_{31})/q_1, \\ q_4 &= 0.25(a_{23}+a_{32})/q_1; \end{aligned} \quad (A.4.17)$$

$$\begin{aligned} q_3 &= \pm 0.5\sqrt{1-a_{11}-a_{22}+a_{33}}, \\ q_1 &= 0.25(a_{13}+a_{31})/q_3, \\ q_2 &= 0.25(a_{23}+a_{32})/q_3, \\ q_4 &= 0.25(a_{12}-a_{21})/q_3; \end{aligned} \quad (A.4.18)$$

$$\begin{aligned} q_2 &= \pm 0.5\sqrt{1-a_{11}+a_{22}-a_{33}}, \\ q_1 &= 0.25(a_{12}+a_{21})/q_2, \\ q_3 &= 0.25(a_{23}+a_{32})/q_2, \\ q_4 &= 0.25(a_{31}-a_{13})/q_2. \end{aligned} \quad (A.4.19)$$

Equations A.4.16–A.4.19 present four different solutions, but the resulting **q**s are identical.

In summary, all the listed quaternion sets are mathematically equivalent. Hence numerical inaccuracies can be minimized by changing between them and using that set for which the divisor component is maximal.

A.4.4 Attitude Transformation in Terms of Quaternions

Representing the attitude of a body in a reference frame by a direction cosine matrix requires knowledge of nine parameters a_{ij}, whereas only four q_i parameters are needed when using the quaternions. (In fact, there exist only six *independent* parameters of the direction cosine matrix and only three independent components of the quaternion vector.) Moreover, the elements of the direction cosine matrix, in contrast to those of the quaternions, are trigonometric functions, which are much more cumbersome to compute.

To obtain an attitude transformation, a *quaternion multiplication* is performed. When dealing with direction cosine matrices, two consecutive attitude transformations are achieved by matrix multiplication of the two individual rotations. These two rotations can be expressed in the quaternion terminology by $[\mathbf{A}(\mathbf{q})]$ for the first rotation and by $[\mathbf{A}(\mathbf{q}')]$ for the second one. The following expression holds for the overall attitude transformation in terms of direction cosine matrices:

$$[\mathbf{A}(\mathbf{q}'')] = [\mathbf{A}(\mathbf{q}')][\mathbf{A}(\mathbf{q})]. \tag{A.4.20}$$

The resulting quaternion $\mathbf{q}''$ can be extracted from $[\mathbf{A}(\mathbf{q}'')]$.

Fortunately, it is much easier to perform a direct quaternion multiplication. Suppose we use the terminology of Section A.4.1. If we define $\mathbf{q}$ and $\mathbf{q}'$ as in Eq. A.4.2 and use the relationships of Eqs. A.4.3, we obtain

$$\begin{aligned}\mathbf{q}'' = \mathbf{q}\mathbf{q}' &= (-q_1 q_1' - q_2 q_2' - q_3 q_3' + q_4 q_4') \\ &+ \mathbf{i}(q_1 q_4' + q_2 q_3' - q_3 q_2' + q_4 q_1') \\ &+ \mathbf{j}(-q_1 q_3' + q_2 q_4' + q_3 q_1' + q_4 q_2') \\ &+ \mathbf{k}(q_1 q_2' - q_2 q_1' + q_3 q_4' + q_4 q_3')\end{aligned} \tag{A.4.21}$$

(see Dalquist 1990). Equation A.4.21 can also be put in matrix multiplication form; the final result is

$$\begin{bmatrix} q_1'' \\ q_2'' \\ q_3'' \\ q_4'' \end{bmatrix} = \begin{bmatrix} q_4' & q_3' & -q_2' & q_1' \\ -q_3' & q_4' & q_1' & q_2' \\ q_2' & -q_1' & q_4' & q_3' \\ -q_1' & -q_2' & -q_3' & q_4' \end{bmatrix} \begin{bmatrix} q_1 \\ q_2 \\ q_3 \\ q_4 \end{bmatrix}. \tag{A.4.22}$$

In other words, if the quaternion components of two successive rotations are known, then Eq. A.4.22 is the matrix vector multiplication that gives the resulting quaternion components of the total rotation.

A.5 Summary

The most efficient mathematical representations of spacecraft attitude are based on the Euler angles, the direction cosine matrix, and the quaternion vector. The analytical transformation techniques for these representations have been explicitly stated for reference and use in the chapters.

References

Dalquist, C. (1990), "Attitude Algebra Software: A Computer Tool for the Quaternion Approach to Satellite Attitude Maneuvers," Paper no. 90-034, American Astronautical Society, San Diego, CA.

Goldstein, H. (1950), *Classical Mechanics*. Reading, MA: Addison-Wesley.

References

Hamilton, W. R. (1866), *Elements of Quaternions*. London: Longmans, Green.
Hildebrand, F. B. (1968), *Methods of Applied Mathematics*. New Delhi: Prentice-Hall.
Hughes, P. C. (1988), *Spacecraft Attitude Dynamics*. New York: Wiley.
Klumpp, A. R. (1976), "Singularity Free Extraction of a Quaternion from a Direction Cosine Matrix," *Journal of Spacecraft and Rockets* 13: 754–5.
Sabroff, A., Farrenkopf, R., Frew, A., and Gran, M. (1965), TRW Report AFFDL-TR-65-115, TRW, Redondo Beach, CA.
Wertz, J. (1978), *Spacecraft Attitude Determination and Control*. Dordrecht: Reidel.

APPENDIX B

Attitude Determination Hardware

B.1 Introduction

Hardware items that are mandatory for realizing almost any spacecraft attitude and orbit control system can be divided into two classes: instrumentation for measuring the attitude of the satellite; and instrumentation for providing forces and torques. The latter category will be treated in Appendix C. Appendix B deals with attitude sensors, but is not an extensive treatment of the subject. There are excellent textbooks and technical papers providing complete treatment of the hardware from both analytical and practical points of view; Wertz (1978) is especially recommended. However, for completeness and for the reader's convenience, a short exposition of the basic principles of satellite attitude hardware will be presented here, together with examples of existing space-proven commercial instruments.

Attitude measurement hardware is used to determine the attitude of the satellite with respect to a defined reference frame. The final product may be, for instance, the Euler angles of the satellite in the orbit reference frame, or (in a different context) the sun vector components in the body axis frame. Attitude determination hardware includes:

(1) earth sensors (in particular, infrared earth sensors);
(2) sun sensors;
(3) star sensors;
(4) rate and rate integrating sensors, based on gyroscopic, laser, or other solid-state principles; and
(5) magnetometers.

The quality of the instruments is responsible for the accuracy that can be achieved in the attitude control system. For instance, there are sun sensors that can measure the direction of the sun with an accuracy of $0.015°$, whereas others have an accuracy of only $0.5°$; however, analytical processing of the two sensors' output is basically the same. The commonly used attitude reference sources are the earth, the sun, and the stars.

The earth is used in two different aspects, optical and magnetic. The more important one is the optical aspect. Unlike the sun (which appears as a small illuminated disk) or the stars (which can be treated as illuminated points), the earth – as seen from nearby space – has a complex appearance that must be adequately modeled for accurate attitude determination. The accuracies that can be achieved with the earth as an attitude source range from $0.02°$ to $0.5°$, depending on the complexity of the hardware and the processing algorithm. The earth's magnetic field, which is known from different sorts of evaluations and measurements, is also used for attitude determination.

The attitude accuracies that can be achieved with the sun as the attitude source are in the region of 0.015° for the best available instruments. For earth-orbiting satellites, the earth and sun sensors are two complementary items that together enable determination of the complete three-axis attitude of the satellite.

For very high attitude accuracies, it is mandatory to use the most accurate attitude sources – the stars. Accuracies in the sub–arc-second range are possible, albeit with instrumentation that is quite complicated and not always reliable.

B.2   Infrared Earth Sensors

The goal of the infrared earth sensor (IRES) is to determine the spacecraft orientation relative to the earth. This sensor operates in one of two principal modes. The first is based on dynamic crossing of the earth's horizon and exact determination of the crossing points; the second mode is based on static determination of the location of the earth's contour inside the instrument's field of view. Both kinds of sensors will be briefly explained and compared.

B.2.1   *Spectral Distribution and Oblateness of the Earth*

In principle, instruments could be based on the earth *albedo,* which is the fraction of reflected radiation from the globe due to all incident energy that falls on its surface. Most of the albedo's energy is in the visual region of the spectrum. Attitude sensors based on the albedo are sometimes used, but they are neither very accurate nor effective because there exists a strong variation in albedo for different refracting surfaces. For instance, the fraction of radiation reflected back into space is very low for some vegetation-covered areas, of the order of 0.06, whereas the reflected ratio for snow-covered surfaces can be as high as 0.8. We must also define the *terminator* – that is, the boundary between day and night on the earth (or any planet). For the albedo, the terminator is not sharply defined; hence we cannot expect to trigger on a well-defined boundary of the horizon, which is the principal requirement for accurate and effective attitude sensing.

The appearance of the earth in the infrared (IR) range is much better for attitude sensing. This is because the energy within that spectrum emitted from all parts of the earth's surface is much more homogeneous, and defines more sharply the profile of the globe (in other words, its *horizon*).

The infrared radiation from the earth is the total thermal radiation from both the surface and the atmosphere. The intensity variations are much smaller than those of the visual albedo. The emitted infrared radiation has a spectral energy distribution that is affected by the temperature of the earth's surface and also by the atmosphere's chemical composition. An approximate spectral distribution of thermal emission from the earth is shown in Figure B.2.1 (overleaf).

In the figure, strong absorption bands are detected that are due to ozone (O_3) and carbon dioxide (CO_2). In fact, in this spectral range, the re-emitted radiated energy is due to the atmosphere above the earth's surface. In the 14.0–16.3 μm (CO_2) spectrum the earth is seen as having a uniform distribution, so this is the most suitable spectral range for attitude determination. Most IR earth sensors use this spectrum

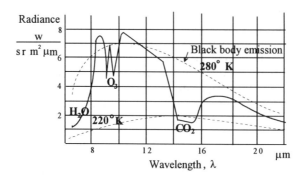

Figure B.2.1 Spectral distribution of thermal emission from the earth; adapted from Lyle, Leach, and Shubin (1971).

for sensing the horizon of the earth; for this reason, earth sensors are often called *horizon sensors*. Because the CO_2 layer is part of the atmosphere extending to an altitude of several tens of kilometers, the sharpness of horizon detection is not absolute. This disadvantage will be discussed in more detail later.

Another important feature related to satellite attitude determination is the earth's oblateness. There are many complex "surface models" used to represent the earth's geometrical shape. In the literature, a relatively simple model is used for purposes of attitude determination. The earth globe is modeled as an ellipsoid (Muller and Jappel 1977); the earth is an ellipsoid rotating about its minor axis, which models the flattening of the globe at the poles. If $R_e = 6{,}378.14$ km is the earth's mean equatorial radius and $R_p = 6{,}356.75$ km is its polar radius, then the flattening is defined as $f = (R_e - R_p)/R_e \approx 0.00335281$.

A simple expression for computing the radius of the earth at different latitudes λ is used in connection with IR sensors that trigger on the atmosphere layer boundary:

$$R = R_e[1 - f\sin^2(\lambda) + k\sin(\lambda)] + h, \tag{B.2.1}$$

where f is the previously defined flatness factor, h is the altitude of the atmosphere, and k is the parametric latitudinal variation of the height of the atmosphere, due to various causes.

B.2.2 Horizon-Crossing Sensors

An infrared horizon-crossing earth sensor (IRHCES) consists of four essential components: a *scanning mechanism,* an *optical system,* a *radiance detector,* and a *signal processing system*. Together, the first three are usually called the *optical sensor head*. The signal processing system is part of the electronic box, which also includes the different power supplies necessary to operate the optical head. The basic principle of a single-scan sensor head is shown in Figures B.2.2 and B.2.3.

Optical System and Scanning Mechanism

The optical system consists of a filter that limits the observed spectral band to the CO_2 range of 14–16 μm, together with a lens that focuses the earth image on the radiance detector, the *bolometer*. Different scanning mechanisms are based on

B.2 / Infrared Earth Sensors

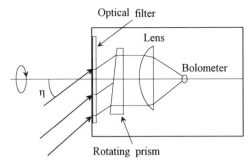

Figure B.2.2 Single-scan horizon sensor principle; adapted from ITHACO (1983) by permission of ITHACO Space Systems.

various principles. For example, some instruments are based on a rotating wedge-shaped prism in front of the lens, as in Figure B.2.2. The image is deflected by η degrees before falling on the bolometer.

The rotating prism scans a ring of the space, and when the field of view crosses the earth, the radiance detector senses its presence (Figure B.2.3). The scanning angle η is optimized according to the geometrical parameters of the system, such as the altitude of the spacecraft and its maximum expected deviation from the nadir direction (the *nadir* is the vector connecting the satellite cm with the earth cm). With a single scanning arrangement, the altitude of the satellite must be known exactly in order to determine both roll and pitch deviations in the reference frame. (With the dual scanning arrangement of Figure B.2.4, knowledge of the altitude is not necessary; moreover, with this arrangement the altitude can be calculated as a byproduct.) The arrangement in Figure B.2.2 is used by ITHACO Space Systems for their conical earth sensor (model no. IPS-6), used in numerous satellites such as *LANDSAT D, P80-1,* and *P78-2* (see ITHACO 1983).

The computation of the roll and yaw Euler angles for small deviations from the nadir is very simple. Figure B.2.3 shows the principle of operation of a single-cone

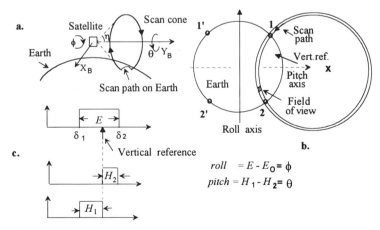

Figure B.2.3 Determination of the roll and pitch attitudes with a single cone; adapted from ITHACO (1983) by permission of ITHACO Space Systems.

arrangement. In Figure B.2.3.a are shown the satellite's $\mathbf{X}_B$ and $\mathbf{Y}_B$ axes; the $\mathbf{Y}_B$ axis is also the axis of rotation of the optical rotating mechanism. The scan cone and its scan path on the earth's surface are also shown. In Figure B.2.3.b, a vertical reference is fixed to the conical optical head. An angular optical encoder in the optical head measures the phase angle between the conical beam horizon crossings and the vertical reference inside the head. With the geometrical definitions in Figure B.2.3.b, the vertical reference is necessary for determining the pitch attitude. In the ES signal processing, the phase angles of the horizon crossing with respect to the vertical reference are labeled δ_1 and δ_2. The first horizon crossing is at point 1, going into the earth atmosphere. The second horizon crossing, at point 2, is going out of the atmosphere. The phase angle difference E between points 1 and 2 is shown in Figure B.2.3.c. This difference is directly proportional to the roll angle. The value of $E = \delta_2 - \delta_1$ depends on the satellite's altitude, and also on the deflection angle η shown in Figure B.2.2. Consequently, a normalizing function E_0 is necessary in order to determine the roll angle from the phase difference E. Finally,

$$\text{roll} = \phi = \frac{\delta_2 - \delta_1}{C_1} - E_0, \tag{B.2.2}$$

where C_1 and E_0 are parameters, the latter depending on the altitude of the satellite.

With no pitch orientation, the two horizon crossings are symmetrically positioned around the vertical optical reference fixed in the optical scanner head. However, with existing pitch inclination the pitch angle is proportional to the difference between H_1 and H_2. Given δ_1 and δ_2 (phase angles measured with respect to the vertical reference of the optical head), we have

$$\text{pitch} = \theta = \frac{-(\delta_2 + \delta_1)}{2C_2} + \frac{90° + C_3}{C_2} + C_r\phi, \tag{B.2.3}$$

where C_2, C_3, and C_r are constants. It is important to mention that, for large roll deviations, the pitch angle depends on the roll orientation.

Some comments related to the geometry of Figure B.2.3 are now necessary.

(1) The pitch and roll angles that the conical ES can measure are limited. As long as the crossing points 1 and 2 in Figure B.2.3.b are located in the right half of the earth's contour as seen from the satellite, there should be no ambiguity in the computation of the roll angle. However, if the satellite's inclination about the $\mathbf{X}_B$ axis brings points 1 and 2 to 1' and 2' on the left half of the globe, then the roll angle will not be defined correctly. With the same geometrical arrangement as in Figure B.2.3, a much larger pitch angle can be measured. Theoretically, with no roll orientation, the satellite can turn about its $\mathbf{Y}_B$ axis without any restriction while still keeping the horizon crossings at points 1 and 2. The pitch angle can be calculated without ambiguity with the aid of the vertical reference of the optical head. With a finite roll orientation, the pitch attitude that can be measured will naturally be limited. For a given satellite altitude, the defined operational pitch-roll range can be optimized via the scanning cone angle η.

(2) The scan field of view is optimized according to two principal arguments. A wider field of view collects more energy, thus increasing the signal-to-noise ratio of the complete IRES assembly. On the other hand, increasing the field of view decreases the sharpness with which the horizon crossing is defined. Fields of view of the order of 1×1 to 2×2 [deg] are commonly used.

B.2 / Infrared Earth Sensors

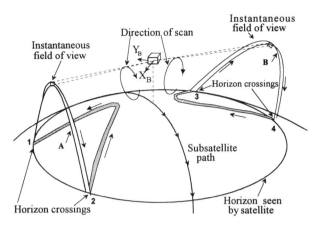

Figure B.2.4 Dual-beam scanning horizon sensor; adapted from Gontin and Ward (1987) by permission of AIAA.

(3) The optical head can be tilted about the body axes for different operational needs – for instance, in order to achieve a desired angular range for attitude determination. With appropriate geometrical positioning within the satellite body, the same optical head can be used at different altitudes – for example, in the transfer orbit stage as well as the final mission orbit.

An arrangement using two scanning paths, as in Figure B.2.4, allows determination of the roll angle with no dependence on the satellite's altitude (see Gontin and Ward 1987). By use of two back-to-back single-cone IR earth sensors, four horizon crossings can be obtained. With the two scan paths (labeled A and B in Figure B.2.4), the attitude determination is independent of altitude information:

$$\text{roll} = \phi = \frac{(\delta_2 - \delta_1) - (\delta_4 - \delta_3)}{4C_1}, \tag{B.2.4}$$

$$\text{pitch} = \theta = \frac{(\delta_4 + \delta_3) - (\delta_2 + \delta_1)}{4C_2} - \frac{90°}{C_3} + C_r\phi, \tag{B.2.5}$$

where δ_3 and δ_4 are defined for the second scan path as in Figure B.2.3. Equations B.2.2–B.2.5 hold only for small deviations of the roll and pitch angles from null. For larger deviations in roll and pitch, the equations must incorporate correcting factors that depend on the altitude and attitude of the satellite.

This configuration, with two back-to-back identical single scanning horizon sensors, has been implemented in numerous satellites. The drawback of such an arrangement is that two complete horizon sensor assemblies are required, increasing the cost of the attitude determination system. To overcome this deficiency, dual-cone optical heads have been designed: the same optical head provides two independent scanning cones, thus achieving the dual-paths arrangement of Figure B.2.4 with only one optical head.

The dual-cone optical head uses two fixed mirrors, which view two different but complementary halves of the earth surface. A rotating mirror scans both fixed mirrors, thus producing the two scanning paths of Figure B.2.4. Such an arrangement is incorporated in the IR earth sensors produced by SODERN – STD 15 (SODERN

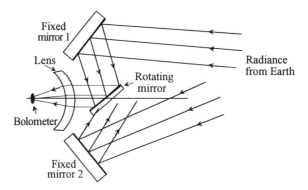

Figure B.2.5 Dual-scan, half-cone design (STD 12); adapted from Desvignes et al. (1985) by permission of SPIE.

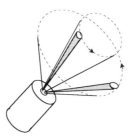

Figure B.2.6 The Barnes dual-cone scanner principle; adapted from Bednarek (1992) by permission of SPIE.

1991a) and STD 16 (SODERN 1991b; see also Pochard 1992). A simplified portrayal of their predecessor, the STE 12 dual-scan optical head, is shown in Figure B.2.5. Another variation of a dual-cone scanner is produced by the Barnes Engineering Division of EDO; the principle of operation is shown in Figure B.2.6. See also Tai and Barnes (1989).

The dual-beam horizon sensor of Figure B.2.4 achieves more accurate attitude determination than the single-beam arrangement of Figure B.2.3. The principal disadvantage is that the angular determination range might be significantly decreased. When the s/c changes its roll attitude, for certain orientations only one of the scanning paths crosses the horizon, while the second path no longer sees the globe. In this case it is common to continue the attitude determination with only one crossing path (and with decreased accuracy).

In fact, direct calculation using Eqs. B.2.2–B.2.5 is not in itself sufficient to achieve a good attitude determination. We must also make corrections for the physical effects mentioned in Section B.2.1, such as the finite equivalent altitude of the atmosphere and its stability with respect to the CO_2 layer, the oblateness of the earth globe, seasonal and geographical earth radiance variations in IR emission, motor-speed variations of the scanning mechanism, the FOV/horizon-crossing angle, and so forth.

According to Figure B.2.4, there are four vector orientations from the ES to the horizon-crossing locations, at points 1, 2, 3, and 4. From a mathematical point of view, three crossings are sufficient to compute the desired pitch and roll angles. This makes the dual-beam principle attractive given the danger of contamination of the crossing edge by the light of a third body, such as the sun or the moon. By using ephemeris data it is possible to predict when the sun or the moon would interfere with one of the scan beams. In this case, data from one of the horizon crossings is disregarded and the attitude determination is calculated on the basis of the three measurements not contaminated by the light of the third body.

The most delicate stage in attitude derivation using IR horizon sensors is the exact location of the horizon. The horizon is not a sharp boundary, owing to the existence of an atmosphere whose equivalent height is finite and whose composition includes different gas layers of various emission wavelengths. The emitted IR energy is dependent on such factors as the seasons, the geographical location on the earth globe, and momentary cloud conditions. Hence, with any technique used to locate the horizon, there exists a specific noise component in determining the horizon crossings. The level of this noise is an important factor when classifying the quality of the entire IR horizon sensor assembly. Irregularities of the IR horizon of the earth, together with the earth's oblateness, are responsible for what is known in the technical literature as *bias attitude error*. Part of the bias errors can be estimated, and the resulting values can be used to correct the attitude. On the other hand, the electronic processing of the individual horizon crossing data is reponsible for *statistical random errors,* expressed in degrees RMS.

Determining the Horizon Crossing

There are two basic techniques for electronically locating crossings of the horizon. In the first technique, energy collected by the bolometer is integrated when the field of view of the optical cone path approaches the earth's atmosphere. A predetermined value of the integrated radiance serves as a locator for the horizon crossing; see Wertz (1978). Alternatively, optical location of the horizon crossing can be based on doubly differentiating the collected radiance energy. This is a very efficient way to locate the horizon; see Fallon and Selby (1990). Formulas for computing pitch and roll from a dual-cone scanner can be found in Tai and Barnes (1989) and Bednarek (1992).

Horizon-Crossing Algorithm

An outline of the algorithm used by an horizon-crossing IRES is as follows.

(1) *Motor-speed correction* – Using motor-speed measurements provided by the instrument electronics, errors due to motor speed are removed.
(2) *Association of horizon crossing to the pertinent scan cone* – If a dual-scan mechanism is used, the horizon crossing must be associated to the correct cone.
(3) *Correction for sensor phase alignment* – The nominal alignment offset of each scan cone relative to the motor phase reference is subtracted from the raw horizon-edge measurement.
(4) *Initial assumption of satellite attitude* – Estimation of initial s/c attitude is necessary to correct for field-of-view/horizon-crossing angle and for other effects that depend on the geographic location of the horizon-crossing edge.

(5) *Correction for earth's oblateness* – The horizon-crossing edge is corrected when the earth radius at the horizon crossing edge is not equal to the earth radius at the equator.

(6) *FOV/horizon correction* – The incident angle of the FOV (field of view) crossing the horizon affects the accuracy of the edge measurement, and the relevant error can be compensated.

(7) *Attitude calculation* – This is performed according to the special formulas used for the relevant IRES assembly.

(8) *Correction for variance in earth radiation* – The earth radiance depends on the geographical location of the crossing edge and also on the season. The correction is performed periodically according to existing atmospheric models.

(9) The satellite yaw attitude has an impact on the roll–pitch calculation.

Step (7) deserves special attention, since it is the heart of the entire algorithm.

Attitude Calculation

The roll and pitch angles are first calculated in the optical sensor head. There are four vectors connecting the center of the sensor to the earth horizon-crossing edges 1–4 in Figure B.2.4. In fact, only three are necessary for computing the roll and pitch angles ϕ and θ; the fourth crossing is redundant data that can nonetheless increase the attitude determination accuracy. If only two earth crossings are available, knowledge of the s/c altitude is necessary in order to calculate its attitude.

We assume for this algebraic analysis that the Euler roll and pitch orientations are to be derived. Since the attitude transformation between the axes of the satellite's body frame and the axes of the sensor's frame is known by definition, the components of the nadir vector in body axes are easily calculated using one of the transformations presented in Appendix A. The definition of the sensor axes is shown in Figure B.2.7.a. In this figure, η is half the cone angle of the earth sensor, whose optical axis is Y_s; δ is the phase angle of the sensor's conical field of view with respect to the optical vertical reference; and δ_1 and δ_2 are the phase angles of the horizon-crossing points 1 and 2 with respect to the optical vertical reference line.

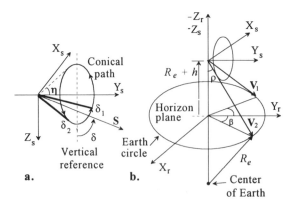

Figure B.2.7 Geometry of scanning, and definition of the horizon-crossing vectors in the sensor's axis frame.

B.2 / Infrared Earth Sensors

In Figure B.2.7.b, $\mathbf{V}_1$ and $\mathbf{V}_2$ are the two vectors joining the center of the sensor with the two horizon-crossing points. The *horizon plane* is the plane containing the two crossing points 1 and 2; it is perpendicular to the line joining the center of the earth with the center of the sensor. This plane intersects the globe in a circle – called the *earth circle* – that contains all the horizon crossing points; a third and fourth crossing (if a second conical FOV is used) will also lie on this circle. In this case we would also have vectors $\mathbf{V}_3$ and $\mathbf{V}_4$ (not shown in the figure). The angle ρ between each of the four vectors and the nadir vector is half the angle by which the earth is seen from the satellite at altitude h. This angle is easily computed from the relation

$$\rho = \sin^{-1}(R_e + H_{eq})/(R_e + h), \tag{B.2.6}$$

where R_e is the radius of the earth, h the altitude of the satellite, and H_{eq} the equivalent height of IR radiation above the earth surface (~ 40 km).

In the reference frame, $\mathbf{V}_1$ can be expressed as

$$\mathbf{V}_1 = -\sin(\rho)\sin(\beta)\mathbf{X}_r + \sin(\rho)\cos(\beta)\mathbf{Y}_r + \cos(\rho)\mathbf{Z}_r, \tag{B.2.7}$$

where the subscript "r" denotes "reference" (see Figure B.2.7.b).

Next we consider the equation of the conical path in the sensor frame. In Figure B.2.7.a, $\mathbf{S}$ is a unit vector representing a momentary direction of the sensor's field of view:

$$\mathbf{S} = \sin(\delta)\sin(\eta)\mathbf{X}_s + \cos(\eta)\mathbf{Y}_s + \cos(\delta)\sin(\eta)\mathbf{Z}_s. \tag{B.2.8}$$

The vector $\mathbf{S}$ is expressed in the sensor axis frame. In order to express it in the reference frame we must perform two transformations, about the pitch (θ) and the roll (ϕ) angular orientations:

$$\mathbf{S}_r = \begin{bmatrix} \cos(-\theta) & 0 & -\sin(-\theta) \\ 0 & 1 & 0 \\ \sin(-\theta) & 0 & \cos(-\theta) \end{bmatrix} \begin{bmatrix} 1 & 0 & 0 \\ 0 & \cos(-\phi) & \sin(-\phi) \\ 0 & -\sin(-\phi) & \cos(-\phi) \end{bmatrix} \begin{bmatrix} \sin(\eta)\sin(\delta) \\ \cos(\eta) \\ \sin(\eta)\cos(\delta) \end{bmatrix}. \tag{B.2.9}$$

With two conical paths we can have four horizon crossings, from which we can calculate four vectors $\mathbf{V}_i$ and four vectors $\mathbf{S}_i$ (which must be equal in the reference frame). If we perform the two vector products then we can find the components of $\mathbf{S}_i$. Only the Z component is important to use because, together with Eq. B.2.7, we find that

$$\mathbf{S}_{rZi} = -\sin(\eta)\sin(\delta_i)\sin(\theta) + \cos(\eta)\sin(\phi)\cos(\theta)$$
$$+ \sin(\eta)\cos(\delta_i)\cos(\phi)\cos(\theta) = \cos(\rho). \tag{B.2.10}$$

If the altitude is known, $\cos(\rho)$ can be calculated from Eq. B.2.6. Next, we need two equations to solve for the desired orientations θ and ϕ, which means that one conical path is sufficient. If the altitude of the satellite is unknown, we need three equations to solve for ϕ and θ.

In our analysis we have used an algebraic approach. Alternatively, the four potential vectors $\mathbf{V}_i$ could be calculated in terms of δ_i, with no reference to θ and ϕ; see Tai and Barnes (1989). The components of the nadir vector in the sensor frame are directly calculated, from which the roll and pitch orientations can be extracted. See also Hablani (1993).

The Infrared Earth Sensor STD 15 is devoted to the attitude measurement of three axis stabilized spacecrafts in geostationary orbit.

Its main characteristics are its light weight and its low power consumption in operation. In addition, it includes new high reliability technologies.

DESIGN CONCEPT

The Infrared Horizon Sensor STD 15 measures the transitions (leading and trailing edges) of the Earth with two traces in normal mode. The four transitions which are detected allow to restitute the attitude deviation of a three axis stabilized spacecraft either with respect to the geocentric direction or a predefined direction whatever the spacecraft altitude is; this latter one is not mandatory to determine these deviations.

The alignment deviations are unambigously delivered in both roll and pitch axes. When this concept is used two sensors are generally necessary to get these informations. This design is one of the main advantages of the infrared Earth Sensor STD 15 when large depointings are envisaged.

The half cone angle of the useful optical path is about 84° the axis of which is oriented in the opposite directions to the Earth center (at nil depointing). This primary scanning cone is intercepted by two flat mirrors symetrically placed at 45° to its axis and are part of the sensor structure.

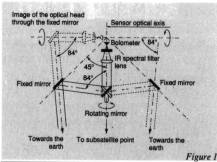

Figure 1

The images of the primary cone through the two fixed mirrors are two half cones the axis of which are perpendicular to the primary cone axis.

The Earth sensor is placed on the satellite such as the half cone axes are perpendicular to the orbit plane. The plane defined by the half cone axes and the primary cone axis are perpendicular to the orbit plane and includes the straight line between the satellite and the Earth center at nil depointing. On figure 2 are represented the two traces on the Earth disc.

An optical encoder linked to the scanning mechanism measures digitally the angle between the Earth path and a reference determined with respect to the reference axes of the satellite. This reference is part of the encoder.

The transition of the trailing and leading edges are detected and their positions are defined by φ_1 and φ_2 for one half cone, φ_3 and φ_4 for the other one. The satellite orientation is defined by the following equations:

- X deviation (pitch) = $\dfrac{(\varphi_3 + \varphi_4) - (\varphi_1 + \varphi_2)}{4}$ + constant

- Y deviation (roll) = $\dfrac{(\varphi_4 - \varphi_3) - (\varphi_2 - \varphi_1)}{4 \times \text{constant}}$

These formulae are valid whatever the altitude.

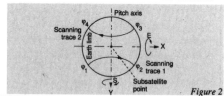

Figure 2

Figure B.2.8 SODERN's STD 15 IR horizon-crossing earth sensor; reproduced by permission of SODERN.

Scanning Rate

The scanning rate of horizon-crossing sensors is important in the design stage of the attitude control loops. It is well known that the maximum gain bandwidth obtainable in a sampled feedback control loop is constrained by the sampling rate of the control processor (Sidi 1980). Hence, a high scanning rate is always pref-

MAIN CHARACTERISTICS

FUNCTIONAL CHARACTERISTICS

Instantaneous field
of view : 1.8 x 1.8 arc degree

Altitude range
– With 4 transitions : 20 000 - 45 000 km
– Usable up to : 125 000 km

Operational field of view
– Acquisition
 . Roll : +14.5 arc degrees (Pitch = 0)
 . Pitch : +12.5 arc degrees (Roll = 0)
   - 15 arc degrees (Roll = 0)

– Normal Mode
 . Roll : ±1.5° arc degree
 . Pitch : ±10° arc degrees

– Transfer orbit mode (starting at 20.000 km)
 . Roll : ±6 arc degrees
 . Pitch : ±6 arc degrees

ACCURACY

Errors $\times 10^{-3}$ Arc.degree	Normal Mode 4 transitions $\omega = 1.25$ rps	Station Keeping mode $\omega = 5$ rps
Constant 10	10	
Periodical:		
• Thermal (30°-60°C)	5	5
• Seasonal Radiance Variations	10	10
• Random Radiance Variations	3	4
Ageing	9	19
• Noise (1 meas) (1 std - 30°C)	5	15

ENVIRONMENTAL CHARACTERISTICS

– Functional temperature range : -25° +55°C
– Storage : -40° +60°C
– Vibrations:
  random (20 - 2000 Hz) : 20 grms

MECHANICAL INTERFACES

– Dimensions:
 . length : 206.5 mm
 . width : 206.5 mm
 . height : 168.0 mm
– Weight : 3.4 kg
– Fixing interface: 4 holes, diameter 4.4 mm

ELECTRICAL INTERFACES

– Power consumption (typical value): 7.5 W
– Power supply range: 22 - 50 Volts

The output data of the sensor correspond to the four transitions Earth-space and Space-Earth. The angular deviations in both roll and pitch axes with respect to the Geocentric axis may be generated at sensor level upon request.

RELIABILITY (MIL HDBK 217E)

(For $\theta = 30$°C)
$\lambda_{30} = 945 - 10\text{-}9$ h^{-1}

LIFETIME (in geostationary orbit)

15 years

Figure B.2.8 (continued)

erable, as long as the accuracy of the horizon sensor will not be compromised. Scanning rates from 1 to 8 revolutions per second are common in today's horizon sensors. Overall accuracies of better than 0.1° (for low orbits) and 0.02° (for geosynchronous orbits) have been achieved in numerous applications. Statistical errors of the order of 0.03°–0.1° (3σ) are common in today's IR horizon sensor assemblies.

B.2.3 IRHCES Specifications

One of the tasks of Appendix B is to give the reader a general idea of the problems encountered in using hardware such as an IRES based on horizon crossings. In addition to fundamental characteristics pertaining to the hardware's principal

B / Attitude Determination Hardware

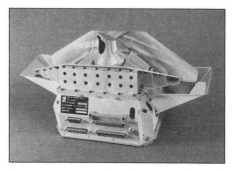

The STD 16 infrared horizon-scanning sensor is designed for attitude control, with respect to the Earth, of three-axis stabilized satellites in low Earth orbits (500 to 1200 km). The operating accuracy is better than 0.1° (3 stand dev.)

This type of sensor could also be used for any elliptic orbits.

This Infrared Earth sensor is derived from the model STD 12 which flew successfully on board the SPOT 1, 2, 3 and ERS 1 and 2 satellites. In particular, the rotating mirror technology, based on dry lubricated bearings, has been retained.

OPERATING PRINCIPLE

The figure opposite shows the satellite at a given point (0) on its orbit around the Earth (0 being the origin of the orthogonal spacecraft body axes OXYZ).

The STD 16 pencil beam is 2° x 2°. The radiation in the 14-16 μm band is selected.

The sensor scans mechanically using a rotating mirror inclined at 60° to the detector-telescope axis. The locus of the scan beam thus formed is a cone with a half-angle of 60° and its axis parallel to the satellite - OZ axis (the Earth center direction). Two plane mirrors canted at 45° transform this scanning cone into two half-cones intercepting the Earth's surface. The axes of the two half-cones are parallel to OX and - OX respectively (and perpendicular to OZ). The half-angles of the scanning half-cones are still 60° (i.e $\alpha = 60°$).

The heavy lines in the figure (labelled trace 1 and trace 2) represent the intersection of the two scanning beams with the Earth's surface.

The detector senses the transitions between the infrared radiance of the Earth and that of space, so the instrument measures the angles $\emptyset_1$, $\emptyset_2$, $\emptyset'_1$ and $\emptyset'_2$ at which the scanning beam is tangential to the Earth. These angles are detected by the mirror shaft angle encoder relative to the satellite reference system (which here is assumed to be identical to the sensor reference system).

The satellite misalignment around OX is given by $(\emptyset_1 + \emptyset_2)/2$ or $(\emptyset'_1 + \emptyset'_2)/2$, or both; while the around misalignment about OY is a function of $(\emptyset'_1 + \emptyset'_2) / (\emptyset_1 + \emptyset_2)$. The geometry of the sensor has been optimized for orbital altitudes in the range 600-1200 km.

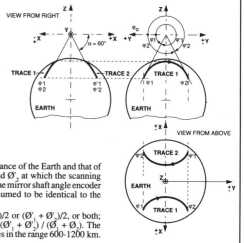

FUNCTIONAL DESCRIPTION

The optical head consists essentially of a rotating mirror, an infrared telescope and two inclined mirrors.

The rotating mirror, made of optically flat machined aluminium, is located in front of the infrared telescope and rotates about the telescope axis at 60 r.p.m. It is driven by a brushless dc motor mounted on dry-lubricated bearings. The telescope consists of a germanium objective lens (focal length : 25 mm, diameter : 20 mm), associated with a 14-16 μm band pass filter, and a germanium-immersed bolometer.

Both the objective lens and the bolometer immersion lens are coated for minimum reflection losses at 15 μm. The two plane inclined mirrors are mounted symmetrically on either side of the telescope axis.

The bolometer bias voltage has been chosen so as to maximize the responsivity in the - 20 to + 45°C temperature range.

After amplification and processing, the four angular data corresponding to the space-Earth and Earth-space transition angles relative to the X-axis are available as 12-bit words from two power C-MOS registers which are available every second for the spacecraft on-board computer. The amplification and processing circuits have been designed for minimal noise. An automatic threshold eliminates practically the influence of local variations in the infrared radiance of the Earth. Other electronic functions performed include : mirror position readout, motor control, and electronic protection against infrared interference from either the Sun or satellite structure.

Figure B.2.9 SODERN's STD 16 IR horizon-crossing earth sensor; reproduced by permission of SODERN.

B.2 / Infrared Earth Sensors

MAIN CHARACTERISTICS

Field
- Instantaneous field of view : 2° x 2°
- Scanning : - rotating mirror speed : 60 r.p.m.
  - cone half-angle : $\alpha = 60°$
  - useful angular extension : 152° (about OY)

Accuracy (about OX and OY)
- Noise equivalent misalignment : 0.015° for 1 standard deviation, for Earth radiance between 0.8 and 3.8 W.m.$^{-2}$sr^{-1}µm^{-1} at 15 µm band for an Earth viewing angle up to 135°.
- Deterministic errors : in order to reduce considerably such errors corrective terms can be calculated on board and applied to the sensor reponse, (more information upon request).

Outputs
- $\varnothing_1$ and $\varnothing_2$, $\varphi'1$, $\varphi'2$: 16 bits word and 1 status word for each transition angle about the X-axis, with reference to + XOY plane.
- Analog outputs for telemetry :
  - 4 Earth radiance
  - bolometer temperature
  - motor drive current

Inputs
- 262 kHz clock frequency (for output data)
- Inhibition gating signals
  - Moon/Sun and Satellite structure interference : upon ground command (2 16-bit word)

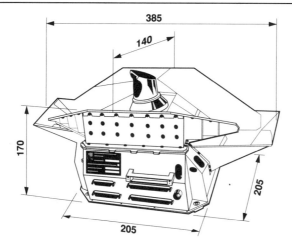

Operating temperature
From - 20°C to + 50° C
Storage temperature
From -40°C to +65°C

Power supply
22 V to 50 V

Power consumption
typical : 7.5 W
end of life : 9.0 W

Vibration acceleration levels
Sinusoidal : max. amplitude of 15 g in the range from 24 to 100 Hz
Random : max. amplitude of 12.4 g (r.m.s. value) in the range from 20 to 2000 Hz.

Lifetime
More than 5 years (low orbit)

Weight
3.7 Kg

Figure B.2.9 (continued)

The Barnes **Dual Cone Scanner** (DCS) is a straightforward modification to the single cone scanning Earth sensor which provides substantially enhanced performance with minimal increase in hardware weight and cost. By substituting a mission-specific optical module, the scanner's field-of-view is split into two halves, 180 degrees out of phase and at two distinct, selectable cone angles. Thus, a single DCS provides the data equivalent of two concentric single cone scanners. The DCS outputs define spacecraft altitude independently of any external reference and provide high accuracy pitch and roll data even when the Sun is directly on the horizon.

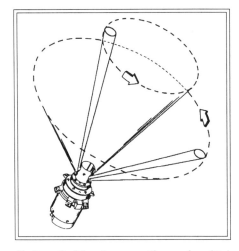

Cone Angles:	Set to meet mission needs 20 to 90 degrees	
Detector:	Pyroelectric	
Accuracy:	± 0.07 degrees (3 sigma)	
Scan rate:	120 rpm	
Weight/Power:	10 Pounds, 10 Watts	
Reliability:	1 Scanner	2 Scanners*
1 Year	0.99+	0.999+
3 Years	0.97+	0.999+
5 Years	0.95+	0.997+
10 Years	0.90+	0.991+

*Both operating. Reliability computed according to MIL-HDBK 217E.

The Barnes Dual Cone Scanner provides error-free, altitude independent attitude data from a single sensor even with the Sun on the Earth's horizon.

Figure B.2.10 Barnes Engineering dual-cone scanner (model no. 13-335); reproduced by permission of EDO Corp.

task of accurate measurement, there exist such prosaic technical characteristics as its weight, dimensions, power consumption, and so on. It is not the purpose of this section to make comparisons between various commercial IRES assemblies; the technical characteristics cited here are for illustrative and reference purposes only.

SODERN's STD 15 is designed for attitude measurement of three-axis–stabilized s/c in geostationary orbits. It can be used in three principal modes: "acquisition" mode, "normal" mode, and "transfer orbit" mode. The technical specifications of STD 15 are summarized in Figure B.2.8 (pp. 338–9). SODERN's STD 16 IRHCES is optimized for low orbits (500–1,200 km); its technical specifications and data are given in Figure B.2.9 (pp. 340–1).

An example of a dual-cone scanner (model no. 13-335) produced by Barnes Engineering (EDO Corp.) is given in Figure B.2.10. Table B.2.1 (page 344) provides basic performance specifications for a 45° half-cone conical earth sensor produced

Dual Cone Scanner Configuration Options

The 30/65 DCS is applicable to low and moderate altitude orbits at any inclination and eccentricity. The geometrical gains and altitude accuracies are substantially better than can be achieved with two 45 degree single cone scanners. As shown at right, two 30/65 DCS sensors provide exceptional coverage of the Earth for high accuracy applications.

In applications where the fields-of-view can clear spacecraft obstructions, a single 45/90 DCS (shown below) provides all of the advantages of both 45 degree and 90 degree conical scanners in a single sensor. Independent knowledge of spacecraft altitude is not required to precisely determine spacecraft attitude. This makes the DCS ideal for applications in which a single sensor must provide fully autonomous attitude data.

A wide variety of other DCS configurations are available to meet specific mission requirements.

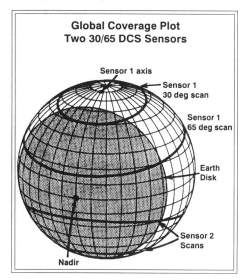

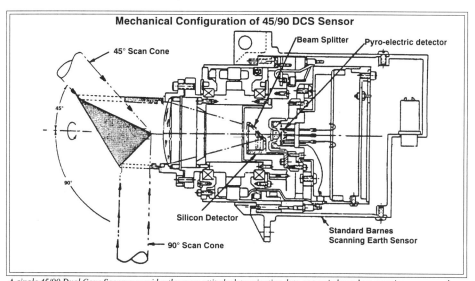

A single 45/90 Dual Cone Scanner provides the same attitude determination data as two independent scanning sensors and eliminates inter-sensor mounting angle biases and Sun interference.

Figure B.2.10 (continued)

by EDO and used in an 880-km sun-synchronous orbit. The spin axis is normal to the orbit plane, and earth horizon data is generated at 4 Hz.

B.2.4 Static Sensors

The layout of an infrared static earth sensor (IRSES) is shown in Figure B.2.11 (overleaf). The optical part of the sensor head projects the earth contour onto

Table B.2.1 *Conical earth sensor performance summary*
(adapted from Bednarek 1992 by permission of SPIE)

Parameter	Numerical Date

Attitude Accuracy (3 sigma):
    Earth Radiance Errors (pitch/roll):
    Alignment Error:
    Noise Equivalent Angle (pitch/roll):
	< 0.12°/0.06°
	0.014°
	0.07°/0.06°

Altitude Accuracy (3 sigma):	680 m
Sun/Moon-on-Horizon Errors:	+/- 0.1°

Maximum Maneuvering Range:
    Pitch:
    Roll:
	
	+/- 180°
	- 30° to + 45°

Spectral Band:	14-16 μm
Instantaneous Field of View:	2.5° circular
Total Weight:	4.85 lbs.
Total Power:	4 watts

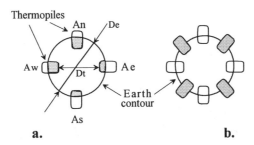

Figure B.2.11 Field configuration of static infrared earth sensor.

a number of radiance detectors (four in the case of Figure B.2.11.a). Similarly to the optical head of an IRHCES, the static ES head consists of a bandpass optical filter in the 14-16 μm (approximately) wavelength band and a focusing lens that projects the earth image onto thermopile surfaces. The earth's disc image is surrounded by thermopiles with uniform temperature sensitivity. In Figure B.2.11.a, D_e is the infrared earth diameter at the specific altitude of the satellite and D_t is the diameter of the circle tangent to the inner side of the thermopile sensitive areas. When the geometrically opposed thermopiles are connected in opposition, the response function near the zero has a constant slope in the range from $-R$ to $+R$, where

$$R = (D_e - D_t)/2. \tag{B.2.11}$$

The null error due to the earth's lack of uniform radiance is

$$2R(L_j - L_k)/(L_j + L_k), \tag{B.2.12}$$

where L_j (L_k) is the average earth radiance in areas common to the jth (kth) thermopiles and the earth. Thus, it is important to minimize the linear field extension $2R$ in order to minimize the null error due to lack of uniformity in the earth's radiance (see Desvignes et al. 1985). The seasonal changes in infrared radiance are a factor that must be taken into consideration, even to the extent of relocating the thermopiles in the optical head.

To increase the reliability of the IRSES, a double set of thermopiles can be used as (for example) in SODERN's STA 03 earth sensor; see Figure B.2.11.b. In the figure, the shaded thermopiles are the second set, to be used in case the first set shows some irregularity in its functioning. For more information, the reader is referred to the technical brochure (SODERN 1977), part of which appears in Figure B.2.12 (pp. 346-7).

Static IR earth sensors are inherently more reliable than horizon-crossing sensors, because no moving parts are used. However, the attitude ranges covered by the instrument are very restricted (a few degrees only) and the linear range might be even smaller (about 0.5°). The total RMS error is typically about 0.03°. The primary deficiency of the static horizon sensor is the very low altitude range (about nominal) in which it can function, so all static sensors are optimized to a specific operational altitude. A static ES cannot be used in both the transfer orbit and the final operational orbit, although sometimes they are used in combination with a less accurate HCES during the transfer orbit stage.

B.3 Sun Sensors

B.3.1 *Introduction*

It is difficult to identify a spacecraft in which sun sensors are not used for attitude determination. The sun has two very important qualities: (1) its luminosity, which remains unaffected by any other bright planet or star; and (2) the smallness of its angular radius (0.267° at 1 AU) compared to the earth globe. This radius is nearly constant for any satellite orbiting the earth.

The applications for sun sensors are numerous, and various types of solar instrumentation have been developed during the last three decades. There are two principal kinds of sun sensors, *analog* and *digital*. Digital sensors are more accurate and versatile, but also more expensive. Analog sun sensors can provide sufficient accuracy for many specific tasks. There are also *sun presence* sensors, which merely indicate if the sun is present in the FOV. For all three types there is a wide variety of implementations, some of which will be explained in this section.

B.3.2 *Analog Sensors*

Analog sun sensors, also called *cosine sun detectors,* are based on silicon solar cells whose output current is proportional to the cosine of the incident sun angle α and on additional physical properties pertinent to the specific cell. With these definitions, the output current amounts to:

$$I(\alpha) = I(0)\cos(\alpha). \tag{B.3.1}$$

In the model of Figure B.3.1 (page 348), secondary effects (e.g., transmission losses, reflections from the cell surface, etc.) are omitted. The output-current behavior of the photocell is shown in Figure B.3.1.b. The ideal output characteristic is a cosine function of α. However, at high incidence angles especially, the output current

Breadboard model optical head.

The STA 03 static infrared horizon sensor has been designed for the attitude control with respect to the Earth, of three axis stabilized geosynchronous satellites. It is used aboard the ESRO telecommunication satellite "OTS" to be launched June 1977.

Its design is such as the sensor main functional characteristics (response functions, linear range, zero accuracy) may be adapted to the mission requirements which are expected for the attitude control system of the next generation of geosynchronous three axis stabilized satellites. The operational accuracy is better than .05 arc degree, 3 standard deviations. For more details see overleaf.

It is made of two separate units: optical head and electronic unit, which are separated.

FUNCTION

In a system of coordinates OXYZ bound to the satellite, O, X and Z are respectively the satellite, the nominal directions of the Earth centre and poles axis.

For a geosynchronous orbit, X and Z are at right angle; in nominal attitude X, Y and Z are respectively the yaw (y), roll (r) and pitch (p) axes.

The STA 03 Earth sensor measures the α_r and α_p misalignments of the Earth centre T with respect to the X axis. The high linearity range is ±1.4°, the output signal remains an increasing function of the misalignment up to ± 2.5°, and does not decrease up to ± 10°.

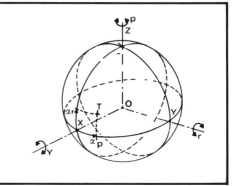

SENSOR DESIGN

OPTICS

The sensor operates according to the radiation balance principle: there is no mechanical movement.

The optical head comprises a single germanium lens, coated for minimum reflection losses, an infrared band pass filter (carbon dioxide atmosphere emission band, 14 to 16 microns), and a set of eight thermopiles placed in the Earth image surface.

In order to avoid the errors due to the Sun and Moon radiations, a visible radiation detector using silicon photocells can be associated.

The thermopiles shape and setting, the optical head housing, the signal processing are designed in order to minimize the overall error due to temperature range, temperature rate of change, components ageing, Earth luminance local variations,...

This optical head is fitted with a reference mirror for sensor alignment on the satellite.

ELECTRONICS

The d.c. signals delivered by the thermopiles are chopped, a.c. amplified, and re-converted in a d.c. output by a phase-locked rectifier. Sun, Moon and Earth output logics are introduced for better performance and simpler operation.

Figure B.2.12 SODERN's static infrared horizon sensor STA 03, optimized for geosynchronous use; reproduced by permission of SODERN.

does not follow the cosine law exactly, as shown by the dashed lines in the figure. For higher accuracies, a lookup table based on laboratory calibration is necessary, but the full range of measurement never attains the desired ±90°; a common practical measurement range is ±80° to ±85°. An example is the Adcole cosine-law analog sensor (model no. 11866) with a conical field of view of 160°; the maximum output current is 0.1 mA. This sun detector was flown on the *OAO-B, OAO-C,* and *ATS-6* satellites.

MAIN CHARACTERISTICS

Response function
Output voltage v versus misalignment α in roll (r) or pitch (p)

Acquisition
Overall field	: α_r, α_p	± 13 arc degrees
Linear range (positive slope)	: α_r, α_p	± 2.5 arc degrees
Slope in the linear range	: dv/dα	4 V/arc degree

Fine pointing
Linear ranges* in	: α_r or α_p	± 1.4 arc degrees
	: $\alpha_r = \alpha_p$	± 1.0 arc degree
Slope in the linear range	: dv/dα	4 V/arc degree
Linearity deviations on	: dv/dα	< 5 %
Seasonal change	: dv/dα	< 1 %

Accuracy
(3 standard deviation values, on roll and pitch channels)
Zero accuracy (noise, components ageing, thermal constraints, Earth radiance lack of uniformity); 3 standard deviation values
Misalignment angle measurement incertitude
— within ± 0.25 arc degree misalignment range : < 0.032 arc degree
— within ±   1 arc degree misalignment range : < 0.057 arc degree
Cross coupling effect up to ± 1 arc degree misalignment: negligible

* Can be extended up to ± 2 arc degrees.

Output
Impedance less than   : 100 Ω
minimum load resistance : 20 kΩ

Time constant
Thermopiles time constant :   < 0,3 sec
Electronics : low pass filter cut off frequency 0.81 Hz
              energy bandwidth   1.3 Hz
Overall :   time constant   < 0.7 sec
              phase shift less than 6° at   0.03 Hz

Power supply
+ and — 15 V, ± 2 %
Tolerated spurious signals . < 300 mV p.t.p.
Power consumption :   < 3 W

Operational temperature
Range   : — 10 to + 40 °C
Rate of change : < 0.5° C/mn

Mechanical stresses
vibration sinusoidal 20 Hz to 2 kHz : 10 g p.t.p.
shocks : 100 g, 5 ms

Dimensions:
Optical head : see figure
Electronic box: 187 × 150 × 80 mm

Weight
optical unit   : 1.45 kg
electronic unit : 1.07 kg

Reliability
failure rate : 5000.10⁻⁹ hr

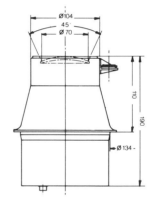
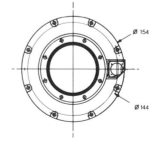

Figure B.2.12 (continued)

One-Axis Sensors

A single solar cell is not by itself very useful, because it measures the angle of the sun's incident rays without identifying their direction. Two cosine detectors

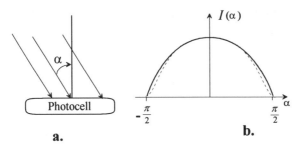

Figure B.3.1 Output-current dependency on sun incidence angle α.

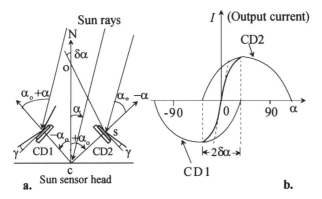

Figure B.3.2 Two-cosine detector arrangement that provides a single-axis almost linear measurement in a predetermined range.

(CD1 and CD2 in Figure B.3.2) are necessary to measure the incidence angle α of the sun in a defined plane in the satellite, thus achieving a one-axis sun measurement.

The simplest possible arrangement is shown in Figure B.3.2. The axes normal to the two photocells, and the normal N to the sensor head, are coplanar. The two photocell axes are inclined by an angle $\pm\alpha_0$ relative to the sensor-head normal axis N (Figure B.3.2.a). The sun rays are inclined relative to the normal N by the angle α, which is to be measured. With the mechanical arrangement shown in the figure, the sun rays deviate by $(\alpha_0 - \alpha)°$ from the normal to the CD2 surface, and by $(\alpha_0 + \alpha)°$ from the normal to the CD1 surface. Thus the output current from CD1 is proportional to $\cos(\alpha + \alpha_0)$; the output current from CD2 is proportional to $\cos(\alpha_0 - \alpha)$. If we subtract the two output currents, the first from the second, the difference will be

$$\Delta I = I_2 - I_1 = K\cos(\alpha_0 - \alpha) - K\cos(\alpha_0 + \alpha)$$
$$= 2K\sin(\alpha_0)\sin(\alpha) = C\sin(\alpha). \tag{B.3.2}$$

In Eq. B.3.2, the output current ΔI of the sensor's optical head is proportional to some constant C that is dependent on the physical properties of the photocells and to the sine of the sun inclination α. In Figure B.3.2.a, $\delta\alpha + \alpha_0 + \beta + 90° = 180°$ for the triangle CSO, from which it follows that the maximum range of angular measurement is

$$\delta\alpha = 90 - \alpha_0 - \gamma, \tag{B.3.3}$$

where γ is the dead zone of each detector unit. Outside the sensor's useful angular limits ($\pm\delta\alpha$ of Figure B.3.2.b), the sun sensor cannot be used at all because of the output-current ambiguity with regard to two different sun inclination angles α.

For comparatively small sun inclinations, the output is nearly linearly proportional to ΔI. The percentage error for increasing sun inclination α will be:

$$\alpha_E = \frac{\alpha - \sin(\alpha)}{\sin(\alpha)} \times 100. \tag{B.3.4}$$

In other words, inclination values of $\alpha = 10$, 20, and 50 degrees will generate percentage errors of $\alpha_E = 0.465$, 2.02, and 13.86, respectively. The analog sun sensor described here has a very good accuracy at null inclination, but clearly the accuracy decreases with increasing α.

In the special case where $\alpha_0 = 45°$, the sensitivity to the scaling factor K can be decreased drastically. To show this, we define $\beta_2 = \alpha_0 + \alpha$ and $\beta_1 = \alpha_0 - \alpha$. Since $\alpha_0 = 45°$, it follows that $\cos(\beta_2) = \sin(\beta_1)$. However,

$$\tan(\alpha) = \tan(45° - \beta_1) = \frac{\tan(45°) - \tan(\beta_1)}{1 + \tan(45°)\tan(\beta_1)}$$
$$= \frac{1 - \tan(\beta_1)}{1 + \tan(\beta_1)} = \frac{\cos(\beta_1) - \sin(\beta_1)}{\cos(\beta_1) + \sin(\beta_1)}.$$

Next, since $\sin(\beta_1) = \cos(\beta_2)$ we have $\tan(\alpha) = (K_1 I_1 - K_2 I_2)/(K_1 I_1 + K_2 I_2)$. If we assume that the scaling factor is identical for the two solar cells, $K_1 = K_2$, then $\tan(\alpha) = (I_1 - I_2)/(I_1 + I_2)$, which is independent of the scale factors of the two solar cells.

Two-Axis Sensors

In order to determine the three components of the sun vector in the body frame, two spatial angles must be measured and hence two single-axis sun sensors are necessary. The second sensor is oriented 90° with respect to the first. In this case, we speak of two-axis sun sensing. We define β as the inclination of the sun vector in the second sensor; now the two measured angles α and β provide the data necessary to determine the sun's vector orientation in the body axis frame.

There exist numerous two-axis sun-sensor optical heads that integrate two single-axis sun sensors as just described, but there are other possibilities, too. Instead of using two single-axis sun sensors, it is common to implement a variation whereby four photocells are located in one optical head (see Figure B.3.3; see also Wertz 1978). The determination principle of sun orientation is shown in Figure B.3.3 (overleaf), which exhibits a two-axis *mask detector*.

Using a simple algorithm based on the sums and differences in individual currents of four photocell detectors, the equivalent currents proportional to the orientation of the sun vector about the $\mathbf{Y}_s$ and $\mathbf{Z}_s$ axes can be calculated. In a restricted region about the null position, the currents I_y and I_z are linearly proportional to the sun orientation about both axes. In principle, this sun detector's optical head has a semispherical FOV (i.e., 2π sr). Here again, since the field of view of an individual detector in Figure B.3.3 is less than 90°, the 2π-sr range cannot be achieved in a practical instrument. Two-axis analog sun sensors based on the principles described here have been produced by Matra Espace S.A.S. (France) and have been flown on several satellites,

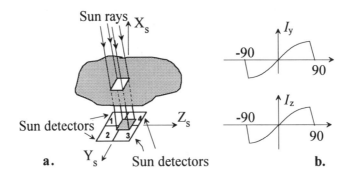

Figure B.3.3 Two-axis mask sun detector.

such as the *OTS* (in 1978), *Marecs-ECS* (1981), *Skynet* and *Spot* (1984), *LSAT* (1985), and others.

The mask sun detector is very effective for *sun acquisition*. In this control task – whereby large angles are attained between the sun direction and the optical sun-sensor axis – the accuracy of sun orientation is irrelevant; only its direction is important. However, the sun angular orientation accuracy at null increases considerably, and is of the order of ±1°.

There exist also analog sun sensors having hard saturation characteristics outside the restricted linear range but with a good accuracy at null (of the order of ±0.1°). Outside the linear range, the output is constant until the edge of the field of view is reached for both positive and negative orientations; see Figure B.3.4. Such analog sun sensors are specially designed to provide attitude information to a solar array orientation control system. Common values of the linear range in such sun sensors are of the order of ±1°. Fields of view range from 20° to 50°. An example is the Adcole (model no. 17470) analog sun sensor.

Characteristics and Specifications

There are five important characteristics necessary to define an analog sun sensor. These characteristics are not independent, so they cannot be prescribed freely. Tradeoffs are necessary when specifying sun-sensor characteristics, depending on the operational constraints of the satellite mission. These characteristics are summarized as follows.

(1) *Field of view* – the maximum deviation of the sun vector from the optical axis of the sensor that can be measured, or sensed. The sun-sensor FOV does not have to be a circular cone. Moreover, there can be different sensitivity ranges for the two directional planes in which the two orientations of the sun vector are measured.

(2) *Linear range* – the angular range in which the sensor output is linearly proportional to the angular deviation of the sun projection in the plane of measurement.

(3) *Linearity error* – the maximum deviation of any point from the best straight line through all points in the linear range of the sensor.

B.3 / Sun Sensors

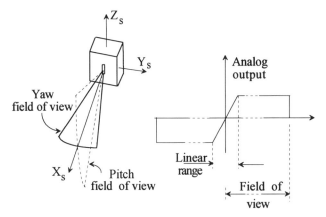

Figure B.3.4 Characteristics of an analog sun sensor with hard saturation outside the linear range.

(4) *Nominal scale factor* – the maximum analog output value in the linear range divided by the total FOV in this range.
(5) *Accuracy* – the minimum unexpected error that cannot be foreseen and compensated.

A summary of analog sun-sensor characteristics is included in Table B.3.1 (pp. 358–62).

B.3.3 Digital Sensors

Analog sun sensors remain popular because they are relatively simple, low-dimensioned, and comparatively cheap. Still, they come with inherent limitations; in particular, they are not accurate enough for high deviations of the sun direction from the sun-sensor optical axis. In such cases, precise measurements can only be achieved with digital sun sensors, which have accuracies of the order of 0.017° inside a large field of view (of the order of 64° × 64°).

The basic digital sun sensor is a single-axis device. Exactly as with their analog counterparts, two single-axis digital sensors are installed with their optical planes positioned 90° apart to obtain a two-axis digital sun sensor. The layout of a digital sun sensor is shown in Figure B.3.5 (overleaf). It consists primarily of an optical head together with an electronics box in which the direction of the sun with reference to the head's axis frame is computed.

The optical head is a slab, with index of refraction n, on whose upper side is a narrow entrance slit for the sun rays. On the lower side of the slab is located the reticle slit pattern, composed of a set of reticle slits laid out on that surface in an array that enables the sun orientation to be expressed in digital code form. In today's digital sun sensors, the *Gray binary coded* reticle pattern is the most commonly used. The Gray code is an equidistant code, which means that one and only one bit changes for each unit distance. In contrast, the disadvantage of a *binary code* is that one or more than one binary bits change for the same unit distance. In this case, if some imperfections exist in the reticle pattern of the binary code and if one of the binary bits is not read

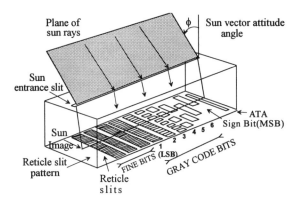

Figure B.3.5 Digital sun sensor (Adcole model no. 17032); reproduced from Wertz (1978) by permission of D. Reidel Publishing Co.

correctly, the change of sun orientation might be erroneously identified as a much larger (multi-unit) distance. This cannot happen with the equidistant Gray code.

For example, suppose there is a transition between the decimal values 19 to 20. In the following table we see that for the transition from 19 to 20, the last three digits are changed in the binary code whereas only the third digit is changed in the Gray code. For the transition from 15 to 16, the situation is even worse: five digits are changed using the binary code; only one is changed with the Gray code.

Decimal	Binary code	Gray code
14	01110	01001
15	01111	01000
16	10000	11000
19	10011	11010
20	10100	11110

The number of bits used depends, naturally, on the prescribed accuracy of the sensor. In Figure B.3.5, there are only six Gray code bits but also three fine bits, for fine interpolation and increase of the total accuracy. As an example, the Adcole two-axis digital sun sensor (model no. 18960) has the following basic characteristics: field of view, $64° \times 64°$; least significant bit size, $0.004°$; accuracy, $0.017°$.

Figure B.3.5 also shows the ATA reticle, which is the automatic threshold adjust voltage. This extra slit is required because the photocells are cosine sun detectors (as explained in Section B.3.2), and for different sun orientations there is a need to provide an adaptive threshold for the different slits of the reticle slit. The ATA slit is narrower by half than all other slits, so that – at any sun orientation – the energy collected by the ATA slit will be half that of any other slit. This energy is compared to that collected from a normal slit; if the latter is more than twice as high then the slit is discriminated as being above the threshold, and the appropriate bit is activated.

The outputs from the Gray code and fine bits are processed in the electronic box, which calculates the orientation of the sun in the sensor's axes frame. A two-axis

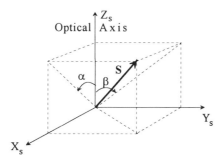

Figure B.3.6 Sensor coordinate system.

digital sun-sensor head consists of two single-axis sensors located 90° apart in the same head plane. A commonly accepted definition for the axis frame of a two-axis solar sensor, whether analog or digital, is shown in Figure B.3.6.

In Figure B.3.6 (see also Table B.3.1, p. 362), the sun vector **S** is projected on the X_s-Z_s plane, creating the angle α. It is also projected on the Y_s-Z_s plane, creating the sun orientation angle β. These are the orientations computed by the two individual sun sensors. The orientation for the digital sun sensor can be calculated as follows:

$$\alpha = \tan^{-1}[A_1 + A_2 N_A + A_3 \sin(A_4 N_A + A_5) + A_6 \sin(A_7 N_A + A_8)] + A_9, \quad (B.3.5)$$

$$\beta = \tan^{-1}[B_1 + B_2 N_B + B_3 \sin(B_4 N_B + B_5) + B_6 \sin(B_7 N_B + B_8)] + B_9. \quad (B.3.6)$$

In these equations, N_A and N_B are the base-10 equivalents of the binary output from each axis. The terms $A_1, ..., A_9$ and $B_1, ..., B_9$ are constants defined in the sensor calibration state and provided to the customer with each individual unit.

As an example of a very accurate two-axis fine digital sensor system (Adcole model 18960), see the partial listing of data characteristics in Figure B.3.7 (pp. 354–6). A summary table of various analog and digital sun sensors, with all relevant specifications included, is given as Table B.3.1 (pp. 358–62).

B.4 Star Sensors

B.4.1 *Introduction*

Stars are the most accurate optical references for attitude determination. The reason stems from the facts that (1) they are inertially fixed bodies, and (2) they are objects of very small size as seen from the solar system. Given these stellar characteristics, star sensors allow attitude determination with accuracies in the sub-arc-second range. The drawbacks are the instrumentation's complexity, elevated price, extensive software requirements, and relatively low reliability as compared to other attitude sensor assemblies.

The number of stars in the sky is very large. In order to use them as attitude references, appropriate techniques to differentiate between them must be implemented. In terms of the necessary hardware, star sensor assemblies are not standard items that can be supplied "off the shelf." In most cases, every space mission places unique demands on the star sensor to be used, which must be defined and designed accordingly. The present section is a short explanation of notions required by the control designer to define the characteristics of a potential star sensor for use in a defined space mission.

SPECIFICATIONS: MODEL NO. 18960

No. of Axes	2
Max. No. of Sensors	2
Field of View: Each Sensor: Total:	$64° \times 64°$ $124° \times 64°$
Least Significant BIT Size:	$0.004°$
Transition Accuracy:	$0.017°$
Sensor Model No. :	19020
Sensor Size:	3.32" x 4.32" x 0.97" (exclusive of connector) 84 mm x 110 mm x 25 mm See Outline Drawing
Sensor Weight:	0.59 lb 270 gr
Electronic Size:	8.13" x 6.19" x 1.19" (exclusive of connector) 206 mm x 157 mm x 30 mm See Outline Drawing
Electronic Weight:	1.62 lb 736 gm
Output:	32 Bits, Serial; 15 Bits/Axis Data (Gray, Natural Binary Mixed), Sensor Select, Sun Presence.
Power Requirements:	+28 +/- 0.56 VDC, 1.8 watts nominal
Transfer Function:	See Figure 1
Sensor Sun Distance:	0.9 to 1.1 AU
Mounting:	See Outline Drawing
Sensor Alignment:	Detachable alignment mirrors, optical axis aligned to one arc minute
Interconnections:	Interconnecting cables to be supplied by customer
Temperature Range:	Sensor Electronics Operating: -10° C to +60° C -10° C to +50° C Non-Operating: -25° C to +60° C -10° C to +50° C
Pressure:	Hard vacuum as encontered in earth orbit
Humidity:	Up to 100%
Acceleration:	25 g
Random Vibration:	$0.31 \text{ g}^2/\text{Hz}$, 20-2000 Hz, 1 minute each axis 14.9 g-rms
EMI:	MIL-STD-461A
Expected Life:	Unlimited
Design Status:	Basic design flown OAO, qualified NASA IEU program
Tests:	Transfer function determined and checked at room and operating temperature limits using solar simulator

Figure B.3.7 Adcole model no. 18960 two-axis fine digital sun-angle sensor system; reproduced by permission of Adcole Corp.

B.4 / Star Sensors

NOTES:

1. The least significant bit size listed is an average of the step size in the angles α (or β) over the field when β (or α) = 0.

2. For a digital sensor , error can be measured only at points where the output changes by one state. Error is defined as the absolute value of the difference between the sun angle calculated from the transfer function and the measured angle at a step. The values quoted are maximum error for α (or β) when β (or α) = 0.

3 Gray code is a special form of a binary code having the property that only one digit changes at a time. Conversion of Gray code to natural binary is accomplished as follows:

   (a) The most significant bit is the same in either code.

   (b) Each succeeding natural binary bit is the complement of the corresponding Gray bit if the preceding natural binary bit is a "1"; or is the same as the corresponding Gray bit if the preceding binary bit is a "0".

   Conversion from natural binary to decimal or base 10 is accomplished by weighing each bit as shown and summing the result.

   Example:

Gray	Natural Binary	Decimal
	$2^7\ 2^6\ 2^5\ 2^4\ 2^3\ 2^2\ 2^1\ 2^0$	
11010011	1 0 0 1 1 1 0 1	157

Figure B.3.7 (continued)

356   B / Attitude Determination Hardware

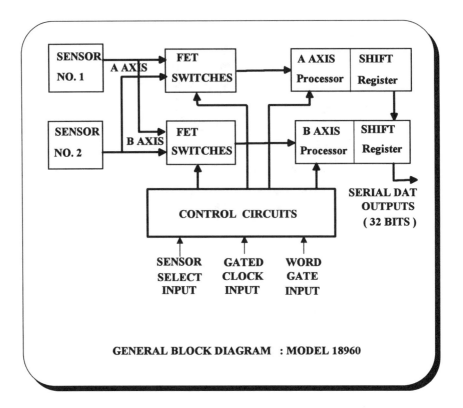

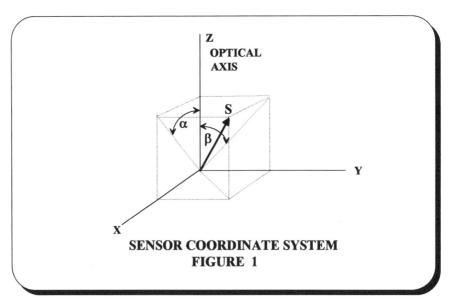

Figure B.3.7 (continued)

B.4.2 Physical Characteristics of Stars

There are two principal characteristics that allow us to differentiate between stars: their *magnitude* and *spectra*. A short explanation follows; see also McCanless, Quasius, and Unruh (1962) and Burnham (1979) for more detailed treatments.

Magnitude m of a Star

An important characteristic of a star is its *brightness* as seen from the earth. This brightness depends on the quantity of light that the star emits, and also on the distance over which that light must travel. The *flux* of light is the energy of light per unit area. Hence, a star's magnitude depends on the flux and the distance of the star. The magnitude is the intensity of light that reaches the earth's vicinity, and is designated by m. The ratio between the magnitudes and the fluxes of two stars is not linearly proportional to the magnitude of the stars. Rather, the relation between fluxes and magnitudes is logarithmic. If the fluxes and the magnitudes of two stars are denoted (respectively) S_2, S_1 and m_2, m_1, then

$$m_2 - m_1 = -2.5 \log_{10}(S_2/S_1). \tag{B.4.0}$$

For example, if the light flux S_1 of a star is higher by a factor of 10^4 than that of another star, then $m_2 - m_1 = -2.5 \log_{10} 10^4 = -10$, which means that the magnitude of the second star is 10 times larger than that of the first. Notice that the result is negative; this indicates that, as flux intensity is larger, the magnitude becomes more negative. Equation B.4.0 assigns differences between magnitudes, but not absolute magnitudes. This can be remedied by choosing a reference magnitude for stars. For instance, in the nineteenth century it was decided to choose the North Star (Polaris), with magnitude $m = 2$, as the reference star. Unfortunately, the magnitude of Polaris varies periodically between $m = 1.95$ and 2.05. Today, Vega is the accepted reference star, with $m = 0$. Table B.4.1 (page 363) shows the magnitudes of some typical stars.

With the help of the star Vega, we can determine $m_0 = -2.5 \log S_0$. There is also another way of defining the magnitude of a star. We can set

$$m = -2.5 \log_{10}(I/I_0),$$

where I denotes the effective irradiance and I_0 the effective standard irradiance.

To help develop our notions of star magnitudes, observe that in Table B.4.1 the brightest star outside the solar system (Sirius) has a magnitude of $m = -1.6$, compared to the sun's magnitude of $m = -26.8$! With the naked eye it is possible to view stars with magnitudes as high as 6 to 6.5. With strong telescopes on the earth's surface (i.e., viewing through the atmosphere surrounding the earth), it is possible to detect stars with magnitudes as "small" as 23 or 24.

Spectra of a Star

In practice, it is common to view (and film) stars in three different spectral ranges of light: ultraviolet, visible, and blue. We must always define in which spectral range the magnitude is measured; for instance, m_V stands for the magnitude in the visible spectrum.

Table B.3.1 *Catalog of sun-angle sensor systems*
(reprinted by permission of Adcole Aerospace Products)

Model number	Field of view per sensor	Maximum number of sensors	Least significant bit size	Accuracy[4]	Transfer function[5]	Output
DIGITAL SYSTEMS FOR SPINNING VEHICLES						
15564	128°[1,2]	1	1.0°	0.50°	A	7-bit parallel Gray TTL
16765	180°[1]	1	0.5°	0.25°	A	9-bit parallel Gray TTL
17083	180°[1]	1	1.0°	0.50°	A	9-bit serial Gray open collector
17126	180°[1]	1	0.5°	0.25°	A	9-bit serial Gray 0–12 V
16151	64°[1,2]	4	0.25°	0.1° $\alpha < 40°$ 0.25° $\alpha > 40°$	A	8-bit serial Gray 0–5 V
17212	128°[1,2]	1	1.0° coarse Analog fine	0.25°	B	7-bit parallel Gray TTL Two analog
18810	128°[1,2]	2	0.25°	0.1°	A	9-bit serial Gray 0–10 V
18656	128°[1,2]	2	1.0° coarse Analog fine	0.1° $\alpha < 40°$ 0.25° $\alpha > 40°$	B	7-bit serial Gray Two analog
15761	5.6°[1]	1	0.02°	0.1°	C	8-bit serial Natural binary TTL
15761	180°[1]	1	1.0°	0.5°	A	8-bit serial Gray TTL
18273	64°[1,2]	2	0.25°	0.1°	A	8-bit serial Gray 0–7 V
TWO-AXIS DIGITAL SYSTEMS						
18273	128° × 128°	5	1.0°[3]	0.5°	D	7-bit/axis serial Gray 0–7 V
15486	128° × 128°	1	1.0°[3]	0.5°	D	7-bit/axis parallel Gray 0–10 V
17115	128° × 128°	3	1.0°[3]	0.5°	D	7-bit/axis parallel Gray TTL
16764	128° × 128°	5	0.5°[3]	0.25°	D	8-bit/axis parallel 3-bit identity Gray TTL
14478	128° × 128°	2	0.5°[3]	0.1° $\theta < 42°$ 0.25° $\theta > 42°$	D	8-bit/axis serial Gray TTL

Size [in / mm]		Weight [lb / g]		Input power	Design status	Remarks
Electronics	Sensor	Electronics	Sensor			
1.7x1.3x2.6 43x33x66		0.32 145		+5.13 vdc 2.0 ma +6.12 vdc 2.5 ma	Flown ESRO IV	Sensor and electronics in one unit
2.9x2.6x3.8 74x66x97		0.81 368		$+12 \pm 0.36$ vdc 6 ma -12 ± 0.36 vdc 1 ma	Flown NRL	Sensor and electronics in one unit
3.3x2.0x3.5 84x51x89	2.0x2.0x1.2 51x51x31	0.98 445	0.31 141	−24.5 vdc 12 ma	Flown AEC-C	
4.1x3.3x1.3 104x84x33	2.0x2.0x1.2 51x51x31	0.54 245	0.25 113	+12 vdc 5.5 ma −12 vdc 4.0 ma	Flown Hawkeye	System includes three 1° FOV earth albedo sensors (0.8"[6]x 5.4" each); 0.59 lb (269 g) total
5.5x4.4x2.7 140x112x69	2.3x1.3x1.0 58x33x25	2.13 966	0.21 95	+28 vdc 10 ma	USAF	LEDs in sensor for ground test Redundant electronics
2.5x2.3x1.4 64x58x37	2.6x1.3x1.0 66x33x25	0.38 172	0.23 104	+15 vdc 4 ma −15 vdc 4 ma	USAF	
4.1x2.3x3.7 104x58x94	2.6x1.3x1.0 66x33x25	1.6 726	0.24 109	+28 vdc 400 mw	Qualified BSE	Redundant sensor and electronics
4.1x2.3x3.7 104x58x94	2.6x1.3x1.0 66x33x25	1.6 726	0.24 109	+28 vdc 400 mw	Qualified RCA SATCOM	Redundant sensor and electronics Analog outputs have sample and hold
4.0x3.5x2.2 102x89x56	1.3x1.3x1.3 33x33x33	1.19 540	0.24 109	+5 vdc 80 ma +8 vdc 4ma −8 vdc 4 ma	Flown AEROS-A AEROS-B	Fine sensor with optical axis parallel to spin axis
4.0x3.5x2.2 102x89x56	1.8x1.7x0.8 46x43x20	1.19 540	0.24 109	+5 vdc 80 ma +8 vdc 4ma −8 vdc 4 ma	Flown AEROS-A AEROS-B	Coarse sensor with optical axis perpendicular to spin axis Common electronics
7.6x4.5x2.5 193x114x64	2.6x1.3x1.0 66x38x25	2.3 1043	0.24 109	+28 vdc 600 mw	Qualified CTS	Command eye has 128° FOV
7.6x4.5x2.5 193x114x64	3.3x1.6x0.7 84x41x18	2.3 1043	0.16 73	+28 vdc 600 mw	Qualified CTS	Sensor for spinning phase shares electronics with five 2-axis sensors
3.0x3.0x2.0 76x76x51	3.3x1.6x0.7 84x41x18	0.79 358	0.18 82	+15 vdc 19.5 ma	Flown ATS-6	
6.5x4.5x2.5 165x114x64	3.3x1.6x0.7 84x41x18	2.11 957	0.25 113	5.0 ± 0.2 vdc 12 ma	Flown ATS-6	
3.5x4.5x1.2 89x114x31	3.2x3.2x0.8 81x81x20	0.65 295	0.57 259	$+12 \pm 0.36$ vdc 8 ma -12 ± 0.36 vdc 2 ma	Flown NTS-1	
5.2x4.5x2.6 132x114x66	3.8x2.8x0.9 97x70x22	1.6 726	0.75 340	+28 vdc 500 mw	Flown DSCS	

continued on next page

Table B.3.1 *Continued*

Model number	Field of view per sensor	Maximum number of sensors	Least significant bit size	Accuracy[4]	Transfer function[5]	Output
TWO-AXIS DIGITAL SYSTEMS (continued)						
17032	64° × 64°	4	0.125°[3]	0.1°	D	9-bit/axis serial Gray 0–7.5V
15671	64° × 64°	1	0.125°[3]	0.1°	D	9-bit/axis parallel Gray TTL
15381	64° × 64°	1	0.004°[3]	0.017°	E	14-bit/axis parallel Natural binary TTL
16467	128° in β 64° in T	1	1.0° in β 0.5° in T	0.5° in β 0.25° in T	F	7-bit/axis parallel Gray TTL
16932	64° × 64°	1	0.004°[3]	0.017°	E	14-bit/axis serial Natural binary TTL
18960	64° × 64°	2	0.004°[3]	0.017°	E	15-bit/axis serial Binary TTL
18970	100° × 100°	6	0.5°	0.25°	G	8-bit/axis parallel 3-bit identity Gray TTL
SINGLE-AXIS DIGITAL SYSTEM						
17061	100° × 100°	1	0.006°[3]	0.05°	E	14-bit serial Natural binary; open collector
TWO-AXIS ANALOG SYSTEMS						
18560	30° cone	1	N.A.	1' at null	±1° linear	±5V
12202	30° cone	N.A.	N.A.	1' at null	±1° linear	±4 ma
18394	180° solid angle	1	N.A.	2° at null	±30° linear	±0.1 ma peak
18980	30° cone	1	N.A.	2' at null	±2° linear	±5V
SINGLE-AXIS ANALOG SYSTEM						
17470	40° × 60°	1	N.A.	6' at null	±1° linear	0–5V 2.5V at null
COSINE-LAW ANALOG SYSTEM						
11866	160° cone	N.A.	N.A.	2 μa	Cosine of angle of incidence	0.1 ma peak

Size [in / mm]		Weight [lb / g]		Input power	Design status	Remarks
Electronics	Sensor	Electronics	Sensor			
7.8x4.5x2.3	3.8x2.8x0.9	2.53	0.6	−24.5 vdc	Flown	
197x114x60	97x71x23	1148	277	560 mw	Nimbus-F	
5.4x4.0x2.1	3.8x2.8x0.9	1.5	0.7	+28 ± 4 vdc	Delivered	
137x102x53	97x71x23	680	318	350 mw	NASA	
7.8x4.5x2.5	3.8x4.1x1.0	3.0	0.82	+28 vdc	Flown	Two-axis analog output
198x114x64	97x104x25	1361	372	61.5 ma	OAO-C	(d-a converter) ±1° about optical axis
4.0x3.5x2.2	2.5x2.1x0.6	1.13	0.31	−24.5 vdc	Flown	Output is sun azimuth
102x89x56	64x53x15	513	141	250 mw	USAF Block V	and elevation in sensor coordinates
7.8x4.5x2.5	3.8x4.1x1.0	2.56	0.82	+28 vdc	Delivered	
198x114x64	97x104x25	1161	372	62 ma	USAF	
8.1x6.2x1.2	3.3x4.3x1.0	1.0	0.75	+28 vdc	Designed for	Weight not final
206x157x30	84x110x25	455	341	1.8 w	NASA-IUE	
3.5x4.5x2.0	3.2x3.2x0.8	1.1	0.40	+28 vdc	New	Dimensions and weight
89x114x51	81x81x20	500	102	500 mw	design	not final
4.0x3.4x2.0	4.3x2.5x1.1	1.14	0.71	+5 vdc 65 mw	USAF	
102x86x51	109x64x28	517	322	+28 vdc 700 mw	Block V	
	2.2x1.4x1.2		0.42	+15 vdc	Designed	
	56x36x31		191	−15 vdc 100 mw	NASA Lewis	
N.A.	2.5x1.2x1.3	None	0.12	None	Flown	Includes sun presence
	64x30x33		55		OAO-B,C ATS-6	sensor with 2 ma output
N.A.	1.9x1.9x1.3	None	0.18	None	Flown	Cluster of five 11866
	48x48x33		82		OAO-B,C ATS-6	cosine-law sensors
	2.7x2.0x1.6		0.38	±10 vdc	New	Dimensions and weight
	69x51x41		171	75 mw	design	not final
	2.7x2.0x1.1		0.26	±15 ± 0.6 vdc	Qualified	Used on CTS for solar
	69x51x28		118	100 mw	CTS	array pointing
N.A.	$0.9^6 \times 0.4$	None	0.01	None	Flown	
	$23^6 \times 10$		4.6		OAO-B,C ATS-6	

continued on next page

Table B.3.1 *Continued*

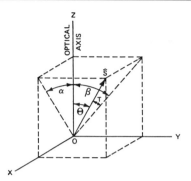

Figure 1. Sensor coordinate system

Notes:

1. The field of view is fan-shaped. An output pulse is provided when the fan crosses the sun, and the digital sun angle is read at this time and stored. The sensor should be mounted so that the plane of the fan is parallel to the spin axis.
2. For fields of view less than 180° the mid-angle of the fan need not be mounted perpendicular to the spin axis.
3. The least significant bit size is an average of the on-axis step sizes over the field of view (FOV).
4. For a digital sensor, error can be measured only at points where the output changes by one step. *Error* is defined as the absolute value of the difference between the sun angle calculated from the transfer function and the measured angle at a step. The values quoted are maximum error for α (or β) when β (or α) equals zero.
5. A wide variety of transfer functions are available for digital sensors; the equations that follow illustrate the principal forms. In these equations, N, N_x, N_y are base-10 equivalents of the binary output number and $k_0, k_1, k_2, \ldots$ are constants; $\alpha, \beta, \Theta,$ and T are the angles defined in Figure 1.
6. Diameter.

Type A $\beta = 0$
         $\alpha = k_0 + k_1 N$

Type B $\beta = 0$
         $X = k_3 + k_4 N + k_5 \tan^{-1}(A_1/A_2)$
         $\alpha = \tan^{-1}\{k_1 X/[1-(k_1^2+k_2)X^2]^{1/2}\}$
         where A_1 and A_2 are the two analog outputs

Type C $\beta = 0$
         $\alpha = 0.0108 + 0.0216N$

Type D $X = k_0 + k_1 N_x$
         $Y = k_0 + k_1 N_y$
         $\alpha = \tan^{-1}\{k_2 X/[k_3-(k_2^2-1)(X^2+Y^2)]^{1/2}\}$
         $\beta = \tan^{-1}\{k_2 Y/[k_3-(k_2^2-1)(X^2+Y^2)]^{1/2}\}$

Type E $\alpha = \tan^{-1}(k_0 + k_1 N_x)$
         $\beta = \tan^{-1}(k_0 + k_1 N_y)$

Type F $T = k_0 + k_1 N_x$
         $\beta = k_2 + k_3 N_y$

Type G $\alpha = k_0 + k_1 N_x$
         $\beta = k_0 + k_1 N_y$

Table B.4.1 *Star magnitudes for extreme cases*

Magnitude m	Star	Name of constellation
0	Vega	α Lyr
-1.6	Sirius	α CMi
-26.8	Sun	-

Different stars have different emission spectra. This is a very important fact, widely used in the design of star sensors. It is convenient to divide the stars into seven principal spectral categories, which are O, B, A, F, G, K, and M. These categories are each subdivided into ten more subgroups, from 0 to 9. The spectra of a star is very much dependent on its surface temperature. An example of different star spectra is shown in Figure B.4.1. The stars may be classified by their visual magnitude m_V and their spectral type. As an example, ten of the brightest stars are characterized in Table B.4.2.

Star detectors collect cosmic energy coming from space in different spectral ranges. This is the reason why the spectra of stellar emissions is important to the design of a

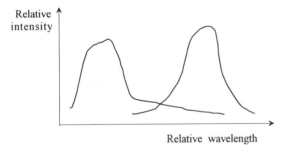

Figure B.4.1 An example of different relative spectra.

Table B.4.2 *Characteristics of the ten brightest stars*
(reproduced from McCanless et al. 1962)

Astronomical Name	Common Name	SHA	DEC	Spectral Type	m_v
α Lyrae	VEGA	81	N39	A0	0.05
α Centauri	RIGIL KENT	141	S61	G2-K3	-0.27
α Bootis	ARCTURUS	147	N19	K0	0.03
β Centauri	AGENA	150	S60	B1	0.69
α Canis Minor	PROCYON	246	N5	F5	0.35
α Canis Major	SIRIUS	259	S17	A1	-1.49
α Argus, Car	CANOPUS	264	S53	F0	-0.77
α Aurigae	CAPELLA	282	N46	G1	0.13
β Orionis	RIGEL	282	S8	B8	0.14
α Eridani	ACHERNAR	336	S57	B5	0.55

Table B.4.3 *Distribution of the stars according to their spectral category*
(reproduced from McCanless et al. 1962)

Type	O	B	A	F	G	K	M	Other
Average Distribution $m_v < +8.5$	<1%	10.5%	22.3%	18.6%	14.3%	31.9%	2.7%	1%
The 20 Bright Stars $m_v < +1.5$	0	6 30%	5 25%	2 10%	2 10%	3 15%	2 10%	0
The 40 Bright Stars $m_v < +2.0$	0	15 38%	10 25%	4 10%	2 5%	6 15%	3 7%	0
The 100 Bright Visual Stars: $m_v < +2.65$	4 4%	32 32%	23 23%	9 9%	7 7%	17 17%	7 7%	1 1%

star sensor. It is important to choose the spectral sensitivity of the detector, so that the star sensor will be optimized for its particular mission; that is, it must be able to detect the stars that have been chosen as references. Note that using star detectors with a special spectral sensitivity is another way of minimizing celestial parasitic disturbances (background noise). It is interesting to view the stellar *distribution* in terms of the spectrum categories; see Table B.4.3.

Celestial Background

The sky is full of stars. When using stars for attitude determination, it is necessary to discriminate the correct and useful ones from the many others that, in this case, constitute a celestial parasitic noise background. On the other hand, for certain missions it is important to use as many stars as necessary in order to allow a continuous attitude determination. The number of stars that can be used within the relevant algorithms is of crucial importance for the attitude accuracies that can be achieved. Table B.4.4 shows the density distributions of stars with different magnitudes. This table is very important in the stage of defining the technical properties of a star sensor.

Table B.4.4 *Target star distribution*
(reproduced from McCanless et al. 1962)

Apparent Magnitude m	Stars Per Square Degree Brighter than m	Square Degrees Per Star Brighter than m	Stars in a 227 Sq. Deg. Field Brighter than m
8	1.0	1.0	227
7	0.34	2.94	77
6	0.12	8.3	27
5	0.04	25.0	9
4	0.013	77.0	3
3	0.0044	227.0	1

Location of Stars on the Celestial Sphere

In order for a star to be useful for attitude determination or guidance purposes, its location must be correctly identified in some predetermined axis frame. All stars outside the solar system are inertially fixed in space. Hence we must define the inertial frame in which all calculations will take place. The commonly accepted inertial frame is defined as follows:

X_I - the vernal equinox direction defined at some reference epoch;
Z_I - axis of rotation of the earth; and
Y_I - completes a right-hand orthogonal axis frame.

Because of the slow precession of the vernal equinox axis, the locations of the stars will also change in this frame, albeit very slowly. Nonetheless, this effect must be taken into consideration.

To determine the location of a star on the celestial map, two angles are used:

(1) the right ascension (RA), measured from the vernal equinox on the equatorial plane of the celestial sphere; and
(2) the declination angle (DEC), defined $\pm 90°$ from the equatorial plane (+ points to the North direction).

These angles are also identified as the longitude and the latitude angles on the celestial map. The longitude ordinates divide the celestial map into 24 hours. An *hour* is divided into *minutes* (') and *seconds* ("); 1 hour is equivalent to 15°. Hence:

$1' = 1 \text{ hr}/60 = 15°/60,$
$1'' = 1 \text{ hr}/3{,}600 = 15°/3{,}600.$

With these definitions, celestial catalogs and maps have been prepared for different vernal equinox epochs – for instance, for the epoch (year) 1950 and more recently for the epoch 2000. Some popular catalogs include *The Catalog of Bright Stars* (Hoffleit 1964), *The Smithsonian Astrophysical Observatory Catalog* (1971), *Burnham's Celestial Handbook* (Burnham 1979), and *Sky Catalogue 2000.0* (Hirshfeld and Sinnott 1990; also available on PC diskette).

When preparing or using star catalogs, we must differentiate between *double* (or multiple) stars, *variable* stars, and so on; see Burnham (1979) for more information. For these denominations, the position angle (PA) denotes the apparent orientation of a pair of double stars. As is well known, stars are agglomerated into constellations. As an example of how stars are cataloged, Table B.4.5 (overleaf) lists some of the double and multiple stars in the constellation Ursa Minor. In this table, the following abbreviations are used:

Dist. – angular separation of the two stars in seconds of arc (the distance of a third or fourth component is given from the primary star);

PA – position angle of the pair in degrees, measured from the brighter to the fainter component;

Yr – year in which the preceding measurements were made (the last two digits only are given; the first two are understood to be "19");

Magn – visual magnitudes of the two stars on the standard scale, to the nearest half magnitude; and

RA, DEC – celestial coordinates (1950 epoch).

Table B.4.5 *List of double and multiple stars*
(adapted from Burnham 1979 by permission of Dover Publications)

Name	Dist	PA	YR	Magn	Notes	RA & DEC
α	18.0	2,128	55	2 - 9	Polaris. (Σ93)	01488n8902
	44.7	83	00	-13	PA slow inc, spect	
	82.7	172	0	-12	F8;Primary cepheid variable (*)	
Σ1583	11.1	285	25	7.5 - 8.5	(0Σ238) relfix,spect A2	11578n8718
β799	1.0	250	67	6.5 - 8.5	PA & dist inc,spect A5	13033n7318
0Σ267	0.2	332	60	9 - 9	Pa measures discordant, spect F5	13244n7615

B.4.3 Tracking Principles

Introduction

In most star-sensor tracking systems, the principle of operation is based on loading a star catalog into the onboard computer, according to the star characteristics enumerated in Section B.4.2. The stars are chosen according to the specific space mission in which the star sensor will be used. The purpose of tracking a star is to measure its direction within the reference frame fixed to the star sensor's optical head.

The first stage in attitude determination is to identify a star, or a set of stars, with reference to the onboard star catalog; the second stage is to track the star(s). The last stage involves processing the acquired data. The major components of a star-sensor attitude determination system are shown in Figure B.4.2.

There are three basic types of star-tracking assemblies: star scanners; fixed-head star trackers; and gimbaled star trackers. The star scanner is used with spinning satellites, where its purpose is to determine the attitude of the spin axis of the satellite. The primary use of fixed-head star trackers is to determine the attitude of three-axis-stabilized satellites. This kind of star sensor provides the most accurate attitude determination. The gimbaled star tracker has the advantage that, when the satellite's attitude is fixed, different parts of the sky can be scanned so that a large quantity of stars can be exploited for tracking and processing. However, imperfections inherent in gimbaled systems reduce the achievable accuracy of attitude determination.

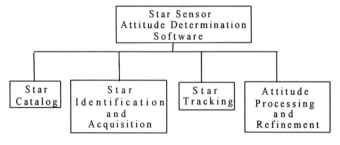

Figure B.4.2 Components of star-sensor attitude determination software.

Hardware Components of a Star-Sensor Assembly

Any star-sensor assembly includes the following basic components: (1) an optical system; (2) a detector for starlight; and (3) electronics for signal processing and attitude determination.

Optical System The optical system consists of a lens system and a stray-light shield. Even with a good sun shade, star sensors are generally inoperable if the sun direction is within 30° to 60° from their optical axes. The earth albedo and moonlight can also compromise the orderly operation of a star sensor. Hence, a good light shield is of utmost importance.

The lens system must meet the primary task of converging the stars' light into the focus of the optical head, where a light detector is located. A secondary task is to provide the desired wavelength filtering of the stars' light, so that discrimination of stars according to their spectral characteristics can be achieved. Since the optical head is exposed to temperature changes, the design of an optical system that is reliable and insensitive to environmental changes is a delicate technical task.

Detectors for Starlight There are two basic kinds of light detectors. The first, which was the only one available until the 1980s, is known as the *image dissector tube* star sensor. It consists of a photocathode located in the focal plane of the optical system, an image dissector, and a photomultiplier; see Figure B.4.3. In this figure, the optical system projects the star image onto the photocathode. An electron beam passes through a small-diameter aperture drilled into the anode and falls on the photocathode and also on the star-field image. The image dissector deflection coils induce this FOV to scan systematically the photocathode, detecting the presence of stars by monitoring the signal output (anode electrical current) and their Cartesian coordinate location on the photocathode, which is recorded from the instantaneous orientation of the deflector's field (Wertz 1978).

With this type of star detector, calibrated accuracies in the arc-second range are possible with sensitivity (on the m_V scale) of up to +14. The FOV aperture of the anode limits the absolute accuracy that can be achieved, and the scanning process suffers from the nonlinearity of the magnetic field responsible for the FOV scan. The need for temperature stability of the photomultiplier within a narrow temperature range also creates technical difficulties.

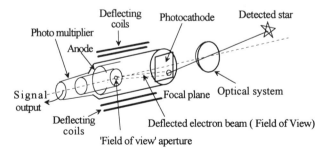

Figure B.4.3 Image dissector tube star detector; adapted from Wertz (1978) by permission of D. Reidel Publishing Co.

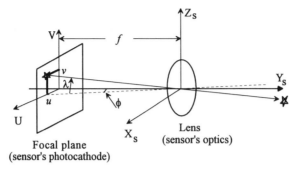

Figure B.4.4 Calculation of the components of the star unit vector in the sensor frame; adapted from Wertz (1978) by permission of D. Reidel Publishing Co.

The star sensor measures the coordinates of the star-direction unit vector in the sensor's three-axis orthogonal frame. Knowing the direction of the sensor's axis frame relative to the axis frame of the satellite, the star vector in the body frame can be calculated by a single transformation.

Let us derive the star vector components in the sensor's axis frame from the readings of the photomultiplier tube. The axis frame of the star sensor, $\mathbf{X}_s, \mathbf{Y}_s, \mathbf{Z}_s$, is shown in Figure B.4.4; the star unit vector is to be expressed in this frame. The direction of the star unit vector is measured in terms of the angles ϕ and λ, where λ is the elevation of the star image above the $\mathbf{X}_s$-$\mathbf{Y}_s$ plane and ϕ is the angle between (a) the projection of the line of sight of the star on the $\mathbf{X}_s$-$\mathbf{Y}_s$ plane and (b) the $\mathbf{Y}_s$ axis. According to Wertz (1978), the components of the star's line-of-sight vector $\mathbf{S}$ in the sensor frame are

$$\mathbf{S} = \begin{bmatrix} -\sin(\phi)\cos(\lambda) \\ \cos(\phi)\cos(\lambda) \\ -\sin(\lambda) \end{bmatrix}. \tag{B.4.1}$$

From Figure B.4.4, we easily find that

$$\tan\phi = u/f \quad \text{and} \quad \tan\lambda = (v/f)\cos(\phi), \tag{B.4.2}$$

where u and v are the two coordinates of the star image on the focal plane (i.e., the image on the photocathode) and f is the focal length.

Because of distortions in the optical system, temperature variations, and various irregular magnetic effects of the deflecting coils, the relationships in Eqs. B.4.2 are not exact. Calibration coefficients must be used in order to increase the accuracy of the derived ϕ and λ angles:

$$\begin{aligned} \phi &= C_0 + C_1 u + C_2 v + C_3 u^2 + C_4 uv \\ &\quad + C_5 v^2 + C_6 u^3 + C_7 u^2 v + C_8 uv^3 + C_9 v^3, \\ \lambda &= D_0 + D_1 u + D_2 v + D_3 u^2 + D_4 uv \\ &\quad + D_5 v^2 + D_6 u^3 + D_7 u^2 v + D_8 uv^3 + D_9 v^3 \end{aligned} \tag{B.4.3}$$

(see Gates and McAloon 1976). In Eqs. B.4.3, the coefficients are also temperature-dependent. Most of them are fitted by laboratory experimentation.

B.4 / Star Sensors

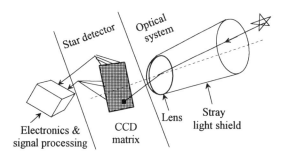

Figure B.4.5 Charge coupled device (CCD) star sensor.

The second type of star detector, available only since the early 1980s, is the *charge coupled device* (CCD) detector. This hardware is very compact and easy to implement in the optical head of a star sensor. Functionally, it replaces the photocathode tubes, which were less durable and consumed more power. The CCD is a solid-state integrated circuit, built as a matrix of photosensitive semiconductor elements called *pixels*. Matrices of the order of 500×500 (rows $\times$ columns) elements are common, providing good angular resolution. The device includes an electronic scanning mechanism that registers any illuminated cell, thus detecting the presence of stellar images and their coordinates in the matrix board.

Using the CCD in a star tracker requires some control of its temperature in order to minimize parasitic noise and bias. Active cooling, performed with Peltier elements, maintains the CCD temperature below 0°C.

The physical arrangement of a CCD-based optical star-sensor head is similar to that of Figure B.4.3, with the exception that the photocathode has been replaced by the CCD matrix board; see Figure B.4.5. With a CCD star sensor it is possible also to obtain light spectrum filtration by choosing the semiconductor material of which the CCD is composed to be sensitive to the desired spectrum only.

The location of the star image on the CCD matrix is easily obtained by reading the digital outputs of the matrix's rows and columns of the CCD pixels on which the starlight is detected. The smaller the pixel size, the higher the attitude accuracy that can be achieved. However, scanning the same FOV then requires a larger number of columns and rows, thus complicating the hardware. The optimum CCD matrix dimensions should be chosen according to mission requirements.

Electronics and Signal Processing Hardware The electronics box incorporates the conventional electronics items necessary to operate the hardware (power supplies, controllers, amplifiers, etc.), as well as a microprocessor unit (CPU) and memory unit needed for signal processing. As we shall see, the computational power required of the CPU depends on such factors as the mission and mode of use of the star sensor, the size of the star catalog used, and the number of stars to be tracked simultaneously.

Signal Processing

The signal processing of star sensors occurs in two stages. First, the target stars are identified with reference to the onboard star catalog. Next, the identified stars are tracked on the CCD focal plane. The acquired data are processed to determine the s/c attitude; it may also be important to derive the attitude rates.

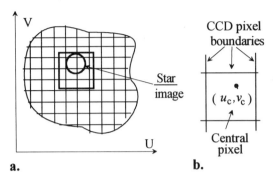

Figure B.4.6 Use of a 3×3 pixel subarray for increasing the resolution of the sensor by the image centroid algorithm; adapted from Stanton and Hill (1980) by permission of AIAA.

Scanning Process and Calculation of Star Coordinates In Figure B.4.4, the coordinates of the star in the focal plane were defined as u and v. A similar definition holds for the CCD-based star sensor, in which the u and v coordinate values are digital outputs. In both cases, the focal plane is scanned in a predetermined sequence; each time a star's image is encountered, its coordinates (i.e., the pixel locations) are sent to the memory part of the electronic processing hardware.

One potential limitation of using CCDs for star tracking is the finite number of sensing elements (pixels) available on a single CCD sensor. Current CCDs have a maximum array size in the range of 500×500 elements. With this CCD size, the resolution of a single pixel is quite low. Suppose we are using a sensor with a FOV of $5° \times 5°$ and a CCD size of 500×500 elements. In this case the angular resolution is only $(5/500)° = 36''$, which is quite low and certainly insufficient for a good star sensor. In order to achieve a resolution higher than that dictated by the detector's pixel-to-pixel spacing, an interpolation algorithm can be used. See Sheela et al. (1991) and Figure B.4.6.

If the dimension of the star's image on the CCD array is greater than that of a single pixel, then more than one pixel will detect its presence. In this case a subarray of 3×3 pixels can be used. By sensing the electrical intensity outputs of each one of the neighboring pixels shown in Figure B.4.6.a, an appropriate algorithm can interpolate the data and calculate the coordinates u_c, v_c of the centroid of the image, as in Figure B.4.6.b. By optically spreading the image over a small array of pixels, stars can be "centroided" quite accurately, typically to 1/20 of a pixel (Strikwerda et al. 1991). With centroidal algorithms, accuracy in the sub-arc region is achievable.

Onboard Star Catalog Selecting stars for inclusion in the onboard star catalog is a complicated task. It depends on the different outputs expected from the star sensor, which in turn depends on the satellite's system and mission requirements.

The number of stars that must be cataloged is a function of the star sensor's field of view. Suppose we wish to have an average of N_s stars continuously inside the FOV, in any direction on the celestial sphere. If the star sensor's FOV is rectangular, $\alpha° \times \beta°$, then the number of stars needed will be at least $[4\pi(180/\pi)^2/\alpha°\beta°]N_s$ stars. For example, if $N_s = 3$, $\alpha = 6°$, and $\beta = 10°$, then the catalog must contain at least

2,064 stars. Having three stars simultaneously in the FOV also requires that the stars be at least quasi-uniformly distributed. However, it is well known that in the direction of the Milky Way, for instance, the concentration of stars is higher than in any other direction. For more about the uniformity of star distribution, see Vedder (1993).

Different optimization criteria can be defined in developing the star catalog, according to the special tasks that the star sensor must perform. In developing the onboard star catalog we can use the two basic star characteristics of magnitude and spectra. We can also choose only that part of the celestial sphere pertinent to the special mission for which the star sensor is designed.

For example, *Polaris* star sensors have been designed for geostationary satellites to serve as the primary attitude means for continuously measuring the s/c yaw angle (Maute, Blancke, and Alby 1989). In most attitude control systems, the yaw angle is measured using sun sensors. However, owing to the unfavorable angle between the nadir and solar directions, the yaw angle can be measured with a sun sensor only during limited portions of the orbit. The Polaris star, whose direction is practically perpendicular to a geostationary orbit, allows a continuous measurement of the yaw angle. Since the Polaris star and its nearby celestial region are always inside the star sensor's FOV, only a limited number of stars are needed for the onboard star catalog. A special star sensor assembly has been developed for this task, the SODERN SED 15. The detection FOV of this sensor is $4.5° \times 6°$, and the range of detectable star magnitudes is from -1 to $+3$, which is sufficient for dealing with a relatively bright star like Polaris.

Star Identification Techniques It is well understood that tracking a star in the sensor axis frame has no meaning unless the star is identified in the star catalog, so that its coordinates in the inertial reference frame are also known; the attitude of the spacecraft can be calculated based on this knowledge. Here we summarize some techniques developed for star identification using a star catalog on board the satellite. A CCD-type star detector will be assumed. First we will assemble some relevant facts and definitions (cf. Wertz 1978) as follows.

(a) The star catalog gives the star position in an inertial reference frame of *celestial coordinates* (CC).

(b) The process of identification begins with the definition of an estimated CC frame (ECC) in which the star sensor is supposed to find the *target star*, and which is as close as possible to the true CC of the target star. The correctness of this ECC frame is very important during the attitude acquisition stage.

(c) Distortion of the ECC may occur if knowledge about the motion of the spacecraft is poor. This means that the angular distances between stars in the ECC frame may be appreciably different from the angular distances between stars measured in the true CC frame.

We are now prepared to discuss the basic star identification techniques as well as their variations.

(1) *Direct match:* With this technique, the ECC frame must be sufficiently close to the star's CC frame in the star catalog. An observation of a star in the star sensor is matched with a catalog star if

$$d(\mathbf{O}, \mathbf{S}) < \epsilon, \tag{B.4.4}$$

where **O** is the observed star unit vector in the ECC and **S** is the catalog star unit vector in the true CC frame; $d(\mathbf{O}, \mathbf{S})$ is the angular distance between both vectors, and ϵ is the error window radius. The star observation is checked against all possible catalog stars until an unambiguous and unique identification is found (see Kosik 1991, Sheela et al. 1991). If this condition is not fulfilled then an additional criterion, such as star magnitude or spectral characterization, may be added to the matching process until a correct identification is achieved.

(2) *Angular separation match:* According to this technique, the angular distances between pairs of sensed stars are compared to the angular distances between pairs of stars in the catalog (Wertz 1978, Kosik 1991). Two stars are selected arbitrarily from a set of measured stars, and the corresponding angular separation between them is calculated as

$$d_m^{12} = \cos^{-1}(\mathbf{S}_1 \cdot \mathbf{S}_2), \tag{B.4.5}$$

where $\mathbf{S}_1$ and $\mathbf{S}_2$ are the directions of the two stars as measured by the star sensor.

Next we search, in a finite region of the catalog around the approximate boresight of the sensor, for a pair of stars (i, j) that fulfills the condition

$$|d(i,j) - d_m^{12}| \le \epsilon, \tag{B.4.6}$$

where $d(i, j)$ is the angular distance calculated for entries i and j in the star catalog. Naturally, if more than one pair of catalog stars meets the condition of Eq. B.4.6 then there is an ambiguity, and the search fails.

It is possible to continue the identification process by selecting more than two stars for comparison purposes. This is called a *polygon match,* in which a pattern of N observed stars is used for identification. The process then requires enormously many more comparisons of angular distances between the set of N measured stars and the similar star distances in the star catalog. For a complete analysis of this approach, see Kosik (1991).

(3) *Phase match:* The phase match technique is used for calculating the direction of the spin axis in spinning satellites. See Wertz (1978) for a complete analysis.

For all these matching techniques, the reliability of the identification process can be increased (and the search time decreased) by using additional star characteristics for identification, such as magnitude and spectra. This additional data must also be loaded into the onboard computer, as is done with many commercial star sensors.

Star Tracking and Computational Loads

Once the target star is identified, it can be used for computing the attitude of the satellite. However, since the s/c is moving in its orbit and its attitude may be intentionally altered, the identified star's image in the focal plane of the star sensor will move. Hence, the star tracker's processor must track the star's image during its motion in the CCD array. In most star sensors, more than one star is identified and tracked simultaneously. The reason is that, since the FOV of the sensor's optical head is limited, previously identified and tracked stars exit the FOV and are lost, so that new stars must be identified and tracked. Current technology enables star-sensor assemblies to track simultaneously five or even more stars. While tracking already identified stars, the sensor must constantly identify new stars.

Considering all the tasks that the sensor must fulfill simultaneously, the computational load can become excessive. This means that tradeoffs must be made between desired sensor characteristics and attendant processing times. For a complete software flowchart containing a *track mode* and also a *search mode* of a star-sensor assembly, see Zwartbol et al. (1985).

Star-Sensor Specifications

As already mentioned, a commercial star sensor is generally not a standard item that can be procured off-the-shelf. Each sensor has special features appropriate for its unique mission, and must be manufactured to well-defined specifications. The most important characteristic features that should be considered when designing or specifying a star sensor are listed as follows.

(1) *Field of view* – The smaller the FOV, the higher the accuracy that can be achieved. However, decreasing the FOV *increases* the number of stars that must be cataloged for successful operation of the sensor.
(2) *Sensitivity to star magnitude* – Note that increasing the range of detectable magnitudes also necessitates an expanded star catalog. However, this does enable a higher attitude accuracy.
(3) *Accuracy* – The accuracy of the sensor must be defined for different modes of operation, such as the *pointing* mode, the *scanning* mode, the *tracking* mode, and so on.
(4) *Noise-equivalent angle* – This is also dependent on the operational mode.
(5) *Update period* – The update period is important because the sensor is part of the attitude control system; long update periods are detrimental to the AOCS.
(6) *Star acquisition time*.
(7) *Tracking capacity*.
(8) *Spatial separation*.
(9) *Star motion* – The accuracy of the star sensor might decrease if the satellite is under angular motion conditions.
(10) *Reliability* – This is a very important factor. If the star sensor is a basic part of the AOCS of a satellite then its reliability must be comparable to other hardware items of the satellite. Since a star sensor is a very delicate instrument, the technical problems of making it reliable are often hard to resolve.
(11) *Power consumption*.
(12) *Weight*.

These characteristics are illustrated in Tables B.4.6 and B.4.7 (overleaf), which list the specifications of two space-proven star sensors.

B.5 Rate and Rate Integrating Sensors

B.5.1 *Introduction*

Rate sensors based on different physical principles are used to measure the angular motion of satellites. All zero–momentum-biased s/c use rate sensors for at least one special task during the satellite's life.

Table B.4.6 *Characteristics of ROSAT precision star tracker, PST1*
(reproduced from Lange et al. 1986 by permission of SPIE)

CHARACTERISTICS	SSST
Field of view	$5.9^\circ \times 4.4^\circ$
Sensitivity	$m_v = 0$ to 6.5
Systematic Errors (bias) pointing scan	< 2 arcsec < 10 arcsec
Noise Equivalent angle (NEA) pointing scan	< 1.0 arcsec < 5.0 arcsec
Magnitude Accuracy	+/- 0.25 m_v
Update Period	1 sec
Star Acquisition Time	< 4 sec
Tracking Capacity	4 targets
Spatial Separation pointing scan	0.1° (x,y) 0.1°(x), 0.3°(y)
Star Motion pointing scan	5 arcsec/sec (x,y) 5 arcmin/sec(y)
Power Consumption	16.4 Watt
Weight Camera Unit Electronic Unit	8 kg 5 kg

Table B.4.7 *CT-601 solid-state star tracker*
(reproduced from McQuerry et al. 1992 by permission of IFAC)

CHARACTERISTICS	SSST
Field of view (deg²)	64
Sensitivity range (M_v)	+1 to +6
Accuracy (arc sec), per coordinate Bias errors Total random error(1σ)	3 5
Update rate (Hz)	10
Acquisition time Full field (sec) Reduced field (sec)	5 0.1
Number of stars tracked simultaneously	1 to 5
Tracking rate (deg/sec) At full performance At reduced performance	0.3 1.0
Data interface	MIL-STD-1553(B)
Maximum power (W at 28 V dc)	<12
Maximum weight without shade (lb)	18
Operating temperature (°C)	-30 to +50

A principal advantage in using a rate sensor is that it can measure the angular rates of the satellite without regard to the s/c angular attitude. For instance, if the attitude of a satellite is determined with the aid of earth and sun sensors, then the rates about the satellite principal axes can be obtained by differentiating the angular position outputs of the attitude sensors. However, once the earth (or sun) leaves the sensor FOV, it is no longer possible – without rate sensors – to measure the satellite's angular rates and thereby control its angular motion.

Another important function of rate sensors is providing angular stability to the satellite. For attitude control of s/c, two measured states must be fed to the ACS: a position error signal and a rate signal, the latter for damping the angular motion (see Chapter 7). In principle, it is possible to obtain the angular rates by differentiating the measured angular position of the satellite. However, the differentiated signals might become very noisy and so inhibit pointing stability of the ACS.

The third (and indispensable) feature of rate sensors is their ability to provide, continuously and accurately, the angular attitude of the satellite via time integration of their outputs. For instance, use of rate integrating gyros allows us to design high-bandwidth and highly accurate attitude control loops that could not be achieved solely with star sensors, whose estimates of s/c attitude are too slow. When using rate sensors for attitude determination, accurate position sensors are needed to provide the initial values used in the integration algorithm (Sections 4.7.4 and 4.7.5), and also to update periodically the attitude during the rate integrating process.

Until recently, the only available rate-sensing instruments were rate gyros (RGs) and rate integrating gyros (RIGs), both based on the gyroscopic stiffness of revolving moments of inertia (see e.g. Swanson 1982). The principal drawback of gyros is their dependence on moving parts – especially the rotor assembly of the electrical motor, which has a limited life expectancy. In the last 10–15 years, new principles have been developed to produce more reliable rate sensors with no moving parts, such as laser gyros, quartz rate sensors, and the hemispherical resonator gyro (Matthews, Baker, and Doyle 1992).

B.5.2 *Rate-Sensor Characteristics*

Rate sensors have inherent noise problems. Various criteria are used to define the qualities of these sensors; the most important are listed as follows.

(1) *Range* – The larger the range of measurement, the higher the noise level of the sensor. Hence, when defining the sensor's desired characteristics it is important to choose the smallest possible range in light of the satellite's attitude control missions. Output ranges of the order of 1°/sec to 100°/sec are possible.

(2) *Bias (constant drift)* – This is one of the most important characteristics of rate integrating gyros. The bias is of crucial importance if the sensor is to be used for attitude determination via rate integration. Sensors with drift levels ranging from 0.03°/hr to 1°/hr are common.

(3) *Output noise* – This is specified per frequency band.

(4) *Scale factor* – This is important when rate integration is performed. Stability of the scale factor has a strong influence on achievable attitude accuracies.

(5) *Linearity* – The linearity is defined over the entire range.

Table B.5.1 *Typical medium-accuracy rate integrating gyro assembly*

	Requirements	Existing RIGA
Life:	10 years	3000 hours of intermittent operation
Minimum Sampling Rate	0.04 sec (25 Hz)	Digital pulse width modulated system pulse frequency =102400 Hz
Range	15 deg/ sec 2 deg/sec	15 deg / sec
Constant drift	1 deg/ hour	1 deg / hour (1 g environment)
Scale factor	1.0 %	1.0 %
Input power maximum	25 Vdc	23.9 V dc

To clarify these characteristics we shall tabulate data for two rate sensors. The RIG described in Table B.5.1 can be used for integration because it has a fairly low constant drift bias of 1°/hr. The RG described in Table B.5.2 cannot be used for rate integrating, owing to its high bias terms. For this reason it is commonly called a "coarse rate gyro."

References

Bednarek, T. (1992), "Dual Cone Scanning Earth Sensor Processing Algorithms," *Small Satellite Technologies and Applications II* (SPIE vol. 1691). Bellingham, WA: International Society of Optical Engineering, pp. 181-91.

Desvignes, F., Doittau, F., Krebs, J., and Tissot, M. (1985), "Optical Sensors for Spacecraft Attitude Measurement with Respect to the Earth," *Infrared Technology and Applications* (SPIE vol. 590). Bellingham, WA: International Society of Optical Engineering, pp. 322-30.

Fallon, J., and Selby, V. (1990), "An Optical Locator for Horizon Sensing," Paper no. 90-32, 13th Annual AAS Guidance and Control Conference (3-7 February, Keystone, CO).

Gates, F., and McAloon, K. (1976), "A Precision Star Tracker Utilizing Advanced Techniques," Paper no. 76-113, AIAA 14th Aerospace Sciences Meeting (January, Washington, DC).

Gontin, R. A., and Ward, K. A. (1987), "Horizon Sensor Accuracy Improvement Using Earth Horizon Profile Phenomenology," Paper no. 87-2598, *Navigation and Control Conference.* Washington, DC: AIAA, pp. 1495-1502.

Hablani, H. (1993), "Modeling of Roll/Pitch Determination with Horizon Sensors: Oblate Earth," AIAA Guidance, Navigation and Control Conference (9-11 August, Monterey, CA). Washington, DC: AIAA, pp. 1133-47.

Hirshfeld, A., and Sinott, R. (1990), *Sky Catalogue 2000.0.* Belmont, MA: Sky Publishing.

Hoffleit, D. (1964), *Catalog of Bright Stars.* New Haven, CT: Yale University Observatory.

ITHACO, Inc. (1983), "Conical Earth Sensor (IPS-6) 12/83," ITHACO Space Products, Ithaca, NY.

Kosik, J. (1991), "Star Pattern Identification Aboard an Inertially Stabilized Spacecraft," *Journal of Guidance, Control, and Dynamics* 14(2): 230-5.

Table B.5.2 *Coarse rate gyro, 2424.0002.00.19*
(TAMAM, Precision Instruments Industries, Electronic Division, Israel Aircraft Industries Ltd.)

NO.	Parameter	Specifications
1	Weight	650 Gr. Max
2	Start Power	55 Watt Max (For 12 sec)
3	Start Current	2.0 Amp. Max
4	Running Power	10 Watt Max.
5	Running Current	0.2 Amp. Max
6	Ready Time	15 Sec Max.
7	Hysteresis	0.15°/sec for X and Y gyros 0.3°/sec for Z gyro
8	Zero Offset	0.6°/sec for X and Y gyros 1.3°/sec for Z gyro
9	Scale Factor	40 mV/°/sec for X and Y gyros 13.3 mV/°/sec for Z gyro
10	Null Voltage Change with Temperature	+/- 0.5°/sec for X and Y gyros +/-1.2°/sec for Z gyro
11	Linearity	0.5% of f.s. up to half scale 2.5% of f.s. from half f.s. up to f.s.
12	Frequency Response	15-20 Hz (-3 db)
13	Input Voltage	25 to 44 Vdc
14	BIT	+8 to +10 Vdc for normal operation -3 to +3 Vdc for failure
15	Range	X and Y gyros -200°/sec Z gyro -600°/sec
16	Cross Coupling	0.5°/sec for X and Y gyros 1.0° /sec for Z gyro
17	Threshold and Resolution	0.05°/sec for X and Y gyros 0.1°/sec for Z gyro
18	Noise	0.1° /sec

Lange, G., Mosbacher, B., and Purll, D. (1986), "The ROSAT Star Tracker," *Instrumentation in Astronomy VI* (SPIE vol. 627). Bellingham, WA: International Society of Optical Engineering, pp. 243–53.

Lyle, R., Leach, J., and Shubin, L. (1971), "Earth Albedo and Emitted Radiation," Document no. SP-8067, NASA, Washington, DC.

Matthews, A., Baker, R., and Doyle, C. (1992), "The Hemispherical Resonator Gyro: New Technology Gyro for Space," 12th FAC Symposium on Automatic Control in Aerospace: Aerospace Control 92 (7–11 September, Ottobrunn, Germany).

McCanless, F., Quasius, G., and Unruh, W. (1962), "Star Tracker Aerospace Reference Study – Stars," Report no. ASD-TDR-62-1056, Honeywell Military Products Group, St. Petersburg, FL.

Maute, P., Blancke, J., and Alby, F. (1989), "Autonomous Geostationary Station Keeping System Optimization and Validation," *Acta Astronautica* 20: 93–101.

McQuerry, J., Wagner, D., Sullivan, M., Deters, R., and Radovich, M. (1992), "A New Generation Stellar Attitude Sensor: The CT-601 Solid-State Star Tracker," 12th IFAC Symposium on Automatic Control in Aerospace: Aerospace Control 92 (7–11 September, Ottobrunn, Germany).

Muller, E., and Jappel, A. (1977), *International Astronomical Union, Proceedings of the Sixteenth General Assembly* (Grenoble 1976). Dordrecht: Reidel.

Pochard, M. (1992), "A Fifteen Years Lifetime Mechanism for an Infrared Earth Sensor" (ESA SP-334), *Proceedings of the Fifth European Space Mechanisms and Tribology Symposium* (28-30 October, ESTEC, Noordwijk, Netherlands). Paris: European Space Agency, pp. 353-8.

Sheela, B., Shekhar, C., Padmanabhan, P., and Chandrasekhar, M. (1991), "New Star Identification Technique for Attitude Control," *Journal of Guidance, Control, and Dynamics* 14(2): 477-80.

Sidi, M. (1980), "On Maximization of Gain-Bandwidth in Sampled Systems," *International Journal of Control* 32(6): 1099-1109.

Smithsonian Institute Staff (1971), *Smithsonian Astrophysical Observatory Star Catalog*, parts I-IV. Washington, DC.

SODERN (1977), "Static Infrared Horizon Sensor STA 03: Description, Operation, Functions and Characteristics," Document no. C.04.1163A, SODERN, Limeil-Brevannes, France.

SODERN (1991a), "Scanning Infrared Earth Sensor STD 15: Description, Operation, Functions and Characteristics," Document no. C.04.1346-05, SODERN, Limeil-Brevannes, France.

SODERN (1991b), "Infrared Horizon Scanning Sensor (500 to 1200 Km) STD 16: Description, Operation, Functions and Characteristics," Document no. C.04.1970-05, SODERN, Limeil-Brevannes, France.

Stanton, R., and Hill, R. (1980), "CCD Star Sensor for Fine Pointing Control of Spaceborne Telescopes," *Journal of Guidance and Control* 3(2): 179-85.

Strikwerda, T., Fisher, H., Kilgus, C., and Frank, L. (1991), "Autonomous Star Identification and Spacecraft Attitude Determination with CCD Star Trackers," *Spacecraft Guidance Navigation and Control Systems* (proceedings of the first international conference organized by ESA at ESTEC, 4-7 June, Noordwijk, Netherlands). Paris: European Space Agency, pp. 195-200.

Swanson, C. (1982), "DRIRU I/ SKIRU - The Application of the DTG to Spacecraft Attitude Control," Paper no. 82-1624, AIAA Guidance and Control Conference (9-11 August, San Diego, CA).

Tai, F., and Barnes, R. (1989), "The Dual Cone Scanner: An Enhanced Performance, Low-Cost Earth Sensor," Paper no. 89-013, 12th Annual AAS Guidance and Control Conference (4-8 February, Keystone, CO). San Diego, CA: American Astronautical Society, pp. 147-69.

Vedder, J. (1993), "Star Trackers, Star Catalogs, and Attitude Determination: Probabilistic Aspects of System Design," *Journal of Guidance, Control, and Dynamics* 16(3): 498-504.

Wertz, J. (1978), *Spacecraft Attitude Determination and Control*. Dordrecht: Reidel.

Zwartbol, T., Van Den Dam, R., Terpstra, A., and Van Woerkom, P. (1985), "Attitude Estimation and Control of Maneuvering Spacecraft," *Automatica* 21(5): 513-26.

APPENDIX C

Orbit and Attitude Control Hardware

C.1 Introduction

The purpose of Appendix C is to introduce the reader to the basic features of the control hardware used to provide translational and angular accelerations to spacecraft.

The level of control forces that can be obtained depends on the source characteristics. For example, ion thrusters can produce forces of the order of tens of millinewtons; liquid propellant thrusters used in spacecraft control provide forces in the range of hundreds of newtons; and solid propellant motor–produced forces are in the range of hundreds of thousands of newtons.

As far as attitude control is concerned, we have seen that torques can be produced with the aid of momentum exchange devices, magnetic torqrods, or solar torque controllers. The torque levels we can obtain with these devices are generally low: in the range of 0.01 to 1 N-m for momentum exchange devices (control moment gyros excluded), of the order of a few centinewton-meters with magnetic torqrods, and tens of micronewton-meters with solar torque controllers. With reaction propulsion means we can achieve torques of about 20–30 N-m, but also much lower torques of the order of 0.1 N-m if needed. Reaction propulsion is also used for momentum dumping – unloading the parasitic angular momentum accumulated in the spacecraft.

The sources of force and torques used for control can be classified as follows:

(1) *propulsion systems,* which can provide translatory and angular accelerations (forces and torques) to the satellite;
(2) *solar radiation pressure,* which can produce forces and torques;
(3) *momentum exchange devices,* which can provide torques and angular momentum; and
(4) *magnetic torqrods,* which can provide only torques.

Two special controllers that are not discussed in this book are (i) atomic energy-based thrusters for translation accelerations, and (ii) control moment gyros, which are used in large inhabited space structures to provide torques in the range of hundreds of newton-meters.

C.2 Propulsion Systems

The task of the propulsion system is to provide forces and torques acting on the body of the spacecraft, thus enabling changes in its translatory and angular velocities. Spacecraft propulsion systems are divided into three categories: cold gas; chemical (solid and liquid); and electrical. The basic equation of propulsion holds for all kinds of propellants.

A rocket engine develops its thrust F by expelling propellant (such as gas molecules, or ions) at a high exhaust velocity V_e relative to the satellite body; see Wertz and Larson (1991). The amount of thrust F can be calculated as follows:

$$F = V_e \frac{dm}{dt} + A_e[P_e - P_a] = V_{ef}\frac{dm}{dt}, \qquad (C.2.1)$$

where P_e and P_a are the gas and ambient pressures (respectively), V_e is the exhaust velocity, V_{ef} is the effective exhaust velocity of the expelled mass with respect to the satellite, dm/dt is the mass flow rate of the propellant, and A denotes the area of the nozzle exit.

The second parameter relevant to the characteristics of the thrust source is the *specific impulse* I_{sp}, a measure of the efficiency with which the propellant mass is converted into thrust energy. The specific impulse is defined as

$$I_{sp} = F/(g\, dm/dt) \text{ sec}, \qquad (C.2.2)$$

where g is the gravitational constant.

To calculate the velocity change per exhausted fuel mass, integration of the acceleration F/m is performed to find, using Eq. C.2.2, that

$$\Delta V = \int_{t_i}^{t_f} \frac{F}{m}\, dt = \int_{t_i}^{t_f} gI_{sp}\frac{1}{m}\frac{dm}{dt}\, dt = gI_{sp}\int_{m_i}^{m_f} \frac{dm}{m}, \qquad (C.2.3)$$

where t_i, t_f and m_i, m_f are the initial (and final) time and masses of the spacecraft, respectively. The solution of Eq. C.2.3, known also as the *rocket equation,* is:

$$m_f = m_i \exp\left\{-\left(\frac{\Delta V}{gI_{sp}}\right)\right\}. \qquad (C.2.4)$$

It follows that the propulsion mass m_p expelled from the satellite's initial mass is

$$m_p = m_i - m_f = m_i\left[1 - \exp\left\{-\left(\frac{\Delta V}{gI_{sp}}\right)\right\}\right]. \qquad (C.2.5)$$

Equation C.2.5 is the solution of the basic propulsion rocket equation, and allows us to calculate the mass of propellant required to change by ΔV the velocity of a satellite with an initial mass m_i. An immediate conclusion is that increasing the specific impulse will decrease the expelled mass of propellant. However, we shall see that high I_{sp} of a potential propellant is not the only factor upon which propulsion selections are based.

Reaction propulsion systems are used for producing forces and torques. Forces are used to increase the linear velocity of the satellite. For this purpose, the propulsion thruster is activated for comparatively large time periods – several tens of minutes. Moreover, since large masses are to be accelerated, high levels of thrust are necessary. Since the thruster must accelerate its own weight also, it is of utmost importance to use propellants with very high specific impulse I_{sp}. The lifting capabilities of a propulsion system are defined as $\int_{t=0}^{\infty} F\, dt$; this is called the system *total impulse,* or simply the *impulse.*

Torques are used in reaction attitude control for which thrusters provide interrupted pulses. In fact, every pulse provides a torque impulse bit TIB, where IB = $\int F\, dt$ and TIB = $\Delta \times$ IB (Δ is the torque arm of the thruster). In order to achieve accurate attitude control, the pulses must be able to provide torque impulse bits as small as possible; hence the impulse bits should likewise be as small as possible. This can be achieved by activating the thruster for very short times (tens of milliseconds or less) if the thrust level F cannot otherwise be decreased. In other situations the same thrusters might be activated for several seconds, so we should anticipate a large ratio of maximum/minimum activation time (Chapter 9). The attitude control system remains active throughout a satellite's life, so the number of thruster activations may become quite large.

A propulsion system is evaluated in terms of various characteristic features. Some have been mentioned already; others will be defined shortly. These basic characteristics include: the thrust level F; the specific impulse I_{sp}; the minimum impulse bit MIB; the maximum number of activations (expected life); the maximum permitted duty cycle of activation; and the total impulse.

C.2.1 Cold Gas Propulsion

Cold gas propulsion is the simplest way of achieving thrust. Such a system consists simply of a tank (with a controllable nozzle) containing pressurized gas. The gas is preferably inert, for example, nitrogen. In order to achieve a reasonable thrust level, the gas must be stored under very high pressure, typically 4,000–10,000 psi. The high pressure requires storage tanks capable of withstanding such pressures. As a result, the tanks are quite heavy. Another significant drawback is the low specific impulse of cold gas propulsion; $I_{sp} = 50$–70 sec is a common average value. Achievable thrust is of the order of 5 N. Table C.2.1 gives a rough idea of the weight of some existing cold gas propulsion systems.

Table C.2.1 *Weight and volume of some cold gas propulsion tanks*

No	Volume [liter]	Weight of Tank (empty) [kg]	Mass of the gas [kg]	Initial Total Weight [kg]	Pressure [PSI]	Technology
1	13	6.7	5	11.7	6000	Titanium & Graphic epoxy
2	11	12	4	16	6000	Titanium
3	9.5	17	4.5	21.5	10000	Titanium

C.2.2 Chemical Propulsion – Solid

Chemical propulsion systems can be subdivided into two basic fuel categories, *solid* and *liquid,* which are the primary systems used in space for achieving thrust. In the technical literature, rockets using solid propellants are called *motors,* whereas rockets using liquid propellants are called *thrusters* or *jets.*

Solid propellant motors are used as the upper stage propulsion system, providing the necessary velocity increment for injection of the spacecraft from the low-altitude initial orbit into the final operational orbit; they are sometimes called apogee kick motors (AKM) or apogee boost motors (ABM). The propellant is a solid chemical material that is a mixture of fuel and oxidant, cast within a metal case ended by a nozzle throat and exit cone. Once ignited, solid propellant motors generally burn until exhausted, since there is no simple physical means to stop the burning within

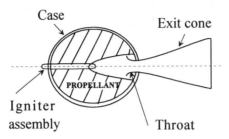

Figure C.2.1 Schematic diagram of a solid propulsion motor.

Table C.2.2 *Performance figures of various solid propellant motors*
(from Wertz and Larson 1991 by permission of Kluwer Academic Publishers)

Name of Motor	Total Mass [kg]	Propellant mass [kg]	Total Impulse [10^6 N-s]	Max Thrust [N]	I_{SP} [s]
STAR 13B	47	41.4	0.16	9608	285.7
IUS SRM-2	2995	2725	8.11	111072	303.8
STAR 75	8066	7496	21.3	242846	288.0
LEASAT PKM	3658	3329	9.26	193200	285.4
IUS SRM-1	10374	9752	28.1	260488	295.5

the motor volume. The I_{sp} of solid propellant motors ranges from 285 to 300 sec. A simplified schematic of a solid motor is shown in Figure C.2.1. Table C.2.2 summarizes the characteristic features of some solid propellant motors; see also Pritchard and Sciulli (1986).

C.2.3 Chemical Propulsion – Liquid

In liquid propulsion systems, we must differentiate between monopropellant and bipropellant fuels. The hardware for bipropellant liquid systems is more complicated, but such systems are characterized by a higher I_{sp}. In either case, the fuel is delivered to the combustion chamber in one of two possible modes.

In the *blowdown* operation mode, pressurized gas is stored in the same tank as the propellant. The drawback here is that the pressure decreases as propellant is consumed, reducing the thrust level accordingly. Typical pressures in this mode are initially 300–400 psi, falling to about 100 psi as the propellant becomes exhausted.

In the *regulated pressure* operation mode, a regulator maintains a constant gas pressure, with the inherent drawback of additional system complexity. A high pressure of about 3,000–4,000 psi is regulated to 200–300 psi at the propellant tank.

Monopropellant Propulsion

The most popular liquid propellant is hydrazine (N_2H_4). Hydrazine vaporizes and decomposes when brought into contact with a suitable catalyst, thus producing hydrogen and nitrogen gases under pressure and so generating propulsion.

C.2 / Propulsion Systems

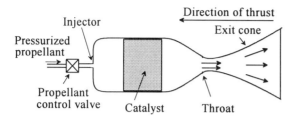

Figure C.2.2 Schematic diagram of a monopropellant thruster engine.

An exemplary catalyst is Shell 405 (manufactured by Shell Oil Co.). A schematic diagram of a typical hydrazine thrust engine is shown in Figure C.2.2.

The thrust is produced in the following stages. (1) When the propellant control valve is in the open condition, the pressurized monopropellant liquid is injected into the catalyst bed at flow rate dm/dt. (2) When in touch with the catalyst, the propellant decomposes according to the chemical reaction formula

$$3N_2H_4 \rightarrow 4NH_3 + N_2 + 36,360,714 \text{ cal.} \tag{C.2.6}$$

A part of the ammonia (NH_3) is further decomposed via the chemical reaction

$$4NH_3 \rightarrow 2N_2 + 6H_2 - 19,956,816 \text{ cal.} \tag{C.2.7}$$

(3) Only part of the ammonia is decomposed, depending on the geometry of the reaction chamber (the fraction of decomposed ammonia is about 2/5). The three elements – ammonia, nitrogen, and hydrogen – are exhausted through the exit nozzle to produce thrust.

With the catalytic thruster just described, a specific impulse $I_{sp} = 235$ is achieved when activated in normal mode (i.e., with nominal pulse width and duty cycles). When very short pulses (near the MIB) are activated, the I_{sp} is reduced by a factor of two or three. Monopropellant thruster engines produce nominal thrust levels in the range of 0.1–500 N.

A hydrazine thruster can be fired for only a limited number of pulses. The limiting factor is the degradation of the catalyst caused by loss of catalytic activity and also physical loss of the catalytic mass. It has also been found that the starting temperature of the catalyst bed is the primary determinant of its catalytic properties. To avoid fast degradation of the catalyst, it is common to keep the bed at high ambient temperatures – about 90°–300°C, depending on the particular thruster, its thrust level, and so on. Maintaining an appropriate catalyst bed temperature can prolong the hydrazine thruster's life by a factor of greater than 10. Currently, preheated hydrazine thrusters may have a life expectancy of 1 million pulses.

The specific impulse achievable with the hydrazine monopropellant can be increased by roughly 25% with a different thruster engine configuration. Instead of using a catalyst for decomposing the hydrazine propellant as in Figure C.2.2, the thruster engine now consists of two decomposition chambers. The first chamber incorporates a low-power heater (about 14 W). The second chamber is a *vortex heat exchanger* of about 400 W, which electrically increases the enthalpy of the decomposition products of hydrazine. In this way $I_{sp} = 300$ sec is achieved, albeit at the expense of greatly increased power consumption. These *electrothermal monopropellant hydrazine* thrusters have been used on the telecommunications satellite *Intelsat V* for North–South station keeping (see Agrawal 1986).

The hydrazine thruster does not by itself constitute a complete propulsion system, which must include at least the following items: propellant tank, pressurizing system, propellant control valves, thrusters, filters, latch valves, pressure transducers, and fill and drain valves.

In general, no two satellites use the same propulsion configuration; each s/c has its own preferred thrust levels. In any given satellite, thrusters with different levels can be used for various tasks. In an integrated propulsion system, thrusters for orbit correction and maneuvers are of the high-thrust (HT) type, of the order of 200–500 N. For attitude control, much lower-thrust level engines are used. Depending on the particular control requirements, low-thrust (LT) engines of the order of 0.1–25 N are used. The weight of a thruster is also of importance. Common LT engines weigh about 0.15 kg; an HT engine's weight is typically about 2 kg.

A schematic diagram of a monopropellant propulsion system is shown in Figure C.2.3. In this system, a blowdown operation mode is assumed. The propulsion system in this figure consists of two identical systems, one of which is fully redundant. The latch valves (LVi) allow simultaneous use of both systems or each one alone. Each system includes at least six LT thrusters to allow 6-DOF attitude control. The number of LT engines need not be limited to six, and neither must they be identical – it all depends on the attitude control philosophy. In Figure C.2.3, each propulsion system also includes one HT engine for orbit control.

Each LT set can be operated by either tank 1 or tank 2. Suppose that leakage is detected in the tank-1 system. In this case, latch valve LV1 is closed, so that the thrusters of set A and those of set B are fed from tank 2. Similarly, if one of the thrusters of the primary set A is leaky then LV2 is closed, and the backup thrusters (set B) can be fed from either tank 1 or tank 2. In other words, this propulsion system is fully redundant. It is called an *integrated* or a *unified* propulsion system because it enables both orbit maneuvering and attitude control (see also Berker 1978).

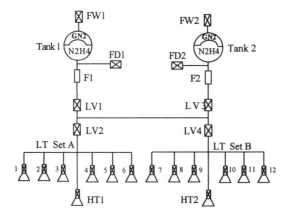

Figure C.2.3 Schematic diagram of a monopropellant propulsion system.

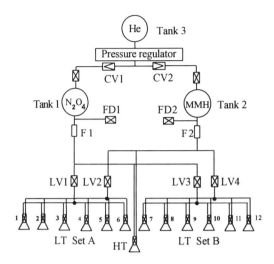

Figure C.2.4 Schematic diagram of a bipropellant propulsion system.

Bipropellant Propulsion

Bipropellant propulsion systems are based on combustion resulting from the contact of two propellants – for instance, monomethyl hydrazine (MMH) and nitrogen tetroxide (N_2O_4), which is used as the oxidant. The primary reason for using such a system is that an augmented specific impulse is achieved, $I_{sp} \approx 320$ or more.

A bipropellant unified propulsion system usually consists of LT thrusters and one or two HT engines, two tanks for the propellants and oxidizer, and an additional tank for the pressurized gas (e.g. helium) used to pressurize the two propellant tanks. To reduce weight, generally no backup tanks are provided. However, a backup set of LT thrusters is used, since their reliability is generally insufficient for an extended operational life. The failure of a single thruster would seriously degrade the entire ACS, so a backup set is absolutely necessary.

A schematic diagram of a bipropellant unified propulsion system is shown in Figure C.2.4. In this system we have two LT sets (A and B) for redundancy. The very high pressure in tank 3 is regulated to a nominal high pressure of about 200 psi supplied to tanks 1 and 2. A system of one-way valves (CV1 and CV2) is used to prevent flow between the two propellant tanks. Only one HT thruster is included in order to decrease the system's weight, which is about 50–60 kg exclusive of propellant mass. Common thrust levels are 10–22 N for LT engines and 400–490 N for the HT engine. Different propellant–oxidant pairs supply varying specific impulses I_{sp}; see the listing in Table C.2.3 (overleaf).

C.2.4 Electrical Propulsion

Electrical propulsion systems are based on accelerating an ionized mass in electromagnetic or electrostatic fields, where the ions leave the thruster nozzle at very high velocities. Specific impulses ranging from 2,000 to 6,000 sec can be achieved. However, such propulsion systems have numerous drawbacks. First, the thrust levels

Table C.2.3 I_{sp} *for different bipropellant pairs*
(from Pritchard and Sciulli 1986 by permission of Prentice-Hall)

Manufacturer	Nominal Thrust (N)	Fuel	Oxidizer	I_{SP} [s]
Bell	4210	MMH	N_2O_4	289
Rocketdyne	445	MMH	N_2O_4	299
M.B.B.	400	MMH	N_2O_4	307
S.E.P.	61668	LH_2	LOX	432

Table C.2.4 *Characteristics of 5-mN and 130-mN thrusters*
(from Poeschel and Hyman 1984 by permission of AIAA)

	Thrust Level [mN]	
	5	130
I_{SP} [s]	2650	3000
Average Power [KW]	0.125	2.6
Thrust efficiency [%]	56	70
Power/thrust [kW/N]	24.4	20.7
Demonstrated lifetime [hr]	>15000	10000
Total Impulse [N-s]	$2.7\ 10^5$	$7\ 10^6$
Propellant	Mercury	

that can be achieved are very low: $5-20 \times 10^{-3}$ N. Next, the exhaust velocity of the ions must be so high (about 40 km/sec) that accelerating voltages of about 1,500 V are required – a complicated and delicate problem in space technology. Finally, the total power input is also high, ranging from 250 W to 550 W. Table C.2.4 details two such thrusters. The expected lifetime of these ion thrusters can exceed 15,000 hours. However, an increase in power consumption has been observed during this lifetime, due to aging of electronic components, as follows: 4% after 8,000 hours of operation; 7% after 15,000 hours of operation.

Given the technical problems of such systems, we must ask if there is any practical reason for using them. A complete ion propulsion system, based on four UK-10 thrusters supplying 0.02 N each and including a propellant mass sufficient to produce a total implse of 800,000 N-sec, weighs only 91 kg. In contrast, the equivalent mass

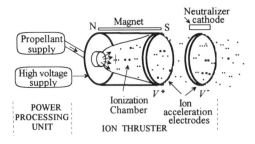

Figure C.2.5 Schematic diagram of an electrostatic ion thruster; adapted from Poeschel and Hyman (1984) by permission of AIAA.

of chemical propellant needed to produce the same total impulse is 286 kg (assuming $I_{sp} = 285$ sec; see Smith 1988).

A simplified diagram of an electrostatic ion thruster is shown in Figure C.2.5. The gaseous propellant (mercury, xenon, or argon vapor) is ionized in the ionization chamber by electron impact, forming a neutral plasma. One side of the ionization chamber is equipped with two electrodes that have an array of aligned apertures. A high voltage is applied to the electrodes in order to extract ions from the discharged plasma and accelerate them to a high velocity, thus forming the ion beam – the thrust beam. Electrons and ions must be injected into the beam in equal numbers to maintain charge neutrality; this is the task of the neutralizer cathode.

C.2.5 Thrusters

The choice of thrusters depends very much on the specific task to be performed by the propulsion system. As already mentioned, a unified propulsion system may consist of several thrusters with different thrust levels. Besides the level of the thrust, some additional characteristics are necessary in order to define a thruster completely. For example, when a thruster is used for fine attitude control, the *time behavior* of its achieved thrust is critical.

Figure C.2.6 shows the time-domain behavior of a commanded thrust. The electrical *on* command applied to the coil of the thruster valve opens that valve for the propulsion liquid, but the thrust force develops only after an inherent initial delay. This *start time* T_s is defined as the time elapsed between the electrical command and the thrust reaching 90% of its maximum value; $T_s \approx 15$ msec for a bipropellant 10-N thruster.

The *shutdown time* (T_{sd}) is the time elapsed between cessation of the *on* command and the final decay of the thrust; $T_{sd} \approx 10$ msec for a bipropellant 10-N thruster. If the *on* command is too short then the thrust will not reach its nominal level F. In this context we can define a *minimum on* command (T_{min}). Another important thruster characteristic is the delay between the *centroid* time of the *on* command and the centroid time of the obtained thrust (T_{cd} in Figure C.2.6).

For fine attitude control, perhaps the most important characteristic is the *minimum impulse bit* (MIB) – a measure of the minimum attitude change that can be commanded to the satellite (see Chapter 9). Lower thrust levels enable lower MIBs. For a bipropellant 10-N thruster, MIBs of 30–40 mN-sec are achievable; for a catalytic monopropellant hydrazine 0.2-N thruster, a MIB of 5 mN-sec is possible.

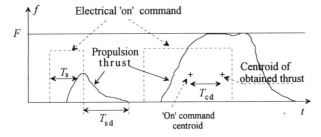

Figure C.2.6 Time-domain behavior of thruster pulse.

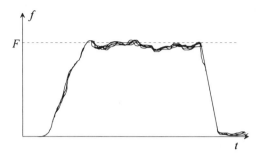

Figure C.2.7 Time histories showing the roughness and repeatability of the output thrust for several activations of the same thruster.

Another property that characterizes thrusters is *roughness,* defined as the irregularity of the thrust about its nominal value (see Figure C.2.7). Typical roughness values are 2%–3% (of the nominal thrust). In the present context, it is also usual to inquire about the *repeatability* of the time histories for consecutive activations of the same thruster. Qualification tests for a thruster record the time behavior for several activations under the same conditions (e.g., temperature, duty cycle, etc.); see Figure C.2.7.

It is also important to note that the specific impulse depends on the impulse bit. The lower the impulse bit, the lower will be I_{sp} – a relation that becomes more pronounced when the thrust does not reach its nominal level. This phenomenon bears significance for the mass of fuel consumed during an attitude control task, which in turn is a major determinant of the satellite's serviceable life. One final thruster feature is the maximum number of activations it can withstand. Contemporary satellites are expected to last more than a decade, so the number of activations is quite large (hundreds of thousands). It's no wonder, then, that thrusters capable of withstanding a *million* activations are now on the market. Table C.2.5 lists technical specifications for two monopropellant thrust engines.

C.3 Solar Pressure Torques

C.3.1 *Introduction*

The literature contains many models purporting to supply a dynamic description of the torques produced by solar pressure on reflecting surfaces. Because of the delicacy of the various relevant parameters, the dynamic equations are often quite complex and so fidelity to nature is far from guaranteed. The purpose of this section is to formulate a simplified model sufficient for the presentation of *solar control torques* in Section 8.6.

C.3.2 *Description*

The basic hardware for realizing solar torques is shown in Figure C.3.1. The complete "solar torque hardware set" consists of two panels (the primary s/c power source) to which are appended the flaps necessary for achieving unbalanced torques

C.3 / Solar Pressure Torques

Table C.2.5 *Catalytic monopropellant hydrazine thrusters*
(by permission of Daimler-Benz Aerospace, MBB/ERNO)

Technical Data	Units	CHT 0.5	CHT 20
Thrust range	N	0.75 - 0.2	24.0 - 7.2
Oper. pressure range	bar	22-5.5	22-5
SSF specific impulse	Ns/kg	2230-2120	2300 - 2180
Minimum impulse bit	Ns	0.015-0.005	0.37 - 0.165
Proof pressure	bar	33/54	33/54
Burst pressure	bar	88/144	88/144
Mass	kg	0.19	0.36
Valve power	Watt	5.0	13.0
Heat power	Watt	2.5	<3.0
Qualification status		qualified in 1977	qualified in 1988
SSF duration total	h	143	4
SSF duration single burn	s	25200	3600
Hot pulse quantity		59000	235000
Off modulation quantity		88000	-
Cold start quantity		311000 at $< 210^0 C$	10000

(to be defined shortly). The geometric dimensions are shown in Figure C.3.2 (overleaf). Our simplified analysis will aim to produce a first-order-approximation dynamic model. Despite its limitations, the model presented in this appendix is suficiently elaborated to give a good understanding of the problems involved in solar torque attitude control.

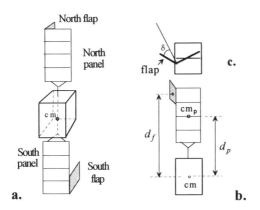

Figure C.3.1 Solar torque hardware as installed on the satellite.

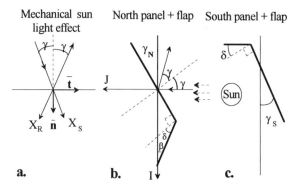

Figure C.3.2 Definition of the panel and flap forces created by solar pressure.

We must first define the following parameters:

P = solar pressure (4.6×10^{-6} N-m^{-2});
S = surface area of flaps or panels;
η = area reflexivity; and
γ = incidence angle of solar pressure.

With these definitions, the force on the pressure center is

$$\mathbf{F} = PS\{\cos^2(\gamma)[1+\eta]\mathbf{n} + \sin(\gamma)\cos(\gamma)[1-\eta]\mathbf{t}\} \quad \text{(C.3.1)}$$

(Lievre 1985; see also Figure C.3.2.a). As seen from this equation, there are force components normal and tangential to the area. Since the panel and the flap are generally manufactured from different materials, the reflexivity coefficients of the panel and the flap will usually differ. The areas of both South and North panels, as well as the areas of both South and North flaps, are assumed to be equal.

Based on Eq. C.3.1, we can compute (with reference to Figure C.3.2.b) the forces along the directions **I** and **J** of the solar inertial frame for all panels and flaps. These forces will be labeled as follows:

FPJN = force on the North panel in the direction **J**,
FPJS = force on the South panel in the direction **J**,
FPIN = force on the North panel in the direction **I**,
FPIS = force on the South panel in the direction **I**;
FFJN = force on the North flap in the direction **J**,
FFJS = force on the South flap in the direction **J**,
FFIN = force on the North flap in the direction **I**,
FFIS = force on the South flap in the direction **I**.

Using Eq. C.3.1 with the assumption that the sun is in the **I–J** plane, we obtain the following results for the panels:

$$\text{FPJN} = PS_p \cos(\gamma_N)[1 + \eta_p \cos(2\gamma_N)], \quad \text{(C.3.2)}$$

$$\text{FPIN} = PS_p \eta_p \cos(90° - 2\gamma_N) = PS_p \eta_p \cos(\gamma_N)\sin(2\gamma_N), \quad \text{(C.3.3)}$$

C.3 / Solar Pressure Torques

$$\text{FPJS} = PS_p \cos(\gamma_S)[1 + \eta_p \cos(2\gamma_S)], \tag{C.3.4}$$

$$\text{FPIS} = PS_p \eta_p \cos(\gamma_S) \sin(2\gamma_S). \tag{C.3.5}$$

To find the forces created by the flaps, we can use the results of Eqs. C.3.2–C.3.5 after correcting for the relations between angles δ, β, and γ in Figure C.3.2.b. For the North flap we have $\beta_N = \gamma_N + \delta - 90°$. For the South flap we have $\beta_S = 90° + \gamma_S - \delta$, from which it follows that

$$\cos(\beta_S) = -\sin(\gamma_S - \delta), \quad \cos(2\beta_S) = -\cos 2(\gamma_S - \delta), \quad \sin(2\beta_S) = -\sin 2(\gamma_S - \delta).$$

Finally, we have the remaining four force equations:

$$\text{FFJN} = PS_f \sin(\gamma_N + \delta)[1 - \eta_f \cos 2(\gamma_N + \delta)], \tag{C.3.6}$$

$$\text{FFIN} = -PS_f \eta_f \sin(\gamma_N + \delta) \sin 2(\gamma_N + \delta), \tag{C.3.7}$$

$$\text{FFJS} = -PS_f \sin(\gamma_S - \delta)[1 - \eta_f \cos 2(\gamma_S - \delta)], \tag{C.3.8}$$

$$\text{FFIS} = PS_f \eta_f \sin(\gamma_S - \delta) \sin 2(\gamma_S - \delta). \tag{C.3.9}$$

The last stage of our computation is to find the torques (about the body center of mass) created by the forces calculated in Eqs. C.3.2–C.3.9:

$$C_I = d_p[\text{FPJN} - \text{FPJS}] + d_f[\text{FFJN} - \text{FFJS}], \tag{C.3.10}$$

$$C_J = d_p[\text{FPIN} - \text{FPIS}] + d_f[\text{FFIN} - \text{FFIS}]. \tag{C.3.11}$$

Since the cm of the flaps is not located on the axis of rotation of the solar panels, there is a small additional torque about this axis that has been ignored in Eq. C.3.10 and Eq. C.3.11. There are two reasons why this assumption is permissible from an engineering point of view: first, the actual distance between the cm of the flaps and the axis of rotation of the panels is small; second, this ignored torque acts about the axis of rotation of the solar panels – which are controlled in *closed-loop mode* relative to the body frame – and hence appears only as a weak disturbance acting on the position control loop of the panels.

Inserting Eqs. C.3.2–C.3.9 into Eq. C.3.10 and Eq. C.3.11 yields the final simplified results, which are nonlinear equations in γ and $\delta\gamma$:

$$C_I = \cos(\epsilon_S)[A_1 \gamma \delta\gamma + A_2 \gamma + A_3 \delta\gamma + A_4], \tag{C.3.12}$$

$$C_J = \cos(\epsilon_S)[B_1 \gamma \delta\gamma + B_2 \gamma + B_3 \delta\gamma + B_4]. \tag{C.3.13}$$

In these equations,

$$\gamma = \tfrac{1}{2}(\gamma_S + \gamma_N) \quad \text{and} \quad \delta\gamma = \gamma_N - \gamma_S. \tag{C.3.14}$$

The following coefficients A_i and B_i are obtained from Eqs. C.3.2–C.3.11:

$$A_1 = -PS_p d_p(1 + 5\eta_p) + PS_f d_f[\tfrac{1}{2}\eta_f \sin(\delta) - \sin(\delta) + \tfrac{9}{2}\eta_f \sin(3\delta)], \tag{C.3.15}$$

$$A_2 = PS_f d_f[2\cos(\delta) - \eta_f 3\cos(3\delta) + \eta_f \cos(\delta)], \tag{C.3.16}$$

$$A_3 = A_4 = B_1 = B_2 = 0, \tag{C.3.17}$$

$$B_3 = PS_f d_p 2\eta_p + PS_f d_f \eta_f \tfrac{1}{2}[\sin(\delta) - 3\sin(3\delta)], \tag{C.3.18}$$

$$B_4 = PS_f d_f \eta_f [\cos(3\delta) - \cos(\delta)]. \tag{C.3.19}$$

In our derivation of solar pressure torque dynamics, existence of the specular reflection effect only was assumed, represented by the area reflexivity coefficient. A more sophisticated model could be elaborated assuming the existence of both specular and diffusing reflection reflexivity effects, as in Wertz (1978). A solar pressure torques model based on both reflexivity effects has also been assayed by Azor (1992). The problem is that in-space tests are required before evaluating any model to be used in the design of an operational satellite's ACS. That is, a purely analytical model might well prove to be erroneous after space tests are performed. For example, Harris and Kyrondis (1990) reported an additional effect of *thermal radiation torque,* produced by the solar panel, that had not been taken into account. This effect turned out to be dependent on the procedure by which the solar panels were manufactured!

C.3.3 *Maximization*

Solar torque capability is clearly dependent on the geometrical characteristics of the panels and flaps, on the flap's deviation angle δ, and on the reflexivity coefficients η_p and η_f of the panels and flaps. The torque capability can be optimized via these parameters without increasing the panel area, which is fixed owing to obvious s/c structural constraints. In Example 8.6.1 – with $S_p = 6$ m^2, $S_f = 1$ m^2, $d_p = 4$ m, and $d_f = 5.5$ m – it was found by cut-and-try methods that the best achievable torque characteristics are procured by $\eta_p = 0.2$, $\eta_f = 0.1$, and $\delta = 15°$ (see Figure C.3.3).

It is important to observe that the torque capabilities decrease with the "maximum permitted power loss" requirement; power losses evolve because, in the process of achieving the solar torques, the solar arrays are moved from their optimal position relative to the sun direction. We also observe in Figure C.3.3 that the produced torques are not symmetrical about the origin. This drawback could theoretically be eliminated by using a more complex flaps geometry to improve flap control (see Duhamel and Benoit 1991).

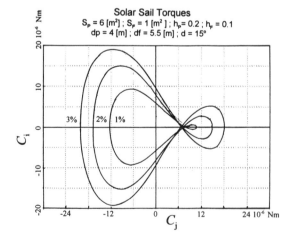

Figure C.3.3 Roll-yaw torque control capability.

C.4 Momentum Exchange Devices

C.4.1 *Introduction*

Momentum exchange devices are controllers that allow changing the distribution of momentum inside the satellite, without altering the total inertial momentum of the entire system, including the device itself. As such, they do not involve any expenditure of fuel. The controller consists of an electrical motor on whose axis is assembled a flywheel designed to increase its angular moment of inertia and so deliver momentum to the satellite's body as part of the ACS. The two basic kinds of momentum exchange devices are momentum wheels and reaction wheels, which are distinguished by their mode of operation.

The *reaction wheel* is used primarily to provide the satellite with sufficient torque for various attitude-maneuvering tasks. Hence it is important for the reaction wheel to provide as much torque as possible in order to achieve fast attitude maneuvers. As is well known (see Section 7.7), a fast attitude maneuver is accompanied by high angular momentum during such maneuvers. The range of achievable torques with reaction wheels is 0.01–1 N-m.

The *momentum wheel* is used primarily to provide the s/c with the momentum bias necessary for inertial attitude stability. As a byproduct, the momentum wheel can also develop torque for controlling the attitude of the satellite's axis that is parallel to the momentum wheel's axis of rotation. The range of angular momentum provided by such wheels is 1–300 N-m-sec. A useful variation is the *double-gimbaled* momentum wheel – a conventional momentum wheel mounted on a double gimbal. With this system, the useful payload of the satellite can be three-axis attitude-controlled in a fine pointing range using only one wheel (Bichler 1991, Auer 1992).

The rotor bearings of momentum exchange devices deserve special mention: we distinguish between mechanical and magnetic bearings. Momentum exchange devices are designed to work without interruption for long time periods (5–12 years). In the space environment of subpressure, lubrication of (mechanical) ball bearings is a major problem that has not been completely solved. Moreover, the ball bearing suffers from excessive friction loading. In recent years, the development of magnetic bearings has taken a decided upturn, with good prospects for the future. Magnetic bearings improve the torque-to-noise ratio by eliminating the parasitic torque noises characteristic of ball bearings (Bosgra and Smilde 1983, Bichler 1991). We proceed with a brief analysis of typical hardware items.

C.4.2 *Simplified Model of a RW Assembly*

A model of reaction wheel (RW) dynamics was introduced in Section 7.3.2, and is reproduced for the reader's convenience as Figure C.4.1 (overleaf). This conventional model incorporates a DC torque motor, where K_M, K_V, I_w, B, and R_M are the usual parameters of the electrical motor. The term I_w is the moment of inertia of the rotor's axis, together with the appended flywheel; B is the viscous damping coefficient of the rotor, and F is the block modeling the dry friction components. Except for this friction block, none of the blocks in Figure C.4.1 exhibit any peculiarities, since they are linear, ideal models. Some probing inside the actual hardware

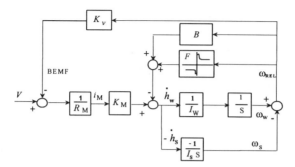

Figure C.4.1 Basic model of a momentum exchange device.

components will aid our understanding of the physical behavior of momentum exchange devices.

Torque Spectra Noise

In practice, a direct-drive motor produces harmonic noises that are a function of the rotor's angular velocity ω_{REL}. The motor is composed of a number p of pole pairs and a number m of phase windings. The commutation of these pairs and windings produces parasitic torque noises with the fundamental and higher harmonic frequencies: $\omega_{REL} pm, \omega_{REL} p(2m), \omega_{REL} p(3m), \ldots$. For example, a motor with 8 pole pairs and 3 phase windings will generate torque noise whose spectral density shows harmonics at $24\omega_{REL}, 48\omega_{REL}, 72\omega_{REL}, \ldots$. Ball-bearing wheels generate another noise due to movement of the retainer. A number of peaks (at wheel-speed-dependent frequencies) are generated: $r, 2r, 3r, \ldots$, where $r = c\omega_{REL}$ and the constant c depends on the bearing's dimensions. These frequencies are all mixed nonlinearly with the wheel-speed harmonics and so introduce subharmonic peaks into the low-frequency range. Of course, this effect is nonexistent in magnetic wheel bearings.

In general, it can be said that the amount of commutation noise increases with the magnitude of the commanded drive-motor current. From this, we conclude that the disturbance noise increases during s/c acceleration. The noise at low wheel speeds is of particular importance, because then the noise lies within the bandwidth of the satellite ACS.

In specifying a reaction wheel assembly, the permitted noise spectrum is defined according to the wheel speed, and a maximum amount of statistical noise is specified in different frequency bands. A sample *reaction wheel torque noise* specification is shown in Table C.4.1 for a RWA with $T_{max} = 0.2$ N-m and $h_{w\,max} = 10$ N-m-sec; spectrum noise measurements of the wheel are also shown.

Dry and Coulomb Friction

The block F in Figure C.4.1 merits special attention. This block is detailed in Figure C.4.2. The *viscous* friction is responsible for the power consumption in the wheel-drive electronics (WDE) of the wheel assembly. For momentum wheels rotating at high velocities (in order to achieve high momentum bias), this power consumption increases notably. The *coulomb* and *stiction* friction torques are responsible for irregularities and finite attitude errors in the ACS. These friction torques must be

Table C.4.1 *Comparison of reaction torque noise specifications with measurements*

Reaction Torque Noise Specifications & Measurement Results				
Frequency [Hz]	Specification Limit [Nm]	Measurement Results [Nm]		
		350 min^{-1}	700 min^{-1}	1050 min^{-1}
0 - 20	7 x 10^{-4}	2 x 10^{-5}	7 x 10^{-5}	2 x 10^{-4}
20 - 50	4 x 10^{-3}	4 x 10^{-5}	1 x 10^{-5}	3 x 10^{-5}
50 - 200	8 x 10^{-3}	9 x 10^{-4}	7 x 10^{-5}	1 x 10^{-4}
(RMS) Total	8.9 x 10^{-3}	9 x 10^{-4}	1 x 10^{-4}	2.3 x 10^{-4}
Motor Current [mA]		36	55	70

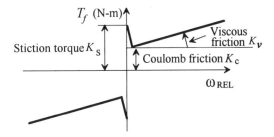

Figure C.4.2 Friction model of a reaction wheel.

dealt with in the design of attitude control loops by using sophisticated nonlinear solutions. It is important to mention that, since momentum wheels are always rotating at an elevated angular speed, there are no velocity sign changes and hence stiction and coulomb friction torques are insignificant.

Torque–Momentum Relationship

The reaction wheel cannot provide the maximum specified torque at all angular velocities; there exists a *torque–momentum* dynamic limitation, as shown in Figure C.4.3. These limitations have a direct impact on the achievable performance in time-optimal attitude control.

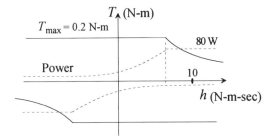

Figure C.4.3 Torque–momentum limitations in a typical reaction wheel assembly.

Tradeoffs on Size of the Flywheel Moment of Inertia

In momentum wheels, there is a tradeoff in sizing the moment of inertia of the flywheel around the rotating axis of the torque motor. The momentum storage capacity is $h_w = I_w \omega(0)$, where $\omega(0)$ is the nominal angular velocity of the flywheel. This simple equation means that it is possible to obtain a desired constant momentum bias by increasing the moment of inertia of the wheel, or by increasing its angular speed. An increase in the moment of inertia is achieved by increasing the weight of the wheel. The rotor and flywheel mass and shape must be optimized so as to obtain a high inertial moment/mass ratio.

Increasing the angular velocity of the wheel leads to an increase in the viscous torques, thus increasing the power consumption in the WDE of the assembly, and also to faster degradation of the ball bearings. Hence, carefully considered tradeoffs are necessary before completing the technical definition of the reaction wheel assembly.

Tachometer

Knowledge of the angular velocity of the wheel is mandatory for many aspects related to the momentum exchange device. First, a momentum wheel is sometimes activated in what is called the *speed mode*. In this mode, the angular velocity must be measured and fed to the electronics for accurate speed control. The angular velocity of momentum wheels is naturally high, so that accurate measurements are comparatively easy. The tachometer is usually a magnetic sensing commutator whose resolution depends on the number of sensors spread along the periphery of the stator of the direct drive motor.

A different use of the measured angular velocity is in *wheel momentum management* of multi-wheel systems, as explained in Section 7.3.5. In these control systems without momentum bias, the four (or more) wheels are controlled to operate under limited angular velocities – close to null, if possible. Here the measurement of momentum wheel angular velocity must be more accurate and the tachometer instrumentation designed accordingly.

C.4.3 *Electronics*

The flywheel and the direct-drive motor can only be controlled through compatible electronics. The basic feedback loops for transforming the drive motor and the flywheel to a reaction wheel assembly were explained in Section 7.3.1. In Figure 7.3.2, use of feedback from the motor current, which is proportional to the motor torque, transforms the motor and flywheel into a reaction wheel system; this is the *torque mode*. If, instead, a constant wheel speed is to be maintained, feedback from the tachometer should be used; the wheel is then operating in the *speed mode*. In other words, the wheel drive electronics is designed to achieve complete fulfillment of the specified characteristics of the reaction or momentum wheel assemblies. A typical WDE scheme is shown in Figure C.4.4.

C.4.4 *Specifications*

Specifying a reaction or momentum wheel assembly is a long and involved process, and an integral part of designing the attitude control system. We have learned

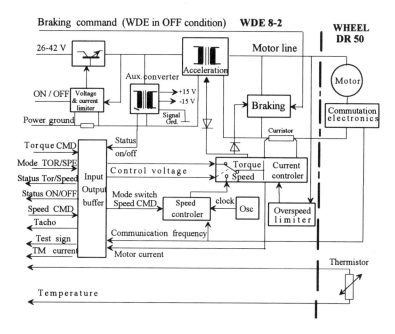

Figure C.4.4 Typical wheel drive electronics block diagram; adapted by permission of TELDIX–Bosch Telecom.

that the most important characteristics concern maximum achievable torques and momentum levels as well as minimum parasitic torque disturbances. We are also faced with more prosaic issues: power consumption, size and weight, conformity to environmental conditions, and so on.

To exemplify these specifications we have chosen the space-proven DRALLRAD momentum and reaction wheels with ball bearings. Table C.4.2 (overleaf) describes the wheel assemblies and Table C.4.3 (page 399) the wheel drive electronics.

C.5 Magnetic Torqrods

C.5.1 *Introduction*

Torqrods (also known as *torque rods, torque bars,* or *magnetotorquers*) are used extensively in the attitude control of spacecraft. They are designed to generate controllable magnetic dipole moments for a variety of applications. For example, we have seen in Section 5.3.4 that they are used for active damping in gravity gradient attitude-stabilized s/c (Section 7.4). In Section 7.5 we described the use of torqrods for magnetic unloading of momentum exchange devices. In general, torqrods are useful because they are substitutes for consumables (such as fuel for thrusters), thus reducing weight. On the other hand, they have their own weight, and it is necessary to examine the associated tradeoffs before deciding whether they are worthwhile from the standpoint of the overall satellite mission.

The torqrod consists of a magnetic core and a coil. When the coil is energized, the torqrod generates a magnetic moment. A serious drawback with torqrods is the

Table C.4.2 *DRALLRAD wheel assembly specifications*
(by permission of TELDIX-Bosch Telecom)

Wheel diameter	cm	20	26	35	50	60*)
Angular momentum range	Nms	1.8...6.5	5.0...20	14...80	50...300	200...1000
Max. reaction torque	Nm	0.2	0.2	0.2	0.3	0.3-0.6
Speed***)	min^{-1}	6000	6000	6000	6000	6000
Loss torque at max. speed**) Nm		≤ 0.012	≤ 0.013	≤ 0.015	≤ 0.022	≤ 0.07
Power consumption:						
-steady state (depending on speed)	W	2...7	2...8	2...10	3...15	10...50
-max. power rating	W	≤ 60	≤ 80	≤ 100	≤ 150	≤ 500
Dimensions:						
-diameter A	mm	203	260	350	500	600
-height B	mm	75	85	120	150	180
Weight	kg	2.7...3.4	3.5...6.0	5.0...8.0	7.5...12	20...37

Environmental conditions
-operating temperature
-vibration (sinusoidal)   suitable for satellites compatible with launchers
-vibration (random)   such as ARIANE or Space Shuttle
-linear acceleration

*)under development
**)with ironless motors (control range +/- 10%)
***)Max. speed of reaction wheels, nominal speed of momentum wheels

fact that the generated moments depend on the earth's magnetic field in an *inverse* cube relation: the earth's field strength is $B = m/R^3$, where $m = 7.96 \times 10^{15}$ Wb/m is the earth's magnetic constant. The moments are also direction-dependent, since the torque is defined as

$$\mathbf{T}_{mag} = \mathbf{M} \times \mathbf{B}, \qquad (C.5.1)$$

where $\mathbf{T}_{mag}$ is the created torque, $\mathbf{B}$ is the earth's magnetic field vector, and $\mathbf{M}$ is the magnetic dipole moment generated by the torqrods.

It is important to develop a sense of the level of attainable torques. With a dipole of $M = 100$ A-m^2 perpendicular to the earth's magnetic field and an altitude of 400 km, the maximum achievable torque will amount to $T_{mag} = 2.56 \times 10^{-3}$ N-m. With the same magnetic dipole but at geostationary altitude, the maximum achievable torque will amount to only 10.46×10^{-6} N-m. Since the perpendicularity condition that we assume does not generally hold, the average torques that we can produce for attitude control are fairly low. Another drawback is that a control system using torqrods must usually include a magnetometer for estimating the earth's magnetic field, a measurement that for the most part is useful only for implementing the control law. Moreover, the applied magnetic dipole moment spoils the magnetometer measurement outputs.

C.5.2 *Performance Curve*

Torqrods can be used in "bang-bang" or proportional attitude control laws. A typical torqrod performance curve is shown in Figure C.5.1. The magnitude of the

C.5 / Magnetic Torqrods

Table C.4.3 *DRALLRAD wheel electronics specifications*
(by permission of TELDIX-Bosch Telecom)

WDE type Application Associated wheels		WDE 1-0 (dual channel) INTELSAT V DR 35	WDE 5-0 (dual channel) TV-SAT/TDF-1 TELE-X, DR 50	WDE 8-2 (single channel) DFS DR 50	WDE 9-9 (dual channel) ROSAT RSR 25
Weight	kg	2.25	3.1	1.9	3.9
Dimensions					
-length	mm	210	205	205	204
-width	mm	160	145	133	184
-height	mm	130	160	118	170
Supply voltages	VDC	+5 ± 15 +50	+50 ± 2% main bus ***	26-42 main bus ***	25-33 digital
Signal interface		digital/analog	digital/analog	digital/analog	digital
Operational Modes		speed command or reaction torque command			
Power consumption at constant speed of the wheels					
-WDE	W	5	5	6	5
-wheels	W	10*	8*	9	12
-total	W	15	13	15	17**
Peak power consumption at max. torque and max. speed of the wheels					
-WDE	W	19	20	30	20
-wheel	W	60*	70*	70	130**
-total	W	79	90	100	150
Environment conditions		suitable for satellites compatible with launchers such as ARIANE or Space Shuttle			

*)One wheel active, second wheel cold redundant
**)Two wheels active simultaneously
***)Isolated DC/DC converters

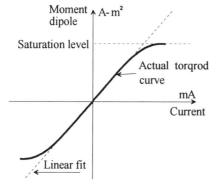

Figure C.5.1 Typical torqrod performance curve.

dipole moment is typically linearly proportional (±4%) to the value of the input current within the specified linear range. The saturation level is defined as the point at which the magnitude of the achieved dipole moment deviates by ±30% from that expected of a straight-line extrapolation of the linear region.

Table C.5.1 *Characteristics of ITHACO torqrods* (by permission of ITHACO Space Systems)

Catalog Number	Moments [A-m²] (Absolute Values)		Linear Voltage [v] (nominal)	Saturation Voltage [V] (nominal)	Resistance[3] at 25°C [ohm] (nominal)	Scale[4] Factor [A-m²/mA]	Mass[2] [Kg] (max)	Length [m] (max)	Diameters[1] [cm] (max)	Number of COILS	Notes
	Linear Moment (minim)	Saturation Moment (minim)	Residual Moment (max)								
TR10CFN	13	15	0.1	11.0	13.9	150	0.18	0.4	0.4	1.8	1	Fiberglass case and 2 mounting blocks
TR10CFR	13	15	0.1	17.0	20.0	270	0.21	0.45	0.39	1.8	2	Fiberglass case, no connector, (pigtail leads), 2 mounting blocks
TR30CFR	35	40	0.2	24.0	28.0	132	0.19	0.95	0.5	2.3	2	Fiberglass case, 3 mounting blocks
TR60CFR	60	70	0.7	10.3	12.6	40	0.25	1.7	0.64	2.6	2	Fiberglass case, 3 mounting blocks
TR65CAR	65	80	0.4	9.2	12.3	39	0.28	1.8	0.64	2.7	2	Aluminium case, Note 5.
TR100CFN	110	130	0.2	10.4	13.0	20	0.21	4.5	0.41	4.9	1	Fiberglass case, 2 mounting blocks
TR100CFR	110	130	1.0	9.6	12.3	106	1.21	3.2	0.87	3.5	2	Fiberglass case, 3 mounting blocks
TR100UPR	110	130	1.0	20.0	24.6	164	0.92	2.1	0.85	2.3	2	Caseless Design for low Weight. Note 5
TR100CAR	110	130	1.0	14.0	19.0	120	0.95	3.6	0.75	3.6	2	Aluminium case, Note 5.
TR140CFR	140	170	1.0	9.5	12.5	110	1.68	5.0	0.94	3.9	2	Fiberglass case, 3 mounting blocks
TR170UPR	170	200	1.2	21.0	27.0	117	1.0	2.8	0.93	2.54	2	Caseless Design for low Weight. Note 5
TR180CAR	180	220	1.0	21.5	29.0	67	0.54	3.5	0.78	3.2	2	Aluminium case, Note 5.
TR230UPR	230	270	1.5	26.0	33.0	127	1.17	4.4	0.92	3.0	2	Caseless Design for low Weight. Note 5
TR480UPR	480	570	3.0	26.5	35.0	116	2.2	8.2	1.3	3.2	2	Caseless Design for low Weight
TR500CFR	500	600	2.0	31.0	40.0	175	2.9	12.25	1.12	6.4	2	Fiberglass case, 3 mounting blocks
TR810UPR	810	970	4.0	27.5	36.2	82	2.5	14.0	1.4	3.2	2	Caseless Design for low Weight. Note 5

1. Not including mounting feet.   2. Mass includes mounting blocks.   3. Copper wire, TC.393%/deg C. Resistance is that of a single coil.   4. Scale factor is that for one coil.
5. Metal case or mounting blocks used. Use with onboard magnetometer closed-loop feedback system is not recommended due to "shorted turn" effects of case or mounting blocks.

C.5.3 *Specifications*

As simple as it appears, even a torqrod must also be carefully designed to exhibit various required characteristics. First, we must define the dipole moment saturation level and the linear range of the torqrod (Section C.5.2).

The principal characteristics can be achieved with different levels of power consumption, reflecting the inversely proportional weight of the torqrod. Torqrods can be produced from different magnetic materials and in different dimensions. For example, ITHACO models TR100CFR and TR100UPR provide practically the same magnetic moments, but need a maximum power of 1.4 W and 3.7 W, respectively; their respective weights are 3.2 kg and 2.1 kg. Torqrods usually are built with *two* coils, the second for redundancy. Table C.5.1 supplies a partial listing of torqrods manufactured by ITHACO Space Systems.

References

Agrawal, B. (1986), *Design of Geosynchronous Spacecraft*. Englewood Cliffs, NJ: Prentice-Hall.

Auer, W. (1992), "A Double Gimballed Momentum Wheel for Precision Three-Axis Attitude Control," IFAC Symposium on Automatic Control in Aerospace: Aerospace Control 92 (7–11 September, Ottobrunn, Germany). Oxford, UK: Pergamon, pp. 581–6.

Azor, R. (1992), "Solar Attitude Control Including Active Nutation Damping in a Fixed-Momentum Wheel Satellite," Paper no. 92-4333, AIAA Guidance and Control Conference (10–12 August, Hilton Head Island, SC). Washington, DC: AIAA, pp. 226–35.

Berker, F. (1978), "MMS Propulsion Model Design," AIAA/SAE 14th Joint Propulsion Conference (25–27 July, Las Vegas, NV).

Bichler, U. (1991), "A Double Gimbaled Magnetic Bearing Momentum Wheel for High-Pointing Accuracy and Vibration Sensitive Space Applications," *Spacecraft Guidance, Navigation and Control Systems* (proceedings of the first international conference organized by ESA at ESTEC, 4–7 June, Noordwijk, Netherlands). Paris: European Space Agency, pp. 393–8.

Bosgra, J., and Smilde, H. (1983), "Experimental and Systems Study of Reaction Wheels, Part I: Measurement and Statistical Analysis of Force and Torque Irregularities," NLF TR 82003 U, ESTEC, Noordwijk, Netherlands.

Duhamel, T., and Benoit, A. (1991), "New AOCS Concepts for ARTEMIS and DRS," *Spacecraft Guidance, Navigation and Control Systems* (proceedings of the first international conference organized by ESA at ESTEC, 4–7 June, Noordwijk, Netherlands). Paris: European Space Agency, pp. 33–9.

Harris, C., and Kyrondis, G. (1990), "Effect of Thermal Radiation Torques on the TDRS," Paper no. 3492-CP, AIAA Guidance and Control Conference (20–22 August, Portland, OR). Washington, DC: AIAA, pp. 1602–14.

Lievre, J. (1985), "Solar Sailing Attitude Control of Large Geostationary Satellite," *IFAC Automatic Control in Space*. Oxford, UK: Pergamon.

Poeschel, R., and Hyman, J. (1984), "A Comparison of Electric Propulsion Technologies for Orbit Transfer," Hughes Research Laboratories, Malibu, CA.

Pritchard, W. L., and Sciulli, J. (1986), *Satellite Communication Systems Engineering*. Englewood Cliffs, NJ: Prentice-Hall.

Smith, P. (1988), "Design and Development of the UK-10 Ion Propulsion Subsystem," GDLR/AIAA/JSASS 20th International Electric Propulsion Conference (3–6 October, Garmisch-Partenkirchen, Germany).

Wertz, J. R. (1978), *Spacecraft Attitude Determination and Control*. Dordrecht: Reidel.

Wertz, J., and Larson, W. (1991), *Space Analysis and Design*. Dordrecht: Kluwer.

Index

absorption characteristic, 41
acceleration
  disturbing, 85
  Keplerian, 29
  linear, 8
  longitudinal, 58, 78
accumulation of momentum, 189
ACS accuracy, 172
action equation of motion, 299
active attitude control, 112
active control, 210, 214
active damping, 117, 126
active nutation control, 135
active nutation damping, 225
active wheel nutation damping, 146
aerodynamic force, 32
air density, 32
altitude of
  apogee, 65
  geostationary orbit, 73
  low-orbit satellite, 28, 72
  orbit, 32
  perigee, 65
amplification of
  position sensor noise, 179
  rate sensor noise, 178
  reaction wheel torque noise, 179
analog sun sensor, 345
angular dynamic equations, 88, 211
angular momentum, 11, 88, 89
angular motion, 88, 90
angular velocity, 88, 101, 162
angular velocity sensors, 175
anomalies, eccentric, true, 18
aphelion, 15
apoapsis, 15
apogee, 15
apogee boost motor (ABM) 60, 381
apsides, 68
argument of perigee, 24, 28, 68
Aries, first point of, 23
ascending node, 23, 36
asymptotes, 17
atmospheric drag, 32, 72
attitude calculation, 336
attitude control, 163
  gravity gradient, 112, 114
  laws, 156

  magnetic, 188, 222
  reaction thrust, 242
  solar torque, 222, 229
  time-optimal, 195
attitude determination hardware, 328
attitude dynamics, 88
attitude error, 76
attitude kinematics, 88, 100
attitude-maneuvering satellite, 112
attitude maneuvers, 152
attitude matrix, 318
attitude sensors, 173, 174
attitude stability, 173
attitude transformations in space, 325
axis of symmetry, 95

ballistic coefficient, 33
bandwidth of attitude control system, 180
bang-bang control, 141, 195
basic attitude control equation, 113, 152
bending modes, 180
bias attitude error, 335
body cone, 98
body coordinate frame, 318
body rates estimation, 158

Canopus, 363
cantilever beam, 297
cantilever natural frequency, 294
Cape Canaveral, 73
Cartesian coordinate system, 24, 27
catalyst, 382
catalytic activity, 383
CCD matrix, 369
central force, 12
celestial background, 364
celestial coordinates, 371
celestial catalog, 365
celestial map, 365
celestial pole, 22
celestial sphere, 23, 365
center of
  attracting body, 43
  earth, 40
  eccentricity circle, 55
central force field, 12
charge coupled device (CCD) detector, 369
chatter, 143, 198, 201

chemical propulsion
  liquid, 382
  solid, 381
circularization of GTO, 75
classical orbit parameters, 24, 27
code
  binary, 352
  Gray, 352
cold gas propulsion, 381
conical section, 14
conservation of angular momentum, 161
conservation of energy, 10
conservative force, 10
constant of gravitation, 9
control hardware, 379
control torques, 113
control torque saturation, 223
coplanar transfer orbit, 69
cosine sun detector, 345
coupling coefficients and matrices, 310
critical inclination, 37

damper
  boom articulation, 123
  external spring boom, 123
  magnetic hysteresis, 123
  point mass, 123
  wheel, 124
damping
  active, 126
  all-magnetic active, 129
  passive, 123, 146
  passive wheel nutation, 144
damping factor, coefficient, 123, 145
dead zone, 140, 285
deflection mode
  in-plane, 309
  out-of-plane, 309
  torsional, 309
denutation, 139
despin, 139
digital sun sensor, 351
direction cosine error matrix, 154,318
direction cosine matrix, 104, 153, 318, 323
direction cosines, 46
discrete control, 273
displacement equations of motion, 300
dissipation function, 293
disturbed Keplerian orbit, 37
disturbing torque, 114, 132
disturbance torques, 122, 274
double stars, 365
drag coefficient, 32
drift rate, 79
dry center of mass, 305
dual-cone optical head, 333

dual-cone scanner, 334
dual-spin stabilization, 132, 148, 210

earth albedo, 329
earth circle, 337
earth
  equatorial plane, 23, 56
  escape velocity, 16
earth sensors, 174 , 329
  noise amplification, 220
earth's magnetic field, 123, 126, 186
eccentric anomaly, 18
eccentricity, 14, 21, 42,
eccentricity circle, 52, 84
eccentricity corrections, 78, 84
eccentricity derivative, 55
eccentricity vector, 43, 52 , 81
eccentricity vector evolution, 55
ecliptic plane, 23, 45
ecliptic pole , 48
eigenaxis rotation, 195
eigenfrequency, 295
eigenvalue, 92 , 299
eigenvector, 92, 299
eigenvector of rotation, 323
electric propulsion, 385
ellipsoid
  of inertia, 93
  of momentum, 94
emission spectra, 363
energy
  kinetic, 9
  potential, 9
energy constant, 16
energy dissipation, 99, 138,149, 244
  rate, 138
energy sink, 138
epoch, 23
equatorial plane, 22, 37, 42, 73
escape velocity (from circular orbit), 16
Euler angle errors, 153
Euler angle rotation, 319
Euler angles, 101, 104, 110
Euler angular rates, 153
Euler axis of rotation, 155, 188, 323
Euler-Hill equations, 57
Euler's moment equation, 95, 145
evolution of the eccentricity vector, 50
evolution of the inclination vector, 43
external disturbances, 275
external torque, 107

field of view (FOV), 329, 350, 373
first point of the Aries, 23
flexibility coefficients, 298
flexibility matrix, 298

Index

flexible solar array, 291
flexural rigidity, 297
flux of light, 357
focus, 8, 15
force-deflection equation, 297
frequency of oscillation, 117
friction, 394
fuel consumption, 64, 70, 136, 143
fuel tank, 302

Gauss planetary equations, 30
generalized coordinates, 293
generalized forces, 293
geocentric inertial system, 22, 23
geocentric latitude, 35
geographical
   latitude, 36, 68, 73
   longitude, 35, 78
geomagnetic equator, 187
geopotential function, 35
geostationary, 28, 42, 73
geostationary orbit corrections, 80
geosynchronous, 42, 73
gravitational
   attraction field, 17
   force, 44
   potential, 34, 37
gravity gradient, 112
   attitude control, 114
   characteristic equation, 114
   moments, 108
   stabilization, 112, 122, 126
   vector, 109
Gray binary code, 351
Greenwich meridian, 43

hardware
   attitude determination, 174, 328
   control, 160, 379
harmonic coefficients, 14
   sectoral, 35
   spherical, 34
   tesseral, 35
   zonal, 35
harmonic motion, 49, 117
Hill equations, 58
Hohmann transfer, 70
horizon-crossing sensor, 330
horizon sensor, 330
hyperbolic orbit, 17

image dissector, 367
immunity to sensor noise, 246
impulsive force, 59
impulsive thrust, 64
inclination, 24, 42
   angle, 28

   circle, 82
   correction, 74, 82
   station keeping, 80, 81
   zeroing, 74
inclination derivatives, 49
inclination drift, 41
inclination vector, 43, 82
   evolution, 43
inertia matrix, 88
inertial coordinate system, 22, 25
inertial frame, 26
inertial measuring unit (IMU), 181
inertial reference frame, 26
influence coefficient, 292
infrared earth sensor, 329
infrared static earth sensor, 343
in-plane deflection mode, 309
integrator, 278
inverse square law of force, 9, 10
ion thruster, 386

Keplerian acceleration, 29
Keplerian orbit, 8, 12
Kepler's laws, 8, 19
Kepler's time equation, 20
Kourou, 73

Lagrange's equations, 45, 293
Lagrange's method, 293
Lagrange's planetary equation, 33
latch valves, 384
launch site, 73, 74
law of areas, 19
Legendre polynomials, 35, 41
linearized attitude dynamics equations of
   motion, 108, 312
line of apsides, 65
line of nodes, 37
liquid slosh, 180, 301
local coordinate system, 25
longitude station keeping, 85
longitudinal acceleration, 56, 78
longitudinal drift rate, 79
lunar pole, 48

magnetic active damping, 129
magnetic attitude control, 185
magnetic control dipole, 126
magnetic field, 123, 126, 128
magnetic moments, 397
magnetic torquers (torqrods), 126, 187, 397
magnetic torque equation, 191
magnetic unloading of momentum, 189
magnetotorquers, 397
magnitude m of a star, 357
major axis, 16, 92

mask detector, 349
mass matrix, 298
massless cantilever beam, 297
mean anomaly, 20
mean longitude, 43
mean motion, 20
mean radius of earth, 35
minimum impulse bit, 245, 284
minimum torque impulse bit, 285
minor axis, 92
modal frequency, 294
modeling liquid slosh, 301
modeling solar panels, 291
modulator
 pulse width, (PW), 273
 pulse width–pulse frequency (PWPF), 266, 270
 pseudo rate (PR), 270
Molniya orbit, 37
moment of inertia, 89
 maximum, 92
 minimum, 92
 principal, 92
 about spin axis, 90
moment of momentum, 11
momentum axis of orbit, 25
momentum bias, 212,
momentum-biased attitude stabilization, 161, 210
momentum accumulation, 165
momentum
 linear, 8,
 angular, 11, 88
momentum-biased satellite, 217
momentum bias stabilization, 150
momentum capacity, 206
momentum dumping, 165, 241
 magnetic, 194
 reaction, 241, 250
momentum exchange device, 107, 160, 393
 control moment gyro, 161
 momentum wheel, 107, 161, 237
 reaction wheel, 161
momentum management, 169
momentum wheel, 161, 211, 393
monopropellant propulsion system, 382
moon's orbit, 45, 48
multi-mass modeling, 296, 308
multi-mass sloshing model, 308

nadir, 214, 331
nadir-pointing stabilized satellite, 120
n-body problem, 39
natural eccentricity radius, 55, 84
natural frequency, 180
natural frequency of oscillation, 300, 301

Newton's laws, 8
 second law of motion, 39
node line, 24,
noise amplification, 178, 220
nonconservative perturbing forces, 28
nonhomogeneity of the earth, 34
nonspinning satellite
 dynamic equations of, 107
 kinematic equations of, 100
nonviscous liquid, 302
North–South station keeping, 81
nozzle throat, 381
nutation
 damper, 146
 destabilization, 99
 instability, 100
 stability, 100,
nutation angle, 98, 133
nutation frequency, 136, 212, 220
nutational motion, 132

oblateness effects of the earth, 34, 329
onboard star catalog, 370
one-vibrating mass model, 302
open-loop gain, 313
operational constraints, 67
optical scanning mechanism, 330, 367
optical sensor head, 349
orbital
 adjustment, 65, 78
 corrections, 80
 frequency, 212
 maneuvers, 64
 period, 32
 plane, 24, 44
 pole, 24, 43, 80
 rate, 117
orbit change
 in-plane, 68, 75
 multi-impulse, 70
 out-of-plane, 75
 single-impulse, 65
orbit coordinates, 25
orbit mechanics, 8
orbit parameters, 24, 32
orbit reference frame, 101, 105
orbits
 altitude, 32
 circular, 15
 coaxial, 71
 coplanar, 71
 elliptical, 15
 equatorial, 24
 geostationary, 42, 73
 geosynchronous, 42, 73
 heliosynchronous, 36

Index 407

hyperbolic, 17
parabolic, 16
sun-synchronous, 36
oscillation frequency, 305
osculating orbit, 29
out-of-plane deflection mode, 309

passive attitude control, 112
passive dampers, 123
passive wheel nutation damping, 144
parabolic trajectory, 78
parasitic disturbing torques, 161
passage time from perigee, 21
pendulum, frequency of oscillation, 302
periapsis, 15
perigee, 15
   argument, 24
   passage, 20
perihelion, 15
perturbation acceleration, 29, 31, 40, 85
perturbed
   orbit, 28
   equation of motion, 29
perturbing body, 40
perturbing forces, 28, 33, 85
perturbing potential function, 37
perturbing third body, 39
photomultiplier, 367
planetary precession, 23
Polaris, 214, 357
pole
   celestial, 22
   ecliptic, lunar, 48
   orbital, 44, 80
polhode, 94
positioning accuracy, 173
potential energy, 10, 14
power loss, 392
precession
   lunar pole, 48
   planetary, 23
precessional motion, 22
prime focus, 15, 18
principal axes, 93
principal axes of inertia, 91
principal moments of inertia, 92
product of inertia, 90, 225, 248
propellant
   control valve, 384
   liquid, 382
   mass, 72
   solid, 381
proper real orthogonal matrix, 319
propulsion, 379
   bipropellant, 385
   chemical, 381
   cold gas, 381
   electric, 385
   liquid, 382
   monopropellant, 382
   solid propellant, 381
propulsion rocket equation, 380
pseudoinverse matrix, 160
pseudo rate (PR) modulator, 270
pulsed controller, 273
pulsed reaction system, 273
pulse width modulation, 265, 273
pulse width-pulse frequency modulation, 265, 267
pulsing mode, 260

quaternion, 104, 322
quaternion error vector, 156
quaternion method, 322
quaternion multiplication, 326
quaternion vector, 104

radiance detector, 330, 344
radiation pressure 41
rate
   gyro, 158, 375
   integrating gyro, 105, 175, 375
rate sensor, 175, 373
reaction control system, 140
reaction thruster attitude control, 260
reaction thruster, 242, 260
   HT, LT, 385
reaction torque, 260, 265
reaction wheel, 161, 206, 393
reciprocity theorem, 298
reference coordinate system, frame, 101
relative acceleration, 58
relative distance, 58
relative motion, 10, 58
restrictions on orbit changes, 69
reticle slit pattern, 351
right ascension, 24, 28, 43
right-handed system, 23
right pseudoinverse transformation, 169
rigid body, 88, 291
   rotation kinetic energy, 90
roll deadbeat limit, 247
roll-yaw attitude control, 237
root locus, 146
rotating frame, 102
rotational axis, 91
rotational kinetic energy, 90, 94
rotational motion, 88, 238

sampled transfer function, 284
sampling frequency, 273
satellite motion, 59

scalar potential function, 34
scanning mechanism, 330
scanning rate, 338
Schmidt trigger, 265,
sectoral harmonic coefficients, 35
secular term
    of inclination derivative, 50
semi-latus rectum, 14
semimajor axis, 16, 65
semiminor axis, 16
sensor noise, 113, 173
    amplification, 178, 266
side force, 308
sidereal angle, 42
sidereal day, 42
signal processing, 330, 369
simulation, 6-DOF, 120
single-mass structural dynamics, 293
single-spin stabilization, 132, 144
slosh dipole, 305
solar
    efficiency, 234
    energy flux, 41
    pressure, 28
    radiation, 41
    torques, 388
    wind, 41
solar control torques, 229, 388
solar flaps, 230
solar panels, 230
solar radiation perturbing function, 53
solar torque capability, 392
space cone, 98
specific angular momentum, 12
specific impulse I_{sp}, 380
spectra of a star, 357
spherical harmonic expansion, 34
stability of rotation, 96
star catalog, 369
star identification, 371
star scanner, 366
star sensor, 353
    assembly, 367
    specification, 373
star tracking, 366
static earth sensor, 343
steady-state error, 278
stellar distribution, 364
stiffness coefficient, 292, 298
stiffness constant, 299
stiffness matrix, 298
structural dynamics, 113, 291, 291
structural model, 295, 296
structural modeling, 291,
sun acquisition, 350

sun sensors, 174, 345
    analog, 345
    digital, 351
    one-axis, 347
    two-axis, 349
sun-synchronous orbit, 36
switching curve, 198

tachometer, 396
terminator, 329
tesseral harmonic coefficients, 35, 37
thermal emission, 329
thermopile, 344
third-body perturbing force, 39
three-axes stabilized, 100
three-body problem, 39
thruster activation time, 263
thrusters, 387
    electrothermal monopropellant, 384
    hydrazine, 383
    ion, 386
time delay, 200
time derivation
    of direction cosine matrix, 104
    of quaternion vector, 104
time-optimal attitude control, 195
time response, 117, 212, 235
time since periapsis passage, 18
torque
    arm, 262
    commands, 141
    control law, 152, 207
    impulse bit, 245, 262
    solar, 234, 388
torque spectra noise, 394
torque impulses, 273
torque wheel, 107
torsional deflection mode, 309
total energy, 10, 14
total energy per unit mass, 14
total impulse, 380
transfer
    geostationary (GTO), 73, 85
    geosynchronous, 73
    Hohmann 70
    orbit, 64, 71
transformation, three-dimensional, 25
transplanetary s/c voyage, 17
true anomaly, 18
two-body problem, 10

unbalanced torque, 232
unified propulsion system, 384
universal constant of gravity, 9

variable stars, 365
variance in earth radiation, 336

Index

variation of parameters, 34
Vega, 357
velocity
 angular, 59
 circular, 51
 radial, 31
 relative, 59
 vector, 8, 30
velocity change, 64, 78
velocity increment, 52
velocity loss, 64, 133
vernal equinox, 23
vibrating mass model, 302

viscosity damping coefficient, 163
visual magnitude, 363

wheel damper, 123
wheel momentum dumping, 250
wheel momentum management, 396
windmill torque, 231
work and energy, 9

yaw error, 217
yaw measurement, 215

zero-bias momentum system, 190